Space Sciences Series of ISSI

Volume 93

Series Editor
International Space Science Institute, Bern, Switzerland

The Space Sciences Series of ISSI books are coherent reports of the findings, discussions, and ideas that result from international scientific workshops regularly held at the International Space Science Institute (ISSI) in Bern, Switzerland. ISSI's main task is to contribute to the achievement of a deeper understanding of the results from space-research missions, adding value to those results through multi-disciplinary research in an atmosphere of international cooperation. The books are reprints of special issues in the Space Science Reviews journal and occasionally of special issues in the Surveys in Geophysics journal.

Benoit Meyssignac • Sonia Seneviratne
Rémy Roca • Graeme L. Stephens • Michael Rast
Editors

Earth's Changing Water and Energy Cycle

Previously published in *Surveys in Geophysics*
Volume 45, Issue 6 December 2024

Editors
Benoit Meyssignac
LEGOS (CNES/CNRS/IRD/UT3)
Université de Toulouse
Toulouse, France

Rémy Roca
LEGOS (CNES/CNRS/IRD/UT3)
Université de Toulouse
Toulouse, France

Michael Rast
Earth Sciences
International Space Science Institute
Bern, Switzerland

Sonia Seneviratne
IAC
ETH Zurich
Zurich, Switzerland

Graeme L. Stephens
California Institute of Technology
Jet Propulsion Laboratory
Pasadena, CA, USA

Spin-off from journal: "Surveys in Geophysics" Volume 45, Issue 6, December 2024

ISSN 1385-7525
Space Sciences Series of ISSI
ISBN 978-94-024-2311-2

© The Editor(s) (if applicable) and The Author(s), under exclusive license to Springer Nature B.V. 2025

This work is subject to copyright. All rights are solely and exclusively licensed by the Publisher, whether the whole or part of the material is concerned, specifically the rights of translation, reprinting, reuse of illustrations, recitation, broadcasting, reproduction on microfilms or in any other physical way, and transmission or information storage and retrieval, electronic adaptation, computer software, or by similar or dissimilar methodology now known or hereafter developed.
The use of general descriptive names, registered names, trademarks, service marks, etc. in this publication does not imply, even in the absence of a specific statement, that such names are exempt from the relevant protective laws and regulations and therefore free for general use.
The publisher, the authors and the editors are safe to assume that the advice and information in this book are believed to be true and accurate at the date of publication. Neither the publisher nor the authors or the editors give a warranty, expressed or implied, with respect to the material contained herein or for any errors or omissions that may have been made. The publisher remains neutral with regard to jurisdictional claims in published maps and institutional affiliations.

Cover illustration: This view of Earth's horizon as the Sun sets over the Pacific Ocean was taken by the Expedition 7 crew onboard the International Space Station in 2003. Image Credit: NASA.

This Springer imprint is published by the registered company Springer Nature B.V.
The registered company address is: Van Godewijckstraat 30, 3311 GX Dordrecht, The Netherlands

If disposing of this product, please recycle the paper.

Contents

**Trends and Variability in Earth's Energy Imbalance
and Ocean Heat Uptake Since 2005** 1
Hakuba, M.Z., Fourest, S., Boyer, T. et al. Trends and Variability in Earth's
Energy Imbalance and Ocean Heat Uptake Since 2005. *Surv Geophys* 45,
1721–1756 (2024). https://doi.org/10.1007/s10712-024-09849-52)

**Observational Assessment of Changes in Earth's Energy
Imbalance Since 2000**..37
Loeb, N.G., Ham, S.H., Allan, R.P. et al. Observational Assessment
of Changes in Earth's Energy Imbalance Since 2000. *Surv Geophys* 45,
1757–1783 (2024). https://doi.org/10.1007/s10712-024-09838-8

Closure of Earth's Global Seasonal Cycle of Energy Storage................................65
Johnson, G.C., Landerer, F.W., Loeb, N.G. et al. Closure of Earth's Global
Seasonal Cycle of Energy Storage. *Surv Geophys* 45, 1785–1797 (2024).
https://doi.org/10.1007/s10712-023-09797-6

**The Global Energy Balance as Represented
in Atmospheric Reanalyses**..79
Wild, M., Bosilovich, M.G. The Global Energy Balance as Represented
in Atmospheric Reanalyses. *Surv Geophys* 45, 1799–1825 (2024).
https://doi.org/10.1007/s10712-024-09861-9

**Assessment of Atmospheric and Surface Energy Budgets
Using Observation-Based Data Products** 107
Mayer, M., Kato, S., Bosilovich, M. et al. Assessment of Atmospheric
and Surface Energy Budgets Using Observation-Based Data Products. *Surv
Geophys* 45, 1827–1854 (2024). https://doi.org/10.1007/s10712-024-09827-x

**North Atlantic Heat Transport Convergence Derived
from a Regional Energy Budget Using Different Ocean Heat
Content Estimates** 135
Meyssignac, B., Fourest, S., Mayer, M. et al. North Atlantic Heat
Transport Convergence Derived from a Regional Energy Budget
Using Different Ocean Heat Content Estimates. *Surv Geophys* 45,
1855–1874 (2024). https://doi.org/10.1007/s10712-024-09865-5

**An Abrupt Decline in Global Terrestrial Water Storage
and Its Relationship with Sea Level Change** 155
Rodell, M., Barnoud, A., Robertson, F.R. et al. An Abrupt Decline in Global
Terrestrial Water Storage and Its Relationship with Sea Level Change. *Surv
Geophys* 45, 1875–1902 (2024). https://doi.org/10.1007/s10712-024-09860-w

**Tropical Deep Convection, Cloud Feedbacks
and Climate Sensitivity**..183
Stephens, G.L., Shiro, K.A., Hakuba, M.Z. et al. Tropical Deep
Convection, Cloud Feedbacks and Climate Sensitivity. *Surv Geophys* 45,
1903–1931 (2024). https://doi.org/10.1007/s10712-024-09831-1

**A Multi-Satellite Perspective on "Hot Tower" Characteristics
in the Equatorial Trough Zone** 213
Pilewskie, J., Stephens, G., Takahashi, H. et al. A Multi-satellite
Perspective on "Hot Tower" Characteristics in the Equatorial Trough Zone. *Surv
Geophys* 45, 1933–1958 (2024). https://doi.org/10.1007/s10712-024-09868-2

**A Geostationary Satellite-Based Approach to Estimate
Convective Mass Flux and Revisit the Hot Tower Hypothesis** 239
Derras-Chouk, A., Luo, Z.J. A Geostationary Satellite-Based Approach
to Estimate Convective Mass Flux and Revisit the Hot Tower Hypothesis.
Surv Geophys 45, 1959–1977 (2024). https://doi.org/10.1007/s10712-024-09856-6

**METEOSAT Long-Term Observations Reveal Changes
in Convective Organization Over Tropical Africa
and Atlantic Ocean**..259
Roca, R., Fiolleau, T., John, V.O. et al. METEOSAT Long-Term
Observations Reveal Changes in Convective Organization Over
Tropical Africa and Atlantic Ocean. *Surv Geophys* 45,
1979–1998 (2024). https://doi.org/10.1007/s10712-024-09862-8

**Lessons Learned from the Updated GEWEX Cloud
Assessment Database**...279
Stubenrauch, C.J., Kinne, S., Mandorli, G. et al. Lessons Learned
from the Updated GEWEX Cloud Assessment Database. *Surv Geophys* 45,
1999–2048 (2024). https://doi.org/10.1007/s10712-024-09824-0

Trends and Variability in Earth's Energy Imbalance and Ocean Heat Uptake Since 2005

Maria Z. Hakuba[1] · Sébastien Fourest[2] · Tim Boyer[3] · Benoit Meyssignac[2] · James A. Carton[4] · Gaël Forget[5] · Lijing Cheng[6] · Donata Giglio[7] · Gregory C. Johnson[8] · Seiji Kato[9] · Rachel E. Killick[10] · Nicolas Kolodziejczyk[11] · Mikael Kuusela[12] · Felix Landerer[1] · William Llovel[11] · Ricardo Locarnini[3] · Norman Loeb[9] · John M. Lyman[8,13] · Alexey Mishonov[3,14] · Peter Pilewskie[7,15] · James Reagan[3] · Andrea Storto[16] · Thea Sukianto[12] · Karina von Schuckmann[17]

Received: 10 December 2023 / Accepted: 17 June 2024 / Published online: 29 July 2024
© The Author(s) 2024

Abstract

Earth's energy imbalance (EEI) is a fundamental metric of global Earth system change, quantifying the cumulative impact of natural and anthropogenic radiative forcings and feedback. To date, the most precise measurements of EEI change are obtained through radiometric observations at the top of the atmosphere (TOA), while the quantification of EEI absolute magnitude is facilitated through heat inventory analysis, where ~90% of heat uptake manifests as an increase in ocean heat content (OHC). Various international groups provide OHC datasets derived from in situ and satellite observations, as well as from reanalyses ingesting many available observations. The WCRP formed the GEWEX-EEI Assessment Working Group to better understand discrepancies, uncertainties and reconcile current knowledge of EEI magnitude, variability and trends. Here, 21 OHC datasets and ocean heat uptake (OHU) rates are intercompared, providing OHU estimates ranging between 0.40 ± 0.12 and 0.96 ± 0.08 W m^{-2} (2005–2019), a spread that is slightly reduced when unequal ocean sampling is accounted for, and that is largely attributable to differing source data, mapping methods and quality control procedures. The rate of increase in OHU varies substantially between -0.03 ± 0.13 (reanalysis product) and 1.1 ± 0.6 W m^{-2} dec^{-1} (satellite product). Products that either more regularly observe (satellites) or fill in situ data-sparse regions based on additional physical knowledge (some reanalysis and hybrid products) tend to track radiometric EEI variability better than purely in situ-based OHC products. This paper also examines zonal trends in TOA radiative fluxes and the impact of data gaps on trend estimates. The GEWEX-EEI community aims to refine their assessment studies, to forge a path toward best practices, e.g., in uncertainty quantification, and to formulate recommendations for future activities.

Keywords Earth's energy imbalance · Ocean heat content · Radiation budget · Reanalysis · Argo · GEWEX

Extended author information available on the last page of the article

Article Highlights

- GEWEX-EEI compares 21 ocean heat content time series from reanalysis, in situ and satellite observations
- We find substantial spread in ocean heat uptake and variable skill in tracking radiometric EEI variability
- Follow-on investigations and recommendations are proposed to reconcile estimates and their uncertainties

1 Introduction

Detection and understanding of climate change rely on research investigations by a broad international science community utilizing a vast array of climate models and Earth observations ranging from in situ measurements made in the deep ocean to satellite measurements made at the top of the atmosphere (TOA). Improvements to our understanding of the state of Earth's climate and projected future changes are communicated to society through several governmental channels such as the climate assessments of the Intergovernmental Panel on Climate Change (e.g., IPCC 2021; Forster et al. 2021; Gulev et al. 2021) or the annual World Meteorological Organization (WMO) State of the Global Climate Reports, which in 2022 (WMO 2023) focused on key climate indicators, such as greenhouse gasses, temperature, sea level rise, ocean heating and acidification, sea ice and glacier melt. Identifying and examining indicators that robustly and comprehensively measure change to the climate and the impact of current and future anthropogenic activities is a crucial aspect of the first Global Stock Take under the Paris Agreement (Peeters 2021; Forster et al. 2023).

The most holistic picture of heat accumulation by the Earth system is gained by quantifying and assessing change in *Earth's radiative Energy Imbalance* (EEI) at the TOA, representing the cumulative effect of radiative forcings and feedbacks. At present, Earth's *heat uptake* or *inventory*, respectively, serves to constrain EEI absolute magnitude over decadal timescales (e.g., von Schuckmann et al. 2016, 2023; Trenberth et al. 2016; Johnson et al. 2016; Meyssignac et al. 2019; Cheng et al. 2022a, b). According to the latest GCOS assessment of *Earth's heat inventory,* the absolute magnitude of EEI for the period 2006–2020 is 0.76 ± 0.2 W m^{-2}, which combines ensemble estimates of ocean heat uptake (OHU), terrestrial as well as atmospheric heat storage, and the heat energy required to melt land and sea ice and evaporate water to increase atmospheric moisture content (von Schuckmann et al. 2023). About 90% of Earth's heat surplus is stored in the ocean; hence, assessments of EEI magnitude and uncertainty largely depend on accurate OHU estimates. To date, standard approaches to estimating OHU (e.g., Hakuba et al. 2021; Cheng et al. 2022a, b) are: (a) to derive ocean heat content (OHC) changes from direct subsurface ocean temperatures observations through hydrographic profiles (e.g., Levitus et al. 2012); (b) to derive the oceans' thermosteric expansion through sea level budget assessment using geodetic observations from space (e.g., Marti et al. 2022); and (c) to estimate the ocean state using global ocean models and reanalyses that assimilate various ocean and atmosphere observations (e.g., Balmaseda et al. 2015; Forget et al. 2015; Zuo et al. 2019).

Since circa 2005, when Argo reached critical spatiotemporal coverage, experts across the globe have been examining Argo profiling float array observations (Riser et al. 2016) along with other ocean in situ observations (e.g., reviewed in Meyssignac et al. 2019; Cheng et al. 2022a, b) to estimate global OHC time series. More recently, geodetic satellite observations have been used for estimating OHU and its variability (Marti et al. 2022; Hakuba et al. 2021). Despite the lack of vertical resolution and requiring independent knowledge of seawater's heat expansion efficiency, the combination of near-global satellite-based estimates of ocean mass and total sea level change has proven successful in matching positive trends in EEI and its year-to-year variability with Clouds and Earth's Radiant Energy System (CERES) Energy Balanced and Filled (EBAF) net radiative flux at the TOA (Marti et al. 2022; Hakuba et al. 2021). Reanalyses that assimilate ocean observations into physically consistent ocean models also provide OHU together with complete ocean state estimates, the latter enabling the study of heat exchanges, distributions and their causes (Storto et al. 2019).

Beyond being a fundamental metric for tracking change in Earth's climate, EEI also represents a target value in global climate model tuning (Cheng et al. 2016a, b; Hourdin et al. 2017; Smith et al. 2015; Schmidt et al. 2023) and serves in constraining Earth's equilibrium climate sensitivity (e.g., Sherwood et al. 2020; Chenal et al. 2022) and climate feedback parameter (Meyssignac et al. 2023a). Improved understanding of EEI and a robust estimate of its uncertainty is vital in reconciling global energy and water cycles using consistent budgets and optimization processes (L'Ecuyer et al. 2015; Roberts et al. submitted to this issue), which is one of the key goals of the Global Energy and Water Exchanges (GEWEX) community (Stephens et al. 2023). EEI monitoring alone is not sufficient for understanding the implications that natural and anthropogenic composition changes impose on Earth's energy budget. Characterizing and attributing EEI changes and their drivers on seasonal (Johnson et al. 2023a, b; Pan et al. 2023) as well as interannual and longer timescales (Cheng et al. 2019a, b; Loeb et al. 2021, 2024, this issue; Stephens et al. 2022) require extensive use of ancillary surface and atmospheric property information as well as radiative transfer and global climate models to help interpret the role of climate forcings and climate feedbacks (e.g., Raghuraman et al. 2021). Although EEI is a global metric of Earth system change, regional investigations of ocean heat distribution and transports (e.g., Meyssignac et al. submitted to this issue; Trenberth et al. 2019) as well as the processes driving heat uptake in the atmosphere and at the surface (Mayer et al. 2024, this issue) are key aspects of holistic EEI assessments. One way to reconcile data from all available observing networks and their scale of variability is through Earth system reanalysis capable of ingesting all observational information, i.e., TOA radiation, OHC, altimetry, gravimetry (Stammer et al. 2016; Storto et al. 2017; de Rosnay et al. 2022).

Due mainly to calibration and retrieval uncertainties that are one order of magnitude larger than EEI itself, EEI absolute magnitude cannot be derived from radiometric observations such as provided by the Clouds and Earth's Radiant Energy System (CERES; Wielicki et al. 1996) and the Solar Radiation and Climate Experiment (SORCE; Rottman 2005), unless adjustments are made to match the global net radiative flux to independent estimates of long-term planetary heat uptake (Loeb et al. 2018). These adjustments do not affect the time variability nor trend in radiometrically derived EEI significantly. Figure 1 contrasts the EEI time series from single-scanner *Terra* data with the *energy balanced* (EBAF) multi-platform (Terra, Aqua, NOAA20) climate data record. Although the *Terra* record suggests unrealistically high EEI absolute magnitude (time series mean), irreconcilable with heat uptake estimates and current knowledge of radiative forcings and feedbacks that shape EEI (Loeb et al. 2009a), both show near-identical time variability and trend over March 2000 to July 2023. The offset correction in EBAF is commensurate with an estimate of EEI absolute magnitude at 0.71 W m^{-2} over 2005–2015 according to Johnson et al. (2016).

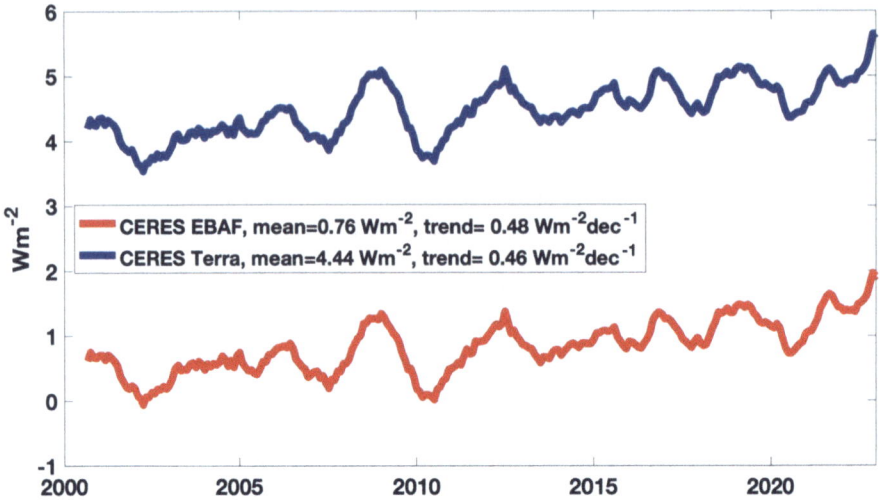

Fig. 1 EEI (12-month running mean) derived from CERES observations. The CERES single-scanner net radiative flux from Terra (blue) provides unrealistically high EEI, irreconcilable with heat uptake estimates and current knowledge of radiative forcings and feedbacks that shape EEI (Loeb et al. 2009a). EBAF combines data from multiple platforms (Aqua, Terra, NOAA20) and is anchored to an independent estimate of EEI from heat uptake analysis (Loeb et al. 2018)

Although satellite measurement absolute accuracy is insufficient to close the TOA energy balance, the unparalleled measurement precision and stability of current CERES and future Libera measurements allows for the study of EEI time variations and trends. As of 2021, the trend in EEI over 2005–2019 is 0.50 ± 0.47 W m^{-2} dec^{-1} (5–95% confidence intervals) according to CERES EBAF analysis, indicating an approximate doubling of EEI in this 14-year period (Loeb et al. 2021; Cheng et al. 2024a). This tendency agrees with global OHU from combined in situ and altimetry measurements (Loeb et al. 2021), as well as satellite-based OHU time series (Marti et al. 2022; Hakuba et al. 2021).

Although comprehensive assessments of EEI and OHU exist (e.g., von Schuckmann et al. 2020; 2023; Meyssignac et al. 2019; Cheng et al. 2019b, 2022a), there is a lack of systematic intercomparison across different methods and products. Methodological differences and assumptions across in situ-based OHC estimates alone are various and range from different data sources, quality control procedures, mapping/interpolation techniques and prior statistics, to mathematical assumptions in the derivation of OHC and OHU from temperature and salinity profiles, and sampling considerations (Boyer et al. 2016; Cheng et al. 2016a, b, 2022b). Furthermore, measurements of OHC (geodetic or in situ) and net radiative flux (satellite) represent inherently different temporal and spatial scales due to different sampling frequencies and processing, e.g., spatial and temporal interpolation, which need to be better understood and quantified. Recognizing the need for a better understanding of the data products available, their discrepancies, uncertainties as well as sources thereof, the WCRP initiated the formation of the GEWEX-EEI assessment working group to intercompare different OHC and OHU estimates in conjunction with other sinks of heat (e.g., atmospheric heat storage) and EEI variability from radiation budget data. GEWEX-EEI aims to (1) facilitate OHC and OHU intercomparison on global and regional scales by the international community, (2) formulate recommendations and best practices to enable "apples-to-apples" intercomparison across products and their uncertainties, (3) improve

EEI central and uncertainty estimates, and (4) improve understanding of EEI variability. In spring 2023, the community met for the first time at a joint WCRP-ESA EEI assessment workshop at ESA-ESRIN, Frascati, Italy, bringing together experts in radiometric remote sensing, satellite altimetry, space gravimetry, ocean in situ measurements and ocean reanalysis/ocean state modeling to assess and intercompare estimates of EEI, their time variability and uncertainties (Meyssignac et al. 2023b). Key findings and recommendations are presented and discussed in Sect. 4.

As part of this EEI assessment effort, the community was asked to share global annual mean OHC series as produced in-house at respective institutions to highlight the spread incurred by different approaches. Based on data available at the time of analysis, this paper intercompares the different OHC estimates, their variability and their trends expressed as OHU in TOA-equivalent W m^{-2} together with uncertainties (Sect. 3.1). Comparison of OHU interannual variability and CERES global net radiative flux (EEI) is presented in Sect. 3.1. Section 3.3 assesses zonal trends in net radiative flux using CERES EBAF data to elucidate regional variations and their co-variability with trends in clear-sky and cloud radiative effects as well as cloud properties.

The critical need to enable seamless climate continuity observations from space has been recognized by the international satellite community (KISS Continuity Study Team 2024) and both GEWEX-EEI and the Global Climate Observing System (GCOS) Working Group on energy cycle closure recommend addressing looming gaps. Section 3.2 demonstrates the impact of data gaps on OHU estimates and EEI trends, highlighting the need for global coverage and seamless continuity to ensure trend estimates are reliable and trend uncertainties remain as low as possible.

2 Methods and Data

2.1 Approach

To intercompare OHU estimates and trends over 2005–2020, we analyze an ensemble of 21 OHC datasets, comprised of ten OHC datasets based on in situ observations, two satellite-based OHC datasets combining satellite altimetry and space gravimetry measurements, three hybrid methods combining in situ measurements with satellite information and six ocean reanalysis products. OHC estimates from in situ data originate from EN4 (Good et al. 2013), the In Situ Analysis System (ISAS20; Gaillard et al. 2016), Scripps Institution of Oceanography (SIO, Roemmich and Gilson 2009), Cheng and Zhu analysis (IAP, Cheng and Zhu 2016; Cheng et al. 2017; Cheng et al. 2024b), NOAA NCEI (Levitus et al. 2012), PMEL (Lyman and Johnson 2008), JAMSTEC (Hosoda et al. 2008), LocalGP (Giglio et al. 2024), JMA (Ishii et al. 2017) and von Schuckmann and Le Traon (2011), hereafter vS&LT. Satellite-based OHC is provided by Jet Propulsion Laboratory (JPL, Hakuba et al. 2021) and Legos-Magellium (Marti et al. 2022). The hybrid estimates originate from CNR-ISMAR (Storto et al. 2022) and PMEL, namely from PMEL-combined (Lyman and Johnson 2014) and RFROM (Lyman and Johnson 2023) analyses. The six reanalysis datasets considered are ECCOv4 (Forget et al. 2015), OCCA2 (Forget 2024), SODA3 (Carton et al. 2018), CIGAR (Storto and Yang 2024), ORAS-5 (Zuo et al. 2019) and the Copernicus Global Reanalysis Ensemble Product (GLORYS2V4, Lellouche et al. 2018; ORAS-5; C-GLORSv7, Storto and Masina 2016). The datasets are summarized in Table 1 and described in more detail in the supplementary information (SI T1).

Table 1 List of all data products considered in analysis and discussion, including product name, reference and keywords describing the source data and methods

Product name	References	Keywords
In situ data-based		
NCEI	Levitus et al. (2012) https://www.ncei.noaa.gov/products/ocean-heat-salt-sea-level	Argo + other; objective analysis (Locarnini 2018)
LocalGP	Giglio et al. (2024) https://doi.org/10.5281/zenodo.10645137	Argo-only, locally stationary Gaussian processes, data-driven decorrelation scales (Kuusela and Stein 2018)
EN4.2.2	Good et al. (2013) https://www.metoffice.gov.uk/hadobs/en4/index.html	ENACT objective analysis, ensemble of XBT and MBT bias correction schemes
ISAS20	Gaillard et al. (2016) https://www.umr-lops.fr/SNO-Argo/Products/ISAS-in-situ-T-S-gridded-fields/Climate-indices	Argo-only, Optimal Interpolation (Bretherton et al. 1976)
PMEL in situ	Lyman and Johnson (2008) https://oceans.pmel.noaa.gov/	Argo + other; objective mapping (Willis et al. 2004)
IAP	Cheng et al. (2017) http://www.ocean.iap.ac.cn/pages/dataService/dataService.html?navAnchor=dataService	Argo + other; OHC reconstruction using Ensemble Optimal Interpolation approach with dynamic ensemble method (Cheng and Zhu 2016)
JMA	Ishii et al. (2017) https://www.data.jma.go.jp/gmd/kaiyou/english/ohc/ohc_global_en.html	Argo + other, 5° regional averaging (Ishii et al. 2003)
SIO	Roemmich and Gilson (2009) https://sio-argo.ucsd.edu/RG_Climatology.html	Argo-only, Optimal estimation
JAMSTEC	Hosoda et al. (2008) https://www.jamstec.go.jp/argo_research/dataset/moaagpv/moaa_en.html	Argo + other; Optimal interpolation (White 1995)
vS&LT	von Schuckmann and Le Traon (2011) https://marine.copernicus.eu/access-data/ocean-monitoring-indicators	Argo + other; box averaging scheme; non-gridded T (Cabanes et al. 2013)
Satellite-based		
JPL	Hakuba et al. (2021) https://doi.org/10.5281/zenodo.5104970	Geodetic, RL06-V2, large GIA and dataset ensemble

Table 1 (continued)

Product name	References	Keywords
Legos-Magellium	Marti et al. (2022) https://earthobservation.magellium.com/project/moheacan/?lang=en	Geodetic, multi-solution, rigorous error budget
CERES	Loeb et al. (2018) https://ceres.larc.nasa.gov/data/	Radiative fluxes at TOA and surface; cloud, atmosphere and surface properties; energy balanced
Hybrid estimates		
PMEL-combined	Lyman and Johnson (2014) https://oceans.pmel.noaa.gov/	Argo+other, OHC local correlations with SSH (Willis et al. 2003, 2004)
PMEL RFROM	Lyman and Johnson (2023) https://www.pmel.noaa.gov/rfrom	Argo+other, random forest regressions, SSH and SST predictors, resolves eddy scales
CNR-ISMAR	Storto et al. (2022) https://journals.ametsoc.org/view/journals/clim/35/14/JCLI-D-21-0726.1.xml?tab_body=supplementary-materials	Multi-platform optimization: satellite, in situ and reanalysis
Ocean reanalysis		
SODA3	Carton et al. (2018) http://www.soda.umd.edu/	GFDL CM2.5, Optimal Interpolation, ERA-5 (SODA3.15.2)
ECCOv4	Forget et al. (2015) https://ecco-group.org/ohc.htm	Ocean state estimate, MITgcm, 4D-Var
OCCA2	Forget (2024) https://doi.org/10.7910/DVN/CAGYQL	ECCOv4, adjustment via objective mapping of Argo-ECCO4 differences
ORAS-5	Zuo et al. (2019) https://www.ecmwf.int/en/forecasts/dataset/ocean-reanalysis-system-5	NEMO, 3D-Var, ERA-I
CIGAR	Storto and Yang (2024) http://cigar.ismar.cnr.it	NEMO, 3D-Var, ERA-5
GLORYS2V4 (in Copernicus ensemble)	Lelluche et al. (2018)	NEMO, 3D-Var, ERA-5
C-GLORSv7 (in Copernicus ensemble)	Storto and Masina (2016)	NEMO, 3D-Var, ERA-5

Full data descriptions are provided in the Supplementary Information (SI T1)

Many of these datasets were provided to the GEWEX-EEI assessment via https://sites.google.com/magellium.fr/eeiassessment/data-records/, while ECCOv4, SIO, JAMSTEC, SODA3, ORAS-5, Copernicus, OCCA2, CIGAR, vS&LT and RFROM data were either directly provided by the data producer or downloaded from the data producer website. Most of the OHC datasets were provided as global annual averages (JPL, Legos-Magellium, ECCO, ORAS-5, CNR-ISMAR, EN4, ISAS20, NCEI, PMEL, PMELc, LocalGP, Ishii, OCCA2, IAP, vS&LT, Copernicus, CIGAR) and/or as monthly gridded data (SIO, ORAS-5, SODA3, JAMSTEC, RFROM, ECCO, Legos, EN4, IAP, Ishii, NCEI). From monthly OHC, we compute annual averages based on bin-averaging of the monthly values. Gridded datasets are globally integrated to obtain global values in J. All OHC products are converted to ZJ and interpolated in time such that each annual mean is centered on the middle of the year. SODA3, JAMSTEC and SIO do not provide OHC datasets but gridded temperature and salinity profiles. For those, we derive the OHC from the temperature and salinity by vertical integration of the specific heat of seawater multiplied by the local density of seawater and the oceanic temperature using the TEOS-10 GSW software (McDougall and Barker 2011) as, e.g., in Melet and Meyssignac (2015).

Some products only cover the ocean down to 2000 m depth (LocalGP, PMEL, PMELc, ISAS, NCEI, EN4, Ishii, SIO, JAMSTEC, RFROM, vS&LT). For consistency across OHU estimates, we add a constant < 2000 m heating rate of 0.06 W m^{-2} ± 0.04 W m^{-2} (Purkey and Johnson 2010; Johnson et al. 2023b). OHU time series are derived from the time derivative of OHC, using centered differences, which applies a light filtering of the data, slightly reducing the noise compared to first differences, and are normalized by Earth's surface area at the TOA: 5.14×10^{14} m^2 at 20 km above the Earth's surface.

The trend and acceleration of OHC are calculated using an ordinary least squares (OLS) estimator. Uncertainties in the trend and acceleration are given by the variance of the estimator which is derived from each dataset's OHC uncertainties. For some of the OHC datasets, at the time of manuscript submission, uncertainty estimates were not available, and the OHU and OHU trend uncertainties are derived from the linear fit. Some other OHC datasets (CNR-ISMAR, EN4, NCEI, PMEL, ECCO, Copernicus, OCCA2, CIGAR, IAP, ORAS-5, vS&LT, ISAS, LocalGP and Ishii) provide annual estimates of uncertainty or ensemble spread without providing the time correlation across annual uncertainties; hence, their trend and acceleration uncertainties ignore any potential temporal correlation effects. A few OHC datasets (JPL and Legos-Magellium) provide a variance–covariance matrix that describes the annual uncertainties and their time correlation.

As part of our analysis, we investigate the impact of ocean sampling discrepancies on the OHU estimates (or OHC trend in W m^{-2}), OHU trends (or OHC acceleration in W m^{-2} dec^{-1}) and OHU correlation (R) with CERES EBAF net radiative flux for 12 gridded OHC datasets made available to us. We apply a restrictive ocean sampling (ROS) mask that covers ocean areas common to all products (SI, Fig. S1), largely masking shelf areas, coastal areas, shallow seas (> 300 m), marginal seas and polar oceans beyond ± 60° latitude, which are generally not sufficiently sampled by Argo profiling floats. In addition, we test the impact of a mask limited to areas covered by satellite altimetry (i.e., limited to ± 66° latitude) to obtain first-order estimates of sampling uncertainty.

2.2 In Situ Observations

Historically, ship-based observations have been the main source of subsurface ocean temperature information. The Argo Program, designed in 1998 (Argo Science Team

1999; Roemmich et al. 2009), was transformational for subsurface ocean observing, enabling high-quality data (Wong et al. 2020) to be obtained nearly anywhere in the ocean, thus reducing geographical and temporal (seasonal) biases of ship-based observation systems. Argo first achieved significant coverage in both hemispheres in 2005 and reached its initial goal of 3000 profiling floats in November 2007. Its present coverage of about 3880 floats is close to the target of 4000 and is becoming more prevalent in marginal seas, seasonally ice-covered regions and the deep ocean below 2000 m (Jayne et al. 2017). Argo's near-global uniform coverage has resulted in a dramatic reduction of the uncertainty of global OHC changes. Other subsurface observing systems contribute significant subsurface temperature data from 2005 to present, and in some cases, they are the main source of data (mainly in areas shallower than 2000 m depth, marginal seas and ice-covered areas). Observations from research ships (mainly conductivity–temperature–depth—CTD—casts) and ships of opportunity (mainly expendable bathythermographs—XBT drops from merchant ships), moored buoys (especially the tropical moored buoy arrays in the Pacific, Atlantic and Indian Oceans), ice-tethered profilers (in the high Arctic), gliders (mainly on and near continental shelves) and even instrumented pinnipeds all augment and extend the observations provided by the Argo array (Abraham et al. 2013; Meyssignac et al. 2019; Cheng et al. 2022a).

To estimate global integrals of OHC, algorithms are developed to grid temperature and/or OHC data, cope with data-sparse regions, and smooth the temporal and spatial fields. These algorithms are generally referred to as "mapping methods" and represent a leading source of uncertainty (Gregory et al. 2004; Boyer et al. 2016) in global OHC estimation, especially in data-sparse regions and eddy-rich regions with large spatiotemporal variability (Wang et al. 2018).

The ten in situ-based datasets used here (NCEI, LocalGP, IAP, Ishii, JAMSTEC, SIO, EN4, PMEL, ISAS and vS&LT) are listed in Table 1 and described in more detail in SI T1.

2.3 Geodetic Observations

The derivation of geodetic OHC is rooted in analysis of Earth's sea level budget (Marti et al. 2022; Hakuba et al. 2021). To obtain global steric sea level change, global mean sea level (altimetry) and ocean mass change (gravimetry) observations are differenced, considering geophysical corrections such as related to glacial isostatic rebound effects (Caron et al. 2018). The steric change is translated into OHC and OHU using estimates of the ocean's expansion efficiency of heat. Full details on the geodetic OHC products used here (JPL and Legos, see Table 1) are provided in SI T1.

2.4 Ocean State Estimates and Reanalysis

Ocean reanalyses combine multiple data sources through data assimilation in a numerical model (Storto et al. 2019). The types of datasets being assimilated as well as the models and assimilation methods vary between reanalyses. In our comparison, we focus on the six datasets described in SI T1 (ECCOv4, OCCA2, SODA3, CIGAR, ORAS-5, Copernicus; Table 1), which continue to be improved and extended.

2.5 Earth Radiation Budget Data

CERES (Wielicki et al. 1996) currently flies multiple instruments on Terra, Aqua, S-NPP and NOAA20 satellite platforms, collecting and processing broadband shortwave (SW) and longwave (LW) radiances since March 2000. Here, we make use of the CERES Energy Balanced and Filled (EBAF) Ed4.2 product (Loeb et al. 2018), providing global solar incoming, and Earth outgoing net, shortwave (SW) and longwave (LW) radiative fluxes, as well as cloud properties (derived from MODIS and VIIRS radiances) at monthly and 1-degree spatial resolution. Detailed descriptions of the data products and a wide range of publications applying the data in climate analyses can be found here: https://ceres.larc.nasa.gov/. The solar irradiances are derived from time-varying instantaneous total solar irradiance measurements from various sources (Loeb et al. 2018). In our comparisons with OHU variability, we consider the period 01/2005–12/2020. In EBAF, the long-term mean EEI (global long-term mean net radiative flux) is adjusted to match with planetary heat uptake derived from largely in situ observations by applying an offset such that its mean value over the period 2005–2015 is consistent with the mean in situ estimate of 0.71 after Johnson et al. (2016). This offset correction anchors the satellite data to the in situ EEI estimate and does not affect the trend or interannual variability of the EBAF time series (see also Fig. 1); thus, temporal variations in radiometric EEI remain independent of those from the OHU data (Loeb et al. 2021).

3 Results

3.1 Global Ocean Heat Content and Heat Uptake Intercomparison

Figure 2 intercompares annual mean OHC time series, highlighting the two satellite-based products (a), the six series from reanalysis (b), the hybrid products that merge in situ with altimetry and other satellite information (c), and ten OHC time series based on in situ observations—four ingesting Argo data primarily (SIO, LocalGP, ISAS and JAM-STEC), and the remaining six using all available ocean temperature data (d). For reference, we also include planetary heat content derived from integrating CERES EBAF net radiative flux (black lines), which itself has been anchored to PMEL-combined OHU and non-oceanic heat uptake estimates and is therefore not independent. Figure 4a combines all OHC time series and provides a summary of the individual OHC long-term trends or OHU, respectively, including trend uncertainties expressed in W m^{-2} (Fig. 4c). The OHU estimates are normalized with respect to Earth's entire surface area at the TOA, approximately 20 km above Earth's surface: 5.14×10^{14} m^2. The OHU derived from OLS trend analysis is comparable but not identical to OHU derived from differencing the last and first years of smoothed OHC time series (e.g., Cheng et al. 2022a, b; von Schuckmann et al. 2023) or deriving the mean OHU from the time-differenced dOHC/dt times series as in Fig. 3e–h (e.g., as done by Loeb et al. 2022). The trend estimate represents the annual rate of OHC change due to both natural and anthropogenic influences over the period of consideration, which means both forced secular and internal variability (e.g., ENSO) contribute to the change observed and ideally require separation. The observed rate of OHC change or OHU, respectively, has been acknowledged as a clear indication of continued warming of the ocean commensurate with observed increase in greenhouse gas emissions, partially

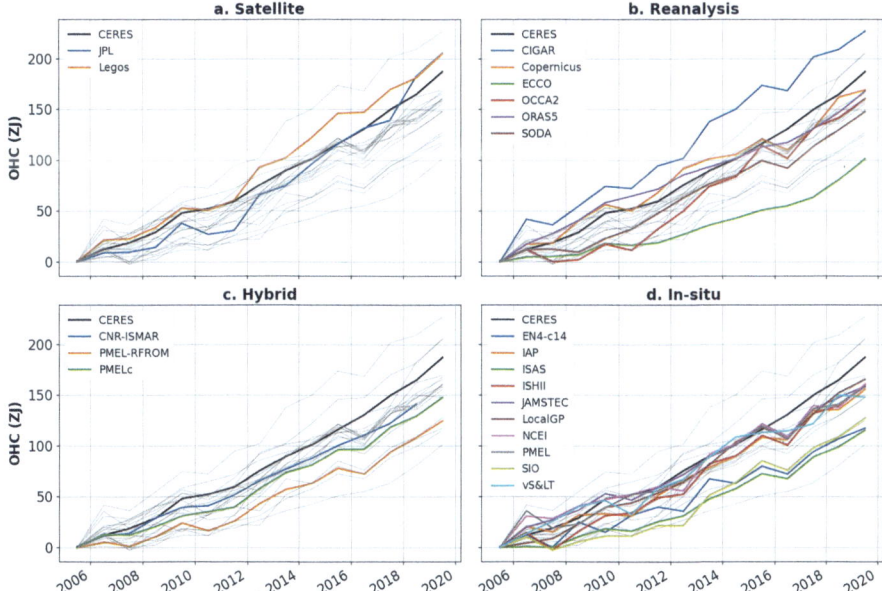

Fig. 2 Annual mean OHC (ZJ) provided by 21 institutions and grouped by overall method and observing system: **a** satellite-based geodetic OHC estimates, **b** OHC time series from ocean reanalyses, **c** hybrid approaches, marrying in situ and altimetry observations (PMELc), machine learning (RFROM) and multi-platform analysis (CNR-ISMAR), **d** OHC derived from in situ observations, considering Argo-only (ISAS, SIO, LocalGP and JAMSTEC) profile data and all available in situ observations (EN4, IAP, Ishii, NCEI, PMEL, vS&LT)

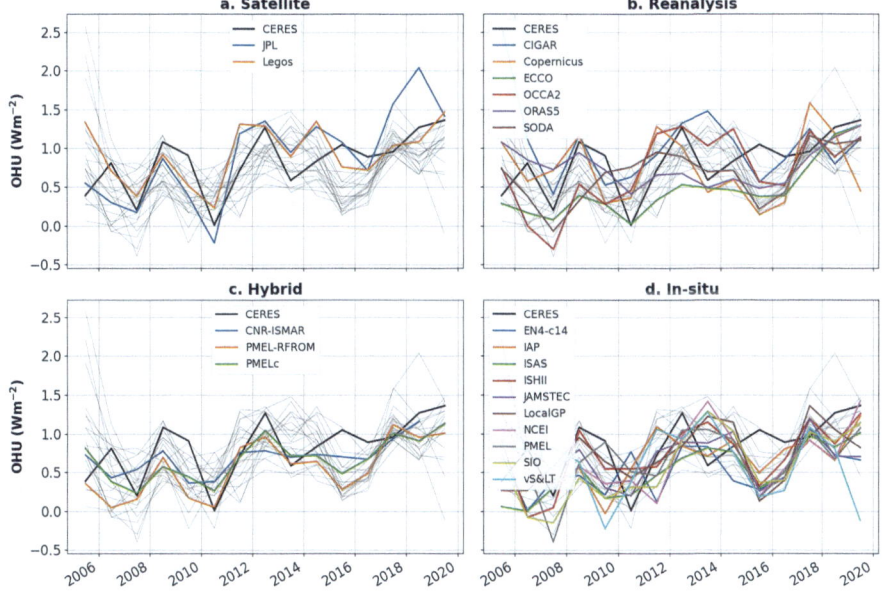

Fig. 3 Same as Fig. 2, but for annual mean ocean heat uptake (OHU) in W m^{-2}

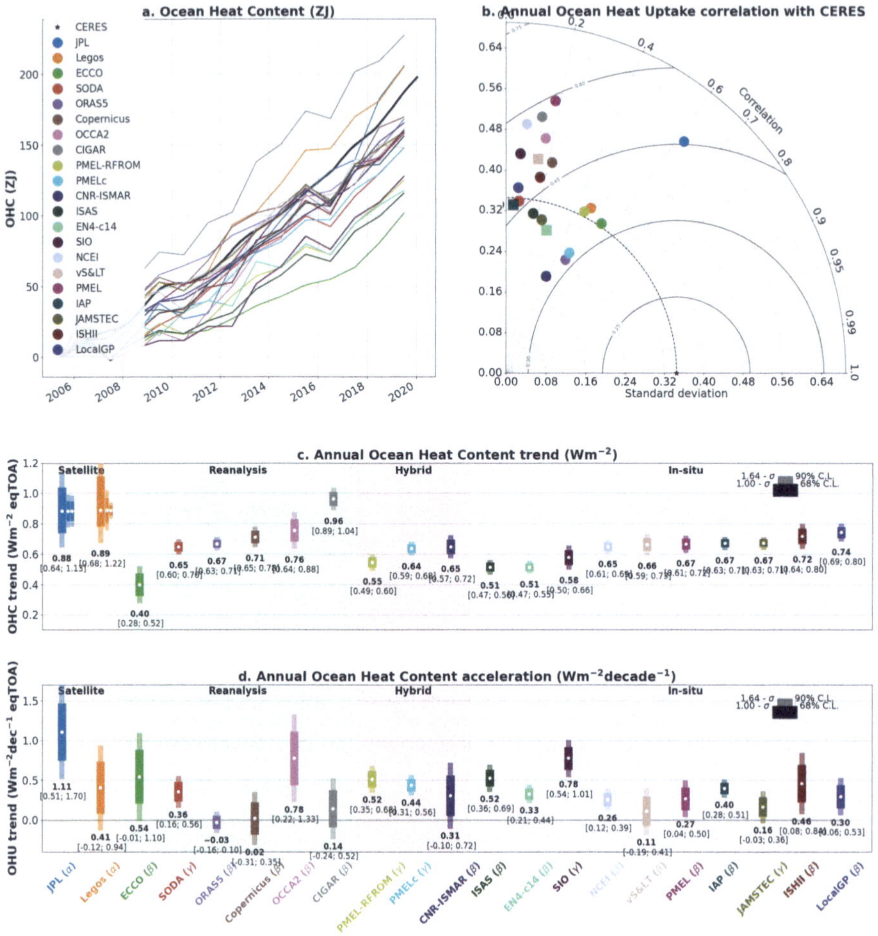

Fig. 4 **a** Ocean heat content (OHC) series (ZJ) derived from 21 data products. **b** Taylor diagram illustrating each detrended OHU time series' standard deviation and correlation coefficient (*R*) with CERES net radiative flux over 2005–2019. Squares indicate a negative correlation. **c** OHC trend or OHU, respectively, including trend uncertainty derived from covariance matrix (α), annual OHC mapping standard errors or ensemble spread (β), and trend line residuals (γ) in terms of 90% and 68% confidence intervals using generalized least squares regression analysis. OHC trends for the JPL and Legos products are calculated three ways left to right using approaches α, β and γ, respectively. (**d**) Same as (**c**) but for OHC acceleration or OHU trend, respectively

compensated by direct aerosol effects, and their radiative forcing (Tokarska et al. 2019; Charles et al. 2020).

The full-column central OHU estimates over 2005–2020 derived from the 21 OHC products vary between 0.40 and 0.96 W m^{-2}, indicating a significant spread with the lowest heating rate obtained from the ECCO ocean state estimate and the largest value derived from CIGAR. The two geodetic OHU estimates agree within uncertainties (0.89 W m^{-2}), exceeding most estimates except for CIGAR. While the reanalysis systems SODA3 and ORAS-5 are in close agreement (0.65–0.67 W m^{-2}), OHU estimates from ECCO (0.40 ± 0.12 W m^{-2}) and CIGAR (0.96 ± 0.12 W m^{-2}) are exceptionally low and high, respectively, representing the lower and upper limit of all estimates provided here. Although not as pronounced as for the reanalysis estimates, the spread across in situ estimates is substantial, from 0.51 (ISAS20 and EN4) to 0.74 (LocalGP) W m^{-2} (amounting to 45% of the lower value) and is significant considering the non-overlapping trend uncertainties at the 90% confidence level in Fig. 3c). In part, the discrepancies across in situ values are associated with different mapping/interpolation techniques, as well as decisions made in the quality control and bias correction of in situ profiles considered (Boyer et al. 2016; Cheng et al. 2016b, 2022b; Tan et al. 2023). Sampling considerations pertaining to the lack of ocean profiles in notoriously under-sampled areas of the ocean, e.g., shallow seas (> 300 m), polar, coastal and shelf areas, yield different coverage areas and ocean volumes considered in the OHU calculations across products, especially between products that use Argo data alone versus products that include profiles from gliders, ice-tethered profilers, XBTs and other in situ observations (e.g., von Schuckmann et al. 2014; Meyssignac et al. 2019; Abraham et al. 2013). Hakuba et al. (2021) found that Argo-only datasets produced larger OHU rates than Argo+other in situ products, which is not strictly the case in the present analysis; however, the primarily Argo-ingesting LocalGP and JAMSTEC products indeed reside at the upper range of in situ-based OHU estimates. Discrepancies in ocean area/volume sampled accounts for some of the spread in OHU across in situ datasets (see Sect. 3.2). Similarly, geodetic observations are constrained to ocean areas equatorward of ±66° latitude given the availability of altimetry data (Marti et al. 2022). Thus, except for ocean reanalyses, and the IAP, Ishii, EN4 and NCEI data products, none of the observed OHC and OHU changes is truly representative of the full global ocean. In Sect. 3.2, we investigate the impact of applying a restrictive ocean sampling mask (Fig. S1) to a subset of twelve gridded OHC dataset and of a mask that limits the calculation of near-global OHC to the satellite sampling for insight on potential sampling uncertainties and their implications for the intercomparison.

From the time derivative (centered differences) dOHC/dt, we derive annual mean OHU time series (Figs. 3, 5, 6). OHU time series are expected to track year-to-year variability in global mean TOA net radiative flux on annual timescales, when the Earth system is near energetic equilibrium and year-to-year heat uptake variations in other Earth system components are assumed to be small (e.g., Loeb et al. 2012, 2021). We assess the agreement in detrended year-to-year variability by providing each products' correlation coefficient with detrended CERES EBAF net radiative flux in the Taylor diagram (Figs. 4b, 5 sub-panel legends) together with each time series' average amplitude expressed as the standard deviation of the time series. Correlation coefficients exceeding 0.44 are obtained for the PMEL-combined (0.47), RFROM (0.44), ORAS-5 reanalysis (0.47), ECCO ocean state estimate (0.55), JPL (0.62) and Legos geodetic OHU (0.46) time series. Correlation coefficients smaller than +0.15 are found for the in situ datasets JAMSTEC, IAP, SIO, LocalGP, NCEI, EN4, vS&LT and the CIGAR and SODA reanalysis. Largest standard deviations exceeding the CERES variability by more than 0.1 W m^{-2} are found for the JPL geodetic OHU time series, PMEL in situ, NCEI, OCCA2, CIGAR and SIO datasets. Eight of the time series produce standard deviations smaller than in CERES EBAF (0.34 W m^{-2}) at this annual timescale. The amplitude and trend in JPL geodetic OHU are sensitive to the estimate of expansion efficiency needed to translate derived steric sea level change to OHC and OHU; hence, the large amplitude in OHU variability in the JPL product is at least partially the result of a smaller expansion efficiency considered (derivation in Hakuba et al. 2021) compared to that in the Legos product (Marti et al. 2022). Further study is needed to find consensus on the magnitude and variability of this critical conversion factor in order to improve both global and regional OHC and OHU estimates from geodetic observations.

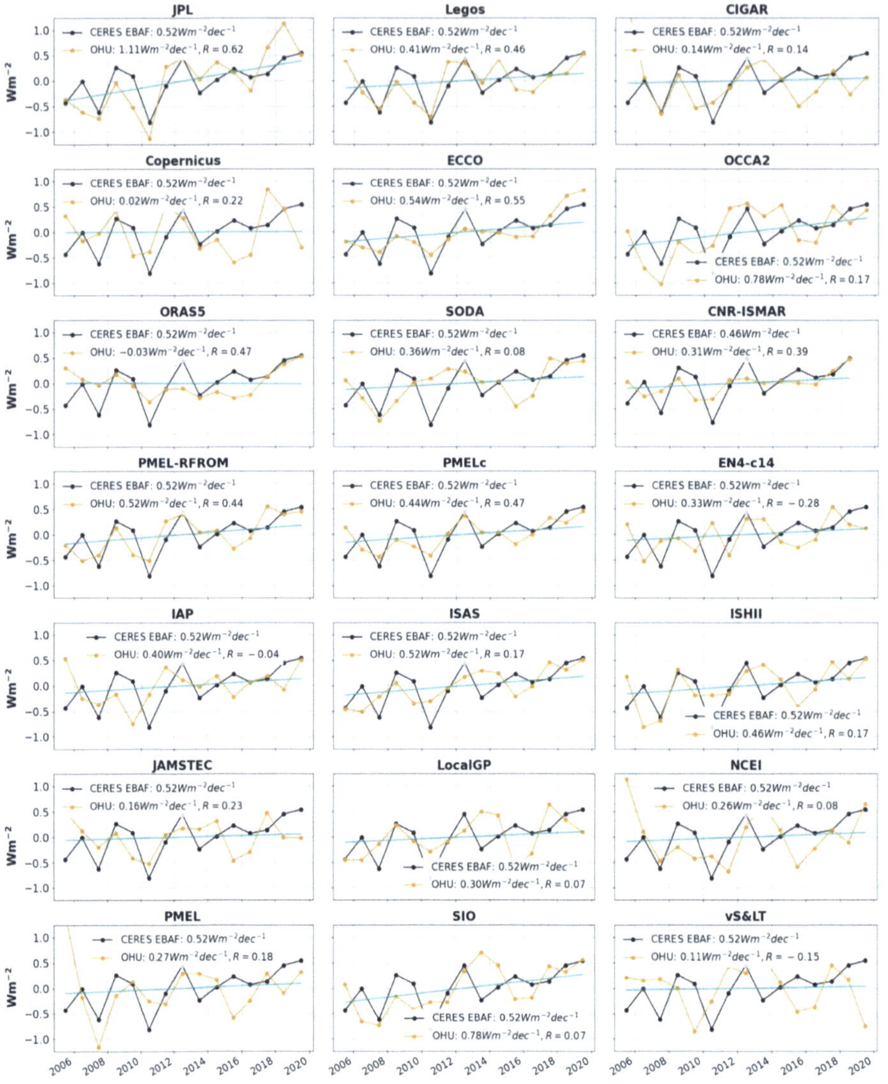

Fig. 5 Annual OHU anomaly time series (dOHC/dt from centered differences, long-term mean removed) derived from 21 OHC products and compared to CERES net radiative flux (black line). Green lines indicate the trend line through the OHU series. Trend magnitude and detrended correlation coefficient are provided in the legend of each sub-figure. Gray shading in the JPL sub-panel indicates the GRACE-FO data gap during which the annual means are based on less than 12 months of data

In terms of co-variability, it is evident that products exploiting temporal and spatial patterns of OHC anomalies and their local correlation with sea surface height (PMEL-combined), and with sea surface height and SST (RFROM), as well as the two satellite-based estimates and the ECCO and ORAS-5 reanalysis exhibit largest correlations (>0.44) with CERES EBAF. All of these products either observe or fill in data-sparse regions based on additional physical knowledge to avoid relaxation of OHC maps back

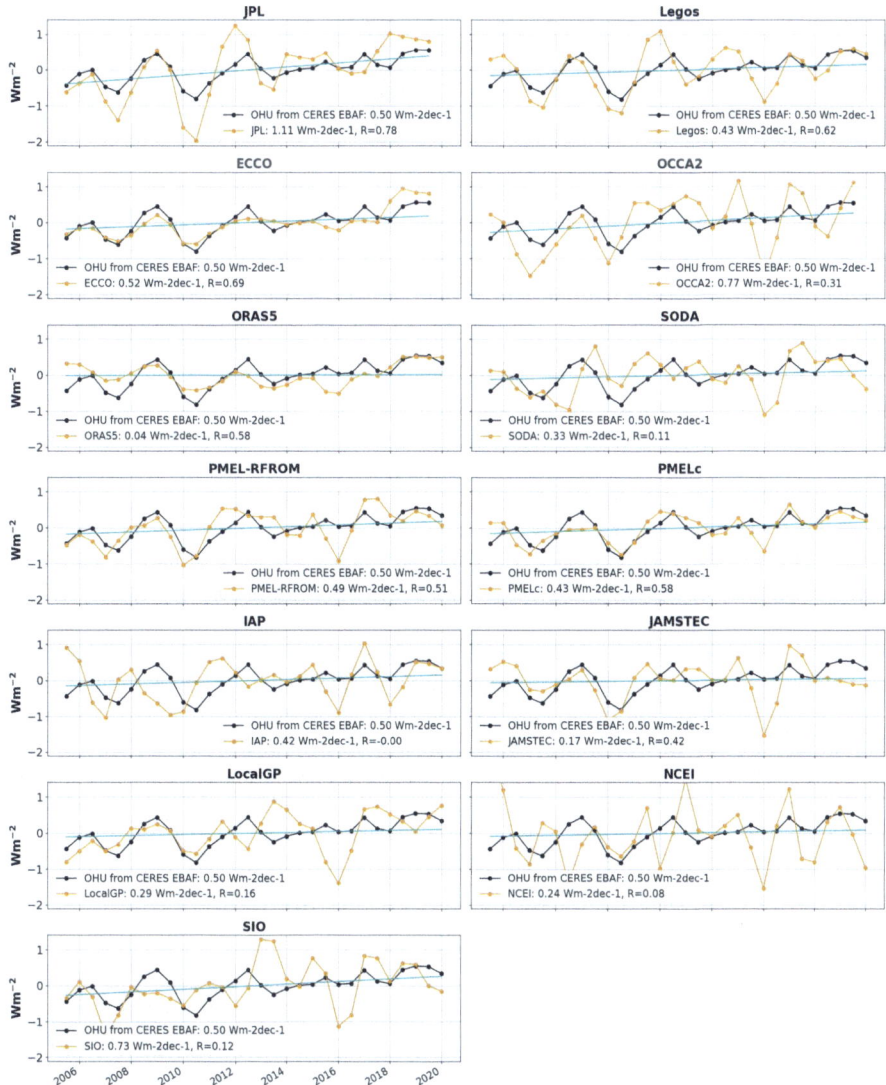

Fig. 6 Same as Fig. 5 but for 13 products comparing annual means at 6-month increments centered mid-year (January–December) and end of year (July–June) akin to analysis in Loeb et al. (2021), their Fig. 1, and reproduced in the panel titled "PMELc." Gray shading in the JPL sub-panel indicates the GRACE-FO data gap during which the annual means are based on less than 12 months of data

to a climatological mean (Durack et al. 2018; Cheng et al. 2019a, b; Lyman and Johnson 2023).

The co-variability of 13 OHU annual mean time series at 6-month increments centered mid-year (January–December) and end of year (July–June) akin to analysis in Loeb et al. (2021) is shown in Fig. 6 and shows that all correlation coefficients slightly increase except for the IAP and NCEI data, by up to +0.19 (JAMSTEC). In line with Loeb et al. (2021), we find good agreement in both correlation and OHU trend between EBAF and the PMEL-combined OHU series and provide an update in the respective panel of Fig. 6. Again, largest correlations exceeding 0.4 are common for reanalysis, satellite and the SSH/SST-informed mapping methods, and are now even more pronounced (up to 0.78 for the JPL product).

In terms of OHU trends, or OHC accelerations (W m^{-2} dec^{-1}, Figs. 4d, 5, 6), respectively, all products yield increase over the observational period, albeit some estimates are not significant at 90% confidence (Legos, ORAS-5, CNR-ISMAR, CIGAR, Fig. 3c). Trends near or below 0.25 W m^{-2} dec^{-1}, and therefore less than 50% of the CERES trend, are obtained for the ORAS-5, CIGAR, Copernicus, vS&LT and JAMSTEC products. Central OHU trend estimates exceeding the CERES trend in net radiative flux originate from SIO, JPL, ECCO and OCCA2 products. Considering the OHU trend uncertainty derived by Loeb et al. (2021) at 0.50 ± 0.47 W m^{-2} dec^{-1} and by data product, these discrepancies are not significant at the 90% confidence level and indicative of accelerated ocean warming or increase in EEI, respectively.

Applying a low-pass filter (Lanczos) with a cutoff period of 3 years as in Marti et al. (2022) removes high-frequency content related to intrinsic ocean variability (Palmer and McNeall 2014) and the mesoscale activity that is visible in altimetry but not in gravimetry, and improves the co-variability of the satellite-based Legos product with CERES net radiative flux (Fig. 7), from 0.62 (annual at 6-month increments) to 0.66. Different smoothing filters and their advantages are under investigation by multiple groups (e.g., Trenberth et al. 2016; Lyman and Johnson 2023; Marti et al. 2022), but their improvement of year-to-year variability, compared to CERES data, is not expected to exceed or fully meet the positive impact of more complete spatiotemporal sampling (e.g., geodetic) or regional filling (e.g., PMEL-combined, RFROM). However, both the "running average" at 6-month increments (Fig. 6) and the low-pass filter provide enhanced R coefficients, ascertaining that variations at timescales shorter than 1–3 years are non-representative of EEI variability, impede the direct comparison with CERES data and ought to be considered, minimized and better understood.

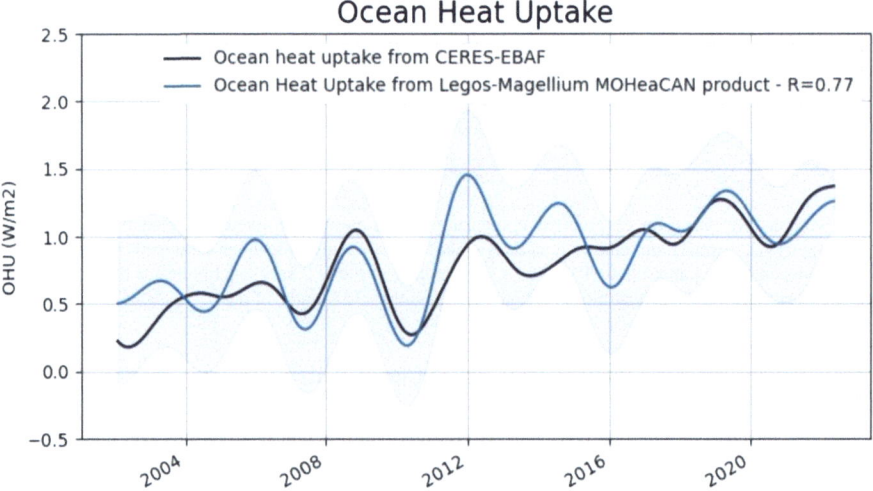

Fig. 7 MOHeaCan (Legos) time series of OHU (blue) and 90% of CERES EBAF net radiative flux (black). Both time series are low-pass-filtered at three-year cutoff time (Lanczos) to remove high-frequency noise related to intrinsic ocean variability (Palmer and McNeall 2014)

Across the 21 different OHC products, we find (1) that CERES net flux and OHU year-to-year variability agrees remarkably well for the satellite-based, two reanalysis, as well as the in situ + satellite hybrid products, suggesting a key to agreement is as complete a spatiotemporal ocean coverage as possible; (2) that datasets that match CERES variability ($R > 0.44$) also agree with a positive trend similar in magnitude, reinforcing the notion of accelerated ocean warming and increase in EEI; (3) that spatial sampling considerations impact the validity of our intercomparison as well as the interpretation of global OHU estimates, OHU trends and correlations with CERES data (see Sect. 3.2); and (4) that smoothing short-term variability in OHU and running averages at the sub-annual scale improve the co-variability with CERES data.

3.2 How Gaps Impact Trend Estimates and Their Accuracy

3.2.1 Ocean Sampling Considerations

To investigate the impact of inconsistent ocean sampling across data products on OHU estimates, OHU trends and their correlations with CERES data, we subsample twelve gridded OHC datasets with the very same restrictive ocean sampling (ROS) mask (SI Fig. 1) prior to computing "global" OHU (orange dots in Fig. 8). Likewise, we apply a mask that limits the near-global OHC calculation to the satellite coverage ($\pm 66°$ latitude, green dots) and compare the results to the unrestricted OHU estimates based on the original gridded data products and their native ocean coverage (blue dots, Fig. 8c). Applying the satellite mask, the OHC trends or OHU, respectively, are reduced in reanalysis products, by about 5% (0.04 W/m^2; relative to SODA3 OHU), while the impact on in situ data, which in many cases do not exceed coverage beyond $\pm 60°$ latitude (except NCEI), is marginal as expected. For reanalyses that much depend on assimilated observations, uncertainties are largest in under-sampled areas, impeding on the interpretation of this result. In addition, a 5% sampling uncertainty is well within OHU uncertainty of most data products (Fig. 4). Subsampling with the ROS mask affects all estimates, reducing satellite-based OHU by 17% (0.15 W m^{-2}; relative to Legos OHU) and OHU from reanalysis by similar

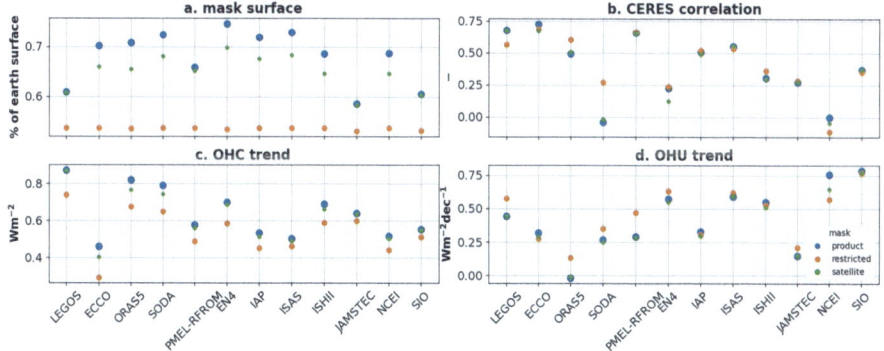

Fig. 8 Restrictive ocean sampling (ROS) mask experiment: **a** Percentage of Earth's surface covered by each gridded product (blue dots), after applying the satellite mask ($\pm 66°$ latitude) (green dots) and after applying the ROS mask (orange dots). **b** Each product's correlation of OHU time series with CERES net radiative flux using different masks, **c** same as (**b**) but for OHC trend or OHU, respectively. **d** Same as (**b**) and (**c**) but for the resulting OHU trend or OHC acceleration, respectively

amounts. In situ-based OHU is at most reduced by 16% for EN4 and by no more than 6% for the Argo-only ISAS20 product. For the in situ OHU estimates, after applying the same ROS mask, the spread across the seven products (EN4, IAP, ISAS, Ishii, JAMSTEC, SIO, NCEI) reduces by 30% from 0.51–0.71 to 0.44–0.60 W m^{-2}, indicating that a common mask is required for the purpose of intercomparison and to reduce systematic differences in OHU across in situ products due to sampling considerations. The remaining discrepancies across Argo-only and across all in situ OHU estimates are dominated by differing source data, mapping techniques, varying baseline climatologies, quality control and bias correction procedures (Boyer et al. 2016; Schuckmann et al. 2014; Meyssignac et al. 2019). At the same time, this experiment highlights the need for as complete a global ocean coverage as possible to represent a true global estimate of OHU, such that no heat is "missed" by under-sampling.

Subsampling with the ROS mask, excluding shallow, coastal and marginal seas, also impacts the OHC acceleration or OHU trend, respectively, in W m^{-2} dec^{-1} (Fig. 8d), as well as the correlation between the OHU time series and CERES net radiative flux (Fig. 8b). The impact on the correlations is most evident for the spatially more complete satellite-based and reanalysis OHU series. For the satellite-based Legos product and ECCO reanalysis, the correlations decrease, potentially indicating the more complete coverage enhances the correlation with CERES; however, the correlations for SODA3 and ORAS-5 increase when applying the ROS mask. Sampling according to the ROS mask, increases the OHU trends for most products overall (except NCEI and ECCO), suggesting a larger OHU trend is observed when the ocean is sampled by the Argo system alone, and could potentially be slightly skewed toward sampling regions of faster warming (Fig. 8d). Masking out the polar regions poleward of ±66° latitude affects neither correlations nor OHU trends significantly in any of the data products, including the reanalysis products, suggesting the polar OHU changes might, currently, play less of a role for tracking global OHU change and the co-variability with CERES EBAF.

Spatial sampling considerations clearly impact the spread between products in our intercomparison of OHC products, the magnitude of "global" OHU estimates, the OHU correlations with CERES net radiative flux and OHU trends in different ways, which require further investigation. Applying the ROS mask reduces spread across products slightly, enabling a more consistent comparison, while at the same time reducing agreement with CERES variability and the OHU magnitude. This suggests more complete ocean volume coverage is instrumental in capturing the "true" global OHU magnitude and change.

3.2.2 Gaps in OHC and EEI Time Series

Here, we investigate the impact of hypothetical data gaps in OHC and EEI climate data records to better understand what implications observing gaps might incur on trend analysis. Currently there is no plan for a follow-on Earth Radiation Budget mission post-Libera, which will be launched in 2027 on JPSS-4 and has a projected lifetime of 5 years (Harber et al. 2023; Hakuba et al. 2024). The in situ ocean observing system (e.g., Argo) is unlikely to suffer from sudden and complete interruptions that would create gaps in a global record, while satellite-based geodetic observations theoretically are.

We quantify the impact of artificial gaps on OHU estimates and EEI trend magnitude and uncertainty, omitting the role of measurement uncertainty. The analysis does not consider (time-varying) calibration uncertainties (e.g., accounted for by Loeb et al. 2009b; Wielicki et al. 2013) nor absolute calibration shifts in the data record after the gap (Loeb et al. 2009b). This analysis therefore solely demonstrates a lower margin of statistical uncertainty introduced by gaps of different lengths (1–25 months) and gap location in the data record.

Our starting point is monthly anomalies of ocean heat content (Fig. 9a) from NOAA NCEI (Levitus et al. 2012). We introduce a gap of at least 1 month and at most 25 months long in the OHC record, varying the gap starting point between the beginning and end of the time series. For each gap ensemble member, we calculate the OHC trend (OHU) and trend uncertainty (95% confidence level, CI95) expressed in W m^{-2}, which is 0.58 W m^{-2} ± 0.06 W m^{-2} for the gap-free record. Figure 9a shows the OHC time series including a 1-year gap (gray line). The red line indicates the linear regression performed to obtain the OHC trend or OHU, respectively, expressed in W m^{-2} with respect to the global surface area. Figure 9b scatters the trend bias in % for all gap-afflicted OHC series against the gap length and is colored by the gap starting year. With increasing gap length, the mean absolute bias (black line) as well as the maximum and minimum biases by gap length (gray envelope) increase, reaching a maximum bias at −9% for a 2-year gap length placed at around 2020. The trend bias expressed in % of the original CI95 is shown in Fig. 9b. It appears that gaps placed toward the end of the record lead to negative biases, while largely positive biases occur with gaps at the beginning of the time series. This is probably related to the flatter increase in OHC at the beginning versus an accelerated increase toward the end of the OHC record. Similarly, but less linearly, the absolute bias in trend uncertainty (95% confidence interval) increases with gap length, reaching maxima near −12% when placed near the middle of the record (Fig. 9d). Positive biases exceeding +6% occur for

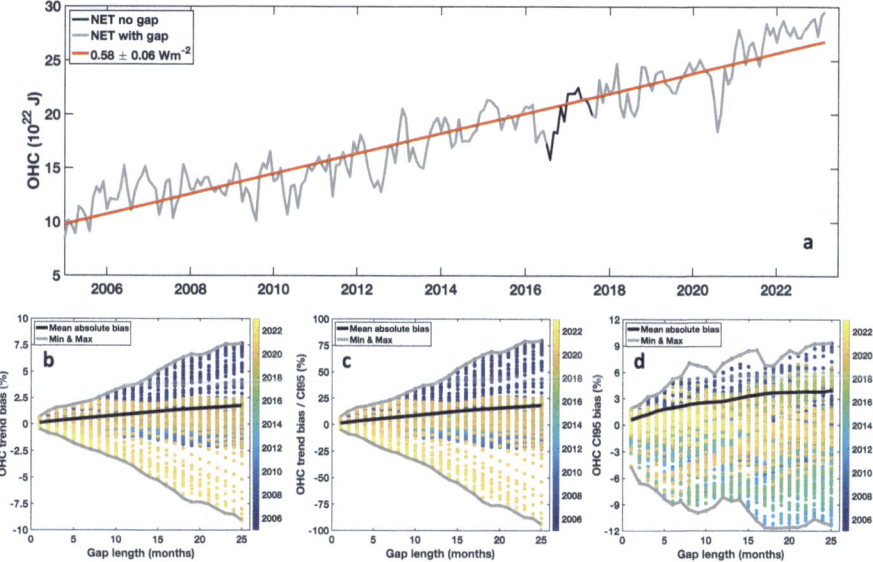

Fig. 9 a OHC monthly anomalies (NOAA NCEI), together with an example gap of 1-year length (black line) and the linear regression trend line (red). **b** OHC trend biases in percent of the original, gap-free trend magnitude (0.58 W m^{-2}) as a function of gap length and gap starting year (color bar). The black line indicates the mean absolute trend bias, the gray line envelopes the maximum and minimum biases incurred by the gaps. **c** Same as (**b**) but for OHC trend biases in % of the original, gap-free trend uncertainty (95% confidence interval, CI95: ± 0.06 W m^{-2}). **d** Same as (**b**) but for the trend uncertainty (CI95) bias in percent

gaps placed at the beginning of the record, a period of modest global OHC increase and variability.

Next, we examine trends in the monthly anomalies of global mean net radiative flux (CERES EBAF Ed.4.2). Same as above, we introduce a gap of at least 1 month and at most 25 months, varying the gap starting point between the beginning and end of the time series. For each gap of different starting times, we calculate the mean absolute bias, maximum and minimum biases in % of the gap-free trend (0.50 ± 0.15 W m^{-2} dec^{-1}). We do the same for the trend uncertainty (CI95). Figure 10a shows the time series including a 6-month gap indicated by the gray line. The red line illustrates imputation of the missing values by linear interpolation as a form of uninformed gap filling. Figure 10b presents the trend biases by gap length with gaps omitted in the trend calculation (no fill, black line in Fig. 10a) and is colored by the gap location starting year. Omitting the gap appears to, on average, introduce trend bias of up to 3% for gaps up to 25 months long, the maximum

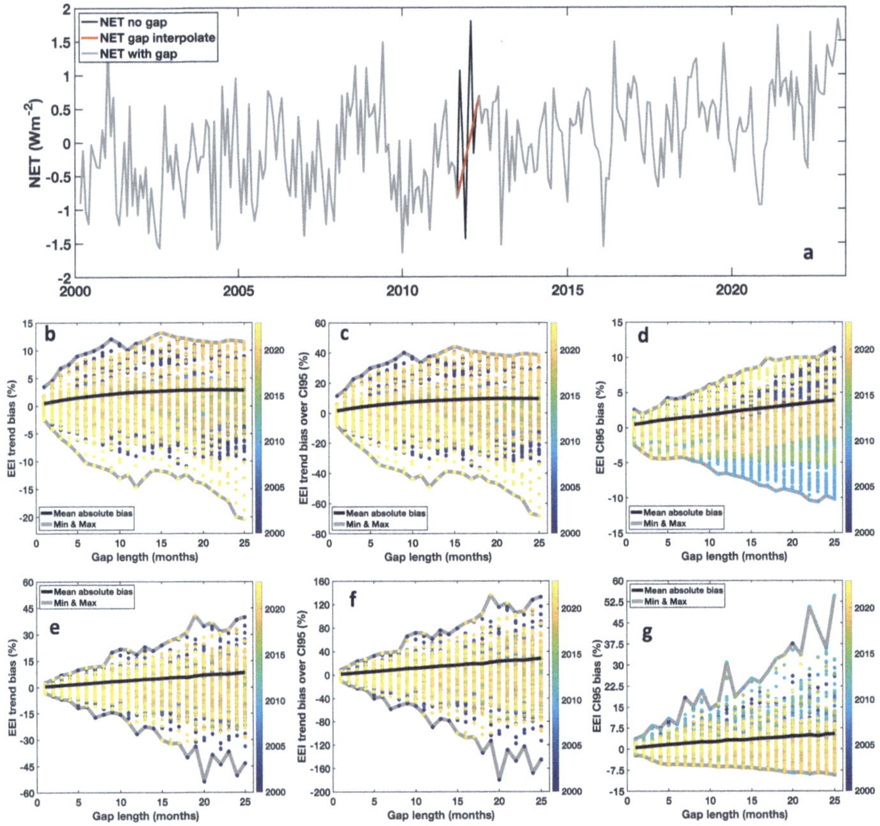

Fig. 10 **a** CERES EBAF EEI monthly anomalies, together with an example gap of 6-month length (black line) and a linear interpolation line across the gap (red). **b** EEI trend biases in percent of the original, gap-free trend magnitude (0.50 W m^{-2}) as a function of gap length and gap starting year (color bar). The black line indicates the mean absolute trend bias, the gray line envelopes the maximum and minimum biases incurred by the gaps. **c** Same as (**b**) but for EEI trend biases in % of the original, gap-free trend uncertainty (95% confidence interval, CI95: ± 0.15 W m^{-2}). **d** Same as (**b**) but for the trend uncertainty (CI95) bias in percent. **e**–**g** Same as (**b**), (**c**), (**d**) but for time series with gaps imputed using linear interpolation (e.g., red line in a)

trend bias incurred is up to 20% for a 25-month gap toward the end of the data record in 2021 (Fig. 10b). Clearly, the longer the gap, the larger the trend bias incurred. This relationship is more pronounced when the gaps are imputed by linear interpolation rather than being omitted (Fig. 10e). The linear interpolation completely disregards the true natural variability during the gap and appears to bias the trend even more, on average by up to 8% for a 25-month gap and at maximum by 54% for a 20-month gap at the beginning of the record. Note that the internal variability (in terms of standard deviation, not shown) is slightly larger in the first half of the record compared to the second, which might explain the sensitivity to gaps at the beginning of the record. Non-informed gap filling can worsen the situation, and gaps of 2 months or longer require more sophisticated, data and physics-informed imputation to not degrade the trend quality as much.

Trend uncertainty (Fig. 10d, g) is sensitive to the gaps as well and shows a near-linear increase with gap length, with enhanced trend accuracy biases toward the end and beginning of the record. Likewise, uninformed gap imputation through linear interpolation increases bias in trend uncertainty even further.

The trend biases normalized by the trend uncertainty derived for the gap-free record (Fig. 10c) indicate that none of the trend biases is significant within trend uncertainty, but 2-year gaps placed toward the end of the record come close to, inducing trend bias of up to 70% of the trend uncertainty. As expected, for the imputed gaps of 15 months and longer, trend biases exceed trend uncertainty (Fig. 10f), implying significant impact of data gaps with uninformed imputation.

Assuming a gap is incurred because of measurement discontinuity or switch to a different measurement platform, then calibration uncertainty and differing instrument characteristics will certainly impede on the quality of trends without overlap or intercalibration capability (Loeb et al. 2009b), potentially making it impossible to tie time series together and derive meaningful trends. This analysis shows that a data gap in a record assumed to be free of measurement uncertainty and inhomogeneities can increase trend uncertainty significantly as a function of gap length and location alone. It also shows that gap imputation can worsen the impact of gaps if done improperly. The gap impacts are more pronounced when identifying trends in EEI from the CERES record, compared to the gap impacts on OHU from the OHC series. This is not surprising but indicates that the impact of gaps is also a function of signal-to-noise ratio; hence, it depends on how reliable the trend is to begin with. The rise in OHC is 20 times larger than the residual standard error, while the EEI trend exceeds the standard error by a factor of six only. The CERES team examined the feasibility of using less accurate imager retrievals to compute radiative fluxes and tie ERB time series before and after a data gap together (RBSP 2018). It was found that this "bridging" method yields uncertainty in net TOA flux that is too large to detect meaningful decade-to-decade changes in EEI. As of now, no comprehensive study exists on potentially suitable methods to bridge gaps and minimize their impact on satellite-derived EEI trends.

3.3 Zonal Trends in Net Radiative Flux

To identify regional patterns of change in net radiative flux (NET), we examine zonal distributions of +20-year trends in NET, absorbed shortwave (SW) and absorbed longwave (LW) radiation. We compare observed global and area-weighted zonal trends in all-sky, clear-sky and cloud radiative effects (CRE), as well as in snow/ice cover and cloud properties. Studies that attribute observed radiative and EEI changes to geophysical processes, forcings and feedbacks require supplemental analysis of climate model experiments (e.g.,

Raghuraman et al. 2021), radiative kernel techniques (Kramer et al. 2021) and radiative perturbation analysis (Loeb et al. 2021, 2024).

In Fig. 11, we show area-weighted zonal trends (March 2000–June 2023, 1° latitude bands) in radiative flux anomalies normalized by the average zonal area. This way, the arithmetic average of area-weighted zonal trends equals the area-weighted global mean trend, enabling relative comparison of trends across latitudes and with the global mean. The radiation data as well as surface, atmospheric and cloud properties are taken from the most recent CERES EBAF Ed.4.2 and SYN Ed.4.1 data records. Figure 11a shows the area-weighted zonal trends for net radiative flux (NET) as well as the clear-sky (b) and cloud radiative effect (CRE, c) components. The all-sky NET trends are positive throughout all zones, implying the increase in EEI is evident across the globe. Zonal peak trends of 0.5 W m^{-2} dec^{-1} (global mean trend is 0.50 ± 0.15) or larger (orange dots), which are statistically significant at the 95% level (purple dots), occur in the deep tropics of the southern hemisphere (SH), tropics and subtropics of the northern hemisphere (NH), and at SH high latitudes poleward of 58° South. Figure 11b and c suggests the global mean NET trend (green line) is almost solely established through a positive NET clear-sky trend (0.48 ± 0.12), partially complemented by a small positive NET CRE trend (0.02 ± 0.13), and largely associated with change in absorbed clear-sky SW radiation (0.35 ± 0.10; Fig. 11e). Near both poles and the NH mid-latitudes, the zonal distribution of clear-sky

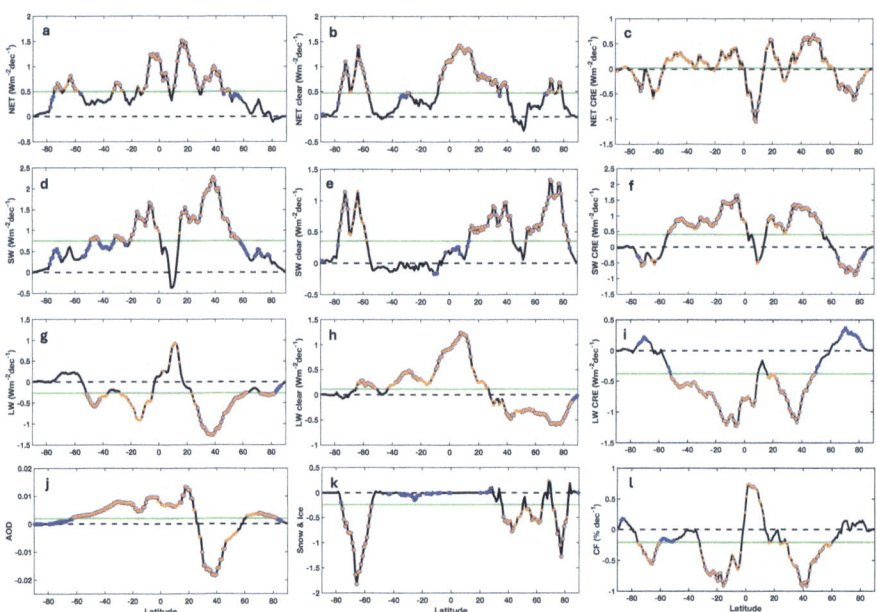

Fig. 11 Area-weighted zonal mean decadal trends in net radiative flux (NET, **a**), clear-sky NET (**b**) and NET cloud radiative effect (CRE, **c**). Decadal trends in shortwave absorbed flux (SW, **d**), clear-sky SW (**e**) and SW CRE (**f**). The same for longwave absorbed flux (LW, **g**), clear-sky LW (**h**) and LW CRE (**i**). Lastly, decadal trends in AOD (**j**), snow and ice cover (**k**), and cloud fraction (**l**). The purple dots indicate zonal mean trends that are significant at the 95% confidence level; the orange dots indicate absolute trends that are equal to or exceed the absolute global mean trend indicated by the green line. The gray shading indicates the zonal trends' 95% confidence interval. Positive trends in **a-i** indicate increase in radiative absorption, i.e., contributing to positive net radiative trend

SW trends appears to primarily correspond to decreases in snow and ice cover (Fig. 11k). Negative changes in aerosol optical depth (AOD, Fig. 11j) appear to add to the positive clear-sky SW trends between 15° and 40° North. Positive clear-sky LW trends are most pronounced in the tropics and shape the NET clear-sky changes in this region, aligning with the expected and regionally enhanced "super greenhouse effect" (Stephens et al. 2016; Raghuraman et al. 2019). Positive trends South of, and negative trends North of 30° N appear to largely cancel in the global mean, resulting in a global clear-sky absorbed LW trend at 0.12 ± 0.10 W m^{-2} dec^{-1}.

The near-zero global NET CRE trend originates from regional cancelations (Fig. 1c) with significant negative trends poleward of 60° North and South, and in the NH tropics that partially, but incompletely, compensate for positive clear-sky trends in these regions, most evidently in the NH polar region. The deep tropical "dip" in CRE trends could be associated with an observed narrowing and intensification of the tropical convection zone (e.g., Wodzicki and Rapp 2016, 2022; Su et al. 2017; Byrne et al. 2018). Significant positive NET CRE trends primarily occur in the NH mid-latitudes. The zonal distribution of NET CRE trends correlates more strongly with SW CRE trends ($R = 0.93$) than with LW CRE trends ($R = -0.69$). On the global scale and by latitude, SW CRE (0.40 ± 0.13 W m^{-2} dec^{-1}) and LW CRE (-0.38 ± 0.06 W m^{-2} dec^{-1}) trends nearly cancel each other, with the overall positive SW CRE dominating the NET CRE changes. Zonal trends in cloud fraction (global trend: -0.20 ± 0.10 W m^{-2} dec^{-1}, Fig. 11l) largely align with the zonal changes in NET CRE ($R = -0.36$), but the zonal trends in cloud optical depth (COD, global trend: 0.02 ± 0.02 W m^{-2} dec^{-1}, Fig. 12a) exhibit a higher correlation with zonal NET CRE trends (-0.63), underscoring the relevance of cloud microphysical change in potentially driving the net radiative changes observed (see also Stephens et al. 2024, this issue). To further investigate the SW and LW CRE cancelation effects in the tropics, we contrast the positive NET CRE trends in the SH tropics (5°–20° South) with the negative CRE changes in the NH tropics (0°–15° North). Figure 12c–h shows three scatter plots per region, each exploring the relationship between LW and SW CRE regional trends, with the dots (regional trends) colored by either cloud fraction (CF, %), cloud optical depth (COD) and cloud top pressure (CTP) trends. It is evident for both tropical regions, even though the resulting zonal NET CRE trends are opposite, that the relationship between regional LW and SW CRE trends is very similar, with the CRE trends being negatively correlated ($R = -0.8$) and dominated by SW CRE effects as indicated by the negative and smaller than 1 slope of the regression line. This means, on average, positive SW CRE trends tend to overcompensate for negative LW CRE trends, and negative SW CRE largely outweigh positive LW CRE trends. In both regions, positive SW CRE trends that outweigh negative LW CRE trends (orange triangle) are largely associated with a decrease in cloud cover, as well as decreased cloud thickness and height. Likewise, the negative SW CRE trends overcompensating for positive LW CRE trends (green triangle) are associated with an increase in cloud cover and clouds getting both thicker and higher. The latter appears to be the case for the deep tropical dip region which coincides with the mean location of the ITCZ (~12° North, e.g., Wodzicki and Rapp 2016). In the NH mid- and high latitudes, the inverse relationship between LW and SW CRE trends is not as pronounced (not shown). The overall positive NET CRE trends in the mid-latitudes is dominated by positive SW CRE trends associated with a decrease in cloud cover and cloud thickness. Likewise, negative NET CRE trends at high latitudes are mostly associated

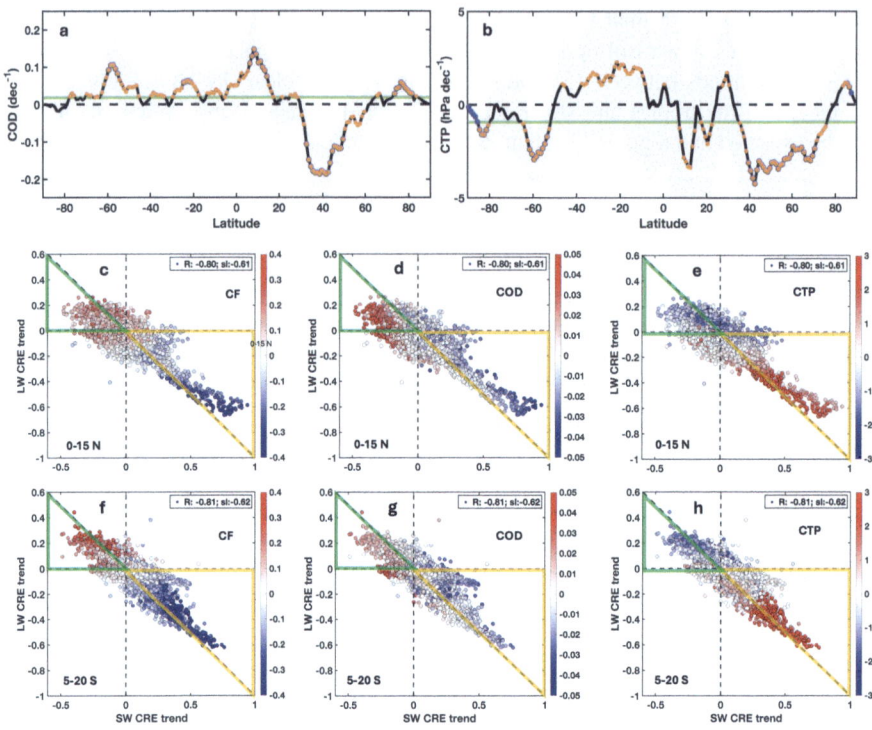

Fig. 12 a and **b** Same as in Fig. 11, but for cloud optical depth (**a**) and cloud top height (**b**). **c–e** Scatter plots of LW versus SW CRE regional +20-year trends in the tropical regions between 0° and 15° North, and are colored by trend in cloud fraction (**c**), cloud optical depth (**d**) and cloud top height (**e**). **f–h** Same as c–e but for tropical regions between 5° and 20° South

with negative SW CRE trends (not shown). The tropical cases suggest that the thinning and thickening of clouds plays an important role in modulating SW and NET CRE changes, which is further investigated for the tropics by Stephens et al. (2024). Zonal and regional NET CRE trends show no clear relationship with changes in cloud height ($R = 0.02$), but more so with cloud cover ($R = -0.36$) and mostly with cloud thickness trends ($R = -0.63$), further underpinning the role of SW CRE and the optical properties of clouds. It is furthermore interesting to note that the zonal distribution of trends in COD and AOD are very much alike ($R = 0.8$) which may suggest that indirect aerosol effects are shaping the changes in COD (e.g., Oreopoulos et al. 2020) and impact the zonal distribution of SW and NET CRE trends. While our trend analysis suggests SW clear-sky trends may be an important contribution to the EEI changes observed, radiative perturbation studies and model analysis find that significant positive contributions by cloud changes are of at least equal importance (Loeb et al. 2021; Raghuraman et al. 2021). Furthermore, Stephens et al. (2022) demonstrate that high-latitude surface changes play a subordinate role in modulating Earth's reflectivity. We recognize the need for radiative perturbation analysis to adequately quantify the relative role of clear-sky and CRE processes driving change in zonal radiative flux. Assessing and intercomparing zonal and regional changes, not only in TOA radiation, but also OHU, is a key recommendation of GEWEX-EEI (Sect. 4).

4 Discussion, Conclusions and Recommendations

Under the umbrella of the GEWEX-EEI assessment, we intercompare 21 OHC time series and OHU estimates (2005–2020) from various sources and institutions, as well as OHU trends and OHU correlations with CERES net radiative flux variability. The goals of this effort are a better understanding of discrepancies and their sources, and to forge a path toward best practices that may enable apples-to-apples comparison across methods, which ultimately improves our estimate of EEI and its uncertainties, a key indicator of global climate change.

Our study shows a significant spread in central OHU estimates (normalized to Earth's TOA surface area) ranging from 0.51 to 0.74 W m^{-2} for in situ-based estimates, generally larger values from satellite-based OHC (0.89 W m^{-2}), and 0.40–0.96 W m^{-2} across six ocean reanalyses. The in situ-based OHC products do not capture the deep ocean below 2000 m, and an estimate of 0.06 ± 0.04 W m^{-2} (Johnson et al. 2023b) for deep OHU has been added to be representative of the full ocean column. Using a subset of gridded OHC products, we demonstrate substantial influence of sampling considerations, with Argo-like sampling generally reducing OHU estimates from satellite and reanalysis data. Assumption of satellite sampling alone ($\pm 66°$ latitude) reduces OHU estimates from reanalysis data only slightly. It is likely that more complete coverage as, for example, provided by satellite observations is beneficial in representing near-global OHU, but neither satellite dataset achieves full global coverage, missing OHU in the polar regions. Fully global data collection would require spacecrafts in polar orbit, as well as enhanced observing systems to more densely sample OHC below sea ice across the full seasonal cycle. Ocean models and reanalysis systems do sample all of the ocean, but are reliant on observational coverage as well, yielding larger uncertainty where direct observations are lacking (e.g., below 2000 m depth and in under-sampled regions; Storto et al. 2019). Clearly, the ocean coverage, both geographically and vertically, plays a substantial role in achieving apples-to-apples comparison, suggesting use of a common mask. However, in terms of portraying the global OHU adequately and its variability with respect to CERES data, it is more complete coverage that should be strived for.

Temporal variability in annual mean OHU at both 12- and 6-month increments is compared against CERES EBAF net radiative flux variability. It stands out that both satellite-based, two reanalyses and the in situ + satellite hybrid products, RFROM and PMEL-combined, exhibit correlations with CERES EBAF net radiative flux of 0.44 or larger, while most of the largely in situ-based OHU series do not agree as well. This may suggest that enhanced spatiotemporal sampling and physically informed regional filling is important for matching the interannual variability and trend found in CERES data.

For the in situ-based OHU fields, quality control choices for the observed data are a critical factor not only in reducing measurement error, but in reducing the representation error, the difference between a point source measurement and the wider spatial area represented in the gridded fields used for the OHU calculations. This also applies to reanalysis products, which rely on external (observations processing centers) and internal quality control procedures (see, e.g., Storto et al. 2016). The mapping method used to calculate uniform gridded fields from heterogeneous measurements in time and space is also crucial to addressing representation error. The representation error for temperature in the ocean is larger than the measurement error (Oke and Sakov 2008). The International Quality-controlled Ocean Database (IQuOD) project brings together the international community working on data rescue, quality control, bias correction and uncertainty quantification for ocean temperature data. The Argo program uniformly monitors measurements from deployed floats, assigning measurement uncertainties and quality flags (Wong et al. 2020). The outcomes of these community-wide efforts will result in a uniformly quality-controlled ocean temperature dataset homogenizing the measurement errors across different in situ-based OHU calculations.

The MapEval4OceanHeat project, an objective assessment of mapping methods used to estimate ocean heat content change, is aiming to improve our understanding of different interpolation/mapping techniques (including the uncertainty estimates some provide), by systematically applying different methods to the very same set of synthetic input data (sampled from a high-resolution model) and comparing the output maps with the actual model fields. While the project is ongoing, a protocol has been released (Giglio et al. 2023) describing which experiments are being conducted.

Deriving OHU from OHC data, and their resulting correlation with CERES, is impacted by assumptions such as the differencing method (e.g., first versus centered differences) and temporal sampling/smoothing (monthly vs. annual vs. low-pass-filtered). For example, most annual mean OHU time series show better agreement with CERES net radiative flux at 6-month intervals than at 12-month intervals. A smoothing filter as applied by Marti et al. (2022) improves the correlations with the Legos product, by suppressing interannual variations in atmospheric heat storage and energy divergence at time resolutions smaller than 2–3 years (Palmer and McNeall 2014). The EEI community has recognized that even though global long-term mean atmospheric heat storage is comparably small, it exhibits significant interannual variability and, in combination with OHU, improves the co-variability with CERES EEI substantially at seasonal timescales (Johnson et al. 2023a, this issue).

A major concern for satellite-derived OHC products is the adequate knowledge of seawater's expansion efficiency of heat, which acts as a scaling factor for OHU variations and significantly affects the magnitude of internal variability and trend in OHU. Hakuba et al. (2021) derived an inverse efficiency of 0.52 ± 0.065 W m^{-2} mm year^{-1} which is significantly larger than the 0.43 W m^{-2} mm year^{-1} derived by Marti et al. (2022). We recommend reassessing the magnitude, variability and uncertainty of global, regional and ocean profile efficiencies, also in light of producing adequate satellite-based regional OHC estimates that would significantly add to our understanding of regional discrepancies and variations in OHC and OHU. For example, Hakuba et al. (2021) found the largest basin-wide discrepancy in steric sea level compared to in situ data in the Indo-Pacific, which has yet to be explained and assessed in terms of OHU. Regional assessments would furthermore help to address a sudden dip in OHU near the year 2016 which is evident in many of the in situ-based OHU time series and yet to be understood.

There is substantial spread not only in OHU central estimates, but the OHU error bars as well (Fig. 4). For example, uncertainties derived for geodetic OHU exceed the error bars derived for in situ-based OHU. While the geodetic community has partially estimated and combined major uncertainties in altimetry and gravimetry observations (e.g., Blazquez et al. 2018; Ablain et al. 2019), the uncertainties for in situ OHC are often not fully quantified or combined. For example, uncertainties due to instrument bias correction or sampling considerations would ideally flow into comprehensive OHU uncertainty estimates. As many before them, Meyssignac et al. (2019) determined OHC trend uncertainties from trend residuals alone, accounting for lag-1 autocorrelation, but state that the trend uncertainties ought to be at least doubled to account for sampling uncertainty. We therefore recommend establishing a roadmap toward comprehensive and consistent uncertainty estimation for OHC, OHU and EEI trend estimates, due to all (known) sources of error and their co-variances, across all approaches discussed—in situ, reanalysis, satellite and hybrid approaches.

Estimating EEI via heat inventory analysis is to date the most viable approach to closing the Earth's energy budget and largely possible due to the unprecedented coverage of the ocean by Argo floats. It is therefore critical to maintain and expand the ocean observing system to improve coverage geographically and vertically into the deepest and notoriously under-sampled layers of the ocean, which Deep Argo regional pilot arrays have demonstrated is feasible (e.g., Jayne et al. 2017), although funding is not yet identified for a global expansion. Additional avenues to monitor EEI independently and directly from space at the TOA are under investigation by several groups around the world (e.g., Schifano et al. 2022; Hakuba et al. 2019), but the feasibility to achieve such unprecedented accuracy requires further study. Such missions would provide additional rapport on EEI magnitude and change from year to year.

The study of zonal trends in net radiation with CERES data has revealed several key regions of change across the tropics, subtropics and at high latitudes. The analysis suggests a primary contribution to positive EEI trends from absorbed SW radiation, in line with previous studies (e.g., Loeb et al. 2021; Stephens et al. 2022). Although SW accumulation is a leading factor, it has to be noted that climate model, kernel and radiative perturbation analyses are consistent in pointing out that LW forcing, although masked by competing climate responses, is the fundamental cause for initiating positive and negative climate responses and feedbacks (Raghuraman et al. 2021; Kramer et al. 2021), which on the one hand mute LW absorption and on the other hand amplify energy uptake driven by positive SW changes. Continued investigation of EEI change and its causes is needed to resolve any disparate conclusions on potential drivers of EEI change and variability (e.g., Stephens et al. 2022; Loeb et al. 2021, 2024; Raghuraman et al. 2021; Kramer et al. 2021).

The gap impact analysis performed ignores the role of measurement uncertainties and potential shifts between non-overlapping parts of a data record, and requires further investigation to include observing system characteristics. The analysis reveals that gaps of any length (between 1 and 25 months) can have a significant impact on deduced trend magnitude and uncertainty, depending on the location of the gap in the data record. The impact is larger for EEI than OHC trends, given the more linear and robust increase in OHC, while EEI trends are more sensitive to the period considered and the interannual variability that substantially shapes the > 20-year EEI record. The gap impacts on trend are at the lower end of impact expected, ignoring intercalibration and bridging issues that would make trend detection likely infeasible for years to come (e.g., Loeb et al. 2009a). It is therefore of utmost importance to prevent data gaps, investigate "bridging" methods and ensure seamless monitoring of EEI change well into the future. Studies to explore observing system requirements, ways to meet them, critical ancillary information and the role of data gaps are likely of great relevance toward a maintained and improved EEI monitoring system and research framework (KISS Continuity Study Team 2024).

The four approaches for estimating OHC and OHU as well as the radiometric observations of TOA radiative flux intercompared here represent the core of today's EEI monitoring system, whereby in situ OHC has been the most vital for constraining EEI magnitude and, together with reanalysis, provides unparalleled insight into the distribution of heat across the ocean volume. Geodetic satellite observations provide enhanced spatiotemporal coverage estimating full-column OHC and confirm the positive EEI trend derived from

TOA radiation measurements. Scanning radiometry not only provides insight into EEI variability at high precision and stability but allows us to study the radiative processes that perturb EEI, e.g., the role of cloud and aerosol radiative effects.

The first GEWEX-EEI Assessment Workshop held in spring 2023, yielded recommendations (Meyssignac et al. 2023b) that in part have been touched upon in this paper and are summarized as follows:

1. Discrepancies among EEI and OHU products, methods and their origin, ought to be systematically assessed and improved upon.
2. Regional, zonal and basin-scale intercomparisons are recommended to better understand global discrepancies and the impact of differing ocean volumes sampled. With respect to regional geodetic OHC analysis, in-depth assessment of expansion efficiency is required.
3. Best practices to enable apples-to-apples comparison—e.g., sampling considerations, uncertainty quantification, OHU derivation—ought to be established and shared with the community.
4. Beyond improving our knowledge of EEI with existing observations, ensuring seamless continuity of these systems and data products should be a priority, as well as efforts to expand those for improved coverage of the ocean, land and cryosphere (see also von Schuckmann et al. 2023).
5. Novel techniques ought to be explored to provide independent and direct measurements of EEI at the TOA.
6. Understanding EEI changes and their attribution is as important as the comprehensive quantification and characterization of EEI and its variability.

The next steps for the GEWEX-EEI assessment involve the intercomparison of OHC and OHU trends at the regional scale and establishing comprehensive and consistent uncertainty quantification of at least the most dominant error sources. While all methods intercompared agree on positive OHU values between 0.40 and 0.96 W m^{-2}, higher confidence and temporal precision is needed to comfortably track changes from year to year, especially if climate mitigation strategies and their impact are to be monitored and understood.

Supplementary Information The online version contains supplementary material available at https://doi.org/10.1007/s10712-024-09849-5.

Acknowledgements This paper is an outcome of the Workshop "Challenges in Understanding the Global Water Energy Cycle and its Changes in Response to Greenhouse Gas Emissions" held at the International Space Science Institute (ISSI) in Bern, Switzerland (26–30 September 2022). We thank the GEWEX Data Analysis Panel (GDAP) and ISSI Bern for support of EEI assessment activities. MZH, supported by the *Libera* project, and FL carried out research at the Jet Propulsion Laboratory, California Institute of Technology, under a contract with the National Aeronautics and Space Administration (80NM0018D0004). BM and SF are supported by the ESA Climate Space Programme under the Cross-ECV project MOTECUSOMA and by the CNRS TOSCA project for the use of Sentinel 6 data. GCJ and JML are supported by NOAA Research and the NOAA Global Ocean Monitoring and Observation Program. AM is supported by the NOAA grant NA19NES4320002 to the Cooperative Institute for Satellite Earth System Studies-CISESS at the University of Maryland/ESSIC. REK was supported by the Met Office Hadley Centre Climate Programme funded by DSIT. DG, MK and TS were supported by NOAA (Awards NA21OAR4310261 and NA21OAR4310258). GF acknowledges support from NASA Awards 80NSSC20K0796, 80NSSC23K0355, 80NSSC22K1697, 1676067 and 1686358. SAS temperature and salinity monthly gridded field products are made freely available by SNO Argo France at LOPS Laboratory (supported by UBO/CNRS/Ifremer/IRD) and IUEM Observatory (OSU IUEM/CNRS/INSU) at https://doi.org/10.17882/52367. The IAP analysis is supported by the National Natural Science Foundation of China (Grant Nos. 42122046, 42261134536) and the new Cornerstone Science Foundation through the XPLORER PRIZE. Argo data were collected and made freely available by the International Argo Program and the national programs that contribute to it http://www.argo.ucsd.edu, http://argo.jcommops.org. The Argo Program is part of the Global Ocean Observing System.

Declarations

Competing interests The authors declare that they have no competing interests.

Open Access This article is licensed under a Creative Commons Attribution 4.0 International License, which permits use, sharing, adaptation, distribution and reproduction in any medium or format, as long as you give appropriate credit to the original author(s) and the source, provide a link to the Creative Commons licence, and indicate if changes were made. The images or other third party material in this article are included in the article's Creative Commons licence, unless indicated otherwise in a credit line to the material. If material is not included in the article's Creative Commons licence and your intended use is not permitted by statutory regulation or exceeds the permitted use, you will need to obtain permission directly from the copyright holder. To view a copy of this licence, visit http://creativecommons.org/licenses/by/4.0/.

References

Ablain M, Meyssignac B, Zawadzki L, Jugier R, Ribes A, Cazenave A et al (2019) Uncertainty insatellite estimate of global mea sea level changes, trend and acceleration. Earth Syst Sci Data Discuss 1–26. https://doi.org/10.5194/essd-2019-10

Abraham J, Reseghetti F, Baringer M, Boyer T, Cheng L, Church J, Domingues C, Fasullo JT, Gilson J, Goni G, Good S, Gorman JM, Gouretski V, Ishii M, Johnson GC, Kizu S, Lyman J, MacDonald A, Minkowycz WJ, Moffitt SE, Palmer M, Piola A, Trenberth KE, Velicogna I, Wijffels S, Willis J (2013) A review of global ocean temperature observations: Implications for ocean heat content estimates and climate change. Rev Geophys 51:450–483. https://doi.org/10.1002/rog.20022

Argo Science Team (1999) Report of the Argo science team meeting (Argo-1). GODAE International Project Office, Melbourne, 27

Balmaseda M, Hernandez F, Storto A, Palmer M, Alves O, Shi L et al (2015) The ocean reanalyses intercomparison project (ORA-IP). J Oper Oceanogr 8(sup1):s80–s97. https://doi.org/10.1080/1755876X.2015.1022329

Blazquez A, Meyssignac B, Lemoine JM, Berthier E, Ribes A, Cazenave A (2018) Exploring the uncertainty in GRACE estimates of the mass redistributions at the Earth surface: implications for the global water and sea level budgets. Geophys J Int 215:415–430. https://doi.org/10.1093/gji/ggy293

Boyer T, Domingues CM, Good SA, Johnson GC, Lyman JM, Ishii M, Gouretski V, Willis JK, Antonov J, Wijffels S, Church JA, Cowley R, Bindoff NL (2016) Sensitivity of global upper-ocean heat content estimates to mapping methods, XBT bias corrections, and baseline climatologies. J Clim 29(13):4817–4842. https://doi.org/10.1175/JCLI-D-15-0801.1

Bretherton F, Davis R, Fandry C (1976) Technique for objective analysis and design of oceanographic experiment applied to MODE-73. Deep-Sea Res Oceanogr Abstr 23:559–582. https://doi.org/10.1016/0011-7471(76)90001-2

Byrne MP, Pendergrass AG, Rapp AD et al (2018) Response of the intertropical convergence zone to climate change: location, width, and strength. Curr Clim Change Rep 4:355–370. https://doi.org/10.1007/s40641-018-0110-5

Cabanes C, Grouazel A, von Schuckmann K, Hamon M, Turpin V, Coatanoan C, Paris F, Guinehut S, Boone C, Ferry N, de Boyer Montégut C, Carval T, Reverdin G, Pouliquen S, Le Traon PY (2013) The CORA dataset: validation and diagnostics of in-situ ocean temperature and salinity measurements. Ocean Sci 9:1–18. https://doi.org/10.5194/os-9-1-2013

Caron L, Ivins ER, Larour E, Adhikari S, Nilsson J, Blewitt G (2018) GIA model statistics for GRACE hydrology, cryosphere, and ocean science. Geo Res Let 45(5):2203–2212. https://doi.org/10.1002/2017GL076644

Carton JA, Chepurin GA, Chen L (2018) SODA3: a new ocean climate reanalysis. J Climate 31:6967–6983. https://doi.org/10.1175/JCLI-D-18-0149.1

Charles E, Meyssignac B, Ribes A (2020) Observational constraint on greenhouse gas and aerosol contributions to global ocean heat content changes. J Clim 33:10579–10591. https://doi.org/10.1175/JCLI-D-19-0091.1

Chenal J, Meyssignac B, Ribes A, Guillaume-Castel R (2022) Observational constraint on the climate sensitivity to atmospheric CO_2 concentrations changes derived from the 1971–2017 global energy budget. J Clim 35:4469–4483. https://doi.org/10.1175/JCLI-D-21-0565.1

Cheng L, Zhu J (2016) Benefits of CMIP5 multimodel ensemble in reconstructing historical ocean subsurface temperature variations. J Clim 29:5393–5416. https://doi.org/10.1175/JCLI-D-15-0730.1

Cheng L, Trenberth KE, Palmer MD, Zhu J, Abraham J (2016a) Observed and simulated full-depth ocean heat-content changes for 1970–2005. Ocean Sci 12:925–935. https://doi.org/10.5194/os-12-925-2016

Cheng L, Abraham J, Goni G, Boyer T, Wijffels S, Cowley R, Gouretski V, Reseghetti F, Kizu S, Dong S, Bringas F, Goes M, Houpert L, Sprintall J, Zhu J (2016b) XBT science: assessment of instrumental biases and errors. Bull Am Meteorol Soc 97(6):924–933. https://doi.org/10.1175/BAMS-D-15-00031.1

Cheng L et al (2017) Improved estimates of ocean heat content from 1960 to 2015. Sci Adv 3:e1601545. https://doi.org/10.1126/sciadv.1601545

Cheng L, Trenberth KE, Fasullo JT, Mayer M, Balmaseda M, Zhu J (2019a) Evolution of ocean heat content related to ENSO. J Clim 32:3529–3556. https://doi.org/10.1175/JCLI-D-18-0607.1

Cheng L, Abraham J, Hausfather Z, Trenberth KE (2019b) How fast are the oceans warming? Science 363:128–129. https://doi.org/10.1126/science.aav7619

Cheng L, von Schuckmann K, Abraham JP et al (2022a) Past and future ocean warming. Nat Rev Earth Environ 3:776–794. https://doi.org/10.1038/s43017-022-00345-1

Cheng L, Foster G, Hausfather Z, Trenberth KE, Abraham J (2022b) Improved quantification of the rate of ocean warming. J Clim 35(14):4827–4840. https://doi.org/10.1175/JCLI-D-21-0895.1

Cheng L, von Schuckmann K, Minière A et al (2024a) Ocean heat content in 2023. Nat Rev Earth Environ 5:232–234. https://doi.org/10.1038/s43017-024-00539-9

Cheng L, Pan Y, Tan Z, Zheng H, Zhu Y, Wei W, Du J, Yuan H, Li G, Ye H, Gouretski V, Li Y, Trenberth K, Abraham J, Jin Y, Reseghetti F, Lin X, Zhang B, Chen G, Mann M, Zhu J (2024b) IAPv4 ocean temperature and ocean heat content gridded dataset. Earth Syst Sci Data Discuss. https://doi.org/10.5194/essd-2024-42

de Rosnay P, Browne P, de Boisséson E, Fairbairn D, Hirahara Y, Ochi K et al (2022) Coupled data assimilation at ECMWF: current status, challenges and future developments. Q J R Meteorol Soc 148(747):2672–2702. https://doi.org/10.1002/qj.4330

Durack PJ, Gleckler PJ, Purkey SG, Johnson GC, Lyman JM, Boyer TP (2018) Ocean warming: from the surface to the deep in observations and models. Oceanography 31(2):41–51

Forget G (2024) Energy imbalance in the sunlit ocean layer. https://doi.org/10.21203/rs.3.rs-3979671/v1

Forget G, Campin J-M, Heimbach P, Hill CN, Ponte RM, Wunsch C (2015) ECCO version 4: an integrated framework for non-linear inverse modeling and global ocean state estimation. Geosci Model Dev 8(10):3071–3104. https://doi.org/10.5194/gmd-8-3071-2015

Forster P, Storelvmo T, Armour K, Collins W, Dufresne J-L, Frame D, Lunt D, Mauritsen T, Palmer M, Watanabe M, Wild M, Zhang H (2021) Chapter 7: the earth's energy budget, climate feedbacks, and climate sensitivity. Climate change 2021: the physical science basis. Contribution of working group I to the sixth assessment report of the intergovernmental panel on climate change

Forster PM, Smith CJ, Walsh T, Lamb WF, Lamboll R, Hauser M, Ribes A, Rosen D, Gillett N, Palmer MD, Rogelj J, von Schuckmann K, Seneviratne SI, Trewin B, Zhang X, Allen M, Andrew R, Birt A, Borger A, Boyer T, Broersma JA, Cheng L, Dentener F, Friedlingstein P, Gutiérrez JM, Gütschow J, Hall B, Ishii M, Jenkins S, Lan X, Lee J-Y, Morice C, Kadow C, Kennedy J, Killick R, Minx JC, Naik V, Peters GP, Pirani A, Pongratz J, Schleussner C-F, Szopa S, Thorne P, Rohde R, Rojas Corradi M, Schumacher D, Vose R, Zickfeld K, Masson-Delmotte V, Zhai P (2023) Indicators of Global Climate Change 2022: annual update of large-scale indicators of the state of the climate system and human influence. Earth Syst Sci Data 15:2295–2327. https://doi.org/10.5194/essd-15-2295-2023

Forget G (2024) Energy Imbalance in the sunlit ocean layer, 11 April PREPRINT (Version 1) available at Research Square. https://doi.org/10.21203/rs.3.rs-3979671/v1

Gaillard F, Reynaud T, Thierry V, Kolodziejczyk N, von Schuckmann K (2016) In-situ based reanalysis of the global ocean temperature and salinity with ISAS: variability of the heat content and steric height. J Clim 29(4):1305–1323. https://doi.org/10.1175/JCLI-D-15-0028.1

Giglio D, Monselesan D, Palmer M (2023) Experimental protocol for MapEval4OceanHeat, an objective assessment of mapping methods used to estimate ocean heat content change (1.0.0). Zenodo. https://doi.org/10.5281/zenodo.10291852

Giglio D, Sukianto T, Kuusela M (2024) Global Ocean Heat Content Anomalies and Ocean Heat Uptake based on mapping Argo data using local Gaussian processes (3.0.0). Zenodo. https://doi.org/10.5281/zenodo.10645137

Good SA, Martin MJ, Rayner NA (2013) EN4: Quality controlled ocean temperature and salinity profiles and monthly objective analyses with uncertainty estimates. J Geophys Res Oceans 118(12):6704–6716. https://doi.org/10.1002/2013JC009067

Gregory JM, Banks HT, Stott PA, Lowe JA, Palmer MD (2004) Simulated and observed decadal variability in ocean heat content. Geophys Res Lett 31:L15312. https://doi.org/10.1029/2004GL020258

Gulev SK, Thorne PW, Ahn J, Dentener FJ, Domingues CM, Gerland S, Gong D, Kaufman DS, Nnamchi HC, Quaas J, Rivera JA, Sathyendranath S, Smith SL, Trewin B, von Schuckmann K, Vose RS (2021) Changing state of the climate system. In: Masson-Delmotte V, Zhai P, Pirani A, Connors SL, Péan C, Berger S, Caud N, Chen Y, Goldfarb L, Gomis MI, Huang M, Leitzell K, Lonnoy E, Matthews JBR, Maycock TK, Waterfield T, Yelekçi O, Yu R, Zhou B (eds) Climate change 2021: the physical science basis. Contribution of working group I to the sixth assessment report of the intergovernmental panel on climate change. Cambridge University Press, Cambridge, pp 287–422. https://doi.org/10.1017/9781009157896.004

Hakuba MZ, Kindel B, Gristey J, Bodas-Salcedo A, Stephens G, Pilewskie P (2024) Simulated variability in visible and near-IR irradiances in preparation for the upcoming Libera mission. AIP Conf. Proc.18 January 2024 2988(1): 050006. https://doi.org/10.1063/5.0183869

Harber D, Catani K, Gieseler J, Haun R, Kruczek N, Sprunk J, Tomlin N, Yung C, Lehman J, Stephens M, Kampe T, Collins S, Peterson J, Latvakoski H, Monte C, Hakuba M, Pilewskie P (2023) The libera mission: bringing next-generation technology to an established climate data record. 15th international conference on new developments and applications in optical radiometry (NEWRAD 2023), 11–15 Sep. 2023, NPL, Teddington, UK

Hakuba MZ, Stephens GL, Christophe B, Nash AE, Foulon B, Bettadpur SV et al (2019) Earth's energy imbalance measured from space. IEEE Trans Geosci Remote Sens 57(1):32–45. https://doi.org/10.1109/TGRS.2018.2851976

Hakuba MZ, Frederikse T, Landerer FW (2021) Earth's energy imbalance from the ocean perspective (2005–2019). Geophys Res Lett 48:e2021GL093624. https://doi.org/10.1029/2021GL093624

Hosoda S, Ohira T, Nakamura T (2008) A monthly mean dataset of global oceanic temperature and salinity derived from Argo float observations. JAMSTEC Rep Res Dev 8:47–59

Hourdin F, Mauritsen T, Gettelman A, Golaz J-C, Balaji V, Duan Q et al (2017) The art and science of climate model tuning. Bull Am Meteorol Soc 98(3):589–602. https://doi.org/10.1175/BAMS-D-15-00135.1

IPCC (2021) Technical summary. In: Masson-Delmotte V, Zhai P, Pirani A, Connors SL, Péan C, Berger S, Caud N, Chen Y, Goldfarb L, Gomis MI, Huang M, Leitzell K, Lonnoy E, Matthews JBR, Maycock TK, Waterfield T, Yelekçi O, Yu R, Zhou B (eds) Climate change 2021: the physical science basis. Contribution of working group I to the sixth assessment report of the intergovernmental panel on climate change. Cambridge University Press, Cambridge, p 2391. https://doi.org/10.1017/9781009157896

Ishii M, Fukuda Y, Hirahara S, Yasui S, Suzuki T, Sato K (2017) Accuracy of global upper ocean heat content estimation expected from present observational data sets. Sola 13:163–167

Ishii M, Kimoto M, Kachi M (2003) Historical ocean subsurface temperature analysis with error estimates. Mon Wea Rev 131:51–73. https://doi.org/10.1175/1520-0493

Jayne SR, Roemmich D, Zilberman N, Riser SC, Johnson KS, Johnson GC et al (2017) The argo program present and future. Oceanography 30:18–28. https://doi.org/10.5670/oceanog.2017.213

Johnson GC, Lyman JM, Loeb NG (2016) Improving estimates of Earth's energy imbalance. Nat Clim Change 6(7):639–640. https://doi.org/10.1038/nclimate3043

Johnson GC, Landerer FW, Loeb NG et al (2023a) Closure of Earth's global seasonal cycle of energy storage. Surv Geophys. https://doi.org/10.1007/s10712-023-09797-6

Johnson GC et al (2023b) Global oceans. Bull Am Meteorol Soc 104:S146–S206. https://doi.org/10.1175/BAMS-D-23-0076.2

KISS Continuity Study Team (2024) Toward a US framework for continuity of satellite observations of Earth's climate and for supporting societal resilience. Earth's Future 12. https://doi.org/10.1029/2023EF003757

Kramer RJ, He H, Soden BJ, Oreopoulos L, Myhre G, Forster PM, Smith CJ (2021) Observational evidence of increasing global radiative forcing. Geophys Res Lett 48:e2020GL091585. https://doi.org/10.1029/2020GL091585

Kuusela M, Stein ML (2018) Locally stationary spatio-temporal interpolation of Argo profiling float data. Proc R https://doi.org/10.1098/rspa.2018.0400

L'Ecuyer TS, Beaudoing HK, Rodell M, Olson W, Lin B, Kato S, Clayson CA, Wood E, Sheffield J, Adler R, Huffman G, Bosilovich M, Gu G, Robertson F, Houser PR, Chambers D, Famiglietti JS, Fetzer E, Liu WT, Gao X, Schlosser CA, Clark E, Lettenmaier DP, Hilburn K (2015) The observed state of

the energy budget in the early twenty-first century. J Clim 28(21):8319–8346. https://doi.org/10.1175/JCLI-D-14-00556.1

Lellouche JM, Greiner E, Le Galloudec O, Garric G, Regnier C, Drevillon M, Benkiran M, Testut CE, Bourdalle-Badie R, Gasparin F, Hernandez O, Levier B, Drillet Y, Remy E, Le Traon PY (2018) Recent updates to the Copernicus Marine Service global ocean monitoring and forecasting real-time 1/12° high-resolution system. Ocean Sci 14:1093–1126. https://doi.org/10.5194/os-14-1093-2018

Levitus S et al (2012) World ocean heat content and thermosteric sea level change (0–2000 m), 1955–2010. Geophys Res Lett 39:L10603. https://doi.org/10.1029/2012GL051106

Locarnini RA, Mishonov AV, Baranova OK, Boyer TP, Zweng MM, Garcia HE, Reagan JR, Seidov D, Weathers K, Paver CR, Smolyar I (2018) World Ocean Atlas 2018, Volume 1: Temperature. A. Mishonov Technical Ed.; NOAA Atlas NESDIS 81 52 pp

Loeb NG, Wielicki BA, Doelling DR, Smith GL, Keyes DF, Kato S, Manalo-Smith N, Wong T (2009a) Toward optimal closure of the earth's top-of-atmosphere radiation budget. J Clim 22(3):748–766. https://doi.org/10.1175/2008JCLI2637.1

Loeb NG, Wielicki BA, Wong T, Parker PA (2009b) Impact of data gaps on satellite broadband radiation records. J Geophys Res 114:D11109. https://doi.org/10.1029/2008JD011183

Loeb N, Lyman J, Johnson G et al (2012) Observed changes in top-of-the-atmosphere radiation and upper-ocean heating consistent within uncertainty. Nat Geosci 5:110–113. https://doi.org/10.1038/ngeo1375

Loeb NG, Doelling DR, Wang H, Su W, Nguyen C, Corbett JG, Liang L, Mitrescu C, Rose FG, Kato S (2018) Clouds and the earth's radiant energy system (CERES) energy balanced and filled (EBAF) top-of-atmosphere (TOA) edition-4.0 data product. J Clim 31(2):895–918. https://doi.org/10.1175/JCLI-D-17-0208.1

Loeb NG, Johnson GC, Thorsen TJ, Lyman JM, Rose FG, Kato S (2021) Satellite and ocean data reveal marked increase in Earth's heating rate. Geophys Res Lett 48:e2021GL093047. https://doi.org/10.1029/2021GL093047

Loeb NG, Mayer M, Kato S, Fasullo JT, Zuo H, Senan R et al (2022) Evaluating twenty-year trends in Earth's energy flows from observations and reanalyses. J Geophys Res Atmos 127:e2022JD036686. https://doi.org/10.1029/2022JD036686

Loeb NG, Ham SH, Allan RP et al (2024) Observational assessment of changes in earth's energy imbalance since 2000. Surv Geophys. https://doi.org/10.1007/s10712-024-09838-8

Lyman JM, Johnson GC (2008) Estimating annual global upper-ocean heat content anomalies despite irregular in situ sampling. J Clim 21:5629–5641. https://doi.org/10.1175/2008JCLI2259.1

Lyman JM, Johnson GC (2014) Estimating global ocean heat content changes in the upper 1800 m since 1950 and the influence of climatology choice. J Clim 27:1945–1957. https://doi.org/10.1175/JCLI-D-12-00752.1

Lyman JM, Johnson GC (2023) Global high-resolution random forest regression maps of ocean heat content anomalies using in situ and satellite data. J Atmos Ocean Technol 40:575–586. https://doi.org/10.1175/JTECH-D-22-0058.1

Marti F, Blazquez A, Meyssignac B, Ablain M, Barnoud A, Fraudeau R, Jugier R, Chenal J, Larnicol G, Pfeffer J, Restano M, Benveniste J (2022) Monitoring the ocean heat content change and the Earth energy imbalance from space altimetry and space gravimetry. Earth Syst Sci Data 14:229–249. https://doi.org/10.5194/essd-14-229-2022

Mayer M, Kato S, Bosilovich M et al (2024) Assessment of atmospheric and surface energy budgets using observation-based data products. Surv Geophys. https://doi.org/10.1007/s10712-024-09827-x

McDougall TJ, Barker PM (2011) Getting started with TEOS-10 and the Gibbs Seawater (GSW) oceanographic toolbox. Trevor J McDougall

Melet A, Meyssignac B (2015) Explaining the spread in global mean thermosteric sea level rise in CMIP5 climate models. J Clim 28(24):9918–9940

Meyssignac B, Boyer T, Zhao Z, Hakuba MZ, Landerer FW, Stammer D et al (2019) Measuring global ocean heat content to estimate the Earth energy imbalance. Front Mar Sci 6:432. https://doi.org/10.3389/fmars.2019.00432

Meyssignac B, Chenal J, Loeb N et al (2023a) Time-variations of the climate feedback parameter λ are associated with the Pacific Decadal Oscillation. Commun Earth Environ 4:241. https://doi.org/10.1038/s43247-023-00887-2

Meyssignac B, Hakuba MZ, Kato S, Boyer T, Benveniste J (2023b) First earth energy imbalance assessment WCRP-ESA workshop summary and recommendations executive brief. ESA Publication. https://doi.org/10.5270/wcrp-esa-eeia-2023.final_report_brief

Meyssignac B, Fourest S, Mayer M, Johnson G, Calafat FM, Ablain M, Boyer T, Cheng L, Desbruyères D, Forget G, Giglio D, Kuusela M, Locarnini R, Lyman J, Llovel W, Mishonov A, Reagan J, Rousseau

V, Beneviste J (submitted to this issue) North Atlantic heat transport convergence derived from a regional energy budget using different ocean heat content estimates

Oke PR, Sakov P (2008) Representation error of oceanic observations for data assimilation. J Atmos Ocean Technol 25(6):1004–1017

Oreopoulos L, Cho N, Lee D (2020) A global survey of apparent aerosol-cloud interaction signals. J Geophys Res Atmos 125:e2019JD031287. https://doi.org/10.1029/2019JD031287

Palmer MD, McNeall DJ (2014) Internal variability of Earth's energy budget simulated by CMIP5 climate models. Environ Res Lett 9(3):034016

Pan Y, Cheng L, von Schuckmann K, Trenberth KE, Li G, Abraham J, Liu Y, Gouretski V, Yu Y, Liu H, Liu C (2023) Annual cycle in upper ocean heat content and the global energy budget. J Clim 36:5003–5026. https://doi.org/10.1175/JCLI-D-22-0776.1

Peeters M (2021) The global stocktake. In: Van Calster G, Reins L (eds) The Paris agreement on climate change. Edward Elgar Publishing, London, pp 326–346

Purkey SG, Johnson GC (2010) Warming of global abyssal and deep southern ocean waters between the 1990s and 2000s: contributions to global heat and sea level rise budgets. J Clim 23:6336–6351. https://doi.org/10.1175/2010JCLI3682.1

Raghuraman SP, Paynter D, Ramaswamy V (2019) Quantifying the drivers of the clear sky greenhouse effect, 2000–2016. J Geophys Res Atmos 124(21):11354–11371

Raghuraman SP, Paynter D, Ramaswamy V (2021) Anthropogenic forcing and response yield observed positive trend in Earth's energy imbalance. Nat Commun 12:4577. https://doi.org/10.1038/s41467-021-24544-4

Radiation Budget Science Working Group (RBSP, 2018) Recommended measurement and instrument characteristics for an earth venture continuity earth radiation budget instrument. https://essp.larc.nasa.gov/EVC-1/evc-1_library.html

Riser SC, Freeland HJ, Roemmich D, Wijffels S, Troisi A, Belbeoch M et al (2016) Fifteen years of ocean observations with the global Argo array. Nat Clim Change 6(2):145–153. https://doi.org/10.1038/nclimate2872

Roberts JB, L'Ecuyer T, Olson B, Haines K, Kato S, Mayer M, Behrangi A, Fourest S, Roca R, Meyssignac B (submitted to this issue) Reconciling global energy and water cycle balances using consistent budgets and optimization approaches

Roemmich D, Gilson J (2009) The 2004–2008 mean and annual cycle of temperature, salinity, and steric height in the global ocean from the Argo Program. Prog Oceanogr 82(2):81–100

Roemmich D, Johnson GC, Riser S, Davis R, Gilson J, Owens WB, Garzoli SL, Schmid C, Ignaszewski M (2009) The Argo program: observing the global ocean with profiling floats. Oceanography 22(2):34–43

Rottman G (2005) The SORCE mission. In: Rottman G, Woods T, George V (eds) The solar radiation and climate experiment (SORCE). Springer, New York, pp 7–25. https://doi.org/10.1007/0-387-37625-92

Schifano L, Berghmans F, Dewitte S, Smeesters L (2022) Optical design of a novel wide-field-of-view space-based spectrometer for climate monitoring. Sensors 22(15):5841

Schmidt GA, Andrews T, Bauer SE, Durack PJ, Loeb NG, Ramaswamy V et al (2023) CERESMIP: a climate modeling protocol to investigate recent trends in the Earth's Energy Imbalance. Front Clim. https://doi.org/10.3389/fclim.2023.1202161

Sherwood S, Webb MJ, Annan JD, Armour KC, Forster PM, Hargreaves JC et al (2020) An assessment of Earth's climate sensitivity using multiple lines of evidence. Rev Geophys. https://doi.org/10.1029/2019RG000678

Smith DM, Allan RP, Coward AC, Eade R, Hyder P, Liu C et al (2015) Earth's energy imbalance since 1960 in observations and CMIP5 models. Geophys Res Lett 42(4):1205–1213. https://doi.org/10.1002/2014GL062669

Stammer D, Balmaseda M, Heimbach P, Köhl A, Weaver A (2016) Ocean data assimilation in support of climate applications: status and perspectives. Ann Rev Mar Sci 8:491–518

Stephens GL, Kahn BH, Richardson M (2016) The super greenhouse effect in a changing climate. J Clim 29(15):5469–5482

Stephens GL, Hakuba MZ, Kato S, Gettelman A, Dufresne JL, Andrews T, Cole JN, Willen U, Mauritsen T (2022) The changing nature of Earth's reflected sunlight. Proc R Soc A 478(2263):20220053

Stephens GL, Shiro KA, Hakuba MZ et al (2024) Tropical deep convection, cloud feedbacks and climate sensitivity. Surv Geophys. https://doi.org/10.1007/s10712-024-09831-1

Stephens G, Polcher J, Zeng X, van Oevelen P, Poveda G, Bosilovich M, Ahn M, Balsamo G, Duan Q, Hegerl G, Jakob C, Lamptey B, Leung R, Piles M, Su Z, Dirmeyer P, Findell KL, Verhoef A, Ek

M, L'Ecuyer T, Roca R, Nazemi A, Dominguez F, Klocke D, Bony S (2023) The first 30 years of GEWEX. Bull Am Meteorol Soc 104(1):E126–E157. https://doi.org/10.1175/BAMS-D-22-0061.1

Storto A, Yang C, Masina S (2016) Sensitivity of global ocean heat content from reanalyses to the atmospheric reanalysis forcing: a comparative study. Geophys Res Lett 43:5261–5270. https://doi.org/10.1002/2016GL068605

Storto A, Masina S (2016) C-GLORSv5: an improved multipurpose global ocean eddy-permitting physical reanalysis. Earth Syst Sci Data 8:679–696. https://doi.org/10.5194/essd-8-679-2016

Storto A, Yang C (2024) Acceleration of the ocean warming from 1961 to 2022 unveiled by large-ensemble reanalyses. Nat Commun 15:545. https://doi.org/10.1038/s41467-024-44749-7

Storto A, Yang C, Masina S (2017) Constraining the global ocean heat content through assimilation of CERES-derived TOA energy imbalance estimates. Geophys Res Lett 44:10520–10529. https://doi.org/10.1002/2017GL075396

Storto A, Alvera-Azcárate A, Balmaseda MA, Barth A, Chevallier M, Counillon F, Domingues CM et al (2019) Ocean reanalyses: recent advances and unsolved challenges. Front Mar Sci 6:418

Storto A, Cheng L, Yang C (2022) Revisiting the 2003–18 deep ocean warming through multiplatform analysis of the global energy budget. J Clim 35:4701–4717. https://doi.org/10.1175/JCLI-D-21-0726.1

Su H, Jiang J, Neelin J et al (2017) Tightening of tropical ascent and high clouds key to precipitation change in a warmer climate. Nat Commun 8:15771. https://doi.org/10.1038/ncomms15771

Tan ZL, Cheng V, Gouretski B, Zhang Y, Wang F, Li Z, Liu JZhu (2023) A new automatic quality control system for ocean profile observations and impact on ocean warming estimate. Deep Sea Res Part I 194:103961. https://doi.org/10.1016/j.dsr.2022.103961

Tokarska KB, Hegerl GC, Schurer AP, Ribes A, Fasullo JT (2019) Quantifying human contributions to past and future ocean warming and thermosteric sea level rise. Environ Res Lett 14(7):074020

Trenberth KE, Fasullo JT, von Schuckmann K, Cheng L (2016) Insights into Earth's energy imbalance from multiple sources. J Clim 29(20):7495–7505. https://doi.org/10.1175/JCLI-D-16-0339.1

Trenberth KE, Zhang Y, Fasullo JT, Cheng L (2019) Observation-based estimate of global and basin ocean meridional heat transport time series. J Clim 32:4567–4583. https://doi.org/10.1175/JCLI-D-18-0872.1

von Schuckmann K, Le Traon PY (2011) How well can we derive global ocean indicators from Argo data? Ocean Sci 7:783–791. https://doi.org/10.5194/os-7-783-2011

von Schuckmann K, Sallée J-B, Chambers D, Le Traon P-Y, Cabanes C, Gaillard F, Speich S, Hamon M (2014) Consistency of the current global ocean observing systems from an Argo perspective. Ocean Sci 10:547–557. https://doi.org/10.5194/os-10-547-2014

von Schuckmann K, Palmer MD, Trenberth KE, Cazenave A, Chambers D, Champollion N et al (2016) An imperative to monitor Earth's energy imbalance. Nat Clim Change 6(2):138–144. https://doi.org/10.1038/nclimate2876

von Schuckmann K, Cheng L, Palmer MD, Hansen J, Tassone C, Aich V, Adusumilli S, Beltrami H, Boyer T, Cuesta-Valero FJ, Desbruyères D, Domingues C, García-García A, Gentine P, Gilson J, Gorfer M, Haimberger L, Ishii M, Johnson GC, Killick R, King BA, Kirchengast G, Kolodziejczyk N, Lyman J, Marzeion B, Mayer M, Monier M, Monselesan DP, Purkey S, Roemmich D, Schweiger A, Seneviratne SI, Shepherd A, Slater DA, Steiner AK, Straneo F, Timmermans M-L, Wijffels SE (2020) Heat stored in the Earth system: where does the energy go? Earth Syst Sci Data 12:2013–2041. https://doi.org/10.5194/essd-12-2013-2020

von Schuckmann K, Minière A, Gues F, Cuesta-Valero FJ, Kirchengast G, Adusumilli S, Straneo F, Ablain M, Allan RP, Barker PM, Beltrami H, Blazquez A, Boyer T, Cheng L, Church J, Desbruyeres D, Dolman H, Domingues CM, García-García A, Giglio D, Gilson JE, Gorfer M, Haimberger L, Hakuba MZ, Hendricks S, Hosoda S, Johnson GC, Killick R, King B, Kolodziejczyk N, Korosov A, Krinner G, Kuusela M, Landerer FW, Langer M, Lavergne T, Lawrence I, Li Y, Lyman J, Marti F, Marzeion B, Mayer M, MacDougall AH, McDougall T, Monselesan DP, Nitzbon J, Otosaka I, Peng J, Purkey S, Roemmich D, Sato K, Sato K, Savita A, Schweiger A, Shepherd A, Seneviratne SI, Simons L, Slater DA, Slater T, Steiner AK, Suga T, Szekely T, Thiery W, Timmermans M-L, Vanderkelen I, Wjiffels SE, Wu T, Zemp M (2023) Heat stored in the Earth system 1960–2020: where does the energy go? Earth Syst Sci Data 15:1675–1709. https://doi.org/10.5194/essd-15-1675-2023

Wang G, Cheng L, Abraham J, Li C (2018) Consensuses and discrepancies of basin-scale ocean heat content changes in different ocean analyses. Clim Dyn 50:2471–2487. https://doi.org/10.1007/s00382-017-3751-5

White WB (1995) Design of a global observing system for gyre-scale upper ocean temperature variability. Prog Oceanogr 36:169–217

Wielicki BA, Barkstrom BR, Harrison EF, Lee RB, Smith GL, Cooper JE (1996) Clouds and the earth's radiant energy system (CERES): an earth observing system experiment. Bull Am Meteorol Soc 77(5):853–868. https://doi.org/10.1175/1520-0477(1996)077%3c0853:catere%3e2.0.co;2

Wielicki BA, Young DF, Mlynczak MG, Thome KJ, Leroy S, Corliss J, Anderson JG, Ao CO, Bantges R, Best F, Bowman K, Brindley H, Butler JJ, Collins W, Dykema JA, Doelling DR, Feldman DR, Fox N, Huang X, Holz R, Huang Y, Jin Z, Jennings D, Johnson DG, Jucks K, Kato S, Kirk-Davidoff DB, Knuteson R, Kopp G, Kratz DP, Liu X, Lukashin C, Mannucci AJ, Phojanamongkolkij N, Pilewskie P, Ramaswamy V, Revercomb H, Rice J, Roberts Y, Roithmayr CM, Rose F, Sandford S, Shirley EL, Smith WL, Soden B, Speth PW, Sun W, Taylor PC, Tobin D, Xiong X (2013) Achieving climate change absolute accuracy in orbit. Bull Am Meteorol Soc 94(10):1519–1539. https://doi.org/10.1175/BAMS-D-12-00149.1

Willis JK, Roemmich D, Cornuelle B (2003) Combining altimetric height with broadscale profile data to estimate steric height, heat storage, subsurface temperature, and sea-surface temperature variability. J Geophys Res 108:3292. https://doi.org/10.1029/2002JC001755

Willis JK, Roemmich D, Cornuelle B (2004) Interannual variability in upper ocean heat content, temperature, and thermosteric expansion on global scales. J Geophys Res 109:C12036 https://doi.org/10.1029/2003JC002260

WMO-No. 1316 (2023) State of the Global Climate 2022. Authors: World Meteorological Organization (WMO)

Wodzicki KR, Rapp AD (2016) Long-term characterization of the Pacific ITCZ using TRMM, GPCP, and ERA-Interim. J Geophys Res Atmos 121:3153–3170. https://doi.org/10.1002/2015JD024458

Wodzicki KR, Rapp AD (2022) More intense, organized deep convection with shrinking tropical ascent regions. Geophys Res Lett 49:e2022GL098615

Wong AP, Wijffels SE, Riser SC, Pouliquen S, Hosoda S, Roemmich D et al (2020) Argo data 1999–2019: two million temperature-salinity profiles and subsurface velocity observations from a global array of profiling floats. Front Mar Sci 7:700

Zuo, H., Balmaseda, M. A., Tietsche, S., Mogensen, K., & Mayer, M. (2019). The ECMWF operational ensemble reanalysis-analysis system for ocean and sea-ice: A description of the system and assessment. *Ocean Science Discussions*, 1–44. https://doi.org/10.5194/os-2018-154

Publisher's Note Springer Nature remains neutral with regard to jurisdictional claims in published maps and institutional affiliations.

Authors and Affiliations

Maria Z. Hakuba[1] · Sébastien Fourest[2] · Tim Boyer[3] · Benoit Meyssignac[2] · James A. Carton[4] · Gaël Forget[5] · Lijing Cheng[6] · Donata Giglio[7] · Gregory C. Johnson[8] · Seiji Kato[9] · Rachel E. Killick[10] · Nicolas Kolodziejczyk[11] · Mikael Kuusela[12] · Felix Landerer[1] · William Llovel[11] · Ricardo Locarnini[3] · Norman Loeb[9] · John M. Lyman[8,13] · Alexey Mishonov[3,14] · Peter Pilewskie[7,15] · James Reagan[3] · Andrea Storto[16] · Thea Sukianto[12] · Karina von Schuckmann[17]

✉ Maria Z. Hakuba
maria.z.hakuba@jpl.nasa.gov

[1] Jet Propulsion Laboratory, California Institute of Technology, Pasadena, CA 91011, USA

[2] LEGOS (CNES/CNRS/IRD/UT3), Université de Toulouse, 31400 Toulouse, France

[3] NOAA/National Center for Environmental Information, Silver Spring, MD 20910, USA

[4] Department of Atmospheric and Oceanic Science, University of Maryland, College Park, MD 20742, USA

[5] Department of Earth, Atmospheric and Planetary Sciences, Massachusetts Institute of Technology, Cambridge, MA 02139, USA

6 Institute of Atmospheric Physics, Chinese Academy of Sciences, Beijing 100029, China

7 Department of Atmospheric and Oceanic Sciences, University of Colorado Boulder, Boulder, CO 80309, USA

8 NOAA/Pacific Marine Environmental Laboratory, Seattle, WA 98115, USA

9 Climate Science Branch, NASA Langley Research Center, Hampton, VA 23681-2199, USA

10 Met Office Hadley Centre, Exeter, EX1 3PB, UK

11 Laboratoire d'Océanographie Physique et Spatiale (LOPS), University Brest CNRS IRD Ifremer, 29280 Brest, France

12 Department of Statistics and Data Science, Carnegie Mellon University, Pittsburgh, PA 15213, USA

13 CIMAR, University of Hawaii, Honolulu, HI 96822, USA

14 ESSIC/CISESS-MD, University of Maryland, College Park, MD 20740-3823, USA

15 Laboratory for Atmospheric and Space Physics, University of Colorado Boulder, Boulder, CO 80303, USA

16 Institute of Marine Sciences, National Research Council, 00133 Rome, Italy

17 Mercator Ocean International, 31400 Toulouse, France

Surveys in Geophysics (2024) 45:1757–1783
https://doi.org/10.1007/s10712-024-09838-8

Observational Assessment of Changes in Earth's Energy Imbalance Since 2000

Norman G. Loeb[1] · Seung-Hee Ham[2] · Richard P. Allan[3] · Tyler J. Thorsen[1] · Benoit Meyssignac[4] · Seiji Kato[1] · Gregory C. Johnson[5] · John M. Lyman[5,6]

Received: 1 October 2023 / Accepted: 1 April 2024 / Published online: 7 May 2024
© The Author(s) 2024

Abstract

Satellite observations from the Clouds and the Earth's Radiant Energy System show that Earth's energy imbalance has doubled from 0.5 ± 0.2 Wm^{-2} during the first 10 years of this century to 1.0 ± 0.2 Wm^{-2} during the past decade. The increase is the result of a 0.9 ± 0.3 Wm^{-2} increase absorbed solar radiation (ASR) that is partially offset by a 0.4 ± 0.25 Wm^{-2} increase in outgoing longwave radiation (OLR). Despite marked differences in ASR and OLR trends during the hiatus (2000–2010), transition-to-El Niño (2010–2016) and post-El Niño (2016–2022) periods, trends in net top-of-atmosphere flux (NET) remain within 0.1 Wm^{-2} per decade of one another, implying a steady acceleration of climate warming. Northern and southern hemisphere trends in NET are consistent to 0.06 ± 0.31 Wm^{-2} per decade due to a compensation between weak ASR and OLR hemispheric trend differences of opposite sign. We find that large decreases in stratocumulus and middle clouds over the sub-tropics and decreases in low and middle clouds at midlatitudes are the primary reasons for increasing ASR trends in the northern hemisphere (NH). These changes are especially large over the eastern and northern Pacific Ocean, and coincide with large increases in sea-surface temperature (SST). The decrease in cloud fraction and higher SSTs over the NH sub-tropics lead to a significant increase in OLR from cloud-free regions, which partially compensate for the NH ASR increase. Decreases in middle cloud reflection and a weaker reduction in low-cloud reflection account for the increase in ASR in the southern hemisphere, while OLR changes are weak. Changes in cloud cover in response to SST increases imply a feedback to climate change yet a contribution from radiative forcing or internal variability cannot be ruled out.

Keywords Earth's energy imbalance · Climate change · Clouds · Satellite · Earth radiation budget

---

✉ Richard P. Allan
r.p.allan@reading.ac.uk

[1] NASA Langley Research Center, Hampton, VA 23681-2199, USA

[2] Analytical Mechanics Associates (AMA), Hampton, VA 23666, USA

[3] Department of Meteorology and National Centre for Earth Observation, University of Reading, Reading RG6 6ET, UK

[4] LEGOS, Université de Toulouse, CNES, CNRS, UPS, IRD, Toulouse 31400, France

[5] NOAA/Pacific Marine Environmental Laboratory, Seattle, WA 98115, USA

[6] CIMAR, University of Hawaii, Honolulu, HI 96822, USA

Article Highlights

- Satellite observations reveal that global mean net flux (NET) at the top-of-atmosphere (or equivalently, Earth's energy imbalance) has doubled during the first twenty years of this century. The increase is associated with a marked increase in absorbed solar radiation (ASR) that is partially offset by an increase in outgoing longwave radiation (OLR)
- While ASR and OLR changes within sub-periods corresponding to the hiatus (03/2000–05/2010), transition-to-El Niño (06/2010–05/2016), and post-El Niño (06/2016–12/2022) vary substantially, NET flux changes are remarkably stable (within 0.1 Wm^{-2} per decade), implying a steady acceleration of climate warming
- The increase in ASR is associated with decreases in stratocumulus and middle cloud fraction and reflection in the Northern Hemisphere, and decreases in middle cloud reflection in the Southern Hemisphere. The cloud changes are especially large in areas with marked increases in sea-surface temperature, such as over the eastern and northern Pacific Ocean
- Continued monitoring of Earth's radiation budget and new and updated climate model simulations are critically needed to understand how and why Earth's climate is changing at such an accelerated pace

1 Introduction

Earth's radiation budget (ERB) describes how radiant energy is exchanged between Earth and space and how it is distributed within the climate system. The balance between incoming solar radiant energy absorbed by Earth and outgoing thermal infrared radiation emitted to space (also called Earth's Energy Imbalance, or EEI) determines whether Earth heats up or cools down (Hansen et al. 2005; Trenberth et al. 2014). A positive EEI is concerning as the extra energy added to the climate system leads to warming of the oceans, land and atmosphere, sea level rise, melting of snow and ice, and shifts in atmospheric and oceanic circulations (von Schuckmann et al. 2016). Approximately 89% of this additional heat is stored in the ocean, while the rest warms the land (5%) and atmosphere (2%) and melts ice (4%) (von Schuckmann et al. 2023).

Multiple lines of evidence show that EEI is increasing. These include an in situ based Earth heat inventory that quantifies how much heat has accumulated in the Earth system and where the heat is stored (von Schuckmann et al. 2023; Minière et al. 2023; Li et al. 2023; Storto and Yang 2024; Cheng et al. 2024), satellite observations of top-of-atmosphere (TOA) radiative fluxes from the Clouds and the Earth's Radiant Energy System (CERES) (Loeb et al. 2021a), and satellite measurements of sea level and ocean mass change (Hakuba et al. 2021; Meyssignac et al. 2023; Marti et al. 2023). In situ based Earth heat inventory observations of global ocean heat content (OHC) and non-ocean components (atmosphere, land and cryosphere) indicate a robust acceleration of Earth system heating since 1960 (von Schuckmann et al. 2023; Minière et al. 2023; Li et al. 2023; Storto and Yang 2024; Cheng et al. 2024). The acceleration rate for 1960–2020 is 0.15 ± 0.05 Wm^{-2} dec^{-1} and 0.30 ± 0.28 Wm^{-2} dec^{-1} for the more recent period between 2002 and 2020 (Minière et al. 2023). The latter is consistent within uncertainty with satellite observations of TOA net flux (Loeb et al. 2021a, 2022). In a comparison of CERES EEI with 18 OHC products derived from in situ, geodetic satellite observations, and ocean reanalyses for 2005–2019, Hakuba et al. (2024, this collection) show that while there is much spread in ocean heat uptake (OHU) and the rate of increase in OHU among the different analyses, the main reason for this spread is inadequate spatial–temporal sampling of the ocean. Datasets with better ocean coverage by filling in data sparse regions with satellite data or physical models (reanalyses) more closely match TOA net flux variability from CERES and show a positive trend in OHU that is similar in magnitude to CERES. It's worth noting that better sampling does not always guarantee better results. Loeb et al. (2022) argue that in the case of ocean reanalyses, achieving reliable temporal fidelity also depends upon model bias and whether new data are introduced/removed from the time series.

Few studies have examined what is driving the EEI increase since 2000. Raghuraman et al. (2021) used Coupled Model Intercomparison Project Phase 6 (CMIP6) (Eyring et al. 2016) simulations from the Geophysical Fluid Dynamics Laboratory Coupled/Atmospheric Model 4.0 (GFDL CM4/AM4) (Zhao et al. 2018; Held et al. 2019) to assess the contributions of internal variability, effective radiative forcing (ERF) and climate feedbacks on the CERES trend. They conclude that the positive EEI trend can only be explained if the simulations account for the increase in anthropogenic radiative forcing and associated climate response since 2000. This is confirmed with four additional CMIP6 models by Hodnebrog et al. (2024), who further showed that effective radiative forcing due to anthropogenic aerosol emission reductions contributes 0.2 ± 0.1 Wm^{-2} dec^{-1} to the trend in EEI. Kramer et al. (2021) used satellite data to infer instantaneous radiative forcing, providing observational evidence that radiative forcing is a major factor behind the EEI trend. Unfortunately, the number of assessments of the observed EEI trend are limited because the CMIP6 protocol ends in 2014. Schmidt et al. (2023) propose a new atmosphere only model intercomparison, CERESMIP, that targets the CERES period using updated sea-surface temperatures (SSTs), forcings and emissions through 2021. These new atmospheric model intercomparison project (AMIP) simulations will greatly expand the number of models available for model–observation comparisons and attribution studies of the EEI trend.

An observation-based partial radiative perturbation (PRP) analysis based upon the methodology of Thorsen et al. (2018) indicates that the CERES trend in EEI since 2000 is manifested in the data through changes in cloud, water vapor, trace gases, surface albedo and aerosols, which combine to increase TOA net downward radiation in excess of a negative contribution from increasing temperature (Loeb et al. 2021a). These changes are a consequence of the combined effects of climate forcing, feedback, and internal variability. To date, there has not been a thorough analysis of how different cloud types contribute to the observed changes in EEI. Loeb et al. (2021a) show that there is a large contribution by clouds to absorbed solar radiation changes and a weaker contribution to outgoing longwave radiation changes of opposite sign, but it does not attribute these to any particular cloud type. Furthermore, Loeb et al. (2021a) note substantial variations in TOA radiation during different sub-periods within the CERES record associated with internal variability.

In the following, we provide an observational assessment of TOA radiation changes that updates prior analyses by considering the period from 2000 to 2022 using CERES data products (Sect. 3.1). We examine the global, zonal and regional variations and trends in TOA radiation both for the entire CERES period and sub-periods corresponding to the hiatus (2000–2010), transition-to-El Nino (2010–2016), and post-El Nino (2016–2022) to highlight TOA radiation changes across periods of markedly different internal variability (Sect. 3.2). We also use the new CERES FluxByCldTyp (FBCT) data product (Sun et al., 2022) to quantify the contribution to TOA radiation changes by different cloud types using a cloud classification scheme based upon cloud types provided in FBCT (Sect. 3.3). Finally, we discuss some of the challenges associated with isolating the underlying processes that contribute to changes in TOA radiation from observations alone (Sect. 4).

2 Data and Methods

2.1 TOA Radiation and Cloud Datasets

Anomalies in TOA radiation components relative to their seasonal cycles are determined from the CERES Energy Balanced and Filled (EBAF) Ed4.2 product (Loeb et al. 2018) for 03/2000–12/2022. The anomalies are determined by differencing the average in a given month from the average of all years of the same month. Throughout the paper, anomalies are defined positive downwards (hence the naming convention "–OLR" to indicate that an increase in OLR corresponds to a loss of energy relative to climatology). Trends are determined from monthly anomalies using least squares linear regression and uncertainties in the trends follow the approach described in Loeb et al. (2022). The EBAF product uses an objective constrainment algorithm (Loeb et al. 2009) to adjust shortwave (SW) and longwave (LW) TOA radiative fluxes within their ranges of uncertainty to anchor global net TOA flux to an in situ estimate of the global mean EEI from mid-2005 to mid-2015 (Johnson et al., 2016). Use of this approach to anchor the satellite-derived EEI does not impact the variability and trends in the data (Loeb et al. 2018). The EBAF product provides two clear-sky fluxes, one for cloud-free portions of a region and a second for the total region. The latter was introduced to provide an observation-based clear-sky flux defined in the same way as climate models (Loeb et al. 2020). Here we only consider clear-sky fluxes for cloud-free areas of a region and use that to compute cloud radiative effect (CRE), defined as the difference between all-sky and clear-sky downward TOA flux. Loeb et al. (2020) show that while the magnitudes of clear-sky fluxes associated with the two definitions can be quite large, differences between their anomalies are relatively small.

TOA radiation changes for different cloud types are evaluated using the CERES FluxbyCldTyp Ed4.1-daily and -monthly products (Sun et al. 2022). The FBCT product has been used previously to generate observation-based cloud radiative kernels to quantify the sensitivity in TOA radiation to perturbations in meteorological conditions (Scott et al. 2020; Oreopoulos et al. 2022; Wall et al. 2022; Myers et al. 2023), to study changes in cloud properties and radiative fluxes by cloud type as a function of convective aggregation (Xu et al. 2023), and to evaluate climate models (Eitzen et al. 2017). FBCT provides CERES Terra and Aqua daytime 1°-regional gridded daily and monthly averaged TOA radiative fluxes and MODIS-derived cloud properties (Minnis et al. 2008, 2011a, 2011b) stratified into 42 cloud types for 6 cloud optical depth and 7 cloud effective pressure intervals, as defined in Rossow and Schiffer (1991). The cloud types are defined from the vantage point of an observer in space that only sees the clouds that are exposed to space. Thus, cloud effective pressure is determined from the topmost portion of a cloudy column and optical depth corresponds to column optical depth (Cole et al. 2011). TOA fluxes are also provided for all-sky and clear-sky conditions. In FBCT, "clear-sky" corresponds to fractional area within a $1° \times 1°$ region (gridbox hereafter) that is not covered by cloud. Since the FBCT uses Terra and Aqua, it only starts in July 2002 onwards. Accordingly, we consider 07/2002–12/2022 to assess changes in cloud fraction by cloud type.

Table 1 Definition of cloud classes used to assess influence of cloud changes on ASR

Cloud class	Cloud top pressure (hPa)	EIS (K)	Latitude range
Stratocumulus (Sc)	>680	>5	60°S–60°N
Stratocumulus-to-cumulus transition (SCT)	>680	0–5	60°S–60°N
Shallow cumulus (Cu)	>680	<0	60°S–60°N
Middle	440–680	–	60°S–60°N
High	<440	–	60°S–60°N
Polar	–	–	90°S–60°S; 60°N–90°N

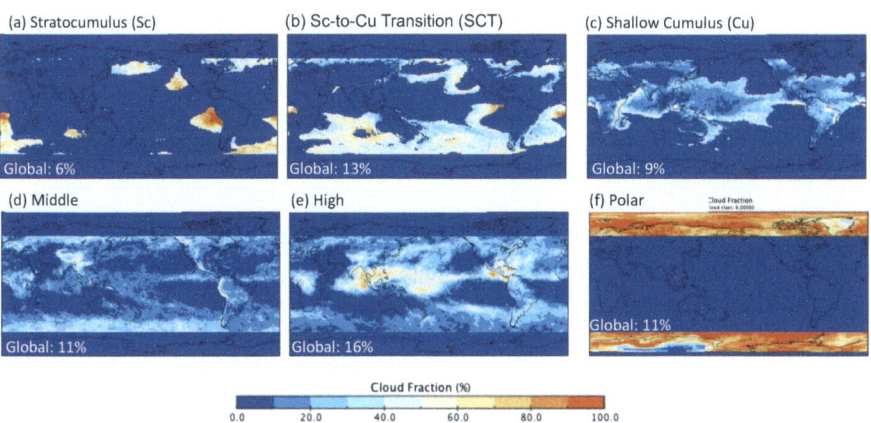

Fig. 1 Cloud fraction by cloud class for September 2002. Global coverages of each cloud class are as indicated

2.2 Changes in TOA Radiation by Cloud Class

To assess the influence of cloud changes on TOA fluxes, we develop a cloud classification scheme using 1°×1° gridded daily mean estimated inversion strength (EIS) parameter (Wood and Bretherton 2006) provided in the SSF1deg Ed4.1-daily product (Doelling et al. 2013) and cloud type information from the FBCT Ed4.1-daily and -monthly products (Sun et al. 2022). EIS is derived from surface pressure, temperature and dew point temperature at 2 m, and temperature and geopotential height at 700 hPa provided in the GEOS-DAS V5.4.1 product (Rienecker et al. 2008).

We first produce a gridded monthly EIS-by-cloud-type dataset from the SSF1deg Ed4.1-daily and FBCT Ed4.1-daily products by sorting gridded daily EIS values into the 42 FBCT cloud types in each gridbox each day and averaging these monthly. The monthly EIS-by-cloud-type data are then used together with the FBCT-monthly product to determine cloud fraction and TOA flux gridbox averages for three low cloud type classes equatorward of 60° (Table 1). The three low cloud classes have cloud effective pressures > 680 hPa with EIS values > 5 K Stratocumulus (Sc), 0–5 K stratocumulus-to-cumulus transition (SCT), and < 0 K cumulus (Cu). This EIS stratification of low clouds is an estimate based upon the regional distribution of annual mean EIS, SW CRE,

SST and vertical velocity at 700 hPa (e.g., see Fig. 1 from Myers and Norris 2015). In regions with EIS > 5 K, SW CRE is strongly negative, indicating that the clouds are bright, SSTs are cooler than surrounding regions, and subsidence is appreciable. These characteristics are consistent with stratocumulus (Wood 2012). Regions with EIS between 0 and 5 K exhibit weaker SW cloud radiative cooling, warmer SSTs, and weaker subsidence, consistent with stratocumulus-to-cumulus transition regimes. Low cloud areas with EIS < 0 K primarily occur in the tropical trade wind region over warm oceans where shallow cumulus typically reside. Middle and high cloud classes equatorward of 60° are defined for cloud effective pressures of 440–680 hPa and < 440 hPa, respectively. A polar cloud class is defined for all clouds poleward of 60°.

The regional distribution of the cloud classes in Table 1 for September 2002 (Fig. 1a–f) shows that three low cloud classes exhibit a smooth transition from Sc off the west coasts of the Americas and southern Africa to SCT mainly over the Southern Oceans and Cu mainly over the tropics. Middle and high clouds are distributed throughout 60°S–60°N, but occur predominantly in the mid-latitudes and tropics, respectively. An important feature of this cloud classification scheme is that the cloud types that can occur in a gridbox vary from month-to-month. In contrast, Scott et al. (2020) assign only one cloud type per region for the entire period to define cloud regimes. Since clouds vary appreciably over short timescales (Oreopoulous et al. 2016), the identified cloud types should be allowed to vary in time to correctly represent TOA flux changes by cloud type.

Global statistics (Table 2) of each cloud category for a 20-year climatology (07/2002–06/2022) show that Sc has a large local area coverage (52%) and exhibits substantial variability, with a monthly SW TOA flux anomaly standard deviation of 4 Wm^{-2}. However, the Sc cloud class accounts for only 7% of the globe, which reduces its global impact. Local cloud fractions for the low cloud types decrease from 52% (Sc) to 20% (Cu); while, SSTs increase from 281 K (Sc) to 300 K (Cu). These general characteristics are consistent with expectation for these cloud types (Wood 2012). Middle clouds have the smallest local fraction (13%) and weakest anomaly standard deviations

Table 2 Local average and monthly anomaly standard deviation in coverage (fraction), SW and OLR TOA fluxes, and SST for clear-sky and the cloud classes in Table 1 for 07/2002–06/2022

	Local fraction (%)		SW TOA flux (Wm^{-2})		OLR TOA flux (Wm^{-2})		SST (K)		Global fraction (%)
	Avg	Stdev	Avg	Stdev	Avg	Stdev	Avg	Std	
Clear	34.1	0.47	53.7	0.36	271.1	0.47	290.0	0.16	34.0
Sc	52.2	1.45	113.9	4.16	242.1	2.09	281.1	0.74	7.0
SCT	40.9	0.71	95.2	1.67	257.1	1.22	289.6	0.40	12.7
Cu	20.2	0.46	97.0	1.06	276.8	0.79	299.6	0.27	8.9
Middle	12.5	0.22	117.1	0.85	234.5	0.68	293.1	0.16	11.1
High	20.4	0.38	125.2	0.97	202.3	1.09	293.1	0.16	18.2
Polar	76.6	1.17	157.6	1.76	198.9	1.31	266.0	0.43	8.1

A "local" average is determined from geodetic-weighted monthly averages of all 1°×1° regions in which a given cloud type is observed. Also provided is the coverage of each clear or cloud class over the entire globe. Here, SSTs are from the CERES SSF1deg Ed4.1 daily product

compared to the other cloud types; while, polar clouds have largest local fraction and average SW flux, but the lowest OLR flux and SST.

2.3 All-Sky TOA Flux Decomposition

The monthly mean all-sky TOA flux over a latitude range (λ_1, λ_2) and longitude range (ϕ_1, ϕ_2) can be expressed in terms of its clear and cloudy column contributions from $1° \times 1°$ regions as follows:

$$\overline{F}_{\text{all}} = \overline{F}_{\text{clr}}^{\text{con}} + \sum_{j=1}^{n} \overline{F}_{j}^{\text{con}} \qquad (1)$$

where $\overline{F}_{\text{clr}}^{\text{con}}$ is the monthly mean clear-sky column flux contribution and $\overline{F}_{j}^{\text{con}}$ is the monthly mean cloud column contribution for cloud class j, and n is the number of cloud classes. These are calculated as follows:

$$\overline{F}_{\text{clr}}^{\text{con}} = \frac{1}{W} \int_{\lambda_1}^{\lambda_2} \int_{\phi_1}^{\phi_2} \left(1 - f_T(\lambda, \phi)\right) F_{\text{clr}}(\lambda, \phi) w_\lambda d\lambda d\phi \qquad (2)$$

$$\overline{F}_{j}^{con} = \frac{1}{W} \int_{\lambda_1}^{\lambda_2} \int_{\phi_1}^{\phi_2} f_j(\lambda, \phi) F_j(\lambda, \phi) w_\lambda d\lambda d\phi \qquad (3)$$

where f_T and F_{clr} are the monthly gridbox total cloud fraction and mean clear-sky flux, respectively, and f_j and F_j are the monthly gridbox cloud fraction and mean flux for cloud class j. The total cloud fraction f_T is equal to the sum of the individual f_j's, and the weights w_λ are geodetic weights whose sum W over the domain is given by:

$$W = \int_{\lambda_1}^{\lambda_2} \int_{\phi_1}^{\phi_2} w_\lambda d\lambda d\phi \qquad (4)$$

This decomposition of all-sky TOA flux represents all-sky TOA flux as the sum of area-weighted clear and cloudy column fluxes. Anomalies and trends in these contribution terms are impacted by area fraction and within column radiative property changes, but the sum is constrained to add to the corresponding all-sky value. We do not correct for non-cloud changes in the cloudy columns, nor do we attempt to remove ERF contributions. We expect that the cloud masking error is smaller than that for CRE since it is confined to the cloudy area only rather than a gridbox-wide difference between clear-sky and total-sky non-cloud contributions (Soden et al. 2008). We plan to extend the methodology to account for cloud masking contributions in the future.

2.4 Validation of MODIS-Based Cloud Fraction Changes

To evaluate MODIS-based cloud fraction changes, Appendix 1 provides a detailed comparison of trends in MODIS cloud fraction by cloud type with those from coincident cloud-aerosol lidar and infrared pathfinder satellite observations (CALIPSO) cloud-aerosol lidar with orthogonal polarization (CALIOP) and CloudSat cloud profiling radar (CPR) data as provided in the CALIPSO-CloudSat-CERES-MODIS (CCCM) RelD1 product (Kato et al. 2010, 2011).

The analysis in Appendix 1 shows that MODIS and CC cloud fraction trends are remarkably similar for each cloud type, providing confidence in the MODIS-based results. Additional comparisons between these and other cloud fraction products are provided in Stubenrauch et al. (2024, this collection), which focuses more on how well the different products agree in their regional cloud fraction distributions than on temporal variability.

3 Results

3.1 Global, Zonal and Regional Changes in TOA Radiation During CERES Period

As noted in Loeb et al. (2021a, 2022), the CERES record indicates that EEI has approximately doubled during the CERES period. During the first decade of CERES observations (03/2000–02/2010), EEI was 0.5 ± 0.2 Wm^{-2} and increased to 1.0 ± 0.2 Wm^{-2} for the most recent decade (01/2013–12/2022) considered here (Table 3). This is the result of a 0.9 ± 0.3 Wm^{-2} ($\approx 0.4\%$) increase in ASR that is partially offset by a 0.4 ± 0.25 Wm^{-2} ($\approx 0.2\%$) increase in outgoing longwave radiation (OLR). The corresponding change in incoming solar irradiance is negligible (0.02 ± 0.09 Wm^{-2}). There is satellite evidence that the increase in EEI began during the decade prior to the CERES period based on a reconstruction of the earth radiation budget experiment (ERBE) record (Liu et al. 2020) and satellite altimetry and space gravimetry measurements (Marti et al. 2023).

Monthly anomalies in global mean TOA radiation show considerable variability superimposed over longer-term trends (Fig. 2a, b). Standard deviations in monthly anomalies for 03/2000–12/2022 are 0.7, 0.5 and 0.7 Wm^{-2} for ASR, –OLR and NET, respectively, and the corresponding trends are 0.71 ± 0.19, -0.26 ± 0.19, and 0.45 ± 0.18 Wm^{-2} per decade (uncertainties given as 2.5–97.5% confidence intervals). Monthly anomalies are consistent across CERES instruments on different platforms to <0.2 Wm^{-2} (Loeb et al. 2018) and trends between Terra and Aqua, the two longest operating missions flying CERES instruments, agree to <0.1 Wm^{-2} per decade (Loeb et al. 2022). Extensive validation of CERES instrument performance using a range of consistency tests involving different vicarious Earth targets and regular scans of the Moon provides further evidence that the CERES instruments are radiometrically stable (Shankar et al. 2023). The trends from CERES observations also agree with independently estimated trends from 0 to 2000 m ocean in situ data to <0.1 Wm^{-2} per decade (Loeb et al. 2021a, 2022).

Analysis of atmospheric climate model simulations with a hierarchy of experiments using the GFDL CM4/AM4 suggest that the large positive ASR trend is due to additive contributions from ERF and climate feedback (radiative response) and the weaker negative trend in outgoing longwave radiation results from compensation between positive ERF and negative climate feedback contributions (Raghuraman et al. 2021; Hodnebrog et al. 2024). Since the ERF contributions add together and the climate feedback contributions offset one another, the model results suggest that ERF is the main driver of the positive trend in NET. However, the

Table 3 Average solar irradiance, ASR, –OLR and Net TOA radiation in Wm^{-2} for the first and most recent decades of CERES observations	Solar irradiance	ASR	–OLR	NET
03/2000–02/2010	340.14	240.7	–240.2	0.53
01/2013–12/2022	340.16	241.6	–240.6	1.05
Difference	0.02	0.9	–0.4	0.52

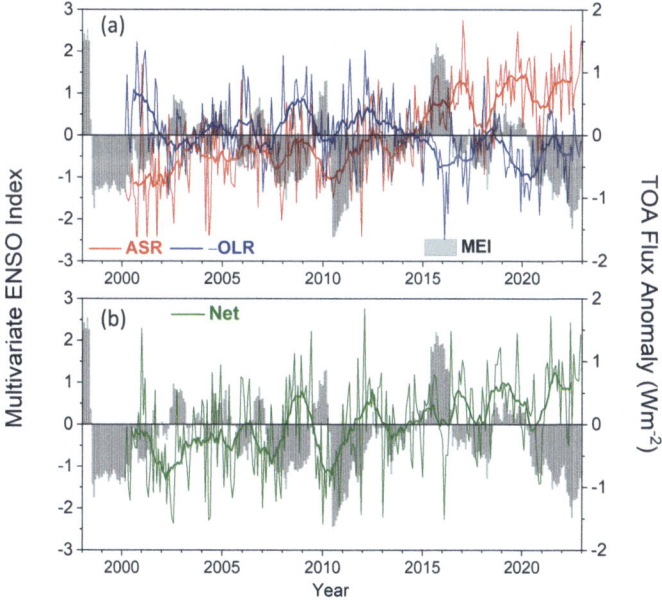

Fig. 2 Global mean all-sky TOA flux anomalies and multivariate ENSO index (MEI) from CERES EBAF Ed4.2 for 03/2000–12/2022. **a** ASR and −OLR; **b** NET

magnitudes of global TOA radiation trends in the climate model simulations are weaker than those in CERES, and there are large discrepancies in regional trend patterns. Furthermore, coupled climate models fail to represent observed SST patterns and associated feedbacks (Andrews et al. 2022; Kang et al. 2023; Olonscheck and Rugenstein 2024), adding to existing questions about the realism of climate model changes during the 21st Century (Trenberth and Fasullo 2009). These, together with substantial updates to SST and forcing datasets, provide additional motivation for further model–observation comparisons (Schmidt et al. 2023).

Zonal average trends for approximately equal-area latitude zones are positive for ASR and NET in the tropics, sub-tropics, and mid-high latitudes of both hemispheres; while, −OLR only shows appreciable negative trends in the NH sub-tropics and NH mid-high latitudes (Fig. 3a–c). Northern and southern hemisphere trends in NET are consistent to 0.06 Wm^{-2} per decade due to a compensation between weak ASR and −OLR hemispheric trend differences of opposite sign (Table 4). Datseris and Stevens

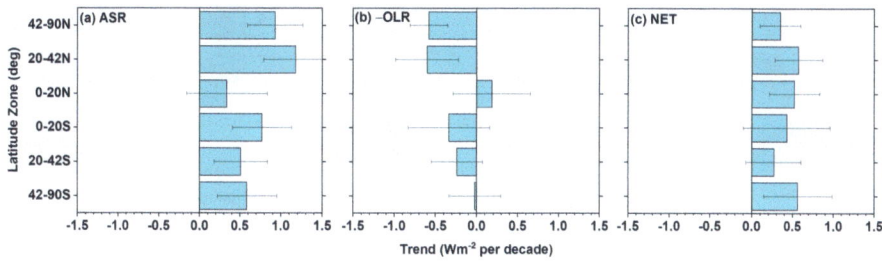

Fig. 3 Zonal mean all-sky TOA flux trends for 03/2000–12/2022. **a** ASR; **b** −OLR; **c** NET

Table 4 Hemispheric and global trends in ASR, –OLR and NET for 03/2000–12/2022 in Wm^{-2} decade^{-1}

	SH	NH	Globe	SH minus NH
ASR	0.62 ± 0.23	0.80 ± 0.22	0.71 ± 0.19	− 0.18 ± 0.36
− OLR	− 0.20 ± 0.21	− 0.33 ± 0.21	− 0.26 ± 0.19	0.13 ± 0.29
NET	0.42 ± 0.26	0.48 ± 0.21	0.45 ± 0.18	− 0.06 ± 0.31

Uncertainties are given as 2.5–97.5% confidence intervals

(2021) also found hemispheric symmetry in reflected SW trends using CERES data for 03/2000–02/2020. Interestingly, GFDL AMIP climate model simulations fall within 0.2 Wm^{-2} per decade of CERES NH trends for ASR, –OLR and NET, but underestimate the ASR trend in the SH by −0.5 Wm^{-2} per decade due to erroneous trends in Antarctic sea ice and Southern Ocean cloud fraction, resulting in a much larger ASR hemispheric contrast (Raghuraman et al. 2021).

Regionally, significant positive trends in CERES ASR occur off both coasts of North America, the Seas of Japan and Okhotsk, over the Arctic Ocean between the Kara and East Siberian Seas, the Southern Ocean to the east of South America, and Antarctica between 60° and 120°E (Fig. 4a). Large positive trends also occur over the equatorial Pacific Ocean, but because interannual variability is so large in this region due to the El Niño-Southern Oscillation (ENSO), the trends do not exceed the 2.5–97.5% confidence interval. Negative trends of –OLR, corresponding to increased thermal infrared emission to space, are appreciable over the NH eastern Pacific Ocean and over much of the Arctic (Fig. 4b). These regions are also associated with strong warming (Fig. 4d).

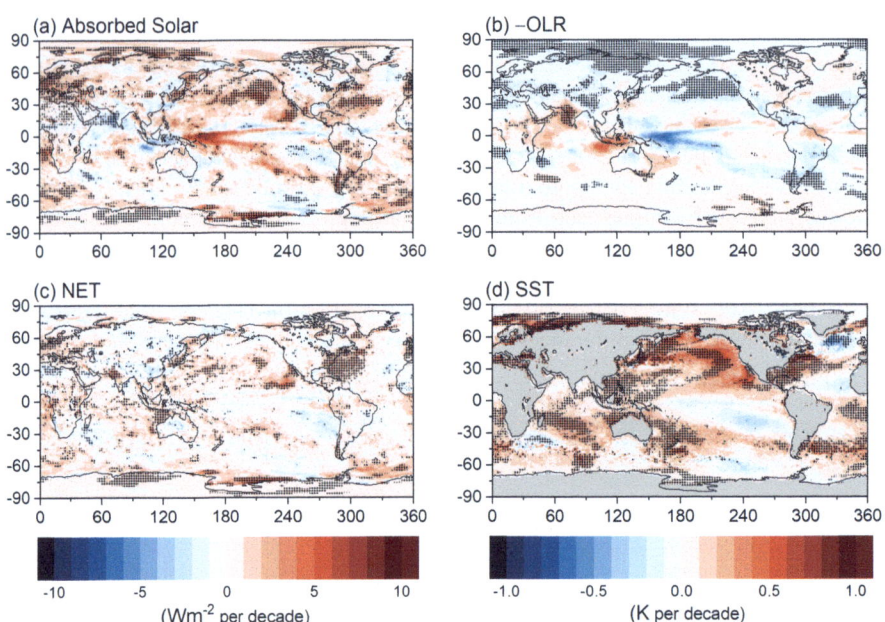

Fig. 4 Regional trends in a ASR, b –OLR, c NET (Wm^{-2} per decade), and d SST (K per decade) for 03/2000–12/2022. Hatching indicates trends significant at 2.5–97.5% confidence level. SSTs are from ECMWF Reanalysis 5 (ERA5) (Hersbach et al. 2020)

Regional net radiation trends are positive over the NH Pacific, Indian and West Atlantic Oceans, but are mainly negative over the marine stratocumulus region off the west coast of South America (Fig. 4c). The similarity between the ASR and SST trend patterns is striking (Fig. 4a, d), particularly over the North Pacific, off the east coast of North America and west coast of South America.

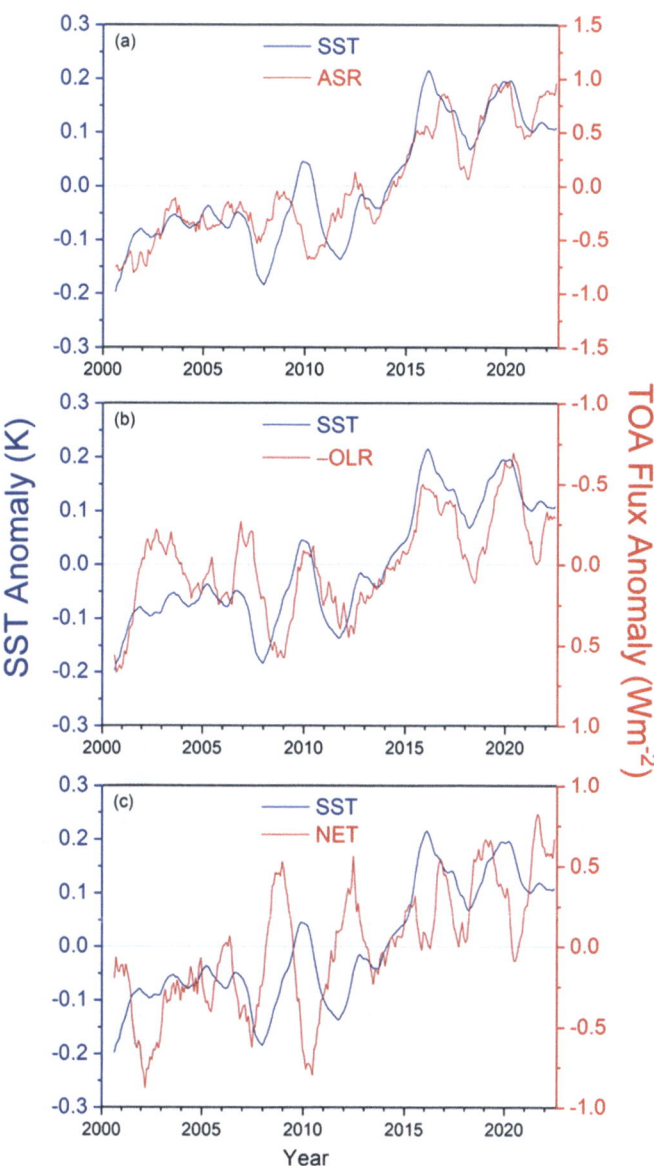

Fig. 5 Twelve-month running average global anomalies in ERA5 SST and CERES **a** ASR, **b** OLR (positive up, since −OLR is displayed with a reversed y-axis), and **c** NET TOA radiation. Period considered: 03/2000–12/2022

Time series of global mean anomalies in SST, ASR, and −OLR also share similar features (Fig. 5a, b). In each case, twelve-month running average anomalies are relatively constant prior to 2010, and then increase sharply (decrease for −OLR) until a maximum is reached during the 2015–2016 El Niño event. The anomalies stay relatively flat after this event, albeit with considerable interannual variability. By comparison, the coherence at interannual timescales between anomalies in SST and NET radiation is much weaker (Fig. 5c) due to compensation between ASR and −OLR changes, but both do show a marked increase for the entire period.

Coupled climate models show a long-term trend in EEI and SST with anthropogenic forcing (Collins et al. 2013; Forster et al. 2021). Results in Fig. 5 confirm that increases in EEI and SST also occur in observations over a 20-year period despite substantial internal variability from heat exchange between the ocean mixed layer—which directly impacts SST—and the ocean layers below. Vertical ocean mixing has been shown to add considerable scatter between TOA radiation and SST trends at decadal timescales (Palmer et al. 2011).

3.2 Changes During the Hiatus, Transition-to-El Niño, and Post-El Niño Sub-Periods

We examine the temporal evolution in SST and TOA radiation for the 3 sub-periods, which we define as follows: (i) "hiatus" (03/2000–05/2010), characterized by a negligible change in the Multivariate ENSO Index (MEI; Wolter and Timlin 1998) (Fig. 6a–d), a slower rate of global warming compared to the longer-term trend (Lewandowsky et al., 2015; Meehl et al. 2013; Trenberth 2015a) and to simulations from coupled climate models (Kosaka and Xie 2013); (ii) "transition-to-El Niño" (06/2010–05/2016), corresponding to the transition between the 2010–2012 La Niña and 2014–2016 El Niño events; and (iii) "post-El Niño" (06/2016–12/2022), corresponding to the transition between the 2014–2016 El Niño and the unusual extended 2020–2022 La Niña (so-called "triple-dip La Niña). During the "transition to El Niño" period, MEI and SST both show rapid increases that exceed the 2.5–97.5% CI (Fig. 6b, d). The SST trend during this period is 0.52 K decade^{-1}, which

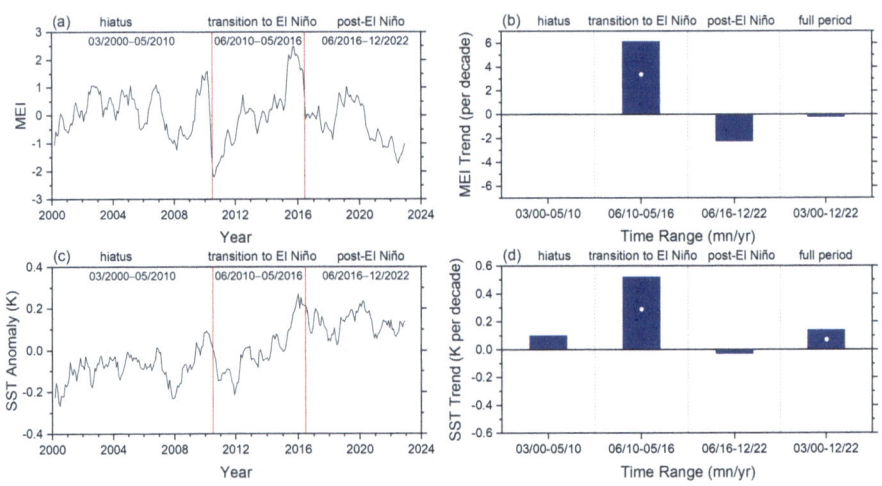

Fig. 6 Monthly time series **a**, **c** and trends **b**, **d** for MEI (top) and anomalies in ERA5 SST (bottom). White circles in **b** and **d** correspond to trends that exceed the 2.5–97.5% CI. Time period 03/2000–12/2022

exceeds the increase during the "hiatus" period by a factor of 5. For the entire period between 03/2000 and 12/2022, the SST trend is 0.14 ± 0.06 K decade^{-1} and the trend in MEI is near zero.

Trends in solar irradiance (SOL) and all-sky reflected SW (–SW, positive downwards), ASR, –OLR, and net radiation (NET) for the three sub-periods and entire time range (Fig. 7a) reveal that despite marked differences among sub-period trends for ASR, –SW and –OLR, reaching 1.3 Wm^{-2} per decade, NET trends remain within 0.1 Wm^{-2} per decade of one another and the trend over the entire period (0.45 Wm^{-2} per decade). During the "hiatus" the –OLR trend is near zero, so that the NET trend is determined by the difference between SOL and –SW. In contrast, all-sky –SW and –OLR both exceed 1 Wm^{-2} per decade in magnitude during the "transition-to-El Niño" period, but their sum (0.26 Wm^{-2} per decade) and the SOL contribution (0.19 Wm^{-2} per decade) add to ≈0.45 Wm^{-2} per decade

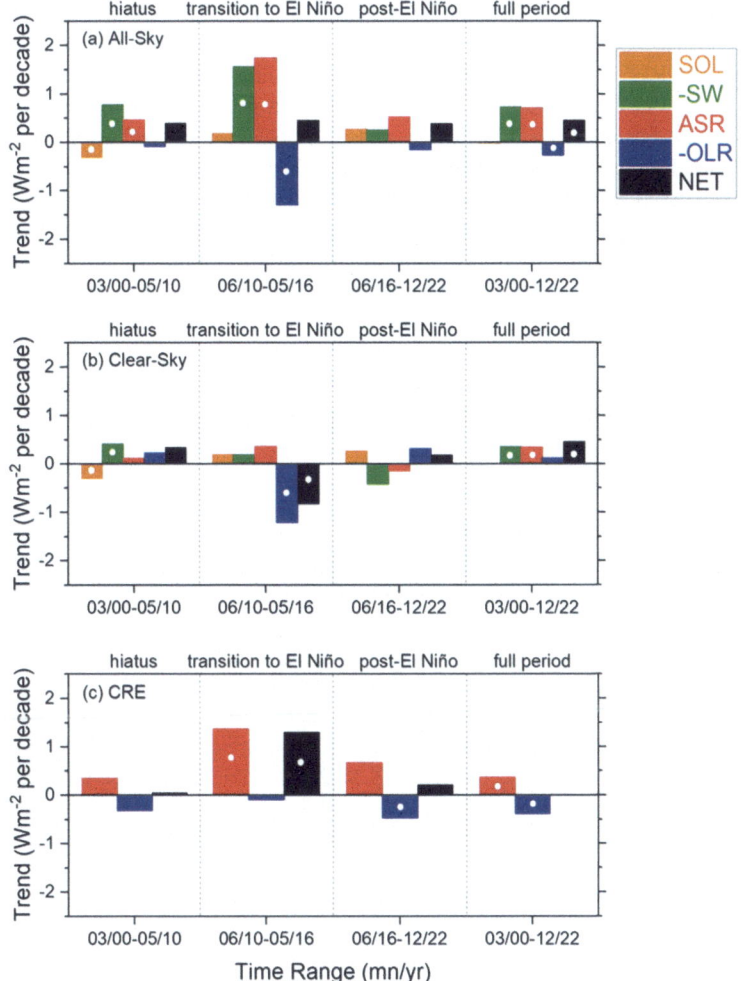

Fig. 7 Trends in solar irradiance (SOL), –SW, ASR, –OLR, and NET TOA flux for **a** all-sky, **b** clear-sky and **c** CRE. White circles indicate trends that exceed the 2.5–97.5% CI

for NET. This period is characterized by a substantial warming, leading to greater thermal emission to space from cloud-free areas (Fig. 7b). There is also a decrease in cloud fraction (not shown) that causes a strong ASR contribution by clouds (Fig. 7c), which compensates for the increased thermal emission. The trend in –OLR during the "post-El Niño" period is small, and SOL and –SW contribute approximately equally to the NET trend. In contrast to the all-sky case, clear-sky NET trends differ by up to -1 Wm^{-2} between sub-periods (Fig. 7b). Changes in clouds compensate for these differences under all-sky conditions, leading to a very similar all-sky NET trend in each sub-period.

It is unclear if the remarkable consistency among all-sky NET trends for the sub-periods occurs by chance or is a robust property of Earth's energy budget. At shorter time scales than those defining these sub-periods, there is substantial interannual variability in NET radiation, as shown in Fig. 2b. Unfortunately, the CERES observational record is too short to test how robust these results are. Nevertheless, it implies a steady acceleration of climate warming since 2000.

It is noteworthy that NET CRE for the full period is near zero (Fig. 7c). Raghuraman et al. (2023) also show a negligible trend in what they describe as the "cloud feedback component of CRE", which is obtained from the difference between CRE and the sum of ERF and cloud masking contributions. The implication is that net cloud feedback is not statistically significant during the CERES period. However, this conclusion assumes the model-derived ERF contribution to CRE is correct. The shortwave ERF contributions are primarily due to greenhouse gas adjustments and the aerosol-cloud indirect effects, both highly uncertain quantities (Smith et al. 2020). Furthermore, in Raghuraman et al. (2023) the model shortwave ERF contribution to CRE exceeds the longwave ERF contribution and accounts for as much as 57% of the total CERES SW CRE. In their observation-based PRP analysis Loeb et al. (2021a) found a significant positive trend in the cloud contribution to NET all-sky TOA flux, but aerosol-cloud indirect effects and greenhouse gas adjustments and were not removed from the cloud contribution. The uncertainty surrounding ERF thus makes it challenging to unambiguously isolate the net effect of clouds during the CERES period.

3.3 TOA Radiation Changes by Cloud Type

The cloud classes (Sect. 2.2) and all-sky TOA flux decomposition (Sect. 2.3) provide a framework to assess TOA radiation changes by cloud type using FBCT. Since the CERES FBCT product uses data from both Terra and Aqua, the time period considered is limited to 07/2002–12/2022. Given that EBAF TOA global trends for this period are very similar to those for the full CERES period (Table 5), we expect results for the shorter period to be representative of the full period. We also find good agreement between EBAF and FBCT all-sky, clear-sky, and CRE trends for 07/2002–12/2022 (Table 5). The reason for the larger clear-sky –OLR difference is unknown. One contributing factor could be because of cloud mask differences as FBCT is a daytime-only product; while, EBAF uses both daytime and nighttime observations.

To illustrate the utility of the all-sky TOA flux decomposition framework, we compare global trends in TOA fluxes for all-sky, clear-sky and CRE alongside cloud fraction-weighted contributions computed using Eqs. (1–4) in Fig. 8. While the trend in net CRE is weak due to compensation between –SW and –OLR components, the trend for the area-weighted cloudy contribution is appreciable due to a large positive trend in –SW and negligible –OLR trend. Without any cloud masking adjustments in the cloudy

Table 5 Global trends in all-sky, clear-sky and CRE from EBAF and FluxbyCldTyp in Wm^{-2} decade^{-1}

	03/2000–12/2022	07/2002–12/2022	
	EBAF all-sky	EBAF all-sky	FBCT all-sky
–SW	**0.73 ± 0.21**	**0.68 ± 0.25**	**0.67 ± 0.26**
–OLR	**−0.26 ± 0.19**	**−0.25 ± 0.22**	−0.20 ± 0.30
NET	**0.45 ± 0.18**	**0.47 ± 0.21**	**0.50 ± 0.23**
	EBAF clear-sky	EBAF clear-sky	FBCT clear-sky
–SW	**0.36 ± 0.11**	**0.32 ± 0.12**	**0.33 ± 0.12**
–OLR	0.12 ± 0.16	0.11 ± 0.19	0.29 ± 0.29
NET	**0.46 ± 0.14**	**0.46 ± 0.16**	**0.65 ± 0.19**
	EBAF CRE	EBAF CRE	FBCT CRE
–SW	**0.37 ± 0.18**	**0.36 ± 0.22**	**0.34 ± 0.20**
–OLR	**−0.38 ± 0.09**	**−0.36 ± 0.11**	**−0.49 ± 0.10**
NET	−0.008 ± 0.19	0.008 ± 0.21	−0.15 ± 0.20

Trends exceeding the 2.5-97.5 confidence interval are indicated in bold

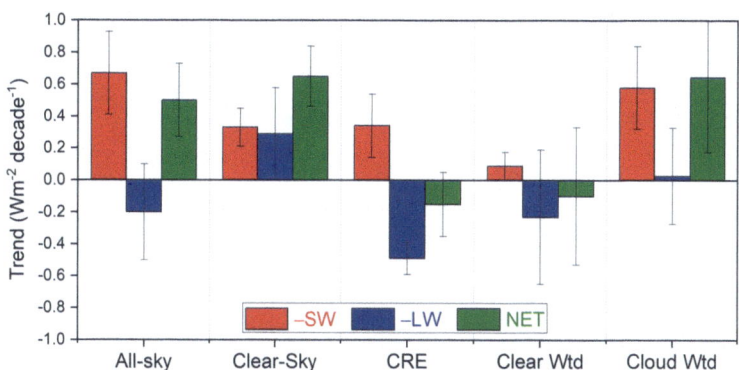

Fig. 8 Trends in all-sky and clear-sky flux, CRE, clear fraction weighted clear-sky column (Clear Wtd) and cloud fraction weighted cloudy column (Cloud Wtd) flux contributions for –SW, –OLR (–LW), and NET TOA flux from FBCT product. Error bars correspond to 2.5–97.5% CI. Time period: 07/2002–12/2022

regions, this result is already comparable to what is obtained using PRP analysis (see Fig. 2 in Loeb et al. 2021a). We expect that after subtracting cloud masking contributions, agreement with the PRP result will improve. After the corrections are made, trends in the –SW, –OLR and NET area-weighted cloudy contribution should decrease because part of the positive –SW trend is impacted by decreases in surface albedo from declining sea-ice coverage during the CERES period, and part of the –OLR trend is associated with reduced emission resulting from increases in water vapor and WMGG above the cloud top (Raghuraman et al. 2023). Results in Fig. 8 show that the all-sky decomposition approach in Sect. 2.3 provides a better framework than CRE for assessing the radiative impacts of cloud changes. The key difference with the CRE approach is that the all-sky decomposition separates changes from clear and cloudy areas whereas the CRE approach can only provide reliable results if there are no changes in cloud-free conditions, which is unrealistic.

TOA radiation and cloud fraction changes by cloud type for different latitude zones (Figs. 9, 10, 11, and 12) provide context for the hemispheric and global trends (Table 4). Since the contribution from each cloud class is an area fraction-weighted quantity over each latitude zone, the sum of all contributions plus the clear-sky contribution is equal to the total all-sky value. Decreases in low and middle cloud fraction and reflection between 20° and 60°N (Figs. 9b, c and 10b, c) and reduced reflection from cloud-free areas between

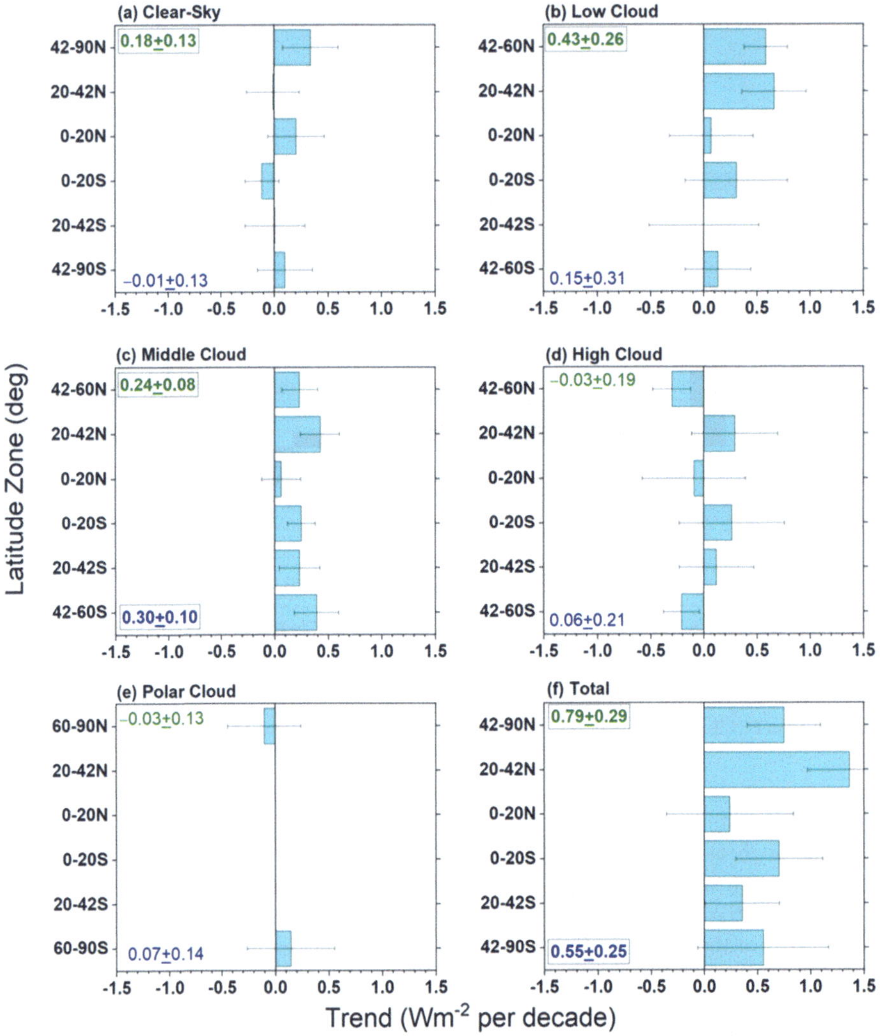

Fig. 9 Contribution to zonal mean −SW trend from **a** clear-sky, **b** low cloud, **c** middle cloud, **d** high cloud, **e** polar cloud, **f** all. Period considered: 07/2002–12/2022. The SH and NH hemispheric average trends for each cloud type are indicated in each figure

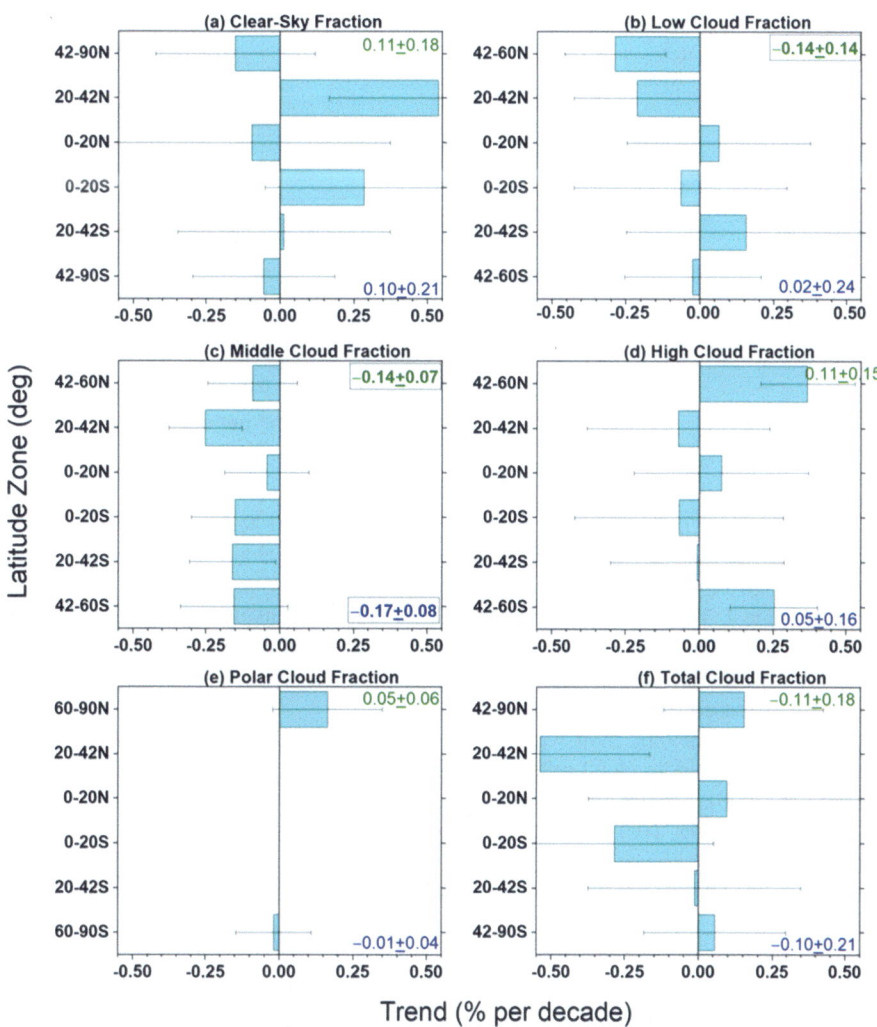

Fig. 10 Same as Fig. 9 but for clear-sky and cloud fraction

42° and 90°N (Fig. 9a) are the primary reasons for the NH ASR increase of 0.8 Wm^{-2} decade^{-1} in Table 4. Low cloud changes are primarily from Sc between 20° and 42°N; while Sc, SCT and Cu all contribute to the low cloud ASR increase between 42° and 60°N (Fig. 11). Regionally, these changes occur over the eastern and northern Pacific and off the east coast of North America, and coincide with large increases in SST (Fig. 4d). Other studies have noted the significant low-cloud response to SST in these regions (Myers et al. 2018; Andersen et al. 2023).

Interestingly, while there is a marked increase in clear-sky fraction in the NH sub-tropics between 20° and 42°N (Fig. 10a), the corresponding ASR trend contribution is near zero (Fig. 9a). This is likely because of a decrease in aerosol optical depth in this latitude range during the CERES period (Zhao et al. 2017; Paulot et al. 2018; Loeb et al. 2021b), which compensates for the increased clear-sky frequency, resulting in a near zero ASR

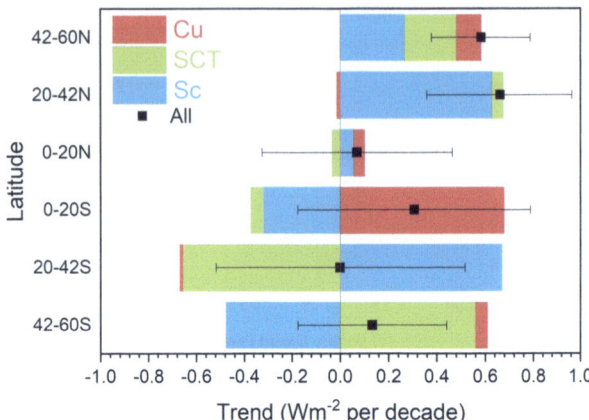

Fig. 11 Zonal low-cloud trends with contribution from Cu, SCT and Sc. Period considered: 07/2002–12/2022

trend contribution. While high clouds contribute little to the overall NH ASR trend, there is a notable increase in high cloud fraction between 42° and 60°N (Fig. 10d) that causes a negative ASR trend (Fig. 9d). Increased thermal emission in cloud-free conditions combined with high cloud changes contribute most to the -0.33 Wm^{-2} $decade^{-1}$ NH $-$OLR change in Table 4. The increase in SST between 20° and 42°N likely contributes to a sharp increase in clear-sky thermal infrared emission ($-$OLR trend of -1.6 Wm^{-2} per decade) (Fig. 12a) while the increase in high cloud thermal emission between 42° and 60°N is associated with increased cloud fraction (Fig. 12d).

The ASR trend of 0.62 Wm^{-2} $decade^{-1}$ in the SH (Table 4) is primarily associated with decreases in middle cloud reflection (Fig. 9c) and a weaker reduction in low-cloud reflection (Fig. 9b). Middle cloud fractions decrease by almost the same amount in each SH latitude zone (Fig. 10c); while, high cloud fraction increases between 42° and 60°S (Figs. 10d), resulting in a weak negative ASR trend contribution to ASR (Fig. 9d). In contrast to the NH, $-$OLR cloud trends in the SH are weak and largely cancel one another.

4 Discussion

A key limitation of relying solely on observations to explain TOA radiation changes is that some of the underlying processes involved are difficult to isolate. For example, there is evidence that anthropogenic aerosol effective radiative forcing is weakening due to a decline in anthropogenic primary aerosol and aerosol precursor emissions (Quaas et al. 2022). Observations can provide estimates of the influence of aerosol-radiation interactions (Bellouin et al. 2005; Subba et al. 2020; Loeb et al. 2021b; Szopa et al. 2021), but the much stronger forcing contribution from aerosol-cloud interactions is more difficult to quantify as both clouds and aerosols are impacted by their environment (e.g., meteorology) in addition to having a two-way interaction between them (Gryspeerdt et al. 2016; McCoy et al. 2020). Furthermore, passive satellite aerosol retrievals are more uncertain in cloudy regions, and cloud retrievals are more uncertain in environments with abundant aerosol (Koren et al. 2007; Loeb and Schuster 2008; Gryspeerdt et al. 2016). This makes it challenging to unambiguously quantify how aerosol and cloud changes separately influence trends in ASR, which we show track closely with trends in SST, particularly over stratocumulus regions off the west coast of North America and over the North Pacific Ocean (see

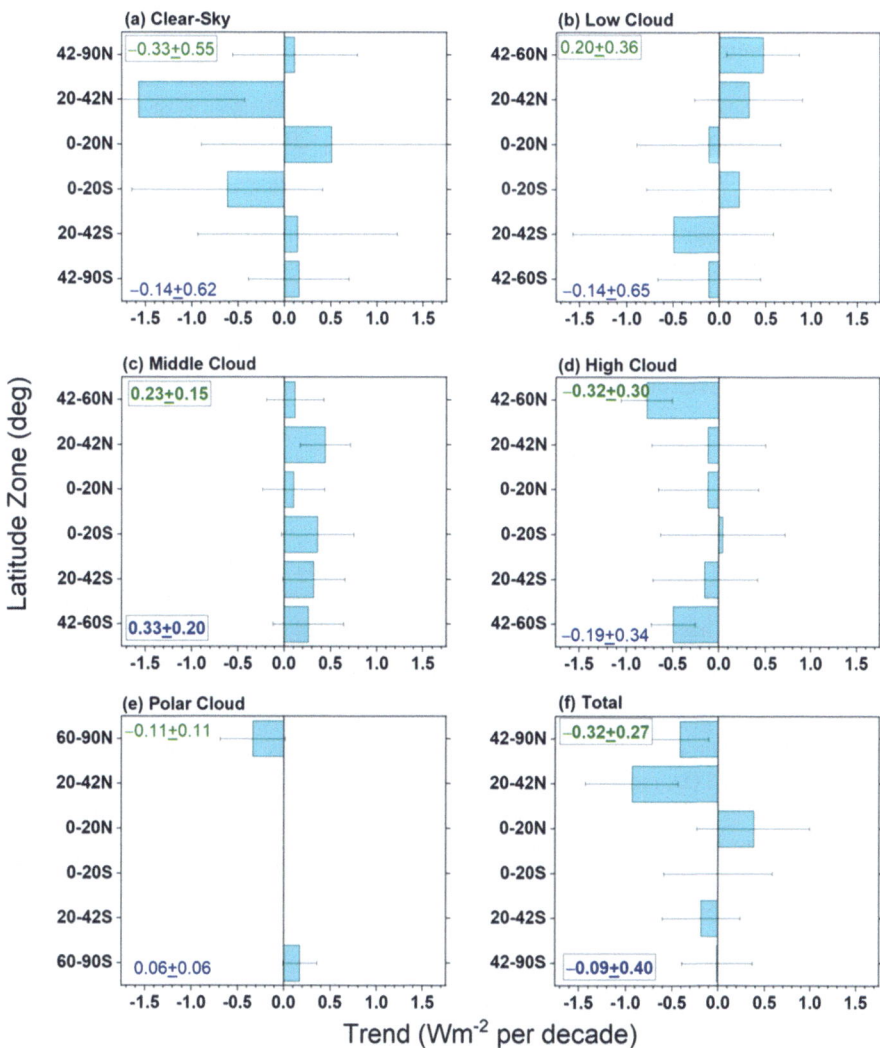

Fig. 12 Same as Fig. 9 but for −OLR

also Andersen et al. 2022; Myers et al. 2018). Establishing causality between observed SST and ASR changes also has its challenges as these share a two-way interaction (Trenberth et al. 2015b).

Nevertheless, progress is being made on the use of satellite observations for studying aerosols. A recent study by Wall et al. (2022) introduces a new method that removes confounding meteorological factors from observed sulfate–low-cloud relationships and narrows the uncertainty in aerosol forcing. Studies by Yuan et al. (2022) and Diamond (2023) use satellite observations to quantify the impact of sulfur regulations for shipping fuel on aerosol indirect forcing. Both studies find evidence for reduced radiative cooling by clouds following new regulations limiting sulfur emissions from the shipping industry by the International Maritime Organization 2020.

A longer TOA ERB observational record and new model output from CERESMIP provides new opportunities to determine how best to use observations and models for improving our understanding of the underlying process related to EEI changes. Current climate model simulations show similar patterns in regional TOA flux changes as observations, but the magnitudes of the changes differ markedly (Loeb et al. 2020), particularly over cloudy extratropical regions (Trenberth and Fasullo 2010; Zelinka et al. 2020). Similarly, the EEI trends from Raghuraman et al. (2021) are systematically lower compared to CERES. Conversely, if we find agreement between trends in TOA radiation in observations and climate model simulations, do they agree for the right reasons? To answer this, it will be necessary to use additional datasets and climate model output describing cloud and aerosol changes. Our comparisons with CC (Appendix 1) provide some confidence that the imager-based cloud changes are realistic. This means that there is some hope that meaningful comparisons between observed and model cloud changes is within reach.

5 Summary and Conclusions

CERES observations show that Earth's energy imbalance (EEI) has doubled from 0.5 ± 0.2 Wm^{-2} during the first 10 years of this century to 1.0 ± 0.2 Wm^{-2} during the past decade. This has led to accelerated increases in global mean temperature, sea level rise, ocean heating, and snow and sea ice melt. The increase in EEI is the result of a 0.9 ± 0.3 Wm^{-2} increase absorbed solar radiation (ASR) that is partially offset by a 0.4 ± 0.25 Wm^{-2} increase in outgoing longwave radiation (OLR). Since most of the energy added to the climate system associated with EEI ends up as heat storage in the ocean, changes in TOA radiation and ocean heat uptake (OHU) derived from in situ ocean data should track one another. Indeed, recently published analyses indicate that when in situ ocean measurements are supplemented with other data to fill in sparsely sampled regions, there is good agreement between variations and trends in OHU and CERES EEI for the Argo period between 2005 and 2019 (Loeb et al. 2021a; Hakuba et al. 2024, this collection).

Regional patterns of CERES ASR, –OLR and SST trends are similar, particularly over the North Pacific, off the east coast of North America and west coast of South America. Time series of global mean anomalies in SST, ASR, and –OLR also share similar features. In each case, twelve-month running average anomalies are relatively constant prior to 2010 ("hiatus" period), increase markedly (decrease for –OLR) prior to the 2015–2016 El Niño event ("transition-to-El Niño" period), and remain relatively flat after this event ("post-El Niño" period). Despite marked differences in global ASR and global –OLR trends between these sub-periods, NET trends remain strikingly within 0.1 Wm^{-2} per decade of one another. Since climate stabilization requires the climate forcing or net radiative imbalance to restore to zero, an increase in Earth's radiative energy imbalance implies an acceleration of climate change rather than a continued, steady heating implied by a constant imbalance (e.g., von Shuckman et al. 2023). However, we note that NET radiation exhibits appreciable internal variability at interannual time scales. A longer observational record is needed to determine how robust these findings are.

We compare global trends in TOA fluxes of CRE alongside an alternate approach that uses the CERES FluxbyCldTyp (FBCT) product to isolate the cloudy and clear-sky contributions to all-sky TOA flux trends. While the trend in net CRE is weak due to compensation between –SW and –OLR components, the trend for the cloudy sky contribution is appreciable due to a large positive trend in −SW (i.e., reduced cloud reflection) and negligible −OLR trend. The latter is comparable to what is obtained using the PRP method and thus provides a better framework than CRE for assessing the radiative impacts of cloud changes. Further refinement would be required to account for cloud masking contributions in cloudy areas. Isolating the cloud contribution also requires removing the contribution from effective radiative forcing (aerosol-cloud indirect effects and greenhouse gas adjustments), which is highly uncertain.

When the cloudy sky contribution is stratified by cloud type, we find that decreases in low and middle cloud fraction and reflection and reduced reflection from cloud-free areas in mid-high latitudes are the primary reasons for increasing ASR trends in the NH. Low cloud changes are primarily from Sc between 20° and 42°N; while Sc, SCT and Cu all contribute to the low cloud ASR increase between 42° and 60°N. In the SH the increase in ASR is primarily from decreases in middle cloud reflection and a weaker reduction in low-cloud reflection. Increased thermal emission in cloud-free conditions combined with high cloud changes contribute most to the increase in OLR.

Climate model AMIP simulations suggest that the larger ASR increase observed during the CERES period is due to additive contributions from effective radiative forcing (ERF) and climate response to warming and it is spatial pattern; while, the weaker OLR change is associated with compensation between increasing ERF from continued emission of well-mixed greenhouse gases and increased infrared cooling to space relating to the radiative response to warming (Raghuraman et al. 2021; Hodnebrog et al. 2024). Model-based attribution of the CERES results are limited in number because the CMIP6 protocol ends in 2014. The new atmospheric model intercomparison project (AMIP) simulations proposed as part of CERESMIP (Schmidt et al. 2023) will provide updated model simulations through 2021 and will use input data sets, greatly expanding opportunities to assess model performance and attribution of the observed EEI trend.

Appendix 1

Cloud fraction trend comparison between MODIS and CC

We compare MODIS-based cloud fraction trends with those from CALIPSO and CloudSat provided in the CCCM RelD1 product (Kato et al. 2010, 2011). The period considered is 01/2008–12/2017. As CALIPSO and/or CloudSat measurements are unavailable ≈20% of the time after 2011, we only include months in which all three instruments provide valid measurements. To ensure consistent spatial sampling, we only use MODIS cloud properties from CERES footprints that are collocated with the CALIPSO and cloudsat (CC) satellite tracks. MODIS cloud fraction is determined for each MODIS pixel using the CERES cloud algorithm (Minnis et al. 2021). The CALIPSO cloud mask is from CALIPSO vertical feature mask (VFM) version 4 product (Vaughan et al. 2009) with a threshold of the cloud-aerosol discrimination (CAD) score ≥ 20 and a horizontal averaging scale for cloud detection ≤ 20 km. Since CALIPSO detects optically thin ice clouds that are often missed by MODIS, we exclude optically thin ice clouds using the following criterion: if the cumulative cloud optical depth (τ) from the top is smaller than 0.3, the CALIPSO cloud layer is removed and treated as clear. For consistency, a τ filtering ($\tau \geq 0.3$) is also applied to

MODIS. We find that the MODIS cloud trends with and without the τ filtering are nearly identical (not shown), meaning that the occurrence of $\tau<0.3$ is small. The CloudSat cloud mask is from the CloudSat 2B-GEORPOF release 5 (R05) product (Sassen and Wang 2008) with a threshold of the cloud mask value ≥ 30 and the radar reflectivity >-25 dBZ. The radar reflectivity condition is considered to minimize the impact of the degradation of the CloudSat cloud profiling radar (CPR) sensor (Mathew Lebsock, personal communication). To combine CC cloud layers we choose the closest CloudSat pixel for a given CALIPSO pixel.

After merging CALIPSO and CloudSat cloud layers, the cloud top height of the uppermost layer is used to assign the cloud type. This is because MODIS usually detects the uppermost cloud layers in the case of multi-layered clouds. The CC cloud top height is converted into the cloud top pressure using pressure profiles of the Global Modeling and Assimilation Office (GMAO)'s Goddard Earth Observing System Data Assimilation System (GEOS-DAS V5.4.1) product (Rienecker et al., 2008).

To evaluate MODIS cloud fraction trends, we compare coincident MODIS and CC during the common period from 01/2008 to 12/2017 for the same cloud types (Fig. 13a–d). Since this comparison is for a much shorter period, these results need not match those in Fig. 10. Furthermore, because −20% of the CC data after 2011 are missing, the trends may not even be representative of 2008–2018. Rather, the intent is to provide an independent assessment of the MODIS results using CC.

Cloud changes inferred from CC are sensitive to the cloud selection criteria applied in the analysis. For example, if we include CALIPSO clouds with small cloud optical depth values (<0.3), high cloud trends become increasingly negative (not shown). In addition, the horizontal averaging scale for CALIPSO cloud detection also impacts the results. If CALIPSO water clouds with cloud top <4 km are detected from a single lidar beam

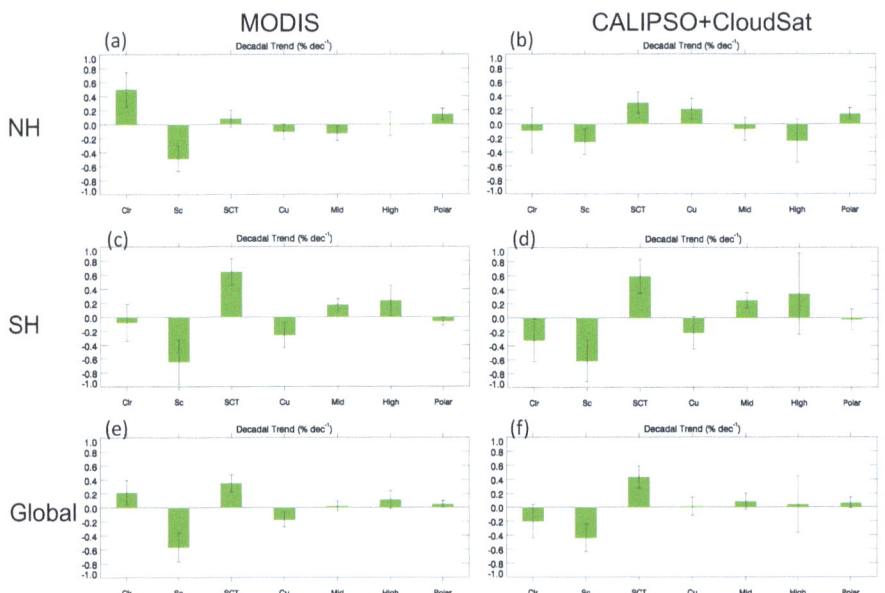

Fig. 13 Clear-sky frequency and cloud fraction trends by cloud type from: **a** MODIS for NH, **b** CC for NH, **c** MODIS for SH, **d** CC for SH, **e** MODIS for globe, and **f** CC for globe using coincident measurements from 01/2008 to 12/2017

(1/3 km resolution) without horizontal averaging, decadal trends of low clouds are reduced relative to that where horizontal averaging is included. We estimate uncertainties in CC cloud fraction trends by combining three factors. The first factor is related to the uncertainty of the linear regression as standard errors ($=\sigma_A$). The second factor is related to the uncertainty related to the τ filtering ($=\sigma_B$). We estimate σ_B as the difference in the decadal trends with and without the τ filtering. The third factor is related to the uncertainty related to the horizontal averaging scales of CALIPSO water clouds below 4 km ($=\sigma_C$). We estimate the value of σ_C as the difference in the decadal trends with 1/3 km scales of clouds and with 1/3, 1, 5, and 20 km averaging scales of water clouds below 4 km. The overall uncertainty is determined by summing the individual contributions in quadrature ($=(\sigma_A^2+\sigma_B^2+\sigma_C^2)^{1/2}$). These are given as error bars in CC cloud trends.

MODIS and CC show remarkably consistent cloud fraction trends for each cloud type in the SH (Fig. 13c, d). Both show a large negative trend in Sc and a large positive trend in SCT, and weaker Cu, Mid, High and Polar cloud trends. The large error bar for CC high clouds is due to a greater sensitivity to our approach used to filter out thin clouds with optical depths <0.3 that are below the MODIS detection threshold (Sect. 2.1). Differences are larger for the clear-sky fraction trend with MODIS showing no trend and CC showing a decrease in clear-sky fraction. With the exception of the Polar cloud case, the NH MODIS and CC cloud trends are generally weaker than those in the SH and show less agreement. Both show a significant decrease in Sc, but the magnitude of the decrease is larger for MODIS. There is a large discrepancy in clear-sky fraction, with MODIS showing an increase and CC showing little change. At the global scale, the main features that stand out are the Sc and SCT trends, which MODIS and CC capture. These comparisons suggest that MODIS is capable of capturing large changes in cloud fraction, but weaker trends are more uncertain.

Acknowledgements We thank the CERES science, algorithm and data management teams and the NASA Science Mission Directorate for supporting this research. This is PMEL contribution number 5554 and CIMAR contribution number 23-404. RPA was funded by funded by the National Centre for Earth Observation Grant Number: NE/RO16518/1. CERES data were obtained from http://ceres.larc.nasa.gov/compare_products.php. ERA5 data are publicly available via the Copernicus Climate Change Service climate (https://confluence.ecmwf.int/display/CKB/). This paper is an outcome of the Workshop "Challenges in Understanding the Global Water Energy Cycle and its Changes in Response to Greenhouse Gas Emissions" held at the International Space Science Institute (ISSI) in Bern, Switzerland (26–30 September 2022).

Open Access This article is licensed under a Creative Commons Attribution 4.0 International License, which permits use, sharing, adaptation, distribution and reproduction in any medium or format, as long as you give appropriate credit to the original author(s) and the source, provide a link to the Creative Commons licence, and indicate if changes were made. The images or other third party material in this article are included in the article's Creative Commons licence, unless indicated otherwise in a credit line to the material. If material is not included in the article's Creative Commons licence and your intended use is not permitted by statutory regulation or exceeds the permitted use, you will need to obtain permission directly from the copyright holder. To view a copy of this licence, visit http://creativecommons.org/licenses/by/4.0/.

References

Andersen H, Cermak J, Zipfel L, Myers TA (2022) Attribution of observed recent decrease in low clouds over the northeastern Pacific to cloudcontrolling factors. Geophys Res Lett. https://doi.org/10.1029/2021GL096498

Andrews T, Bodas-Salcedo A, Gregory JM, Dong Y, Armour KC, Paynter D et al (2022) On the effect of historical SST patterns on radiative feedback. J Geophys Res. https://doi.org/10.1029/2022JD036675

Bellouin N, Boucher O, Haywood J, Reddy MS (2005) Global estimate of aerosol direct radiative forcing from satellite measurements. Nature 438:1138–1141. https://doi.org/10.1038/nature04348

Cheng LJ et al (2024) New record ocean temperatures and related climate indicators in 2023. Adv Atmos Sci. https://doi.org/10.1007/s00376-024-3378-5

Cole J, Barker HW, Loeb NG, von Salzen K (2011) Assessing simulated clouds and radiative fluxes using properties of clouds whose tops are exposed to space. J Clim 24:2715–2727. https://doi.org/10.1175/2011JCLI3652.1

Collins M et al (2013) Long-term climate change: projections, commitments and irreversibility. In: Stocker TF, Qin D, Plattner G-K, Tignor M, Allen SK, Boschung J, Nauels A, Xia Y, Bex V, Midgley PM (eds) Climate change 2013: the physical science basis. Contribution of working group i to the fifth assessment report of the intergovernmental panel on climate change. Cambridge University Press, Cambridge

Datseris G, Stevens B (2021) Earth's albedo and its symmetry. AGU Adv. https://doi.org/10.1029/2021AV000440

Diamond MS (2023) Detection of large-scale cloud microphysical changes within a major shipping corridor after implementation of the International Maritime Organization 2020 fuel sulfur regulations. Atmos Chem Phys 23:8259–8269. https://doi.org/10.5194/acp-23-8259-2023

Doelling DR, Loeb NG, Keyes DF, Nordeen ML, Morstad D, Nguyen C, Wielicki BA, Young DF, Sun M (2013) Geostationary enhanced temporal interpolation for CERES flux products. J Atmos Oceanic Tech 30(6):1072–1090. https://doi.org/10.1175/JTECH-D-12-00136.1

Eitzen ZA, Su W, Xu K-M, Loeb N, Sun M, Doelling D, Bodas-Salcedo A (2017) Evaluation of a general circulation model by the CERES flux-by-cloud type simulator. J. Geophys. Res. Atmos. 122:10655–10668. https://doi.org/10.1002/2017JD027076

Eyring V, Bony S, Meehl GA, Senior CA, Stevens B, Stouffer RJ, Taylor KE (2016) Overview of the coupled model intercomparison project phase 6 (CMIP6) experimental design and organization. Geosci Model Dev 9:1937–1958. https://doi.org/10.5194/gmd-9-1937-2016,2016

Forster P et al (2021) The Earth's energy budget, climate feedbacks, and climate sensitivity. In: Masson-Delmotte V, Zhai P, Pirani A, Connors SL, Pan C, Berger S, Caud N, Chen Y, Goldfarb L, Gomis MI, Huang M, Leitzell K, Lonnoy E, Matthews JBR, Maycock TK, Waterfield T, Yeleki O, Yu R, Zhou B (eds), Climate change 2021: The physical science basis. Contribution of working group I to the sixth assessment report of the intergovernmental panel on climate change. Cambridge University Press, Cambridge, pp 923–1054. https://doi.org/10.1017/9781009157896.009

Gryspeerdt E, Quaas J, Bellouin N (2016) Constraining the aerosol influence on cloud fraction. J Geophys Res 121:3566–3583. https://doi.org/10.1002/2015JD023744

Hakuba MZ, Frederikse T, Landerer FW (2021) Earth's energy imbalance from the ocean perspective (2005–2019). Geophys Res Lett. https://doi.org/10.1029/2021GL093624

Hakuba MZ et al (2024) Trends and variability in Earth's energy imbalance and ocean heat uptake since 2005. Surveys in Geophysics (submitted, this collection)

Hansen J, Nazarenko L, Ruedy R, Sato M, Willis J, Del Genio A, Koch D, Lacis A, Lo K, Menon S, Novakov T, Perlwitz J, Russell G, Schmidt GA, Tausnev N (2005) Earth's energy imbalance: confirmation and implications. Science 308:1431–1435. https://doi.org/10.1126/science.1110252

Held IM et al (2019) Structure and performance of GFDL's CM4.0 climate model. J Adv Model Earth Syst 11(11):3691–3727

Hersbach H, Bill B, Berrisford P, Hirahara S, Horanyi A, Munoz-Sabater J et al (2020) The ERA5 global reanalysis. Q J R Meteorol Soc 146:1999–2049. https://doi.org/10.1002/qj.3803

Hodnebrog Ø, Myhre G, Jouan C, Andrews T, Forster PM, Jia H, Loeb NG, Olivié DJL, Paynter D, Quaas J, Raghuraman SP, Schulz M (2024) Recent reductions in aerosol emissions have increased Earth's energy imbalance. Nature Comm Earth Environ. https://doi.org/10.1038/s43247-024-01324-8

Johnson GC, Lyman JM, Loeb NG (2016) Improving estimates of Earth's energy imbalance. Nat Clim Change 6(7):639–640. https://doi.org/10.1038/nclimate3043

Kang SM, Ceppo P, Yu Y, Kang I-S (2023) Recent global climate feedback controlled by Southern Ocean cooling. Nat Clim Change 16:775–780. https://doi.org/10.1038/s41561-023-01256-6

Kato S, Sun-Mack S, Miller WF, Rose FG, Chen Y, Minnis P, Wielicki BA (2010) Relationships among cloud occurrence frequency, overlap, and effective thickness derived from CALIPSO and CloudSat merged cloud vertical profiles. J Geophys Res. https://doi.org/10.1029/2009JD012277

Kato S et al (2011) Improvements of top-of-atmosphere and surface irradiance computations with CALIPSO-, CloudSat-, and MODIS-derived cloud and aerosol properties. J Geophys Res 116:D19209. https://doi.org/10.1029/2011JD016050

Koren I, Remer LA, Kaufman YJ, Rudich Y, Martins JV (2007) On the twilight zone between clouds and aerosols. Geophys Res Lett. https://doi.org/10.1029/2007GL029253

Kosaka Y, Xie S-P (2013) Recent global-warming hiatus tied to equatorial Pacific surface cooling. Nature 501:403–407. https://doi.org/10.1038/nature12534

Kramer RJ, He H, Soden BJ, Oreopoulos L, Myhre G, Forster PM, Smith CJ (2021) Observational evidence of increasing global radiative forcing. Geophys Res Lett. https://doi.org/10.1029/2020GL091585

Lewandowsky S, Risbey JS, Oreskes N (2015) On the definition and identifiability of the alleged "hiatus" in global warming. Sci Rep. https://doi.org/10.1038/srep16784

Li Z, England MH, Groeskamp S (2023) Recent acceleration in global ocean heat accumulation by mode and intermediate waters. Nat Comm 14:6888. https://doi.org/10.1038/s41467-023-42468-z

Liu C, Allan RP, Mayer M, Hyder P, Desbruyères D, Cheng L, Xu J, Xu F, Zhang Y (2020) Variability in the global energy budget and transports 1985–2017. Clim Dynam 55:3381–3396. https://doi.org/10.1007/s00382-020-05451-8

Loeb NG, Schuster GL (2008) An observational study of the relationship between cloud, aerosol and meteorology in broken low-level cloud conditions. J Geophys Res 113:D14214. https://doi.org/10.1029/2007JD009763

Loeb NG, Wielicki BA, Doelling DR, Smith GL, Keyes DF, Kato S, Smith NM, Wong T (2009) Towards optimal closure of the Earth's top-of-atmosphere radiation budget. J Climate 22:748–766

Loeb NG, Doelling DR, Wang H, Su W, Nguyen C, Corbett JG, Liang L, Mitrescu C, Rose FG, Kato S (2018) Clouds and the Earth's radiant energy system (CERES) energy balanced and filled (EBAF) top-of-atmosphere (TOA) edition 4.0 data product. J Clim 31:895–918. https://doi.org/10.1175/JCLI-D-17-0208.1

Loeb NG, Rose FG, Kato S, Rutan DA, Su W, Wang H, Doelling DR, Smith WL, Gettelman A (2020) Toward a consistent definition between satellite and model clear-sky radiative fluxes. J Clim 33(1):61–75. https://doi.org/10.1175/JCLI-D-19-0381.1

Loeb NG, Wang H, Allan RP, Andrews T, Armour K, Cole JNS et al (2020) New generation of climate models track recent unprecedented changes in earth's radiation budget observed by CERES. Geophys Res Lett. https://doi.org/10.1029/2019GL086705

Loeb NG, Johnson GC, Thorsen TJ, Lyman JM, Rose FG, Kato S (2021a) Satellite and ocean data reveal marked increase in Earth's heating rate. Geophys Res Lett. https://doi.org/10.1029/2021GL093047

Loeb NG, Su W, Bellouin N, Ming Y (2021b) Changes in clear-sky shortwave aerosol direct radiative effects since 2002. J Geophys Res. https://doi.org/10.1029/2020JD034090

Loeb NG, Mayer MM, Kato S, Fasullo JT, Zuo H, Senan R, Lyman JM, Johnson GC, Balmaseda M (2022) Evaluating twenty-year trends in Earth's energy flows from observations and reanalyses. J Geophys Res. https://doi.org/10.1029/2022JD036686

Marti F, Rousseau V, Ablain M, Fraudeau R, Meyssignac B, Blazquez A (2023) Monitoring the global ocean heat content from space geodetic observations to estimate the Earth energy imbalance, State Planet Discuss. [preprint]. https://doi.org/10.5194/sp-2023-26

McCoy DT, Field P, Gordon H, Elsaesser GS, Grosvenor DP (2020) Untangling causality in midlatitude aerosol-cloud adjustments. Atmos Chem Phys 20:4085–4103

Meehl GA, Hu A, Arblaster JM, Fasullo J, Trenberth KE (2013) Externally forced and internally generated decadal climate variability associated with the Interdecadal Pacific Oscillation. J Clim 26:7298–7310. https://doi.org/10.1175/JCLI-D-12-00548.1

Meyssignac B et al (2023) How accurate is accurate enough for measuring sea-level rise and variability. Nat Clim Change 13:796–804. https://doi.org/10.1038/s41558-023-01735-z

Minière A, von Schuckmann K, Sallée J-B, Vogt L (2023) Robust acceleration of Earth system heating observed over the past six decades. Sci Rep 13:22975. https://doi.org/10.1038/s41598-023-49353-1

Minnis P et al (2008) Cloud detection in non-polar regions for CERES using TRMM VIRS and Terra and Aqua MODIS data. IEEE Trans Geosci Remote Sens 46:3857–3884. https://doi.org/10.1109/TGRS.2008.2001351

Minnis P et al (2011a) CERES Edition-2 cloud property retrievals using TRMM VIRS and Terra and Aqua MODIS data—Part I: algorithms. IEEE Trans Geosci Remote Sens 49:4374–4400. https://doi.org/10.1109/TGRS.2011.2144601

Minnis P et al (2011b) CERES Edition-2 cloud property retrievals using TRMM VIRS and Terra and Aqua MODIS data—Part II: examples of average results and comparisons with other data. IEEE Trans Geosci Remote Sens 49:4401–4430. https://doi.org/10.1109/TGRS.2011.2144602

Minnis P et al (2021) CERES MODIS cloud product retrievals for edition 4—Part I: algorithm changes. IEEE Trans Geosci Remote Sens 59:2744–2780. https://doi.org/10.1109/TGRS.2020.3008866

Myers T, Norris JR (2015) On the relationship between subtropical clouds and meteorology in observations and CMIP3 and CMIP5 models. J Clim 28:2945–2967. https://doi.org/10.1175/JCLI-D-14-00475.1

Myers TA, Mechoso CR, Cesana GV, DeFlorio MJ, Waliser DE (2018) Cloud feedback key to marine heatwave off Baja California. Geophys Res Lett 45(9):4345–4352. https://doi.org/10.1029/2018GL078242

Myers T, Zelinka MD, Klein SA (2023) Observational constraints on the cloud feedback pattern effect. J Clim 36:6533–6545. https://doi.org/10.1175/JCLI-D-22-0862.1

Olonscheck D, Rugenstein M (2024) Coupled climate models systematically underestimate radiation response to surface warming. Geophys Res Lett. https://doi.org/10.1029/2023GL106909

Oreopoulos L, Cho N, Lee D, Kato S (2016) Radiative effects of global MODIS cloud regimes. J Geophys Res Atmos 121(5):2299–2317. https://doi.org/10.1002/2015JD024502

Oreopoulos L, Cho N, Lee D, Lebsock M, Zhang Z (2022) Assessment of two stochastic subcloud generators using observed fields of vertically resolved cloud extinction. J Atmos Tech 39:1229–1244. https://doi.org/10.1175/JTECH-D-21-0166.s1

Palmer MD, McNeall DJ, Dunstone NJ (2011) Importance of the deep ocean for estimating decadal changes in Earth's radiation balance. Geophys Res Lett 38:L13707. https://doi.org/10.1029/2011GL047835

Paulot F, Paynter D, Ginoux P, Naik V, Horowitz LW (2018) Changes in the aerosol direct radiative forcing from 2001 to 2015: observational constraints and regional mechanisms. Atmos Chem Phys 18(2018):13265. https://doi.org/10.5194/acp-18-13265-2018

Quass J et al (2022) Robust evidence for reversal in the aerosol effective climate forcing trend. Atmos Chem Phys. https://doi.org/10.5194/acp-2022-295

Raghuraman SP, Paynter D, Ramaswamy V (2021) Anthropogenic forcing and response yield observed positive trend in Earth's energy imbalance. Nat Comm. https://doi.org/10.1038/s41467-021-24544-4

Raghuraman SP, Paynter D, Menzel R, Ramaswamy V (2023) Forcing, cloud feedbacks, cloud masking, and internal variability in the cloud radiative effect satellite record. J Clim 36:4151–4167. https://doi.org/10.1175/JCLI-D-22-0555.1

Rienecker MM et al (2008) The GOES-5 data assimilation system—documentation of versions 5.0.1, 5.1.0, and 5.2.0. In: NASA technical report series on global modeling and data assimilation, vol 27, NASA/TM-2008–105606, p 97

Rossow WB, Schiffer RA (1991) ISCCP cloud data products. Bull Amer Meteor Soc 72:2–20. https://doi.org/10.1175/1520-0477(1991)072,0002:ICDP.2.0.CO;2

Sassen K, Wang Z (2008) Classifying clouds around the globe with the CloudSat radar: 1-year of results. Geophys Res Lett 35:L04805. https://doi.org/10.1029/2007GL032591

Schmidt GA, Andrews T, Bauer SE, Durack P, Loeb NG, Ramaswamy V et al (2023) CERESMIP: a climate modeling protocol to investigate recent trends in the Earth's energy imbalance. Front Clim. https://doi.org/10.3389/fclim.2023.1202161

Scott RC, Myers TA, Norris JR, Zelinka MD, Klein SA, Sun M, Doelling DR (2020) Observed sensitivity of low-cloud radiative effects to meteorological perturbations over the global oceans. J Clim 33(18):7717–7734. https://doi.org/10.1175/JCLI-D-19-1028.1

Shankar M, Loeb NG, Smith N, Smith N, Daniels JL, Thomas S, Walikainen D (2023) Evaluating the radiometric performance of the Clouds and the Earth's Radiant Energy System (CERES) instruments on Terra and Aqua over 20 years. IEEE Trans Geosci Rem Sens. https://doi.org/10.1109/TGRS.2023.3330398

Smith CJ, Kramer RJ, Myhre G, Alterskjær K, Collins W, Sima A, Boucher O, Dufresne J-L et al (2020) Effective radiative forcing and adjustments in CMIP6 models. Atmos Chem Phys 20:9591–9618. https://doi.org/10.5194/acp-20-9591-2020

Soden BJ, Held IM, Colman R, Shell KM, Kiehl JT, Shields CA (2008) Quantifying climate feedbacks using radiative kernels. J Clim 21:3504–3520. https://doi.org/10.1175/2007JCLI2110.1

Storto A, Yang C (2024) Acceleration of the ocean warming from 1961 to 2022 unveiled by large-ensemble reanalyses. Nat Comm 15:545. https://doi.org/10.1038/s41467-024-44749-7

Stubenrauch C et al (2024) Lessons learned from the updated GEWEX Cloud Assessment database. Surv Geophys (Accepted, this collection)

Subba T, Gogoi MM, Pathak B, Bhuyan PK, Babu SS (2020) Recent trend in the global distribution of aerosol direct radiative forcing from satellite measurements. Atmos Sci Lett. https://doi.org/10.1002/asl.975

Sun M, Doelling DR, Loeb NG, Scott RC, Wilkins J, Nguyen LT, Mlynczak P (2022) Clouds and the Earth's Radiant Energy System (CERES) FluxByCldTyp Edition 4 data product. J Atmos Oceanic Technol 39(3):303–318. https://doi.org/10.1175/JTECH-D-21-0029.1

Szopa S et al (2021) Short-lived climate forcers. In: Masson-Delmotte V, Zhai P, Pirani A, Connors S, Péan C, Berger S, Caud N, Chen Y, Goldfarb L, Gomis M, Huang M, Leitzell K, Lonnoy E, Matthews J, Maycock T, Waterfield T, Yelekçi O, Yu R, Zhou B (eds). Climate change 2021: the physical science basis. Contribution of working group I to the sixth assessment report of the intergovernmental panel on climate change, ch 6, Cambridge University Press, Cambridge

Thorsen TJ, Kato S, Loeb NG, Rose FG (2018) Observation-based decomposition of radiative perturbations and radiative kernels. J Clim 31:10039–10058. https://doi.org/10.1175/JCLI-D-18-0045.1

Trenberth KE (2015) Has there been a hiatus? Science 349:691–692. https://doi.org/10.1126/science.aac9225

Trenberth KE, Fasullo JT (2009) Global warming due to increasing absorbed solar radiation. Geophys Res Lett 36:L07706. https://doi.org/10.1029/2009GL037527

Trenberth KE, Fasullo JT (2010) Simulation of present day and 21st century energy budgets of the southern oceans. J Clim 23:440–454

Trenberth KE, Fasullo JT, Balmaseda M (2014) Earth's energy imbalance. J Clim 27:3129–3144. https://doi.org/10.1175/JCLI-D-13-00294.1

Trenberth KE, Zhang Y, Fasullo JT, Taguchi S (2015) Climate variability and relationships between top-of-atmosphere radiation and temperatures on Earth. J Geophys Res Atmos 120:3642–3659. https://doi.org/10.1002/2014JD022887

Vaughan M et al (2009) Fully automated detection of cloud and aerosol layers in the CALIPSO lidar measurements. J Atmos Oceanic Technol 26:2034–2050. https://doi.org/10.1175/2009JTECHA1228.1

von Schuckmann K, Palmer MD, Trenberth KE, Cazenave A, Chambers D, Champollion N, Hansen J, Josey SA, Loeb N, Mathieu P-P, Meyssignac B, Wild M (2016) An imperative to monitor Earth's energy imbalance. Nat Clim Change 6:138–144. https://doi.org/10.1038/nclimate2876

von Schuckmann K et al (2023) Heat stored in the Earth system 1960–2020: where does the energy go? Earth Syst Sci Data 15:1675–1709. https://doi.org/10.5194/essd-15-1675-2023

Wall CJ, Norris JR, Possner A, McCoy DT, McCoy IL, Lutsko NJ (2022) Assessing effective radiative forcing from aerosol-cloud interactions over the global ocean. PNAS. https://doi.org/10.1073/pnas.2210481119

Wolter K, Timlin MS (1998) Measuring the strength of ENSO events—how does 1997/98 rank? Weather 53:315–324. https://doi.org/10.1002/j.1477-8696.1998.tb06408.x

Wood R (2012) Stratocumulus clouds. Mon Weather Rev 140:2373–2423. https://doi.org/10.1175/MWR-D-11-00121.1

Wood R, Bretherton CS (2006) On the relationship between stratiform low cloud cover and lower-tropospheric stability. J Clim 19:6425–6432. https://doi.org/10.1175/JCLI3988.1

Xu K-M, Zhou Y, Sun M, Kato S, Hu Y (2023) Observed cloud type-sorted cloud property and radiative flux changes with the degree of convective aggregation from CERES data. J Geophys Res Atmos 128:e2023JD039152

Yuan T, Song H, Wood R, Wang C, Oreopoulos L, Platnick SE, von Hippel S, Meyer K, Light S, Wilcox E (2022) Global reduction in ship-tracks from sulfur regulations for shipping fuel. Sci Adv. https://doi.org/10.1126/sciadv.abn7988

Zelinka MD, Myers TA, McCoy DT, Po-Chedley S, Caldwell PM, Ceppi P et al (2020) Causes of higher climate sensitivity in CMIP6 models. Geophys Res Lett. https://doi.org/10.1029/2019GL085782

Zhao B, Jiang JH, Gu Y, Diner D, Worden J, Liou K-N et al (2017) Decadal-scale trends in regional aerosol particle properties and their linkage to emission changes. Environ Res Lett 2017(12):054021

Zhao M et al (2018) The GFDL global atmosphere and land model AM4.0/LM4.0: 1. Simulation characteristics with prescribed SSTs. J Adv Model Earth Syst 10(3):691–734. https://doi.org/10.1002/2017MS001208

Publisher's Note Springer Nature remains neutral with regard to jurisdictional claims in published maps and institutional affiliations.

Closure of Earth's Global Seasonal Cycle of Energy Storage

Gregory C. Johnson[1] · Felix W. Landerer[2] · Norman G. Loeb[3] · John M. Lyman[1,4] · Michael Mayer[5,6] · Abigail L. S. Swann[7] · Jinlun Zhang[8]

Received: 12 January 2023 / Accepted: 19 June 2023 / Published online: 18 July 2023
This is a U.S. Government work and not under copyright protection in the US; foreign copyright protection may apply 2023

Abstract

The global seasonal cycle of energy in Earth's climate system is quantified using observations and reanalyses. After removing long-term trends, net energy entering and exiting the climate system at the top of the atmosphere (TOA) should agree with the sum of energy entering and exiting the ocean, atmosphere, land, and ice over the course of an average year. Achieving such a balanced budget with observations has been challenging. Disagreements have been attributed previously to sparse observations in the high-latitude oceans. However, limiting the local vertical integration of new global ocean heat content estimates to the depth to which seasonal heat energy is stored, rather than integrating to 2000 m everywhere as done previously, allows closure of the global seasonal energy budget within statistical uncertainties. The seasonal cycle of energy storage is largest in the ocean, peaking in April because ocean area is largest in the Southern Hemisphere and the ocean's thermal inertia causes a lag with respect to the austral summer solstice. Seasonal cycles in energy storage in the atmosphere and land are smaller, but peak in July and September, respectively, because there is more land in the Northern Hemisphere, and the land has more thermal inertia than the atmosphere. Global seasonal energy storage by ice is small, so the atmosphere and land partially offset ocean energy storage in the global integral, with their sum matching time-integrated net global TOA energy fluxes over the seasonal cycle within uncertainties, and both peaking in April.

Keywords Earth · Climate · Global energy · Seasonal cycle

✉ Gregory C. Johnson
gregory.c.johnson@noaa.gov

[1] NOAA/Pacific Marine Environmental Laboratory, Seattle, WA 98115, USA
[2] Jet Propulsion Laboratory, California Institute of Technology, Pasadena, CA 91109, USA
[3] NASA Langley Research Center, Hampton, VA 23681, USA
[4] CIMAR, University of Hawaii, Honolulu, HI 96822, USA
[5] Research Department, ECMWF, 53175 Bonn, Germany
[6] Department of Meteorology and Geophysics, University of Vienna, 1090 Vienna, Austria
[7] Departments of Atmospheric Sciences and Biology, University of Washington, Seattle, WA 98195, USA
[8] Applied Physics Laboratory, University of Washington, Seattle, WA 98105, USA

Article Highlights

- Earth's global seasonal energy storage budget closes when ocean storage is limited to 50 m below the surface mixed layer maximum depth
- Global ocean seasonal heat storage is maximum in April, consistent with its larger Southern Hemisphere area and massive thermal inertia
- Global atmospheric seasonal energy storage is maximum in June, land in September, consistent with larger Northern Hemisphere land area

1 Introduction

In recent decades, the ocean, atmosphere, and land have all been warming substantially and ice (at sea and on land) has been melting, primarily owing to the buildup of greenhouse gases in the atmosphere from human activities. The amounts of energy required for these changes have been estimated globally (e.g., von Schuckmann et al. 2020). To estimate the time history of heat energy storage in the ocean, measurements of ocean temperature have been mapped on their own over time or assimilated in ocean models, with the results of the latter being known as reanalyses. Energy storage in the atmosphere is dominated by changes in temperature and moisture content (the latter necessitating phase changes which require substantial energy). In the atmosphere measurements assimilated into models producing reanalyses are the primary means for assessing the time history of atmospheric energy storage. Over long time scales, land warming has been estimated mostly from borehole temperature profiles (Beltrami et al. 2002), but for shorter time scales, land process models of varying complexity are driven with atmospheric reanalysis fields to estimate the time history of changes in land heat energy storage (e.g., Fasullo and Trenberth 2008; McKinnon and Huybers 2016). Sea ice areas are measured by satellites (e.g., Parkinson and Cavalieri 2008), and sea ice volumes in recent years may be estimated using satellite observations of sea ice thickness (e.g., Kwok and Rothrock 2009; Laxon et al. 2013). Sea ice volumes in all seasons may be estimated also using reanalysis sea ice models that are often constrained by satellite observations of sea ice concentration via data assimilation (e.g., Zhang 2014; Zhang and Rothrock 2003). Land ice mass changes can be derived from satellite gravimetry observations (e.g., Velicogna 2009).

On decadal and longer time scales, ocean heat storage dominates increases in energy storage in the climate system, accounting for about 90% of the total energy taken up in recent decades according to observational estimates (von Schuckmann et al. 2020). However, observational estimates of ocean warming are noisy to varying extents (Trenberth et al. 2016) even in recent years, with month-to-month ocean heat content anomaly (OHCA) estimates using Argo data alone or all available in situ data having about three times the variance of top of the atmosphere (TOA) radiation budgets from satellite measurements (Lyman and Johnson 2023). Even on annual time scales, variations in the rates of 0–2000 m ocean heat storage rates generally have larger variances than TOA net energy fluxes and the two independent measurements are at best correlated at less than 0.8 (e.g., Johnson et al. 2016). However, ocean warming estimates and TOA energy flux measurements both show a substantial and statistically significant increase in energy uptake by Earth's climate system in recent years, with the rate roughly doubling from 2005 to 2019 according to these independent measurement systems (Loeb et al. 2021).

Over six months, the change of globally integrated seasonal cycles of energy storage may exceed 60 ZJ (e.g., Pan et al. 2023). Hence, energy moves in and out of the global ocean on seasonal time scales at an order of magnitude larger than the average rate of long-term uptake of energy by the global ocean over the past few decades, which is on the order of 10 ZJ yr^{-1} (e.g., Loeb et al. 2021). The percentage contributions to the seasonal cycle of globally averaged energy storage in the climate system from the atmosphere and land are larger than their percentage contributions to the long-term energy increase, but the ocean still dominates, with a seasonal cycle that is larger than that of the net flux at the TOA as shown by Fasullo and Trenberth (2008) and McKinnon and Huybers (2016). However, in both of these previous studies, the sums of the individual components' (ocean's, atmosphere's, land's, and ice's) global seasonal cycles of energy storage do not agree within uncertainties with the net TOA energy storage. That is to say, the seasonal cycles of energy storage do not close. In both of those studies, this disagreement was attributed to poorly quantified measured or completely absent estimates of ocean heat storage in sparsely measured regions (e.g., the marginal seas and the Arctic Ocean) in the OHCA maps analyzed.

Here, we revisit the seasonal cycle of energy storage in light of a longer time series of Argo data and new high resolution ($1/4° \times 1/4° \times 7$ day) maps of OHCA that use Argo and other in situ OHCA data to train a random forest regression with satellite sea surface height, sea surface temperature, location, and time as predictors (Lyman and Johnson 2023). These maps include estimates of OHCA in the Arctic and most marginal seas (with exceptions being the Black and Caspian seas). Their global integrals are also substantially less noisy than maps made with in situ temperature profiles only, and their time derivative is generally better correlated with global integrals of CERES TOA net energy fluxes on time scales from monthly to annual, with or without the seasonal cycle included (Lyman and Johnson 2023). Nonetheless, with these maps, despite their coverage of high latitudes and most marginal seas, here we find lack of closure very similar to previous studies when using 0–2000 m depth integrals of OHCA. However, we can close Earth's seasonal global energy budget when integrating from the surface to 50 m deeper than the maximum local monthly 95th percentile mixed layer depth taken from Johnson and Lyman (2022). That depth range encompasses the portion of the ocean where energy is likely to be exchanged with the atmosphere on seasonal time scales.

2 Data and Methods

Weekly OHCA values constructed using Argo and other in situ ocean temperature data to train random forest regressions using location, time, satellite sea surface height anomaly maps, and satellite sea surface temperature maps (Lyman and Johnson 2023) are locally depth integrated from the surface to 2000 m and then globally integrated for comparison with previous studies. They are also locally depth integrated from the surface to 50 m deeper than the maximum monthly climatological 95th percentile mixed layer depth from Johnson and Lyman (2022) and then globally integrated. Where that mixed layer depth is not available (mostly in ice covered high latitudes), the integration is from the surface to 130 m, which should be sufficiently deep to encompass the ocean diabatic seasonal cycle in heat storage in those regions. In both cases, the resulting OHCA values are smoothed with a one-month half-power loess filter and interpolated to mid-months. The results presented below are very similar if the interpolation is performed on the unsmoothed time series, or if the smoothing varies between 0.5 and 1.5 months. Once the interpolation is increased to > 1.5 months, the fidelity of the seasonal cycle of OHCA begins to be compromised sufficiently that the closure achieved below is reduced.

Monthly means of total atmospheric energy are obtained using output from the fifth generation European Re-Analysis (ERA5; Hersbach et al. 2020). ERA5 combines a wealth of atmospheric observations (remotely sensed and in situ) with a dynamical model using four-dimensional variational data assimilation to provide physically consistent fields of atmospheric state and flux quantities with ~ 30 km spatial resolution on 137 vertical levels. The ERA5 archive provides monthly means of all output fields, which facilitates computation of monthly mean atmospheric energy content. Atmospheric total energy encompasses internal and potential energy (which can be combined to give sensible heat), latent heat, and kinetic energy, the latter being small (see, e.g., Peixoto and Oort 1992). In this study, we compute total energy on the native horizontal grid, integrate vertically using model level data, and from this compute global averages. Uncertainties of atmospheric state quantities like temperature and moisture from reanalyses are deemed relatively small since they are well constrained by a large number of observations, especially in the study period of the present work. Even in the Arctic, where observations are relatively sparse, the systematic differences in the seasonal cycle of atmospheric energy storage from three state-of-the-art reanalyses are fairly small, much smaller compared to uncertainties in energy storage in other components of the system (Mayer et al. 2019).

Monthly land energy flux estimates are obtained from the Community Land Model version 5 (Lawrence et al. 2019; https://doi-org.cuucar.idm.oclc.org/10.1029/2018MS0015 83), run to produce a historical reanalysis following the TRENDY Protocol (Friedlingstein et al. 2020), forced with CRU TS version 4.03 fields (Harris and Jones 2020; Harris et al. 2020) prior to 1954 and JRA55 fields (Kobayashi et al. 2015) from 1955 through 2020, both at 0.5° spatial resolution. Greenland and Antarctica are omitted from the area integration, as those regions are largely covered by ice sheets, and the energy required to melt those ice sheets is estimated from the time variation of satellite gravity measurements as detailed below. The global integrals are integrated in time to obtain a cumulative land energy storage estimate and then interpolated to mid-months.

Monthly global sea ice volumes (Arctic and Antarctic) are estimated using a coupled global sea ice–ocean model, the Global Ice–Ocean Modeling and Assimilation System (GIOMAS) (Zhang 2014; Zhang and Rothrock 2003). The ocean model component of GIOMAS is from the Parallel Ocean Program developed at the Los Alamos National Laboratory (Smith et al. 1992). The sea ice model component employs the multicategory ice thickness and enthalpy distribution model (Zhang and Rothrock 2003). It has 8 sub-grid categories each for ice thickness, ice enthalpy, and snow depth, representing variable ice thicknesses up to 28 m. GIOMAS assimilates satellite observations of sea ice concentration. Sea ice volume changes are converted to energy storage changes assuming an average density of 910 kg m^{-3} and a heat of fusion of 3.3×10^5 J kg^{-1} for sea ice.

Monthly land ice mass variations are estimated from the Gravity Recovery and Climate Experiment (GRACE) and GRACE follow-on satellite data records (Landerer et al. 2020) using the JPL (Jet Propulsion Laboratory) Mascon data (Watkins et al. 2015; Wiese et al. 2019). Here, we sum ice mass changes over Greenland and its peripheral glaciers and Antarctica. The ice mass variations are then converted to energy changes following Slater et al. (2021, their Eq. 1 with $\Delta T = 10$ °C).

Globally integrated TOA monthly averages of net energy flux from the Clouds and the Earth's Radiant Energy System (CERES) Energy Balanced and Filled (EBAF) Edition 4.1 product (Loeb et al. 2018) are integrated in time to convert them from energy flux into energy storage values and then interpolated to mid-months. The CERES EBAF product provides monthly mean TOA shortwave (SW), longwave (LW), and net radiative fluxes on a $1°\times 1°$ grid from March 2000 onward. Net radiative fluxes are calculated from the difference between spatially and temporally averaged monthly solar irradiances and the sum of outgoing SW and LW fluxes. The EBAF product uses an objective constraint algorithm (Loeb et al. 2009) to adjust SW and LW TOA fluxes within their ranges of uncertainty to remove the inconsistency between the 10-year average (07/2005–06/2015) global net TOA energy flux and energy storage in the earth–atmosphere–land–ice climate system, determined primarily from OHCA data (Johnson et al. 2016). Use of this approach to anchor the satellite-derived Earth energy imbalance to the in situ value does not affect the temporal variability in the data.

A quadratic function of time and six harmonics (with annual, semiannual, triannual, quarterly, five per year, and bimonthly periods) are fit to each time series starting in January 2008 and ending in December 2020. The harmonics are then used to re-construct the seasonal cycle. The quadratic and annual harmonics are also removed from the time series to create a residual time series. The variances of the residuals for each calendar month, together with the number of degrees of freedom, are used to construct 5–95% confidence limits for each of the monthly means. Hence, the confidence limits include only information about the variances from the average seasonal cycle for the 13 years analyzed, and neither any systematic nor possibly some random sources of error in the datasets. Therefore, they should be considered as a lower bound for the actual confidence limits. When summing or differencing estimates, the confidence limits are assumed to be independent. Hence, they are added in quadrature (taking the square root of the sum of the squares). Again, this assumption makes those sums and differences a lower bound on the actual confidence limits.

3 Results

We discuss the seasonal cycles in energy storage in the various components of Earth's climate system roughly in order of amplitude from the largest to the smallest. We then compute their sum and compare it to the seasonal cycle of time-integrated net TOA energy flux, to assess whether or not the budget is closed. First, we present the seasonal cycle when integrating OHCA to 2000 m (Fig. 1), as previous studies (Fasullo and Trenberth 2008; McKinnon and Huybers 2016) have done, although the budget did not close in those previous studies, and does not close here.

The ocean integrated from 0 to 2000 m has the largest seasonal cycle of heat storage of the energy reservoirs in Earths' climate system (Fig. 1, dark blue line with circles). In the global average, the ocean absorbs 63.5 ZJ of heat energy from late August (near the start of austral spring) to mid-March, a time period of around 7 months. It discharges that energy more rapidly, over a period of about 5 months. The standard deviation of the seasonal cycle of ocean heat storage is 22.1 ZJ.

The atmosphere (Fig. 1, orange line with pluses) and land (Fig. 1, green line) have comparably sized seasonal cycles of energy storage. The atmosphere takes up 13.9 ZJ of energy from mid-December through late July, a period of 7 months. Like the ocean, it discharges that energy over a period of 5 months. The standard deviation of the seasonal cycle of atmospheric energy storage is just under 4.8 ZJ. The land takes up 13.5 ZJ of heat from late February to early September, a period 6 months, and discharges that

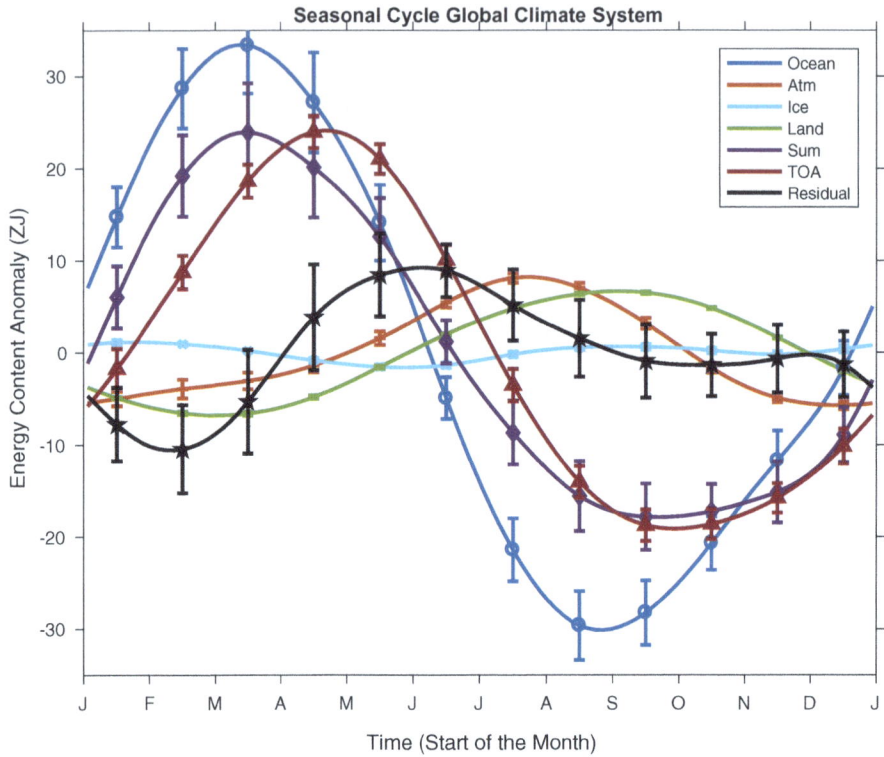

Fig. 1 Seasonal cycles of energy storage for reservoirs in Earth's climate system estimated from monthly values from January 2008 through December 2020 as described in the text. Ocean heat storage is estimated by integrating from the surface to 2000 m. The residual is the sum of energy storage in the various reservoirs minus that at the top of the atmosphere (TOA). Error bars show 5–95% confidence intervals

heat over an approximately equal period of time. The standard deviation of the seasonal cycle of land heat storage is just over 4.8 ZJ.

The cryosphere (Fig. 1, light blue line with crosses) accounts for the smallest portion of the seasonal energy storage, gaining 2.8 ZJ of energy in total from late May through late January, a period of 8 months, and losing it in four months. The standard deviation of the seasonal cycle of energy storage in the cryosphere is only 0.8 ZJ. However, the energy in the semiannual harmonic is more evident in ice energy storage than in the other components of Earth's climate system, with a local maximum in energy storage in early September and a local minimum in mid-November.

Because of the opposing phasing of seasonal energy storage in the ocean versus the atmosphere and land reservoirs of Earth's climate system, their sum (Fig. 1, purple line with diamonds, also including the smaller energy reservoirs of sea ice and ice sheets) has an amplitude smaller than that of the ocean. When integrating ocean heat storage from the surface to 2000 m, the seasonal cycle of the sum of energy storage in Earth's climate system gains 41.8 ZJ of energy from late September through mid-March, a period of around 6 months, and discharges it again over 6 months. The standard deviation of the seasonal cycle of the sum of energy in Earth's climate system is 15.3 ZJ.

The seasonal cycle of the integral of net energy flux through the TOA (Fig. 1, red line with triangles) is similar in amplitude to the seasonal cycle of the sum of energy storage in Earth's climate system, but lags it by about a month. The seasonal cycle of the integral of net energy flux through the TOA gains 43.2 ZJ of energy from late September through late April, a period of about 7 months. Its standard deviation is 15.4 ZJ. Largely because of this mismatch in phase, the residual of the integral of the TOA energy flux minus the sum of storage in the ocean, atmosphere, land, and ice reservoirs (Fig. 1, black line with pentagrams) gains 9.8 ZJ of energy from mid-February though early July, with substantial energy in some of the higher harmonics evident, and a standard deviation of 5.8 ZJ, larger than that of either the land or the atmosphere. Furthermore, the residual is statistically significantly different from zero for five months of the year, so the global seasonal energy budget does not close within uncertainties.

Limiting the depth of integration of OHCA from the surface to just 50 m deeper than the maximum climatological 95th percentile monthly mixed layer depth (Fig. 2) from Johnson and Lyman (2022) changes this result and allows the global seasonal energy budget (Fig. 3) to close within uncertainties. In the global average, the ocean (Fig. 3, dark blue line with circles) absorbs 56.1 ZJ of heat energy from early October (near the start of austral spring) to mid-April, a time period of about 7 months. It discharges that energy more rapidly, over a period of about 5 months. The standard deviation of the seasonal cycle of ocean heat storage is 19.6 ZJ.

The reduction in amplitude and slight change in phase of ocean heat storage when limiting the integration from the surface to just 50 m deeper than the maximum climatological 95th percentile monthly mixed layer depth means that the seasonal cycle of the sum of the energy storage reservoirs (Fig. 3, purple line with diamonds) gains 40.6 ZJ of energy from mid-November through mid-April, with a standard deviation of 14.6 ZJ, pretty closely matching

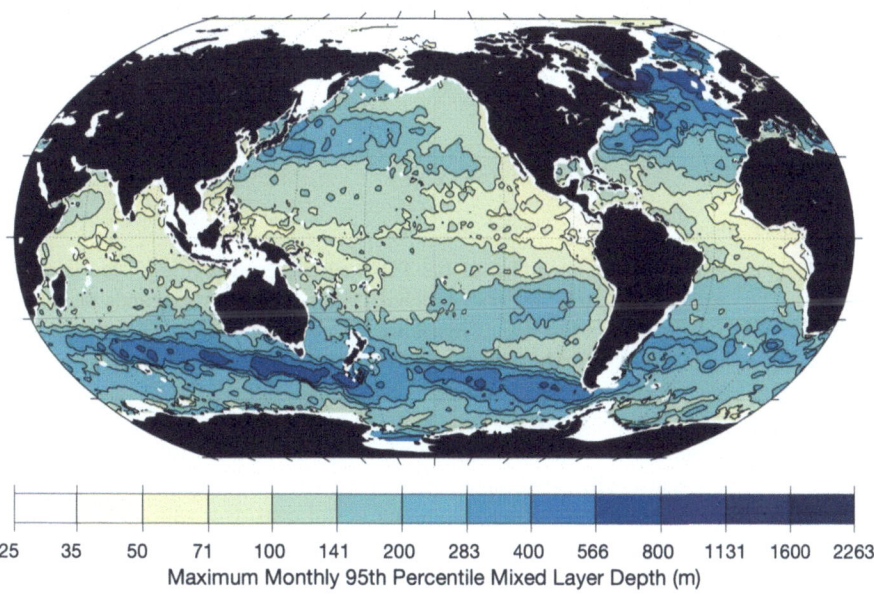

Fig. 2 Maximum monthly climatological 95th percentile mixed layer depth from the Global Ocean Statistical Mixed Layer (GOSML) climatology (Johnson and Lyman 2022)

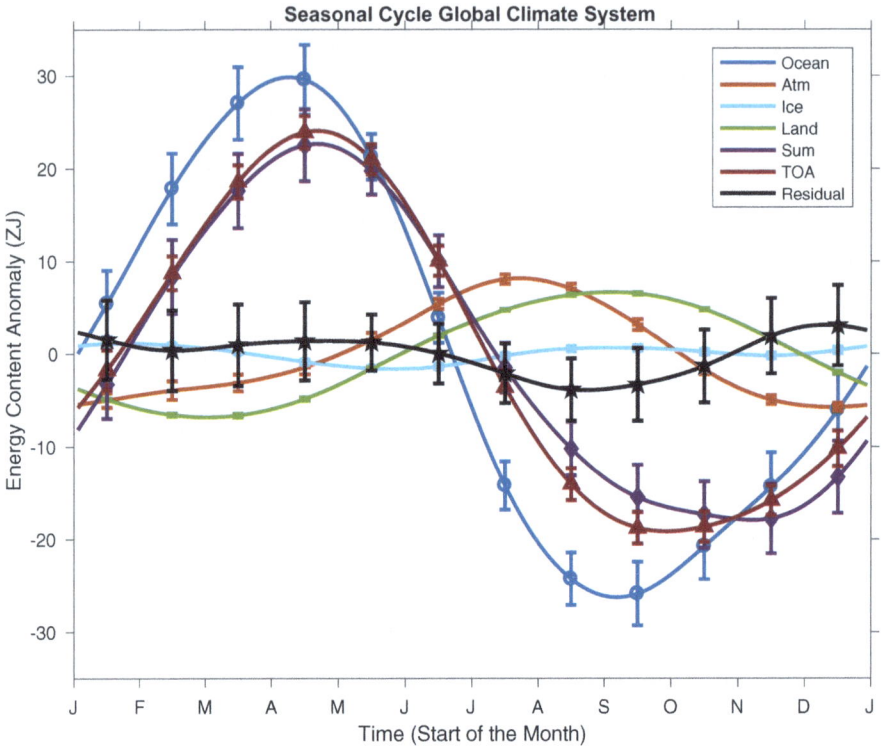

Fig. 3 Seasonal cycles of energy storage for reservoirs in Earth's climate system estimated from monthly values from January 2008 through December 2020 as described in the text. Ocean heat storage is estimated by integrating from the surface to 50 m below the maximum monthly 95th percentile mixed layer depth from Johnson and Lyman (2022), or from the surface to 130 m where those values are not mapped, primarily in ice-covered high-latitude regions where that depth of integration should encompass the seasonal cycle of ocean diabatic heat energy storage. The residual is the sum of energy storage in the various reservoirs minus that at the TOA. Error bars show 5–95% confidence intervals

the seasonal cycle of the time-integrated global TOA net energy flux (Fig. 3, red line with triangles). The residual of the time-integrated global TOA net energy flux minus the sum of the global energy storage reservoirs (Fig. 3, black line with pentagrams) now gains just 7.0 ZJ of energy from mid-August to mid-December, with substantial variance in semi-annual and shorter period harmonics. The standard deviation of seasonal cycle of the residual is 2.1 ZJ, much less than that for any of the other terms in the budget except for the ice volume reservoir (Fig. 3, light blue line with crosses). Furthermore, only the monthly value of the residual for August is barely statistically significantly different from zero at 5–95% confidence limits. Given that there are 12 months in a year, this result is still consistent with a budget that is closed within uncertainty.

4 Discussion and Conclusions

Previous analyses (Fasullo and Trenberth 2008; McKinnon and Huybers 2016) have noted that the seasonal cycle of the globally integrated energy storage by Earth's climate system was not closed, and attributed that to either errors in estimates or absence of estimates of the ocean's seasonal cycle in 0–2000 m temperature at high latitudes. Here, we attempt a similar exercise with near-global OHCA maps that extend well into the high latitudes, yet still get a similar result (Fig. 1) to those previous studies. As in those studies, the seasonal cycle in 0–2000 m ocean heat storage amplitude is too large and peaks too early in the year for seasonal energy budget closure, resulting in a monthly residual that has more variance over the year than the seasonal cycles of either land or atmospheric energy storage.

However, if we limit the local depth of OHCA integration to from the ocean surface to 50 m below the maximum monthly 95th percentile mixed layer depth (Fig. 2) from Johnson and Lyman (2022), the seasonal energy budget does close (Fig. 3), with the seasonal cycle of the sum of the globally integrated energy storage reservoirs (ocean, atmosphere, land, and ice) agreeing within uncertainties to the time-integrated TOA global net energy flux. This procedure is well justified, since diabatic seasonal heating is limited to slightly deeper than the deepest seasonal mixed layer (Moisan and Niiler 1998). Furthermore, seasonal adiabatic heave of isotherms is found in many locations below the seasonal thermocline. For example, prominent annual Rossby waves extend well below the permanent thermocline in the tropics of all three oceans (Brandt and Eden 2005; Johnson 2011; Kessler and McCreary 1993; Nagura 2018; Zanowski et al. 2019). We varied the depth added to the maximum monthly 95th percentile mixed layer depth from 0 to 100 m without an appreciable change in results (not shown). In fact, simply integrating in depth from the surface to 290 m gave a very similar result of budget closure (also not shown), so the finding seems fairly robust.

When OHCA is globally integrated from 50 m below the maximum depth of the seasonal mixed layer to 2000 m, this isotherm heave results in a substantial seasonal adiabatic signal (the difference between the seasonal cycles of OHCA in Figs. 2 and 3). This deep heave signature has a cooling effect in April–September and a warming effect in November–March. A recent global analysis of the seasonal cycle in ocean temperature using the mean of ocean temperature maps from seven different groups shows the global integral of this deep dynamically driven heave signal propagating down to at least 1000 m (Pan et al. 2023, their Fig. 8a), and our maps (not shown) confirm that this propagation extends to at least 2000 m.

The only way to eliminate this adiabatic heave signal would be to observe (and integrate over) the full ocean depth globally, as is proposed with the Deep Argo mission (Roemmich et al. 2019). Of course, for ocean heat gain over years to decades, integrating as deep as possible is desirable, since the ocean is gaining heat from the surface to the 2000 m sampling limit of core Argo floats (Wijffels et al. 2016), well below the deepest seasonal mixed layer, and even in the Antarctic Bottom Waters from 4000 to 6000 m (Desbruyeres et al. 2016; Purkey and Johnson 2010), where Deep Argo floats can sample.

The ocean, with its great mass, large heat capacity, and low albedo, dominates the seasonal cycle of energy storage within Earth's climate system. Because the ocean's surface area is much larger in the Southern Hemisphere than in the Northern Hemisphere, and it has substantial thermal inertia, its seasonal heat storage (Fig. 3, dark blue line with circles, integrated from the surface to 50 m below the depth of the maximum climatological 95th percentile monthly mixed layer depth) peaks in mid-April, well after the austral summer solstice in late December. Because of its low albedo and massive heat capacity, the ocean is the largest seasonal energy reservoir in Earth's climate system, with a seasonal standard deviation of 19.6 ZJ.

The land (Fig. 3, green line) and atmosphere (Fig. 3, orange line with pluses) both have seasonal energy storage cycles with standard deviations of 4.8 ZJ, and so are reservoirs with very similar magnitudes in terms of Earth's seasonal energy cycling. Hence, they each have ~¼ the standard deviation, or ~6% of the variance, of the ocean's seasonal cycle of heat energy storage. Land area is much larger in the Northern Hemisphere than in the Southern Hemisphere, the atmosphere heats and cools much more seasonally over the land than over the ocean, and the atmosphere has relatively little thermal inertia. Hence, atmospheric energy storage peaks in mid-July, shortly after the boreal summer solstice. In contrast, seasonal energy storage by the land, with its considerably larger thermal inertia, peaks in mid-September, well after that of the atmosphere.

The seasonal cycle of energy storage in the cryosphere (Fig. 3, light blue line with crosses) has a standard deviation of 0.8 ZJ, making it by far the smallest of the seasonal energy reservoirs in Earth's climate system. Hence it has ~4% the standard deviation, or ~0.2% of the variance, of the ocean's seasonal cycle of heat energy storage. There is considerable semi-annual energy in the seasonal cycle, mostly because the rates of growth and decay of sea ice in both hemispheres are not symmetrical, although there is considerable cancellation of the hemispheric contributions in the global integral, as for the other reservoirs of energy.

The cooling periods for the globally integrated TOA, ocean, and atmosphere seasonal cycles of energy storage are all noticeably shorter, about five months, than their warming periods of about seven months. For the TOA, a likely contributing factor to this asymmetry is the planetary albedo: The globally averaged albedo (not shown) is higher than average during November through January, when highly reflective mid-latitude regions (clouds) and high latitude regions (Antarctic ice) in the Southern Hemisphere are illuminated, which induces asymmetry in the seasonal cycle of absorbed solar radiation. Additionally, TOA outgoing longwave radiation (not shown) has a relatively narrow maximum centered on July–August, with phasing and amplitude likely associated with Northern Hemisphere land heating, and a broader minimum centered on December. The Southern Hemisphere OHCA seasonal cycle (not shown) is highly symmetric, with ~99.99% of its variance in the annual harmonic, and has about 1.9 times the variance of the Northern Hemisphere OHCA seasonal cycle, primarily owing to the larger area of the ocean in the Southern Hemisphere. The Northern Hemisphere OHCA seasonal cycle has a cooling period of 5.6 months, probably owing to the greater fraction of land there. In boreal fall and winter, the atmosphere over land cools more quickly than that over the ocean, enhancing atmospheric land-to-ocean energy transport (e.g., Trenberth and Fasullo 2013) and exchange of energy from the ocean to the atmosphere via turbulent fluxes (e.g., Cayan 1992)—thus likely accelerating, and shortening, the oceanic cooling period. The sum of the two largely opposing OHCA seasonal cycles in each hemisphere accentuates the asymmetry in the global OHCA seasonal cycle. For the atmosphere, one possible reason for the shorter cooling period could be Northern Hemisphere snow cover on land: Once there is a thin snow cover in boreal fall, ground heat flux is suppressed and atmospheric cooling accelerates, while snow cover in boreal spring decelerates warming, and hence the buildup of energy in the warming period takes longer.

The seasonal cycles of globally integrated energy storage in Earths' climate system are substantial with respect to the rates of long-term energy accumulation in the reservoirs of that climate system. For instance, the global seasonal cycle of OHCA has rates of heat

accumulation about an order of magnitude greater than the decadal increase in global OHCA. This attribute makes the seasonal cycle of global energy storage in the various reservoirs an important benchmark for climate models.

The sum of the seasonal cycle of energy storage in all the components of Earth's climate system (Fig. 3, purple line with diamonds) overlaps with the time integral of net TOA energy flux (Fig. 3, red line with triangles) in all months within their 5–95% confidence limits, meaning that the budget closes (Fig. 3). The residual (Fig. 3, black line with pentagrams, sum minus TOA) has noticeable semi-annual and higher energy, but overall has a very small global seasonal cycle compared with all other seasonal climate system energy storage reservoirs except ice. The standard deviation of the seasonal cycle of the global residual is only 2.1 ZJ, considerably less than the mean 5–95% confidence interval of the residual, which is ± 3.4 ZJ. Furthermore, the variance of the seasonal cycle of the global residual is only 2% of the variance of the seasonal cycle of either the time-integrated net TOA energy flux or the seasonal cycle of the sum of energy storage in Earth's climate system. Only one monthly mean, the minimum found in August, is barely statistically significantly different from zero at 5–95% confidence. This marginal discrepancy of one month out of 12 is consistent with a closed budget at 5–95% confidence limits. This result suggests that our choice of depth integrating ocean heat content from the surface to 50 m below the maximum 95% monthly climatology mixed layer depth is capturing the seasonal ocean heat storage, as expected.

Acknowledgements GCJ and JML are supported by NOAA Research and the NOAA Global Ocean Monitoring and Observation Program. ALSS acknowledges support from Office of Biological and Environmental Research of the US Department of Energy Regional and Global Model Analysis Program award DE-SC0021209 to the University of Washington. JZ is supported by NASA Cryosphere Program (NNX-17AD27G and 80NSSC20K1253) and NSF Office of Polar Programs (NNA-1927785). MM's contribution was supported by Austrian Science Fund P33177. FWL is supported by the Jet Propulsion Laboratory, California Institute of Technology, under a contract with the National Aeronautics and Space Administration (80NM0018D0004). PMEL Contribution Number 5440.

Declarations

Conflict of interest The authors have no financial or other conflicting interests regarding this paper.

Open Access This article is licensed under a Creative Commons Attribution 4.0 International License, which permits use, sharing, adaptation, distribution and reproduction in any medium or format, as long as you give appropriate credit to the original author(s) and the source, provide a link to the Creative Commons licence, and indicate if changes were made. The images or other third party material in this article are included in the article's Creative Commons licence, unless indicated otherwise in a credit line to the material. If material is not included in the article's Creative Commons licence and your intended use is not permitted by statutory regulation or exceeds the permitted use, you will need to obtain permission directly from the copyright holder. To view a copy of this licence, visit http://creativecommons.org/licenses/by/4.0/.

References

Beltrami H, Smerdon JE, Pollack HN, Huang SP (2002) Continental heat gain in the global climate system. Geophys Res Lett. https://doi.org/10.1029/2001gl014310

Brandt P, Eden C (2005) Annual cycle and interannual variability of the mid-depth tropical Atlantic ocean. Deep-Sea Res Part I-Oceanogr Res Papers 52:199–219. https://doi.org/10.1016/j.dsr.2004.03.011

Cayan DR (1992) Variability of latent and sensible heat fluxes estimated using bulk formulae. Atmos Ocean 30:11–42. https://doi.org/10.1080/07055900.1992.9649429

Desbruyeres DG, Purkey SG, McDonagh EL, Johnson GC, King BA (2016) Deep and abyssal ocean warming from 35 years of repeat hydrography. Geophys Res Lett 43(19):10356–10365. https://doi.org/10.1002/2016gl070413

Fasullo JT, Trenberth KE (2008) The annual cycle of the energy budget. Part I: global mean and land-ocean exchanges. J Clim 21(10):2297–2312. https://doi.org/10.1175/2007jcli1935.1

Friedlingstein P et al (2020) Global carbon budget 2020. Earth Syst Sci Data 12:3269–3340. https://doi.org/10.5194/essd-12-3269-2020

Harris I, Osborn TJ, Jones P et al (2020) Version 4 of the CRU TS monthly high-resolution gridded multivariate climate dataset. Sci Data 7:109. https://doi.org/10.1038/s41597-020-0453-3

Harris IC, Jones, PD (2020) CRU TS4.03: climatic research unit (CRU) time-series (TS) version 4.03 of high-resolution gridded data of month-by-month variation in climate (Jan. 1901–Dec. 2018). Centre for environmental data analysis, 22 January 2020. https://doi.org/10.5285/10d3e3640f004c578403419aac167d82

Hersbach H, Bell B, Berrisford P et al (2020) The ERA5 global reanalysis. Q J R Meteorol Soc 146:1999–2049. https://doi.org/10.1002/qj.3803

Johnson GC (2011) Deep signatures of southern tropical Indian ocean annual Rossby waves. J Phys Oceanogr 41(10):1958–1964. https://doi.org/10.1175/jpo-d-11-029.1

Johnson GC, Lyman JM (2022) GOSML: a global ocean surface mixed layer statistical monthly climatology: means, percentiles, skewness, and kurtosis. J Geophys Res-Oceans. https://doi.org/10.1029/2021jc018219

Johnson GC, Lyman JM, Loeb NG (2016) Improving estimates of Earth's energy imbalance. Nature Clim Change 6(7):639–640. https://doi.org/10.1038/nclimate3043

Kessler WS, McCreary JP (1993) The annual wind-driven Rossby-wave in the Subthermocline Equatorial Pacific. J Phys Oceanogr 23(6):1192–1207. https://doi.org/10.1175/1520-0485(1993)023%3c1192:tawdrw%3e2.0.co;2

Kobayashi S, Ota Y, Harada Y, Ebita A, Moriya M, Onoda H, Onogi K, Kamahori H, Kobayashi C, Endo H, Miyaoka K, Takahashi K (2015) The JRA-55 reanalysis: general specifications and basic characteristics. J Meteorol Soc Jpn Ser II 93(1):5–48. https://doi.org/10.2151/jmsj.2015-001

Kwok R, Rothrock DA (2009) Decline in Arctic sea ice thickness from submarine and ICESat records: 1958–2008. Geophys Res Lett 36:L15501. https://doi.org/10.1029/2009gl039035

Landerer FQ, Flechtner FM, Save H, Webb FH, Bandikova T, Bertiger WI, Bettadpur SV, Byun SH, Dahle C, Dobslaw H, Fahnestock E, Harvey N, Kang Z, Kruizinga GLH, Loomis BD, McCullough C, Murböck M, Nagel P, Paik M, Pie N, Poole S, Strekalov D, Tamisiea ME, Wang F, Watkins MM, Wen H-Y, Wiese DN, Yuan D-N (2020) Extending the global mass change data record: GRACE follow-on instrument and science data performance+ Geophys Res Lett 47(12). https://doi.org/10.1029/2020GL088306

Lawrence DM, Fisher RA, Koven CD, Oleson KW, Swenson SC, Bonan G et al (2019) The community land model version 5: description of new features, benchmarking, and impact of forcing uncertainty. J Adv Model Earth Syst 11:4245–4287. https://doi.org/10.1029/2018MS001583

Laxon WS, Giles KA, Ridout AL, Wingham DJ, Cullen RWR, Kwok R, Schweiger A, Zhang J, Haas C, Hendricks S, Krishfield S, Kurtz N, Farrell S, Davidson M (2013) CryoSat-2 estimates of Arctic sea ice thickness and volume. Geophys Res Lett 40:732–737. https://doi.org/10.1002/grl.50193

Loeb NG, Wielicki BA, Doelling DR, Smith GL, Keyes DF, Kato S, Smith NM, Wong T (2009) Towards optimal closure of the earth's top-of-atmosphere radiation budget. J Climate 22:748–766. https://doi.org/10.1175/2008JCLI2637.1

Loeb NG, Doelling DR, Wang H, Su W, Nguyen C, Corbett JG, Liang L, Mitrescu C, Rose FG, Kato S (2018) Clouds and the earth's radiant energy system (CERES) energy balanced and filled (EBAF) top-of-atmosphere (TOA) edition 4.0 data product. J Climate 31:895–918. https://doi.org/10.1175/JCLI-D-17-0208.1

Loeb NG, Johnson GC, Thorsen TJ, Lyman JM, Rose FG, Kato S (2021) Satellite and ocean data reveal marked increase in earth's heating rate. Geophys Res Lett. https://doi.org/10.1029/2021gl093047

Lyman JM, Johson GC (2023) Global high-resolution random forest regression maps of ocean heat content anomalies using in situ and satellite data. J Atmos Ocean Tech. https://doi.org/10.1175/JTECH-D-22-0058.1

Mayer M, Tietsche S, Haimberger L, Tsubouchi T, Mayer J, Zuo H (2019) An improved estimate of the coupled arctic energy budget. J Clim 32(22):7915–7934. https://doi.org/10.1175/JCLI-D-19-0233.1

McKinnon KA, Huybers P (2016) Seasonal constraints on inferred planetary heat content. Geophys Res Lett 43(20):10955–10964. https://doi.org/10.1002/2016gl071055

Moisan JR, Niiler PP (1998) The seasonal heat budget of the North Pacific: net heat flux and heat storage rates (1950–1990). J Phys Oceanogr 28(3):401–421. https://doi.org/10.1175/1520-0485(1998)028%3c0401:tshbot%3e2.0.co;2

Nagura M (2018) Annual Rossby waves below the pycnocline in the indian ocean. J Geophys Res-Oceans 123(12):9405–9415. https://doi.org/10.1029/2018jc014362

Pan Y et al (2023) Annual cycle in upper ocean heat content and the global energy budget. J Clim. https://doi.org/10.1175/JCLI-D-22-0776.1

Parkinson CL, Cavalieri DJ (2008) Arctic sea ice variability and trends, 1979–2006. J Geophys Res-Oceans 113(C7):C07003. https://doi.org/10.1029/2007jc004558

Peixoto JP, Oort AH, Lorenz EN (1992) Physics of climate, vol 520. American Institute of Physics, New York

Purkey SG, Johnson GC (2010) Warming of global abyssal and deep southern ocean waters between the 1990s and 2000s: contributions to global heat and sea level rise budgets. J Clim 23(23):6336–6351. https://doi.org/10.1175/2010jcli3682.1

Roemmich D, Alford MH, Claustre H, Johnson K, King B et al (2019) On the future of argo: a global, full-depth, multi-disciplinary array. Front Mar Sci. https://doi.org/10.3389/fmars.2019.00439

Slater T, Lawrence IR, Otosaka IN, Shepherd A, Gourmelen N, Jakob L, Tepes P, Gilbert L, Nienow P (2021) Review article: earth's ice imbalance. Cryosphere 15:233–246. https://doi.org/10.5194/tc-15-233-2021

Smith RD, Dukowicz JK, Malone RC (1992) Parallel ocean general circulation modeling. Physica D-Nonlinear Phenomena 60(1–4):38–61. https://doi.org/10.1016/0167-2789(92)90225-c

Trenberth KE, Fasullo JT (2013) Regional energy and water cycles: transports from ocean to land. J Clim 26(20):7837–7851. https://doi.org/10.1175/JCLI-D-13-00008.1

Trenberth KE, Fasullo JT, von Schuckmann K, Cheng L (2016) Insights into earth's energy imbalance from multiple sources. J Clim 29:7495–7505. https://doi.org/10.1175/JCLI-D-16-0339.1

Velicogna I (2009) Increasing rates of ice mass loss from the Greenland and Antarctic ice sheets revealed by GRACE. Geophys Res Lett 36(4):L19503. https://doi.org/10.1029/2009gl040222

von Schuckmann K et al (2020) Heat stored in the earth system: where does the energy go? Earth Syst Sci Data 12(3):2013–2041. https://doi.org/10.5194/essd-12-2013-2020

Watkins MM, Wiese DN, Yuan D-N, Boening C, Landerer FW (2015) Improved methods for observing Earth's time variable mass distribution with GRACE using spherical cap mascons. J Geophys Res Solid Earth 120:2648–2671. https://doi.org/10.1002/2014JB011547

Wiese DN, Yuan D-N, Boening C, Landerer FW, Watkins MM (2019) JPL GRACE and GRACE-FO mascon ocean, ice, and hydrology equivalent HDR water height RL06M CRI filtered version 2.0, ver. 2.0, PO.DAAC, CA, USA. https://doi.org/10.5067/TEMSC-3MJ62

Wijffels S, Roemmich D, Monselesan D, Church J, Gilson J (2016) Ocean temperatures chronicle the ongoing warming of Earth. Nat Clim Chang 6(2):116–118. https://doi.org/10.1038/nclimate2924

Zanowski H, Johnson GC, Lyman JM (2019) Equatorial Pacific 1,000-dbar velocity and isotherm displacements from Argo data: beyond the mean and seasonal cycle. J Geophys Res: Oceans 124(11):7873–7882. https://doi.org/10.1029/2019JC015032

Zhang JL (2014) Modeling the impact of wind intensification on Antarctic sea ice volume. J Clim 27(1):202–214. https://doi.org/10.1175/jcli-d-12-00139.1

Zhang JL, Rothrock DA (2003) Modeling global sea ice with a thickness and enthalpy distribution model in generalized curvilinear coordinates. Mon Weather Rev 131(5):845–861. https://doi.org/10.1175/1520-0493(2003)131%3c0845:mgsiwa%3e2.0.co;2

Publisher's Note Springer Nature remains neutral with regard to jurisdictional claims in published maps and institutional affiliations.

The Global Energy Balance as Represented in Atmospheric Reanalyses

Martin Wild[1] · Michael G. Bosilovich[2]

Received: 6 March 2024 / Accepted: 2 September 2024 / Published online: 21 September 2024
© The Author(s) 2024

Abstract

In this study, we investigate the representation of the global mean energy balance components in 10 atmospheric reanalyses, and compare their magnitudes with recent reference estimates as well as the ones simulated by the latest generation of climate models from the 6th phase of the coupled model intercomparison project (CMIP6). Despite the assimilation of comprehensive observational data in reanalyses, the spread amongst the magnitudes of their global energy balance components generally remains substantial, up to more than 20 Wm^{-2} in some quantities, and their consistency is typically not higher than amongst the much less observationally constrained CMIP6 models. Relative spreads are particularly large in the reanalysis global mean latent heat fluxes (exceeding 20%) and associated intensity of the global water cycle, as well as in the energy imbalances at the top-of-atmosphere and surface. A comparison of reanalysis runs in full assimilation mode with corresponding runs constrained only by sea surface temperatures reveals marginal differences in their global mean energy balance components. This indicates that discrepancies in the global energy balance components caused by the different model formulations amongst the reanalyses are hardly alleviated by the imposed observational constraints from the assimilation process. Similar to climate models, reanalyses overestimate the global mean surface downward shortwave radiation and underestimate the surface downward longwave radiation by 3–7 Wm^{-2}. While reanalyses are of tremendous value as references for many atmospheric parameters, they currently may not be suited to serve as references for the magnitudes of the global mean energy balance components.

Keywords Global energy balance · Reanalysis · Earth radiation budget · Surface energy balance · Earth's energy imbalance · Energy and water cycles

---

✉ Martin Wild
martin.wild@env.ethz.ch

[1] ETH Zurich, Institute for Atmospheric and Climate Science, 8092 Zurich, Switzerland

[2] Global Modeling and Assimilation Office, NASA Goddard Space Flight Center, Greenbelt, MD 20771, USA

Article Highlights

- The global mean energy balance components of 10 different reanalyses are compared to reference estimates and state-of-the-art climate models from CMIP6
- The spread in the global energy balance components amongst the reanalyses is substantial (exceeding 20 Wm^{-2} in some quantities) and typically not smaller than amongst the CMIP6 models
- The spread amongst the reanalyses is particularly large in global mean latent heat fluxes and associated intensity of the global water cycle, as well as in the representation of the Earth energy imbalance (EEI)
- Compared to reference estimates, the reanalyses tend to overestimate the global mean surface downward shortwave radiation, which is compensated by an underestimation of the surface downward longwave radiation

1 Introduction

The Earth's energy balance fundamentally determines the climatic conditions on our planet. While the energy balance at the top-of-atmosphere (TOA), consisting of the shortwave and longwave radiation fluxes in an out of the climate system, determines the overall heat uptake in the climate system, the energy balance at the surface governs the thermal changes in our environments and defines the radiative energy that drives the evaporative flux and with it the global water cycle (e.g., Kiehl and Trenberth 1997; Wild et al. 1998, 2013; Hatzianastassiou et al. 2004; Trenberth et al. 2009; Bosilovich et al. 2011; Stephens et al. 2012; Allan et al. 2014; L'Ecuyer et al. 2015; Hakuba et al. 2019; Wang et al. 2022). An accurate knowledge of the magnitude of these energy fluxes is therefore essential for an adequate quantification of the state of climate and climate change.

Atmospheric reanalyses are widely used as references for various climate parameters, as they assimilate comprehensive amounts of in situ and space-based weather observations from the Global Observing System (GOS) into a numerical model, which enables an observationally constrained representation of the three-dimensional atmospheric structure (e.g., Bosilovich et al. 2013). Reanalyses are thus much more constrained by observations than "free-running" climate models. While reanalyses provide well-accepted and widely used reference estimates for quantities like geopotential heights, sea level pressure or upper air temperature and humidity fields, their ability to adequately represent the global energy balance components is less comprehensively assessed. The representation of different global mean energy balance components in earlier and individual reanalyses have been evaluated by Allan et al. (2004) in the ERA-40 reanalysis, by Trenberth et al. (2009) in ERA-40 and early versions of the NCEP and JRA reanalyses, by Berrisford et al. (2011) in the ERA-Interim reanalysis, by Bosilovich et al. (2011) and (Roberts et al. 2012) in the MERRA reanalysis, and by Bosilovich et al. (2015) and (Stamatis et al. 2022) in the MERRA-2 reanalysis. Net surface energy fluxes using TOA radiation measurements combined with atmospheric energy transports and tendencies from reanalyses have been investigated by Trenberth and Solomon (1994) and Liu et al. (2017).

In the present study, we will focus in the following on the global energy balance as represented in 10 different reanalyses. We will cover all radiative energy balance components under both all-sky and clear-sky conditions, as well as the non-radiative surface energy balance components of sensible and latent heat. We will focus on the consistency in the magnitudes of these various energy balance components across the different reanalyses,

and compare them to independent reference estimates as well as to the respective quantities in state-of-the-art climate models participating in the 6th phase of the Coupled Model Intercomparison Project (CMIP6) as analyzed in Wild (2020).

2 Data

Data from 10 different reanalysis products are used in this study (Table 1). These include the reanalyses MERRA-2 (Gelaro et al. 2017), MERRA-2 AMIP (Collow et al. 2017) ERA5 (Hersbach et al. 2020), JRA-55 (Kobayashi et al. 2015), NCEP-R2 (Kanamitsu et al. 2002), 20CRv3 (Slivinski et al. 2019), ERA20C (Poli et al. 2016), ERA20CM (Hersbach et al. 2015), MERRA (Rienecker et al. 2011) and JRA-3Q (Kosaka et al. 2024). While MERRA, MERRA-2, ERA5, JRA-55, JRA-3Q and NCEP-R2 consider full data assimilation (i.e., as many observing platforms as the systems can assimilate including radiance assimilation), ERA20C and 20CRv3 only assimilate surface observations, whereas MERRA-2 AMIP as well as ERA20CM do not include any data assimilation except the evolution of sea surface temperatures and sea-ice extent, which is prescribed according to observations in all reanalysis products considered here. MERRA-2 AMIP as well as ERA20CM use identical physical models as their counterparts MERRA-2 and ERA20C, respectively. The only reanalysis that forces global mass conservation in the presence of water and/or dry mass assimilation is MERRA-2. Regionally, the water vapor and mass increments have significant magnitudes in the respective budgets (Takacs et al. 2016).

The values presented in this study are long-term global annual averages over the period 2001–2010, which is covered by all 10 reanalyses considered here. They are thus representative for the first decade of the twenty-first century. NCEP-R2 did not provide any clear-sky fluxes. While all other reanalyses provided net clear-sky shortwave and longwave radiation at the surface, none of them explicitly stored the related clear-sky downward components, which can be directly compared to surface observations. The clear-sky downward shortwave and longwave components were therefore inferred from the available quantities by combining the clear-sky surface net shortwave radiation with the surface albedo, and the clear-sky surface net longwave radiation with the upward longwave radiation, respectively.

With respect to reference estimates, we refer to published satellite-derived data from CERES-EBAF for the TOA and surface radiative components under all-sky and clear-sky conditions (Loeb et al. 2018; Kato et al. 2018), plus to independent estimates derived by Wild et al. (2015) and L'Ecuyer et al. (2015) for the all-sky radiative and non-radiative surface energy balance components, as well as to the clear-sky surface and atmospheric radiative estimates derived by Wild et al. (2019). Generally, global reference estimates of the TOA fluxes are afflicted with smaller uncertainties than their atmospheric and surface counterparts, since they can be directly measured from space, whereas the surface and atmospheric estimates must rely to some degree on modeling in addition to the available direct observations. Thus, the various surface and atmospheric energy balance estimates published over the years differed substantially. Their consistency has however improved in recent years, yet still not reaching the level of accuracy of the TOA estimates (Wild 2017).

In addition, the results of Wild (2020) covering the representation of the global mean energy balance components in up to 40 CMIP6 climate models are used for comparison. These stem from "historical-all-forcings" experiments and represent the global energy balance components as simulated in the CMIP6 models at the beginning of the twenty-first century.

Table 1 Global annual means of various energy balance components under clear-sky and all-sky conditions at the TOA, within the atmosphere and at the surface, representative for the first decade of the twenty-first century (2001–2010), as calculated in 10 individual reanalyses, together with reference estimates

Energy balance component	Reference Estimates Wm^{-2}	MERRA-2 Wm^{-2}	M2AMIP Wm^{-2}	ERA5 Wm^{-2}	JRA-55 Wm^{-2}	NCEP-R2 Wm^{-2}	20CRv3 Wm^{-2}	ERA20C Wm^{-2}	ERA20CM Wm^{-2}	MERRA Wm^{-2}	JRA-3Q Wm^{-2}
TOA											
SW down	340^a, 340^b, 340^c	340.4	340.4	340.4	341.3	341.3	340.3	340.4	340.4	341.3	341.3
SW up all-sky	−99^a, −100^b, −102^c	−106.0	−104.2	−97.6	−100.0	−104.8	−98.2	−100.5	−99.2	−99.5	−96.8
SW absorbed all-sky	241^a, 240^b, 238^c	234.4	236.2	242.8	241.3	236.5	242.1	239.9	241.2	241.8	244.6
SW up clear-sky	−53^a, −53^b	−51.4	−51.1	−51.4	−55.3	DNR	−57.4	−51.2	−51.4	−52.3	−51.2
SW absorbed clear-sky	287^a, 287^b	289.0	289.3	289.1	286.1	DNR	282.9	289.2	289.1	289.0	290.1
SW CRE	−46^a, −47^b	−54.6	−53.1	−46.3	−44.8	DNR	−40.8	−49.3	−47.8	−47.2	−45.5
LW up (OLR) all-sky	−240^a, −239^b, −238^c	−238.4	−238.9	−242.1	−251.4	−243.3	−227.1	−240.9	−240.8	−242.5	−250.0
LW up (OLR) clear-sky	−268^a, −267^b	−267.3	−266.2	−264.0	−265.9	DNR	−260.4	−264.0	−265.0	−268.2	−266.3
LW CRE	28^a, 28^b	28.9	27.3	21.9	14.5	DNR	33.2	23.1	24.2	25.7	16.3
Net CRE	−18^a, −19^b	−25.8	−25.8	−24.4	−30.3	DNR	−7.6	−26.2	−23.7	−21.5	−29.3
Imbalance	0.7^a	−4.0	−2.8	0.7	−10.1	−6.8	14.9	−1.0	0.4	−0.6	−5.4
Atmosphere											
SW absorbed all-sky	80^b, 74^c, 77^d	71.5	71.5	79.0	77.5	75.7	77.4	77.6	77.4	72.7	78.4
SW absorbed clear-sky	73^b, 73^d	70.0	69.9	76.5	71.3	DNR	74.1	74.4	74.5	70.2	74.4
SW CRE	7^b, 4^d	1.5	1.6	2.5	6.2	DNR	3.3	3.2	2.9	2.5	4.0
LW net all-sky	−183^b, −180^c, −187^d	−176.6	−177.7	−184.2	−189.9	−186.6	−170.4	−182.2	−180.9	−178.7	−189.3
LW net clear-sky	−183^b, −184^d	−183.3	−182.8	−181.4	−180.3	DNR	−182.4	−178.9	−179.3	−182.8	−181.2
LW CRE	0^b, −3^d	6.8	5.1	−2.8	−9.6	DNR	12.0	−3.3	−1.6	4.1	−8.1
Net CRE	7^b, 1^d	8.2	6.7	−0.2	−3.4	DNR	15.2	0.0	1.3	6.5	−4.1

Table 1 (continued)

Energy balance component	Reference Estimates W m^{-2}	MERRA-2 W m^{-2}	M2AMIP W m^{-2}	ERA5 W m^{-2}	JRA-55 W m^{-2}	NCEP-R2 W m^{-2}	20CRv3 W m^{-2}	ERA20C W m^{-2}	ERA20CM W m^{-2}	MERRA W m^{-2}	JRA-3Q W m^{-2}
Surface											
SW down all-sky	185[b], 186[c], 187[d]	186.0	188.2	188.1	189.4	187.5	··193.0	187.8	189.3	192.7	189.6
SW up all-sky	−25[b], −22[c], −23[d]	23.2	23.6	24.3	25.6	26.7	28.2	25.4	25.5	23.6	23.4
SW absorbed all-sky	160[b], 164[c], 164[d]	162.8	164.6	163.8	163.8	160.8	164.7	162.4	163.8	169.1	166.2
SW down clear-sky*	247[b], 244[d]	250.1	250.8	244.2	248.3	DNR	244.6	248.6	247.9	249.5	246.1
SW up clear-sky*	33[b], 30[d]	−31.2	−31.4	−31.6	−33.5	DNR	−35.8	−33.7	−33.3	−30.6	−30.4
SW absorbed clear-sky	214[b], 214[d]	218.9	219.4	212.6	214.8	DNR	208.8	214.9	214.6	218.8	215.7
SW CRE	−54[b], −50[d]	−56.1	−54.7	−48.8	−50.9	DNR	−44.1	−52.5	−50.7	−49.7	−49.5
LW down all-sky	342[b], 341[c], 344[d]	336.4	337.1	339.5	338.1	340.9	341.1	337.6	336.8	334.9	339.5
LW up all/clear-sky	−398[b], −399[c], −398[d]	−398.2	−398.4	−397.4	−399.6	−397.6	−397.9	−396.3	−396.7	−398.6	−400.2
LW net all-sky	−56[b], −58[c], −54[d]	−61.8	−61.2	−57.9	−61.5	−56.7	−56.7	−58.7	−59.9	−63.7	−60.7
LW down clear-sky	314[b], 314[d]	314.3	315.0	314.8	314.0	DNR	319.9	311.2	311.0	313.2	315.1
LW net clear-sky	−84[b], −84[d]	−83.9	−83.4	−82.6	−85.6	DNR	−78.0	−85.1	−85.7	−85.4	−85.1
LW CRE	28[b], 30[d]	22.1	22.2	24.7	24.1	DNR	21.3	26.4	25.8	21.7	24.4
Net CRE	−26[b], −20[d]	−34.0	−32.5	−24.1	−26.8	DNR	−22.8	−26.2	−25.0	−28.0	−25.1
Net radiation	104[b], 106[c], 110[d]	101.0	103.4	105.9	102.3	104.1	108.0	103.6	103.9	105.4	105.5
Latent heat flux	−82[b], −81[c]	86.2	85.0	85.0	93.2	94.1	85.6	82.9	83.0	75.9	89.3
Sensible heat flux	−21[b], −25[c]	19.3	19.4	16.9	20.2	7.3	15.5	19.1	19.0	18.0	20.6
Surface Imbalance	0.6[b], 0.5[c]	−4.5	−1.0	4.0	−11.2	2.7	6.9	1.6	1.9	11.4	−4.4

For related statistics on all-reanalyses means, inter-reanalyses spreads and standard deviations as well as respective reference estimates see Table 2. Not all reanalyses provide all energy balance components listed above, missing components are marked with DNR (Did Not Report). Units W m^{-2}. Reference estimates from Loeb et al. (2018) ([a]), Wild et al. (2015/2019) ([b]), L'Ecuyer et al. (2015) ([c]) and Kato et al. (2018) ([d])

*Approximation only

3 Results

3.1 Shortwave Components

Table 1 presents global annual means of various TOA, atmospheric and surface energy balance components as estimated by the 10 reanalyses for the period 2001–2010, as well as recent reference estimates. As can be inferred from the first data row of Table 1, the reanalyses MERRA-2, M2AMIP, ERA5, ERA20C and ERA20CM consider the solar constant as 1361 Wm^{-2}. (The solar constant refers to the measured incoming shortwave radiation at the TOA per m^{-2} perpendicular to the incoming beam, which is four times higher than the same quantity per square meter on the Earth's sphere given in Table 1.) The value of 1361 Wm^{-2} is in line with the current best estimate of the solar constant based on the Solar Radiation and Climate Experiment (SORCE, Kopp and Lean 2011). However, the earlier reanalyses considered in this study (MERRA, JRA-55 and NCEP-R2) used an older and slightly higher estimate for the solar constant of 1365 Wm^{-2} considered as best estimate at the time of their production. The recent JRA-3Q reanalysis also still uses the older estimate of 1365 Wm^{-2}.

Figure 1 illustrates the global annual mean net shortwave radiation at the TOA (i.e., the total absorption of solar radiation in the climate system, upper panel), within the atmosphere (middle panel) and at the Earth's surface (lower panel) as represented in 10 reanalyses (red bars). Figure 1 and all subsequent bar-chart figures further include the mean over all 10 (9 for clear-sky) reanalyses (pink bars), the CMIP6 multi-model mean as given in Wild (2020) (green bars), observational references (black bars, taken from CERES-EBAF (Loeb et al. 2018) for the TOA fluxes and from Wild et al. (2015, 2019) for the surface, atmospheric and clear-sky fluxes, as well as a mean over the most recent reanalyses (blue bars). The latter consists of the most recent reanalyses provided by each institution performing reanalyses, namely the four reanalyses MERRA-2, ERA5, NCEP-R2 and JRA-3Q. This gives an indication how the contemporary generation of reanalyses represents to global mean energy balance components in comparison with previous generations and reference estimates.

Table 2 provides a summary of the related statistics in terms of all-reanalyses means, standard deviations and spreads (as defined here by the difference between the largest and smallest values). For comparison, Table 2 also contains the same statistics for the set of more than 30 CMIP6 climate models, which have been taken from Wild (2020), as well as the means over the 4 most recent reanalyses mentioned above (MERRA-2, ERA5, NCEP-R2 and JRA-3Q). Generally, the means over the recent 4 reanalyses are very similar to the means over all 10 reanalyses considered here, and within 2 Wm^{-2} for most of the components (Table 2).

The spread amongst the 10 different reanalyses in their global mean TOA, atmospheric and surface shortwave absorption estimates is fairly similar, between 7.5 Wm^{-2} (atmospheric absorption) and 10.2 Wm^{-2} (TOA absorption). Interestingly, the standard deviations of the reanalysis estimates for these components are not smaller than the corresponding ones of the CMIP6 models, or even slightly larger in case of the TOA and atmospheric absorption (Table 2). This applies also for various other energy balance components listed in Table 2 in terms of their climatological global means. Specifically, in 20 out of the 36 variables in Table 2 the standard deviation amongst the reanalysis estimates is larger than the comparable CMIP6 estimates. This suggests that the representation of the global energy balance components is generally not more consistent

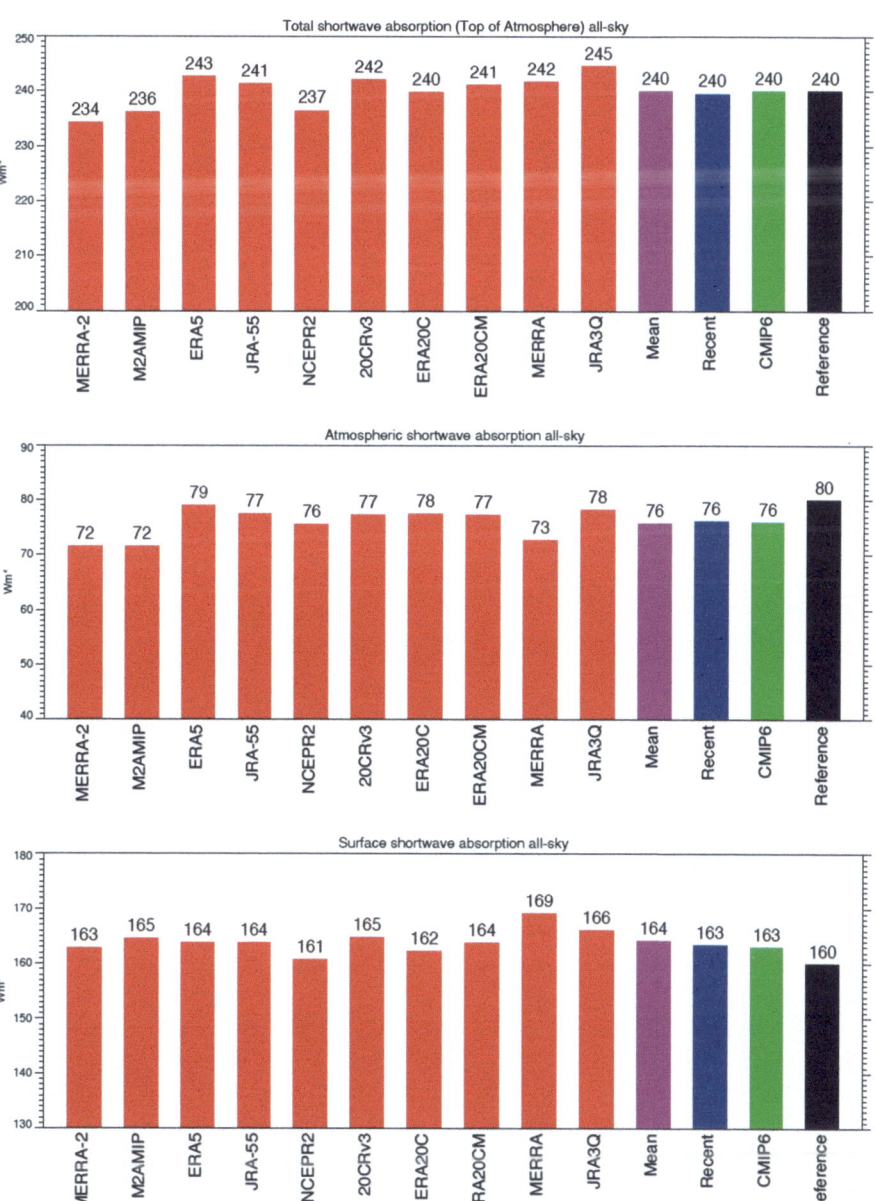

Fig. 1 Global annual mean shortwave all-sky radiation budgets representative for the period 2001–2010 as simulated by 10 different reanalyses (red bars), plus the mean over all 10 reanalyses (pink bar), the mean over the 4 most recent reanalyses (blue bar), the CMIP6 multi-model mean (green bar) and a reference estimate (black bar): shortwave radiation absorbed in the total climate system (TOA, upper panel), within the atmosphere (middle panel) and at the surface (lower panel). TOA reference estimates from the CERES-EBAF dataset (Loeb et al. 2018; Kato et al. 2018), atmospheric and surface reference estimates from Wild et al. (2015). Numbers above bars rounded to integers in Wm^{-2}

Table 2 Global annual means of various energy balance components at the TOA, within the atmosphere and at the surface under all-sky and clear-sky conditions from reanalyses, CMIP6 climate models and reference estimates, representative for the first decade of the twenty-first century (2001–2010)

Energy balance component	Reference Estimates Wm^{-2}	#Reanal (#CMIP6 Models)	Reanal mean Wm^{-2}	Reanal spread Wm^{-2}	Reanal stdev Wm^{-2}	CMIP6 mean Wm^{-2}	CMIP6 spread Wm^{-2}	CMIP6 stdev Wm^{-2}	Recent Reanal mean Wm^{-2}
TOA									
SW down	340^a, 340^b, 340^c	10 (37)	340.8	1.0	0.5	340.2	5.3	0.9	340.9
SW up all-sky	−99^a, −100^b, −102^c	10 (38)	−100.7	9.3	3.2	−100.6	13.1	2.7	−101.3
SW absorbed all-sky	241^a, 240^b, 238^c	10 (37)	240.1	10.2	3.3	239.5	14.5	2.9	239.6
SW up clear-sky	−53^a, −53^b	9 (37)	−52.5	6.3	2.3	−53.0	7.7	1.9	−51.3
SW absorbed clear-sky	287^a, 287^b	9 (37)	288.2	7.2	2.3	287.3	7.1	1.8	289.4
SW CRE	−46^a, −47^b	9 (37)	−47.7	13.8	4.2	−47.8	19.2	3.6	−48.8
LW up (OLR) all-sky	−240^a, −239^b, −238^c	10 (40)	−241.5	24.2	6.6	−238.3	15.6	2.8	−243.4
LW up (OLR) clear-sky	−268^a, −267^b	9 (38)	−265.2	7.8	2.3	−262.4	12.5	2.6	−265.8
LW CRE	28^a, 28^b	9 (38)	23.9	18.7	5.9	24.1	10.4	2.3	22.4
Net CRE	−18^a, −19^b	9 (37)	−23.8	22.6	6.6	−23.6	13.5	3.3	−26.5
Imbalance	0.7^a	(37)	−1.5	25.0	6.7	1.1	4.5	0.8	−3.9
Atmosphere									
SW absorbed all-sky	80^b, 74^c, 77^d	10 (37)	75.9	7.5	2.9	76.0	8.9	2.0	76.1
SW absorbed clear-sky	73^b, 73^d	9 (36)	72.8	6.6	2.4	72.8	8.6	1.8	73.6
SW CRE	7^b, 4^d	9 (36)	3.1	4.7	1.4	3.2	4.0	1.1	2.7
LW net all-sky	−183^b, −180^c, −187^d	10 (37)	−181.6	19.4	6.1	−182.1	17.2	4.2	−184.2
LW net clear-sky	−183^b, −184^d	9 (33)	−181.4	4.4	1.6	−180.9	15.1	3.0	−182.0
LW CRE	0^b, −3^d	9 (33)	0.3	21.6	7.1	−1.3	9.8	2.9	−1.4
Net CRE	7^b, 1^d	9 (33)	3.4	19.3	6.3	1.9	10.0	2.6	1.3

Table 2 (continued)

Energy balance component	Reference Estimates W m^{-2}	#Reanal (#CMIP6 Models)	Reanal mean W m^{-2}	Reanal spread W m^{-2}	Reanal stdev W m^{-2}	CMIP6 mean W m^{-2}	CMIP6 spread W m^{-2}	CMIP6 stdev W m^{-2}	Recent Reanal mean W m^{-2}
Surface									
SW down all-sky	185 [b], 186 [c], 187 [d]	10 (38)	189.2	6.9	2.2	187.4	20.8	4.5	187.8
SW up all-sky	−25 [b], −22 [c], −23 [d]	10 (37)	−25.0	5.1	1.6	−23.9	9.4	2.0	−24.4
SW absorbed all-sky	160 [b], 164 [c], 164 [d]	10 (37)	164.2	8.3	2.3	163.4	12.1	3.0	163.4
SW down clear-sky*	247 [d], 244 [d]	9 (37)	247.8	6.6	2.4	244.8	15.4	2.8	246.8
SW up clear-sky*	33 [b], 30 [d]	9 (36)	32.4	5.4	1.8	30.2	12.7	2.3	31.0
SW absorbed clear-sky	214 [b], 214 [d]	9 (36)	215.3	6.8	2.7	214.6	11.0	2.2	215.8
SW CRE	−54 [b], −50 [d]	9 (36)	−50.8	12.0	3.5	−51.2	20.4	4.0	−51.5
LW down all-sky	342 [b], 341 [c], 344 [d]	10 (38)	338.2	6.3	2.0	343.8	20.3	5.2	339.1
LW up all-sky/clear-sky	−398 [b], −399 [c], −398 [d]	10 (37)	−398.1	3.9	1.2	−399.9	11.7	3.0	−398.3
LW net all-sky	−56 [b], −58 [c], −54 [d]	10 (37)	−59.9	7.0	2.3	−56.2	14.0	3.6	−59.3
LW down clear-sky	314 [b], 314 [d]	9 (33)	314.3	8.9	2.6	318.0	22.5	5.1	314.7
LW net clear-sky	−84 [b], −84 [d]	9 (33)	−83.9	7.7	2.5	−81.7	16.1	3.5	−83.9
LW CRE	28 [b], 30 [d]	9 (33)	23.6	5.1	1.9	25.5	7.5	2.2	23.7
Net CRE	−26 [b], −20 [d]	9 (33)	−27.2	11.2	3.8	−25.4	15.3	3.6	−27.7
Net radiation	104 [b], 106 [c], 110 [d]	10 (37)	104.3	6.9	2.0	107.2	13.1	3.1	104.1
Latent heat flux	−82 [b], −81 [c]	10 (38)	−86.0	18.2	5.3	−85.3	18.0	3.5	−88.7
Sensible heat flux	−21 [b], −25 [c]	10 (39)	−17.5	13.3	3.9	−20.1	13.2	2.7	16.0
Surface Imbalance	0.6 [b], 0.5 [c]	10 (36)	0.7	22.6	6.4	1.5	1.2	0.3	−0.6

Given are means over up to 10 Reanalyes and 40 CMIP6 models together with their inter-model spreads as well as their standard deviations. "Reanal mean" is an average over all 10 (9) reanalyses considered in this study. "Recent Reanal mean" corresponds to an average over the most recent reanalyses (*MERRA-2, ERA5, NCEP-R2 and JRA-3Q*). Reanalyses results from the present study, CMIP6 results from Wild (2020). Reference estimates from Loeb et al. (2018) ([a]), Wild et al. (2015/2019) ([b]), L'Ecuyer et al. (2015) ([c]) and Kato et al. (2018) ([d])

*Approximation only

amongst the reanalyses than amongst the climate models. This might be somewhat surprising, given the fact that the atmospheric structure in the reanalyses is strongly constrained by the assimilated observations (except for the AMIP-type reanalysis runs of ERA20CM and MERRA-2 AMIP), in contrast to the "free-running" CMIP6 models, where not even the SSTs are observationally constrained but instead calculated by the coupled atmosphere–ocean ensemble modeling systems. Note, however, that climate models are typically tuned to match some of the observational reference quantities on a global mean level (Hourdin et al. 2017). This is particularly the case for their global mean TOA fluxes, which are usually tuned to match the satellite-based reference values from CERES-EBAF (Loeb et al. 2018). Reanalyses may also be tuned, however, while climate models can be integrated and reintegrated until the tuning converges on climate time scales, reanalysis systems in data assimilation mode can only be tuned for brief periods because of the computational demand. This may result in a less effective tuning, and may partly explain the similar or higher standard deviations of the reanalysis global mean energy balance components compared to the CMIP6 models. Also, changing observing systems can introduce spurious changes over time in reanalyses, which is not the case in free-running climate models.

In Table 2, one can further also see that the total spreads in the majority of the variables are still larger amongst the CMIP6 models than amongst the reanalyses, but this can be expected due to the many more CMIP6 models (33–40) than reanalyses (9–10) considered in this analysis.

The spread and standard deviation amongst the different reanalyses with respect to their global mean clear-sky net shortwave radiation at the TOA (i.e., the total solar absorption in the climate system under cloud-free conditions) are lower than their all-sky equivalents (Table 2, Fig. 2 upper panel). However, within the atmosphere and at the surface, the spreads and standard deviations of their clear-sky shortwave absorption are almost as large as the ones of their all-sky counterparts (Table 2, Fig. 2 middle and lower panels). This suggests that uncertainties in the partitioning of the shortwave absorption between atmosphere and surface cause similar discrepancies under clear-sky and all-sky conditions. The standard deviations in the shortwave clear-sky budgets of the reanalyses are also similar to the CMIP6 climate models (Table 2). The slightly larger standard deviation in the atmospheric clear-sky absorption in the reanalyses compared to the climate models might again be somewhat surprising, given the fact that the humidity, an essential absorber of shortwave radiation in the cloud-free atmosphere, is observationally constrained in the reanalyses, in contrast to the climate models. This may suggest that differences in the formulation of the radiation codes used in the reanalyses and climate models could be more relevant for the discrepancies in the radiative fluxes than differences in the physical atmospheric structure entering the radiation codes. Discrepancies could further also be enhanced by different treatments of absorbing aerosols or ozone in the reanalyses and climate models, as well as potential differences in the observational humidity inputs in the various reanalyses.

For the 9 reanalyses which provide both all-sky and clear-sky budgets, the global mean cloud radiative effect (CRE) can be diagnosed at the TOA, within the atmosphere and at the surface, as the difference between the global mean all-sky and clear-sky estimates (Tables 1, 2). The standard deviations of the global mean shortwave CREs amongst the reanalyses are again similar to the CMIP6 models (slightly larger at the TOA and in the atmosphere, while slightly smaller at the surface, Table 2).

With respect to the agreement with independent reference values, both the all-reanalyses means and the recent-reanalyses means are close to the reference estimates for most of their shortwave components, i.e., within 2 Wm^{-2} (Table 2).

At the TOA, the absorbed all-sky solar radiation averaged over all 10 reanalyses is in close agreement with the reference estimates from CERES-EBAF (Loeb et al. 2018), although favored by a small compensation between a slight overestimation of TOA clear-sky absorption and a slight overestimation of the reflectivity in cloudy conditions (i.e., a slightly too strong SW CRE (Table 2)). This applies even more so for the 4 recent-reanalyses mean (Table 2). As noted in Bosilovich et al. (2015), a low bias in the TOA all-sky shortwave absorption is evident in MERRA-2, on the order of 6 Wm^{-2}, whereas the recent JRA-3Q shows a high bias of similar magnitude in this quantity (Table 1, Fig. 1).

As already mentioned in Sect. 2, reference values for the shortwave absorption in the atmosphere and at the surface are less well established, since these quantities cannot be directly measured from satellites. Compared to our best estimate of 80 Wm^{-2} for the all-sky shortwave atmospheric absorption (Wild et al. 2015), the all- and recent-reanalyses means are, at 75.9 and 76.1 Wm^{-2} somewhat low, but still within the range of the observational references (Table 2). However, the entire MERRA reanalysis family (MERRA, MERRA-2 and MERRA-2 AMIP) tends to calculate a too transparent atmosphere for solar radiation under all-sky conditions compared to all reference estimates (Table 1). The spread in all-sky atmospheric absorption remains substantial also amongst the most recent reanalyses, ranging from 71.5 Wm^{-2} in MERRA-2 to 79.0 Wm^{-2} in ERA5 (Table 1).

Under cloud-free conditions, the atmospheric shortwave absorption in the all- and recent-reanalyses means nearly match the reference values of 73 Wm^{-2} (Table 2). Note also that the two reference estimates for this quantity in Table 2 (Kato et al. 2018; Wild et al. 2019) perfectly agree, despite their entirely different and independent derivations. With respect to the individual reanalyses, the MERRA family also systematically underestimates the global mean clear-sky shortwave atmospheric absorption, by 3 Wm^{-2}, whereas ERA5 overestimates this quantity by a similar amount.

When comparing the shortwave fluxes at the surface to the reference estimates, the all-reanalyses mean tends to overestimate the all-sky downward shortwave components, on the order of 3 Wm^{-2}, which is reinforced by the strong overestimation of the 20CRv3 and the older MERRA reanalyses, on the order of 7 Wm^{-2} (Tables 1, 2, Fig. 3 upper panel). This overestimation is partly caused by an overestimation of this quantity under cloud-free conditions (Table 1), and an underestimated all-sky absorption in the atmosphere (Fig. 1 middle panel) in some of the models. This applies again particularly to the MERRA family, which shows excessive global mean downward clear-sky shortwave fluxes (Table 1), in line with the overly low atmospheric clear-sky shortwave absorption mentioned above, as well as a somewhat high TOA clear-sky shortwave absorption (Fig. 2 upper and middle panels). In the case of MERRA-2 under all-sky conditions, the excessive clear-sky insolation is partly compensated by an excessive shortwave reflectance in cloudy areas (i.e., a too strong shortwave CRE, Table 1) as mentioned above. The excessive surface insolation due to an overly transparent atmosphere has been a long-standing issue over the history of climate model development (Wild et al. 1995; Wild 2008). It has been partly attributed to the lack of water vapor absorption in the atmosphere due to deficiencies in the related spectroscopic absorption coefficients and in the formulations of the near-infrared water vapor continuum used in the radiation codes (Morcrette 2002; Paynter and Ramaswamy 2012, 2014; Pincus et al. 2015; Radel et al. 2015). Deficiencies in shortwave atmospheric absorption have also been noted in the CMIP5 models in DeAngelis et al. (2015), which have been alleviated in the some of the CMIP6 models (Pendergrass 2020; Wild 2020).

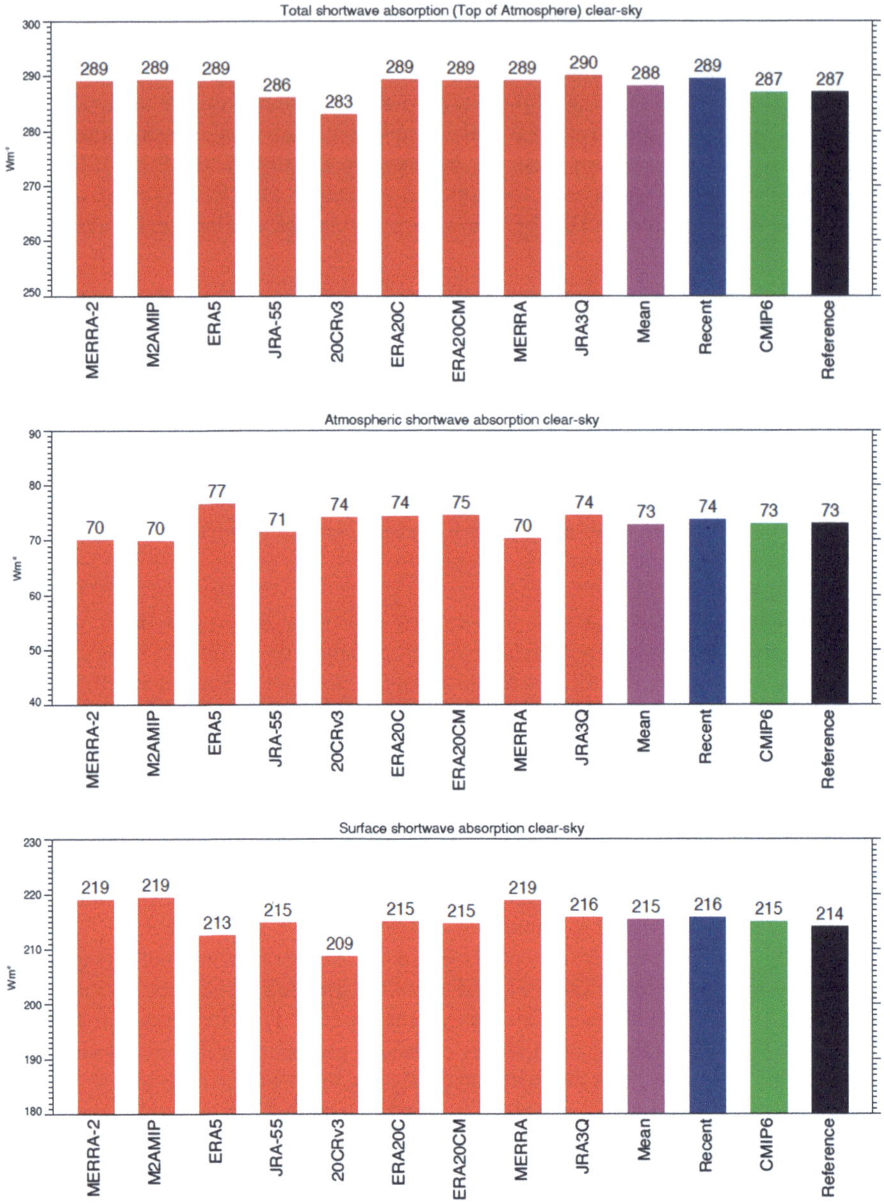

Fig. 2 as Fig. 1, but for clear-sky shortwave budgets of 9 different reanalyses. TOA reference estimates from the CERES-EBAF dataset (Loeb et al. 2018), atmospheric and surface reference estimates from Wild et al. (2019)

Generally, compared to the CMIP6 multi-model means, the shortwave components in the all- and recent-reanalyses means are not closer to the reference estimates neither under all-sky nor under clear-sky conditions.

3.2 Longwave Components

Figures 4 and 5 illustrate the global annual means of the net longwave radiation at the surface (lower panel), within the atmosphere (middle panel) and at the TOA (outgoing longwave radiation, OLR, upper panel), under all-sky and clear-sky conditions, respectively, as given by the different individual reanalyses, their all- and recent-reanalyses means, the CMIP6 multi-model means and the reference estimates. As in the shortwave, several of the longwave components including the longwave CREs determined by the reanalyses show similar or larger standard deviations and spreads than the ones simulated by the CMIP6 models (Table 2). This applies for example to the global mean all-sky OLR, with a spread amongst the reanalyses of as much as 24.2 Wm^{-2}, corresponding to 10% of the absolute magnitudes, and a standard deviation of 6.6 Wm^{-2}. This is partly due to the exceptionally low all-sky OLR in the 20CRv3 reanalysis (Fig. 4). Also in the TOA longwave CRE, the spread and standard deviation is substantial, at 18.7 and 5.9 Wm^{-2}, respectively, reinforced by the particularly low and high TOA longwave CRE in JRA-55 and 20CRv3, respectively. This causes the reanalysis spread to be considerably larger than amongst the many more CMIP6 models (Table 2). This similarly applies to the longwave CRE in the atmosphere, which shows, at 21.6 Wm^{-2}, a much larger spread amongst the 9 reanalyses than amongst the 33 CMIP6 models with a spread of 9.8 Wm^{-2}.

The longwave clear-sky components, on the other hand, show generally a higher consistency in the reanalyses compared to the CMIP6 models. This is likely due to the observationally constrained temperature and water vapor profiles in the former, which are particularly relevant for the determination of the longwave clear-sky fluxes.

With respect to the longwave reference estimates, at the TOA, both all- and recent-reanalyses means slightly overestimate and underestimate the global mean OLR under all-sky and clear-sky conditions, respectively, compared to the CERES-EBAF estimates. This leads to a too weak global mean TOA longwave CRE (Table 2) in the all- and recent-reanalyses means, on the order of 4 and 6 Wm^{-2}, respectively. As mentioned above, the spread amongst the individual reanalyses in this quantity is, however, substantial.

At the surface, a noteworthy feature is the somewhat low global mean downward longwave radiation in the all- and recent-reanalyses means compared to the reference estimates. Specifically, compared to our best estimate (Wild et al 2013, 2015), the all-reanalyses mean is low by 4 Wm^{-2}, while by 3 Wm^{-2} for the recent-reanalysis mean (Table 2, Fig. 3 lower panel). Also, all individual reanalyses shown in Fig. 3 (lower panel) calculate a global mean downward longwave radiation that is up to 5 Wm^{-2} lower than the lowest reference estimate, or at best equal to the lowest reference estimate (NCEP-R2 and 20CRv3 at 341 Wm^{-2}). In some of the reanalyses, the underestimation is also evident under clear-sky conditions (Table 2). The underestimation of the downward longwave radiation, similarly as the overestimation of the downward shortwave radiation discussed in the previous section, is a long-standing issue in numerical models of weather and climate (Wild et al. 1995, 2001). This underestimation has been partly related to uncertainties in the formulation of the longwave water vapor continuum (Iacono et al. 2000; Paynter and Ramaswamy 2011; Wild et al. 2001). Over generations of weather and climate models, the underestimation of the downward longwave radiation has compensated for the overestimation of the downward shortwave radiation, leading to a superficially correct surface net radiation in the global mean due to error compensation (Wild et al. 1995, 2013; Wild 2008). The situation has gradually improved over time (Wild 2020), but seems to remain an issue in some of the reanalyses.

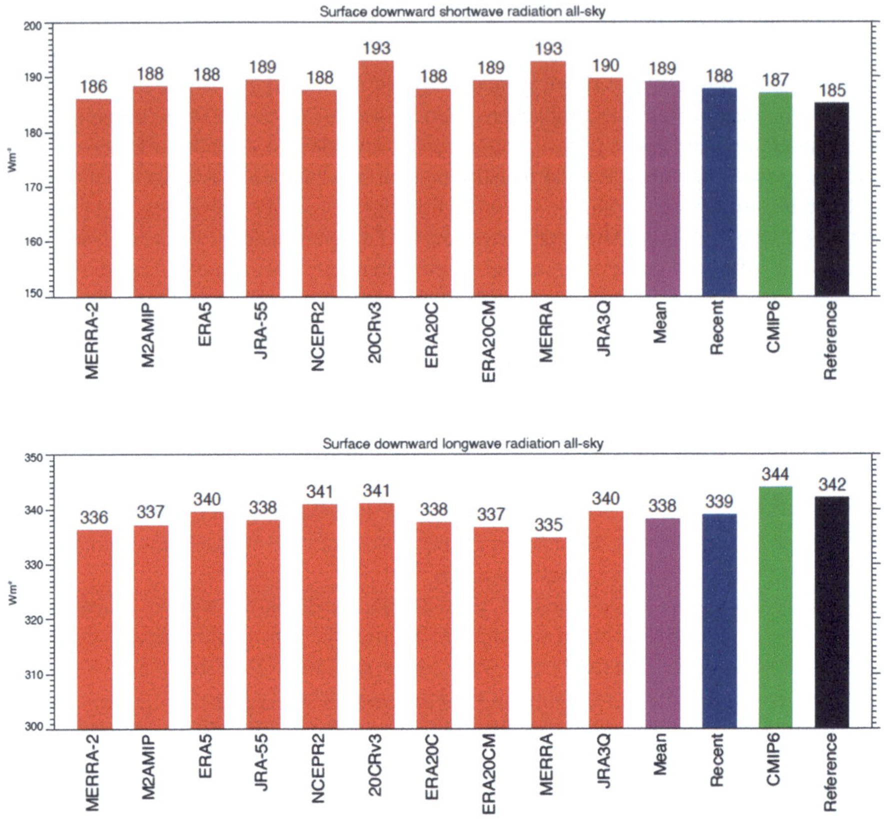

Fig. 3 Global annual mean all-sky downward shortwave (upper panel) and longwave (lower panel) radiation at Earth's surface representative for the period 2001–2010 as simulated by 10 different reanalyses (red bars), plus the mean over all 10 reanalyses (pink bar), the mean over the 4 most recent reanalyses (blue bar), the CMIP6 multi-model mean (green bar) and a reference estimate (black bar). Reference estimates from Wild et al. (2015). Numbers above bars rounded to integers. Units Wm^{-2}

3.3 Surface Net Radiation and Non-radiative Fluxes

Figure 6 displays the global mean surface net radiation (upper panel) as well as the non-radiative fluxes of surface latent and sensible heat (middle and lower panel, respectively) of the 10 reanalyses. In both all- and recent-reanalyses means, the global mean surface net radiation (also known as surface radiation balance) perfectly matches our best estimate of 104 Wm^{-2} (Wild et al. 2015) (Table 2). However, as noted above in Sects. 3.1 and 3.2, this overall good agreement is partly caused by an excessive downward solar radiation, which is compensated by a too weak downward longwave radiation. Thereby, also the individual reanalyses calculate magnitudes of global mean surface net radiation which are mostly within a few Wm^{-2} of our estimate. The largest compensation between overestimated downward shortwave and underestimated downward longwave radiation is found in the older MERRA reanalysis, with counteracting biases on the order of 8 Wm^{-2}. The same tendency, albeit with smaller biases is also found in JRA-55, JRA-3Q, M2AMIP, ERA20CM and ERA5, while NCEP2 shows hardly any biases in this respect. On the other

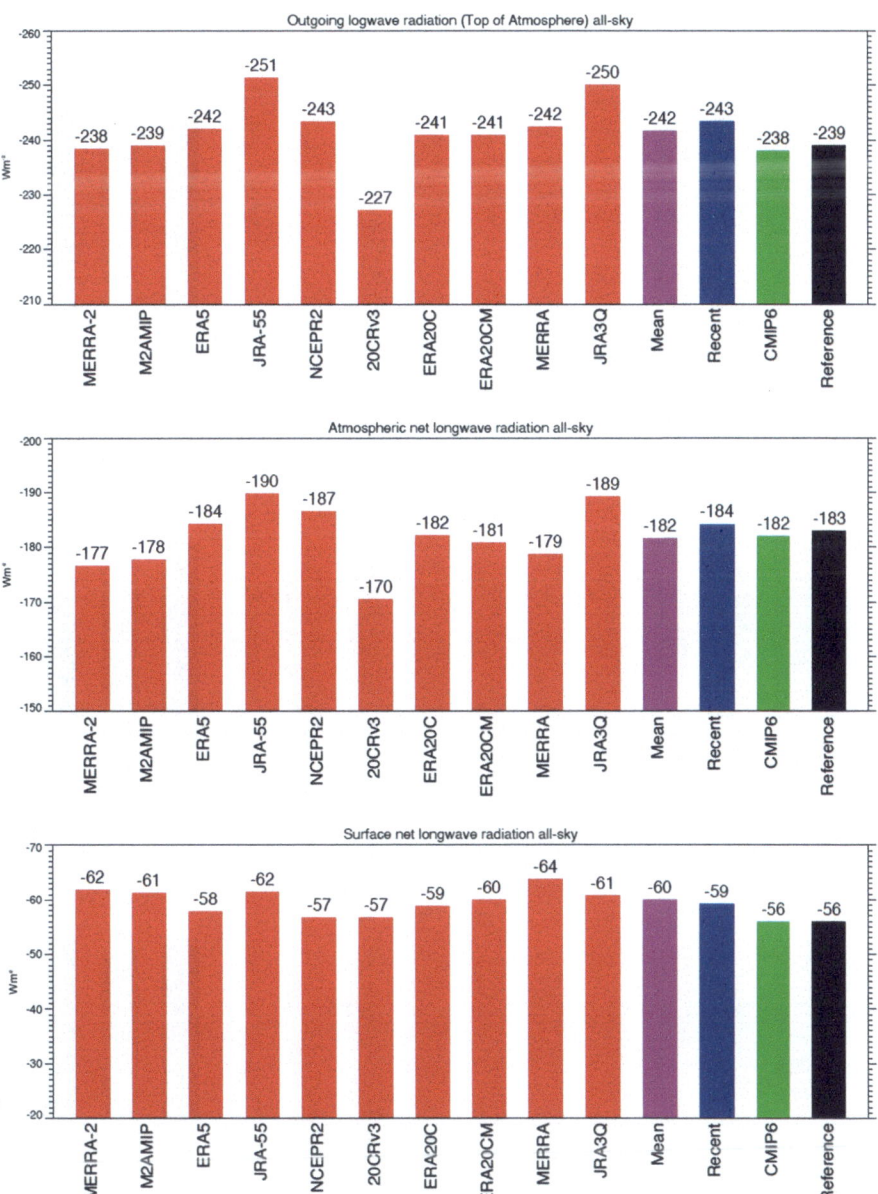

Fig. 4 Global annual mean longwave all-sky radiation budgets representative for the period 2001–2010 as simulated by 10 different reanalyses (red bars), plus the mean over all 10 reanalyses (pink bar), the mean over the 4 most recent reanalyses (blue bar), the CMIP6 multi-model mean (green bar) and a reference estimate (black bar): longwave radiation emitted to space (upper panel), net longwave radiation within the atmosphere (middle panel) and at the surface (lower panel). TOA reference estimates from the CERES-EBAF dataset (Loeb et al. 2018), atmospheric and surface reference estimates from Wild et al. (2015). Numbers above bars rounded to integers. Units Wm^{-2}

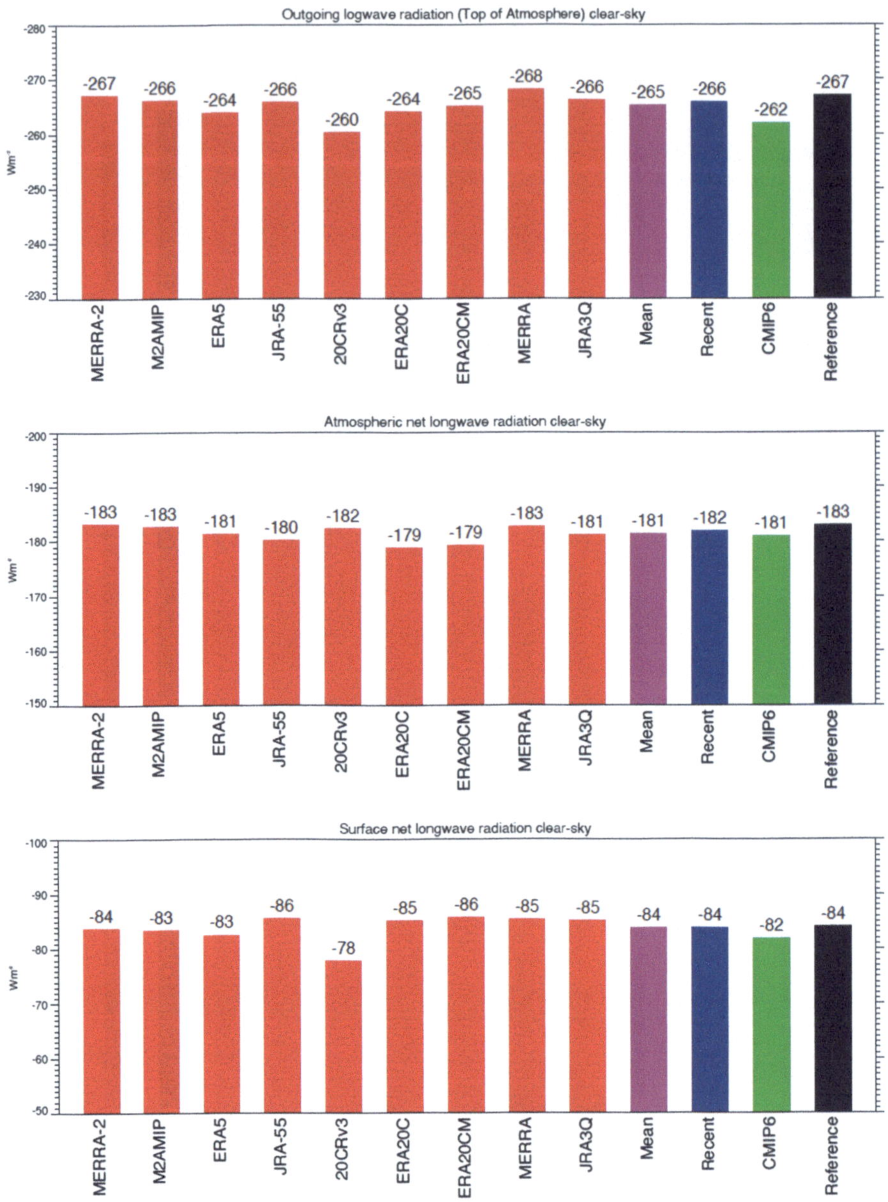

Fig. 5 As Fig. 4, but for clear-sky longwave budgets of 9 different reanalyses. TOA reference estimates from the CERES-EBAF dataset (Loeb et al. 2018), atmospheric and surface reference estimates from Wild et al. (2019)

hand, the somewhat low surface net radiation in MERRA-2 (101 Wm^{-2}) is primarily a consequence of a too low downward longwave radiation (336 Wm^{-2}). At the other end of the spectrum, the somewhat high surface net radiation of 108 Wm^{-2} in 20CRv3 is primarily induced by a too high downward solar radiation (193 Wm^{-2}), while the downward longwave radiation is, at 341 Wm^{-2}, in line with the reference estimates.

Despite the reasonably similar global mean surface net radiation and thus radiative energy available for evaporation at the surface in the 10 reanalyses, the global mean latent heat fluxes largely vary across the different reanalyses, in a range of as much as 18 Wm^{-2} (Fig. 6 middle panel). This spread is not smaller than the spread in the CMIP6 models, despite the much larger number of CMIP6 models than reanalyses considered here (Table 2). As a consequence, also the representation of the global water cycle largely differs in the various reanalyses. Since the latent heat flux varies across the reanalyses by 21%, this implies that the global mean precipitation in the different reanalyses also varies by a similar amount (i.e., by more than 20%). A similarly large spread in global mean precipitation in earlier generation reanalyses can be inferred from Table 1 of Bosilovich et al. (2008). Still, it is noteworthy that several of the recent reanalyses show global mean latent heat flux values near 85 Wm^{-2}, in reasonable agreement with the reference estimates (Fig. 6 middle panel, Table 1).

The much larger spread and standard deviation in the latent heat flux than in the surface net radiation also indicates that the Bowen ratio (ratio between sensible and latent heat flux) varies considerably across the different reanalyses. Accordingly, also the global mean sensible heat flux in the different reanalyses shows a substantial spread, of 13.3 Wm^{-2}, corresponding to 75% of its absolute value (Table 2, Fig. 6 lower panel). This is largely caused by the very low sensible heat flux in NCEP2, whereas the other reanalyses tend to cluster around 19 Wm^{-2} in their global mean sensible heat fluxes. Due to the lack of widespread long-term direct observations of sensible heat fluxes and the considerable uncertainties in estimates from bulk parameterizations (Yu 2019), this quantity is poorly constrained and the available global mean reference estimates differ considerably (Trenberth et al. 2009; Berrisford et al. 2011; Wild et al. 2015). A detailed assessment of the turbulent surface fluxes in MERRA can be found in Roberts et al. (2012).

3.4 TOA and Surface Imbalance

To keep the Earth's climate system in equilibrium, the shortwave radiation absorbed by the climate system should match the outgoing longwave radiation at the TOA, i.e., the radiation balance at the TOA should be zero. However, due to anthropogenic climate change, a positive imbalance (nonzero TOA radiation balance) is expected to occur, which causes an accumulation of energy in the climate system (e.g., von Schuckmann et al. 2016; Hakuba et al. 2019; Hakuba et al. 2021; von Schuckmann et al. 2023). This imbalance (also known as Earth Energy Imbalance EEI) is estimated to be +0.8 Wm^{-2} over the period 2006–2018, primarily inferred from measurements of the energy accumulation in the global oceans (Forster et al. 2021). However, in contrast to coupled atmosphere–ocean GCMs, in reanalyses the energy accumulation is not only governed by the TOA imbalance, but also influenced by the prescribed SSTs and the assimilated observational data, which can induce additional energy sources. Still, the examination of the imbalance in reanalyses is relevant, considering that they aim at reproducing the physical structure of the atmosphere as realistic as possible, and as such may also be thought to have a potential to realistically capture the imbalance. However, in both all- and recent-reanalyses means the imbalance at the TOA is, at −1.5 and −3.9 Wm^{-2}, respectively, of opposite sign compared to any published TOA imbalance reference estimate (Table 2). This suggests that current reanalyses may not be able to serve as alternative pathways to better constrain the magnitude of the TOA imbalance, despite their incorporation of comprehensive observational data. With respect to the individual reanalyses, only ERA5 and ERA20CM simulate TOA imbalances, at 0.7

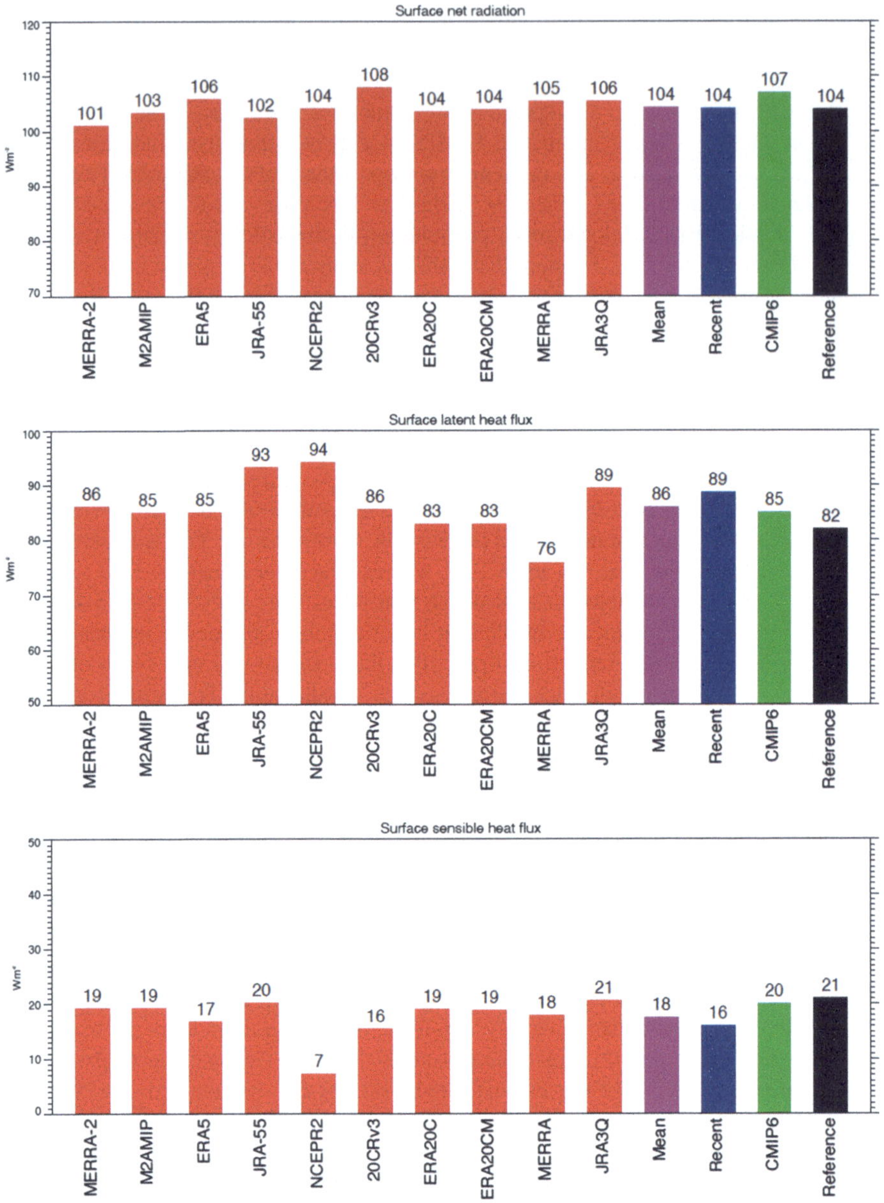

Fig. 6 Global annual mean surface net radiation (upper panel), latent heat fluxes (middle panel) and sensible heat fluxes (lower panel) representative for present-day climate as calculated by 10 different reanalyses (red bars), plus the mean over all 10 reanalyses (pink bar), the mean over the 4 most recent reanalyses (blue bar), the CMIP6 multi-model mean (green bar) and a reference estimate (black bar) from Wild et al. (2015). Numbers above bars rounded to integers. Units Wm^{-2}

and 0.4 Wm^{-2}, respectively, which are compatible with reference estimates and associated uncertainty ranges (Fig. 7 upper panel, Table 1, see also discussion in Hersbach et al. (2020), Sect. 6.1).

Also, the temporal change of the TOA imbalance is not adequately represented in the reanalyses (Fig. 8). While the annual imbalance in the CERES satellite data shows a dramatic increase over the first two decades of the twenty-first century as pointed out by Loeb et al. (2021) (thick black line in Fig. 8), only the JRA-3Q shows a slight increase, yet of an order of magnitude smaller than the CERES reference trend. All other reanalyses which cover the first two decades of the twenty-first century show a negative trend (linear trend magnitudes given in Fig. 8) and thus fail to capture the increasing energy imbalance in the climate system. The lack of a significant positive global mean trend in the ERA5 reanalysis and a related AMIP version is also documented in Table 1 of Loeb et al. (2022). Further with respect to ERA5, the lack of increase has been related to the TOA shortwave reflectance, which does not show the substantial decrease seen in CERES-EBAF, possibly related to the deficient representation of the decline in aerosol (Hodnebrog et al. 2024; Liu et al. 2020). Finally, climate models also underestimate the CERES trend by half (Schmidt et al. 2023), and seemingly the data assimilation in reanalysis systems does not appear to overcome this general modeling deficiency.

The lower panel of Fig. 7 shows the annual mean energy imbalance at the Earth's surface of the different reanalyses. This surface imbalance is deduced here from the difference between the surface net radiation and the sum of the surface sensible and latent heat fluxes. This roughly corresponds to the net energy flux into the oceans, because the amount of energy going into land surface and the melting of snow and ice are much smaller in comparison. The surface imbalance is thus quantitatively closely related to the TOA energy imbalance discussed above, since the energy storage in the atmosphere is small. In contrast to the CMIP6 models, where all 36 models show a positive surface imbalance as expected with increasing greenhouse-gas forcing (Wild 2020), 4 out of the 10 reanalyses show a negative surface imbalance (Fig. 7 lower panel). Both spread and standard deviation across the 10 reanalyses, at 22.6 and 6.4 Wm^{-2}, respectively, are massively larger (more than an order of magnitude) than across the 36 CMIP6 models, at 1.2 and 0.3 Wm^{-2}, respectively. There is also limited correlation in the reanalyses between their TOA and surface imbalances, which points to the effect of observational assimilation, where the correction of state variables (temperature and water vapor) by observations each day causes subsequent variations in the energy balance (see also Hersbach et al. 2020).

4 Discussion and Conclusions

In this study, the global mean energy balance components of 10 atmospheric reanalyses, representative for the period 2001–2010, have been intercompared and related to available reference estimates. Figure 9 provides an illustrative summary of the magnitudes of the global mean energy balance components according to the all-reanalyses mean and the corresponding values from the CMIP6 multi-model mean as well as the reference estimates from Kato et al. (2018) and Wild et al. (2015, 2019). Depicted are the values for both for all-sky and clear-sky conditions (Fig. 9 upper and lower panel, respectively).

The majority of the reanalyses considered in this study incorporate a comprehensive observational data assimilation (including prescribed observational SSTs), whereas two reanalysis models are only constrained by observed SSTs. To this end, it is interesting to compare the energy balance components of MERRA-2 and MERRA-2 AMIP, where the

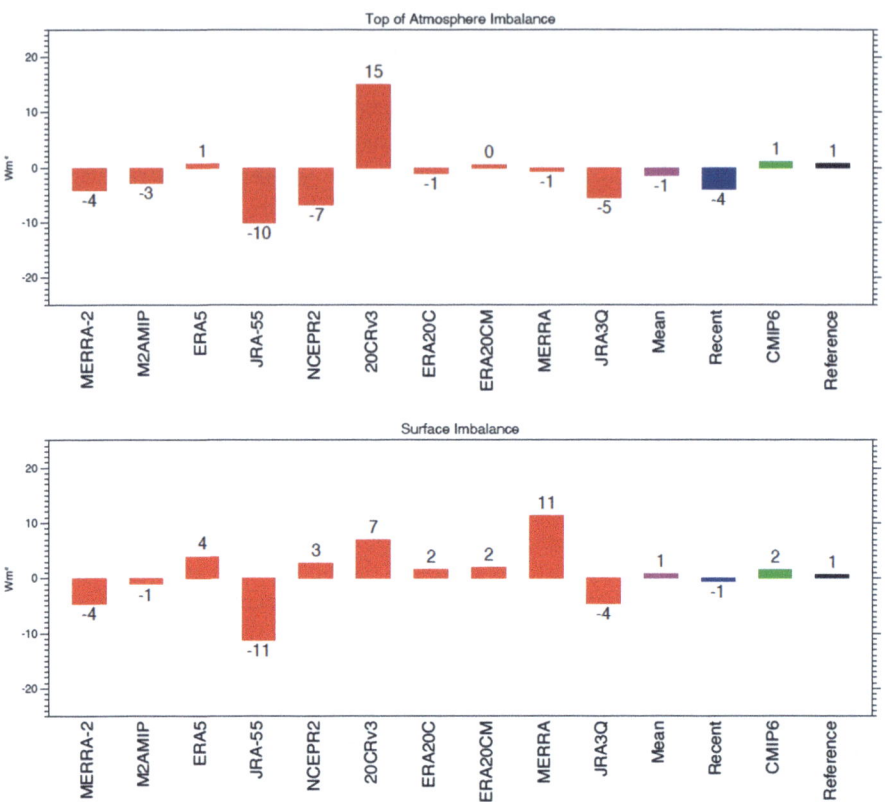

Fig. 7 Global annual mean energy imbalance at the TOA (upper panel) and at the Earth's surface (lower panel) for the period 2001–2010 estimated by 10 different reanalyses (red bars), plus the mean over all 10 reanalyses (pink bar), the mean over the 4 most recent reanalyses (blue bar), the CMIP6 multi-model mean (green bar) and a reference estimate (black bar). TOA energy imbalance determined as difference between absorbed shortwave radiation in the climate system (Fig. 1 upper panel) and the longwave emission to space (Fig. 4 upper panel). Surface imbalance determined as difference between surface net radiation (Fig. 6 upper panel) and the sum of surface sensible and latent heat fluxes (Fig. 6 middle/lower panels). Reference estimates from Forster et al. (2021). Numbers above bars rounded to integers. Units Wm^{-2}

same model is once run in full assimilation mode (MERRA-2), and once only constrained by the evolving SSTs (MERRA-2 AMIP). As can be inferred from Table 1 (columns 2 and 3), for most quantities the differences between the two realizations over the identical period 2001–2010 are within 2 Wm^{-2} for their global means. Exceptions are the surface downward shortwave radiation, the surface net radiation and the surface imbalance, where the differences amount to 2.2, 2.4 and 3.5 Wm^{-2}, respectively. The differences in the surface downward shortwave and net radiation come primarily from a stronger shortwave cloud radiative effect in the assimilated realization, which reduces the surface insolation, in better agreement with the reference estimates. In this case, the assimilation of observational data seems to have been beneficial, whereas for most other energy balance quantities the impact of the assimilation has been negligible in the global mean. Differences in the clear-sky fluxes between the two realizations tend to be particularly small and are typically within a few tenths of one Wm^{-2}. The surface imbalance in the assimilated realization is even less realistic and more negative than in the realization constrained by SSTs only. A similar

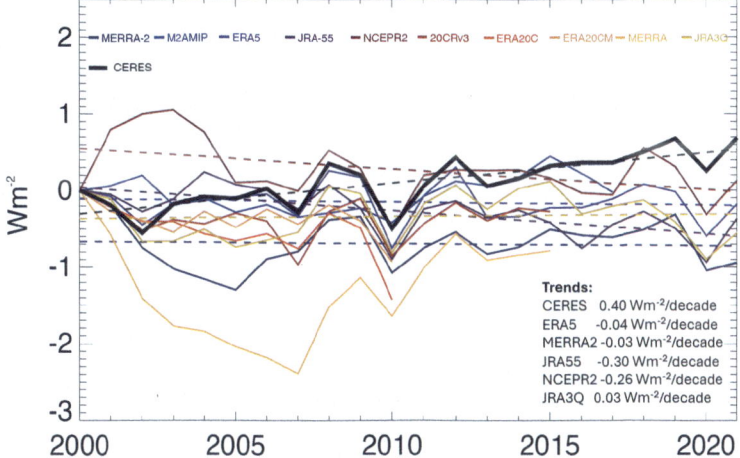

Fig. 8 Temporal evolution of the global annual mean TOA radiation balance (Earth energy imbalance EEI) as derived from the CERES-EBAF satellite dataset (Loeb et al. 2018) (black line) and as calculated in 10 different reanalyses (colored lines) over the first two decades of the twenty-first century. Shown are annual anomalies with respect to the beginning of the twenty-first century. Unit on vertical axis is Wm^{-2}. Linear trend magnitudes are given for CERES and for those 5 reanalyses which cover the entire first two decades of the twenty-first century

comparison can be made between ERA20C and ERA20CM, where ERA20C assimilates surface observations in addition to SSTs, whereas ERA20CM is only constrained by SSTs. Except for the net CRE at the TOA, all quantities in Table 1 differ by less than 2 Wm^{-2} between ERA20C and ERA20CM.

These comparisons suggest that the assimilation of observations does not substantially modify the overall representation of the global energy balance in reanalyses. This fits to the general picture portrayed here, in that the representation of the global energy balance components in reanalyses, despite the assimilation of observational data, is not obviously improved compared to the less constrained CMIP6 climate models. This generally also applies for the most recent reanalyses considered in this study (MERRA-2, ERA5, NCEP-R2 and JRA-3Q). We also showed that the consistency between the magnitudes of the global mean energy balance components in the different reanalyses is not necessarily higher than in the CMIP6 models, despite the additional observational constraints. This indicates that the magnitude of the global energy balance components in reanalysis systems, as in climate models, may depend more on the formulation of the specific parameterization schemes (i.e., the radiation codes for the radiative components) than on degree of observational constraints applied to the inputs to these schemes. In this respect, the reanalysis energy balance components are similarly prone to some of the well-known deficiencies in climate models. Specifically, we noted the tendency of an overestimated surface downward shortwave radiation globally, compensated by an underestimated surface downward longwave radiation, a long-standing issue in climate models, to be similarly present in the reanalyses. This compensational effect between the overestimated downward shortwave and underestimated downward longwave radiation, however, only works on a global annual mean basis, and does not apply on regional, seasonal and diurnal scales, thereby likely inducing biases on any of these scales. This should apply similarly to both reanalyses

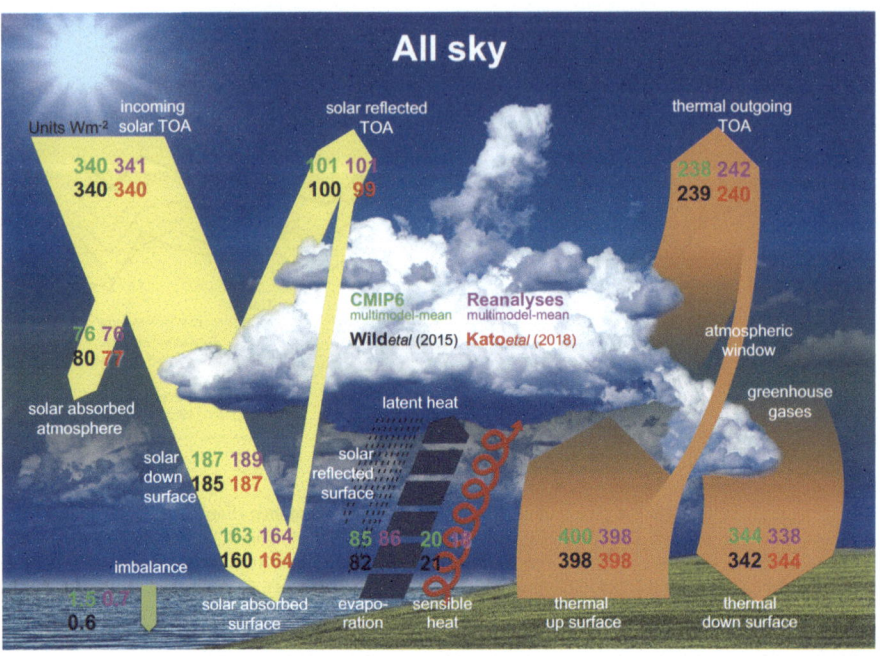
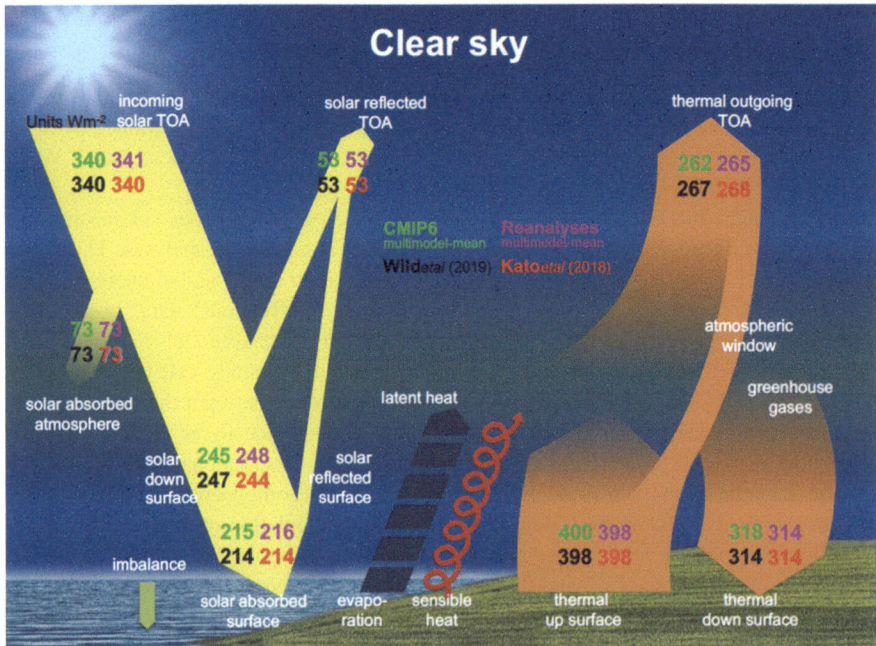

Fig. 9 Magnitudes of the different global annual mean energy balance components at the beginning of the twenty-first century under "all-sky" (upper panel) and "clear-sky" (lower panel) conditions, as simulated in the all-reanalyses mean (upper right (pink) values) and in the CMIP6 multi-model mean (upper left (green) values), and estimated by Wild et al. (2015, 2019) (lower left (black) values) and Kato et al. (2018) (lower right (red) values). Units Wm^{-2}

and climate models. Noteworthy is also the substantial spread in the intensity of the global water cycle between some of the reanalyses, as indicated by their largely diverging global mean surface latent heat fluxes. Also, current reanalyses cannot provide additional constraints on the magnitude of the Earth energy imbalance nor on its temporal evolution, since their absolute magnitudes as well as the sign of their trends are mostly unrealistic.

This does not put in question in any way the tremendous value of reanalyses for a wide range of applications. Reanalysis fields, such as geopotential height, upper air temperature, velocity or humidity fields have been most successfully used over many years as reliable references in countless studies. With respect to the different global energy balance components, however, it is not recommended to rely on currently available reanalyses as references.

Acknowledgements This paper is an outcome of the Workshop "Challenges in Understanding the Global Water Energy Cycle and its Changes in Response to Greenhouse Gas Emissions" held at the International Space Science Institute (ISSI) in Bern, Switzerland (26–30 September 2022). We are most grateful to the different institutions which engage in the enormous task of performing reanalyses. Research on the global energy balance at ETH Zurich is supported by the Swiss National Science Foundation, Grant No. 200020_188601, and at GSFC by the NASA Energy and Water cycle Studies (NEWS) program. We would like to thank also Richard Allan and 3 additional anonymous reviewers for their very constructive and helpful comments.

Funding Open access funding provided by Swiss Federal Institute of Technology Zurich.

Declarations

Conflict of interest The authors have no relevant financial or non-financial interests to disclose.

Open Access This article is licensed under a Creative Commons Attribution 4.0 International License, which permits use, sharing, adaptation, distribution and reproduction in any medium or format, as long as you give appropriate credit to the original author(s) and the source, provide a link to the Creative Commons licence, and indicate if changes were made. The images or other third party material in this article are included in the article's Creative Commons licence, unless indicated otherwise in a credit line to the material. If material is not included in the article's Creative Commons licence and your intended use is not permitted by statutory regulation or exceeds the permitted use, you will need to obtain permission directly from the copyright holder. To view a copy of this licence, visit http://creativecommons.org/licenses/by/4.0/.

References

Allan RP, Liu CL, Loeb NG, Palmer MD, Roberts M, Smith D, Vidale PL (2014) Changes in global net radiative imbalance 1985–2012. Geophys Res Lett 41(15):5588–5597. https://doi.org/10.1002/2014gl060962

Allan RP, Ringer MA, Pamment JA, Slingo A (2004) Simulation of the earth's radiation budget by the european centre for medium-range weather forecasts 40-year reanalysis (era40). J Gerontol Ser A Biol Med Sci 109(D18):D18107. https://doi.org/10.1029/2004jd004816

Berrisford P, Kallberg P, Kobayashi S, Dee D, Uppala S, Simmons AJ, Poli P, Sato H (2011) Atmospheric conservation properties in era-interim. Q J R Meteorol Soc 137(659):1381–1399. https://doi.org/10.1002/Qj.864

Bosilovich MG, Akella S, Coy L, Cullather R, Draper C, Gelaro R, Kovach R, Liu Q, Molod A, Norris P, Wargan K, Chao W, Reichle R, Takacs L, Vikhliaev Y, Bloom S, Collow A, Firth S, Labow G, Partyka G, Pawson S, Reale O, Schubert SD, Suarez M (2015) Merra-2: Initial evaluation of the climate. NASA/TM–2015–104606 43:145

Bosilovich MG, Chaudhuri AH, Rixen M (2013) Earth system reanalysis: progress, challenges, and opportunities. Bull Am Meteorol Soc 94(8):Es110–Es113. https://doi.org/10.1175/Bams-D-12-00191.1

Bosilovich MG, Chen JY, Robertson FR, Adler RF (2008) Evaluation of global precipitation in reanalyses. J Appl Meteorol Clim 47(9):2279–2299. https://doi.org/10.1175/2008jamc1921.1

Bosilovich MG, Robertson FR, Chen JY (2011) Global energy and water budgets in merra. J Clim 24(22):5721–5739. https://doi.org/10.1175/2011jcli4175.1

Collow, Marquardt AB, Mahanama SP, Bosilovich MG, Koster RD, Schubert SD (2017) An evaluation of teleconnections over the united states in an ensemble of amip simulations with the merra-2 configuration of the geos atmospheric model. NASA/TM-2017–104606 47:68

DeAngelis AM, Qu X, Zelinka MD, Hall A (2015b) An observational radiative constraint on hydrologic cycle intensification. Nature 528(7581):249. https://doi.org/10.1038/nature15770

Forster P, Storelvmo T, Armour K, Collins W, Dufresne J-L, Frame D, Lunt DJ, Mauritsen T, Palmer MD, Watanabe M, Wild M, Zhang H (2021) The earth's energy budget, climate feedbacks, and climate sensitivity. In: V M-D, Zhai P, Pirani A et al. (eds) Climate change 2021: The physical science basis. Contribution of working group 1 to the sixth assessment report of the intergovernmental panel on climate change. Cambridge University Press, Cambridge, , United Kingdom and New York, NY, USA, pp 923–1054. https://doi.org/10.1017/9781009157896.009

Gelaro R, McCarty W, Suárez MJ, Todling R, Molod A, Takacs L, Randles CA, Darmenov A, Bosilovich MG, Reichle R, Wargan K, Coy L, Cullather R, Draper C, Akella S, Buchard V, Conaty A, da Silva AM, Gu W, Kim GK, Koster R, Lucchesi R, Merkova D, Nielsen JE, Partyka G, Pawson S, Putman W, Rienecker M, Schubert SD, Sienkiewicz M, Zhao B (2017) The modern-era retrospective analysis for research and applications, version 2 (merra-2). J Clim 30(14):5419–5454. https://doi.org/10.1175/Jcli-D-16-0758.1

Hakuba MZ, Frederikse T, Landerer FW (2021b) Earth's energy imbalance from the ocean perspective (2005–2019). Geophys Res Lett 48(16):e2021GL93624. https://doi.org/10.1029/2021GL093624

Hakuba MZ, Stephens GL, Christophe B, Nash AE, Foulon B, Bettadpur SV, Tapley BD, Webb FH (2019) Earth's energy imbalance measured from space. Ieee T Geosci Remote 57(1):32–45. https://doi.org/10.1109/Tgrs.2018.2851976

Hatzianastassiou N, Matsoukas C, Hatzidimitriou D, Pavlakis C, Drakakis M, Vardavas I (2004) Ten year radiation budget of the earth: 1984–93. Int J Climatol 24(14):1785–1802. https://doi.org/10.1002/joc.1110

Hersbach H, Bell B, Berrisford P, Hirahara S, Horányi A, Muñoz-Sabater J, Nicolas J, Peubey C, Radu R, Schepers D, Simmons A, Soci C, Abdalla S, Abellan X, Balsamo G, Bechtold P, Biavati G, Bidlot J, Bonavita M, De Chiara G, Dahlgren P, Dee D, Diamantakis M, Dragani R, Flemming J, Forbes R, Fuentes M, Geer A, Haimberger L, Healy S, Hogan RJ, Hólm E, Janisková M, Keeley S, Laloyaux P, Lopez P, Lupu C, Radnoti G, de Rosnay P, Rozum I, Vamborg F, Villaume S, Thépaut JN (2020) The era5 global reanalysis. Q J R Meteorol Soc 146(730):1999–2049. https://doi.org/10.1002/qj.3803

Hersbach H, Peubey C, Simmons A, Berrisford P, Poli P, Dee D (2015) Era-20cm: a twentieth-century atmospheric model ensemble. Q J R Meteorol Soc 141(691):2350–2375. https://doi.org/10.1002/qj.2528

Hodnebrog O, Myhre G, Jouan C, Andrews T, Forster PM, Jia HL, Loeb NG, Olivie DJL, Paynter D, Quaas J, Raghuraman SP, Schulz M (2024) Recent reductions in aerosol emissions have increased earth's energy imbalance. Commun Earth Environ 5(1):166. https://doi.org/10.1038/s43247-024-01324-8

Hourdin F, Mauritsen T, Gettelman A, Golaz JC, Balaji V, Duan QY, Folini D, Ji DY, Klocke D, Qian Y, Rauser F, Rio C, Tomassini L, Watanabe M, Williamson D (2017) The art and science of climate model tuning. Bull Am Meteor Soc 98(3):589–602. https://doi.org/10.1175/Bams-D-15-00135.1

Iacono MJ, Mlawer EJ, Clough SA, Morcrette JJ (2000) Impact of an improved longwave radiation model, rrtm, on the energy budget and thermodynamic properties of the ncar community climate model, ccm3. J Gerontol Ser A Biol Med Sci 105(D11):14873–14890

Kanamitsu M, Ebisuzaki W, Woollen J, Yang SK, Hnilo JJ, Fiorino M, Potter GL (2002) Ncep-doe amip-ii reanalysis (r-2). Bull Am Meteor Soc 83(11):1631–1643. https://doi.org/10.1175/Bams-83-11-1631(2002)083%3c1631:Nar%3e2.3.Co;2

Kato S, Rose FG, Rutan DA, Thorsen TJ, Loeb NG, Doelling DR, Huang X, Smith WL, Su WY (2018a) Surface irradiances of edition 40 clouds and the earth's radiant energy system (ceres) energy balanced and filled (ebaf) data product. J Clim 31(11):4501–4527

Kiehl JT, Trenberth KE (1997) Earth's annual global mean energy budget. Bull Am Meteor Soc 78(2):197–208

Kobayashi S, Ota Y, Harada Y, Ebita A, Moriya M, Onoda H, Onogi K, Kamahori H, Kobayashi C, Endo H, Miyaoka K, Takahashi K (2015) The jra-55 reanalysis: general specifications and basic characteristics. J Meteorol Soc Jpn 93(1):5–48. https://doi.org/10.2151/jmsj.2015-001

Kopp G, Lean JL (2011) A new, lower value of total solar irradiance: evidence and climate significance. Geophys Res Lett 38:L01706. https://doi.org/10.1029/2010gl045777

Kosaka Y, Kobayashi S, Harada Y, Kobayashi C, Naoe H, Yoshimoto K, Harada M, Goto N, Chiba J, Miyaoka K, Sekiguchi R, Deushi M, Kamahori H, Nakaegawa T, Tanaka TY, Tokuhiro T, Sato Y, Matsushita Y, Onogi K (2024) The jra-3q reanalysis. J Meteorol Soc Jpn 102(1):49–109. https://doi.org/10.2151/jmsj.2024-004

L'Ecuyer TS, Beaudoing HK, Rodell M, Olson W, Lin B, Kato S, Clayson CA, Wood E, Sheffield J, Adler R, Huffman G, Bosilovich M, Gu G, Robertson F, Houser PR, Chambers D, Famiglietti JS, Fetzer E, Liu WT, Gao X, Schlosser CA, Clark E, Lettenmaier DP, Hilburn K (2015) The observed state of the energy budget in the early twenty-first century. J Clim 28(21):8319–8346. https://doi.org/10.1175/Jcli-D-14-00556.1

Liu CL, Allan RP, Mayer M, Hyder P, Desbruyères D, Cheng LJ, Xu JJ, Xu F, Zhang Y (2020) Variability in the global energy budget and transports 1985–2017. Clim Dyn 55(11–12):3381–3396. https://doi.org/10.1007/s00382-020-05451-8

Liu CL, Allan RP, Mayer M, Hyder P, Loeb NG, Roberts CD, Valdivieso M, Edwards JM, Vidale PL (2017) Evaluation of satellite and reanalysis-based global net surface energy flux and uncertainty estimates. J Gerontol Ser A Biol Med Sci 122(12):6250–6272. https://doi.org/10.1002/2017jd026616

Loeb NG, Doelling DR, Wang HL, Su WY, Nguyen C, Corbett JG, Liang LS, Mitrescu C, Rose FG, Kato S (2018b) Clouds and the earth's radiant energy system (ceres) energy balanced and filled (ebaf) top-of-atmosphere (toa) edition-4.0 data product. J Clim 31(2):895–918. https://doi.org/10.1175/Jcli-D-17-0208.1

Loeb NG, Johnson GC, Thorsen TJ, Lyman JM, Rose FG, Kato S (2021c) Satellite and ocean data reveal marked increase in earth's heating rate. Geophys Res Lett 48(13):e2021GL093047. https://doi.org/10.1029/2021GL093047

Loeb NG, Mayer M, Kato S, Fasullo JT, Zuo H, Senan R, Lyman JM, Johnson GC, Balmaseda M (2022a) Evaluating twenty-year trends in earth's energy flows from observations and reanalyses. J Geophys Res-Atmos 127(12):e036686. https://doi.org/10.1029/2022JD036686

Morcrette JJ (2002) Assessment of the ecmwf model cloudiness and surface radiation fields at the arm sgp site. Mon Weather Rev 130(2):257–277

Paynter D, Ramaswamy V (2012) Variations in water vapor continuum radiative transfer with atmospheric conditions. J Geophys Res Atmos 117:D16310. https://doi.org/10.1029/2012jd017504

Paynter D, Ramaswamy V (2014) Investigating the impact of the shortwave water vapor continuum upon climate simulations using gfdl global models. J Gerontol Ser A Biol Med Sci 119(18):10720–10737. https://doi.org/10.1002/2014jd021881

Paynter DJ, Ramaswamy V (2011) An assessment of recent water vapor continuum measurements upon longwave and shortwave radiative transfer. J Geophys Res Atmos 116:D20302. https://doi.org/10.1029/2010jd015505

Pendergrass AG (2020) The global-mean precipitation response to co-induced warming in cmip6 models. Geophys Res Lett 47(17):e2020GL089964. https://doi.org/10.1029/2020GL089964

Pincus R, Mlawer EJ, Oreopoulos L, Ackerman AS, Baek S, Brath M, Buehler SA, Cady-Pereira KE, Cole JNS, Dufresne JL, Kelley M, Li JN, Manners J, Paynter DJ, Roehrig R, Sekiguchi M, Schwarzkopf DM (2015) Radiative flux and forcing parameterization error in aerosol-free clear skies. Geophys Res Lett 42(13):5485–5492. https://doi.org/10.1002/2015gl064291

Poli P, Hersbach H, Dee DP, Berrisford P, Simmons AJ, Vitart F, Laloyaux P, Tan DGH, Peubey C, Thépaut JN, Trémolet Y, Hólm EV, Bonavita M, Isaksen L, Fisher M (2016) Era-20c: An atmospheric reanalysis of the twentieth century. J Clim 29(11):4083–4097. https://doi.org/10.1175/Jcli-D-15-0556.1

Radel G, Shine KP, Ptashnik IV (2015) Global radiative and climate effect of the water vapour continuum at visible and near-infrared wavelengths. Q J R Meteorol Soc 141(688):727–738. https://doi.org/10.1002/qj.2385

Rienecker MM, Suarez MJ, Gelaro R, Todling R, Bacmeister J, Liu E, Bosilovich MG, Schubert SD, Takacs L, Kim GK, Bloom S, Chen JY, Collins D, Conaty A, Da Silva A, Gu W, Joiner J, Koster RD, Lucchesi R, Molod A, Owens T, Pawson S, Pegion P, Redder CR, Reichle R, Robertson FR, Ruddick AG, Sienkiewicz M, Woollen J (2011) Merra: Nasa's modern-era retrospective analysis for research and applications. J Clim 24(14):3624–3648. https://doi.org/10.1175/Jcli-D-11-00015.1

Roberts JB, Robertson FR, Clayson CA, Bosilovich MG (2012) Characterization of turbulent latent and sensible heat flux exchange between the atmosphere and ocean in merra. J Clim 25(3):821–838. https://doi.org/10.1175/Jcli-D-11-00029.1

Schmidt GA, Andrews T, Bauer SE, Durack PJ, Loeb NG, Ramaswamy V, Arnold NP, Bosilovich MG, Cole J, Horowitz LW, Johnson GC, Lyman JM, Medeiros B, Michibata T, Olonscheck D, Paynter D, Raghuraman SP, Schulz M, Takasuka D, Tallapragada V, Taylor PC, Ziehn T (2023) Ceresmip:

a climate modeling protocol to investigate recent trends in the earth's energy imbalance. Front Clim 5:1298599. https://doi.org/10.3389/fclim.2023.1298599

Slivinski LC, Compo GP, Whitaker JS, Sardeshmukh PD, Giese BS, McColl C, Allan R, Yin XG, Vose R, Titchner H, Kennedy J, Spencer LJ, Ashcroft L, Bronnimann S, Brunet M, Camuffo D, Cornes R, Cram TA, Crouthamel R, Dominguez-Castro F, Freeman JE, Gergis J, Hawkins E, Jones PD, Jourdain S, Kaplan A, Kubota H, Le Blancq F, Lee TC, Lorrey A, Luterbacher J, Maugeri M, Mock CJ, Moore GWK, Przybylak R, Pudmenzky C, Reason C, Slonosky VC, Smith CA, Tinz B, Trewin B, Valente MA, Wang XL, Wilkinson C, Wood K, Wyszynski P (2019) Towards a more reliable historical reanalysis: Improvements for version 3 of the twentieth century reanalysis system. Q J R Meteorol Soc 145(724):2876–2908. https://doi.org/10.1002/qj.3598

Stamatis M, Hatzianastassiou N, Korras-Carraca MB, Matsoukas C, Wild M, Vardavas I (2022b) Inter-decadal changes of the merra-2 incoming surface solar radiation (ssr) and evaluation against geba & bsrn stations. Appl Sci-Basel 12(19):10176. https://doi.org/10.3390/app121910176

Stephens GL, Li JL, Wild M, Clayson CA, Loeb N, Kato S, L'Ecuyer T, Stackhouse PW, Lebsock M, Andrews T (2012) An update on earth's energy balance in light of the latest global observations. Nat Geosci 5(10):691–696. https://doi.org/10.1038/Ngeo1580

Takacs LL, Suárez MJ, Todling R (2016) Maintaining atmospheric mass and water balance in reanalyses. Q J R Meteorol Soc 142(697):1565–1573. https://doi.org/10.1002/qj.2763

Trenberth KE, Fasullo JT, Kiehl J (2009) Earth's global energy budget. Bull Am Meteor Soc 90(3):311. https://doi.org/10.1175/2008bams2634.1

Trenberth KE, Solomon A (1994) The global heat-balance - heat transports in the atmosphere and ocean. Clim Dyn 10(3):107–134. https://doi.org/10.1007/s003820050039

von Schuckmann K, Minière A, Gues F, Cuesta-Valero FJ, Kirchengast G, Adusumilli S, Straneo F, Ablain M, Allan RP, Barker PM, Beltrami H, Blazquez A, Boyer T, Cheng LJ, Church J, Desbruyeres D, Dolman H, Domingues CM, García-García A, Giglio D, Gilson JE, Gorfer M, Haimberger L, Hakuba MZ, Hendricks S, Hosoda S, Johnson GC, Killick R, King B, Kolodziejczyk N, Korosov A, Krinner G, Kuusela M, Landerer FW, Langer M, Lavergne T, Lawrence I, Li YH, Lyman J, Marti F, Marzeion B, Mayer M, MacDougall AH, McDougall T, Monselesan DP, Nitzbon J, Otosaka I, Peng J, Purkey S, Roemmich D, Sato K, Sato K, Savita A, Schweiger A, Shepherd A, Seneviratne SI, Simons L, Slater DA, Slater T, Steiner AK, Suga T, Szekely T, Thiery W, Timmermans ML, Vanderkelen I, Wjiffels SE, Wu TH, Zemp M (2023) Heat stored in the earth system 1960–2020: Where does the energy go? Earth Syst Sci Data 15(4):1675–1709. https://doi.org/10.5194/essd-15-1675-2023

von Schuckmann K, Palmer MD, Trenberth KE, Cazenave A, Chambers D, Champollion N, Hansen J, Josey SA, Loeb N, Mathieu PP, Meyssignac B, Wild M (2016) An imperative to monitor earth's energy imbalance. Nat Clim Change 6(2):138–144. https://doi.org/10.1038/Nclimate2876

Wang QY, Zhang H, Yang S, Chen Q, Zhou XX, Xie B, Wang YY, Shi GY, Wild M (2022) An assessment of land energy balance over east asia from multiple lines of evidence and the roles of the tibet plateau, aerosols, and clouds. Atmos Chem Phys 22(24):15867–15886. https://doi.org/10.5194/acp-22-15867-2022

Wild M (2008) Short-wave and long-wave surface radiation budgets in gcms: a review based on the ipcc-ar4/cmip3 models. Tellus A 60(5):932–945. https://doi.org/10.1111/J.1600-0870.2008.00342.X

Wild M (2017) Towards global estimates of the surface energy budget. Curr Clim Change Rep 3(1):87–97. https://doi.org/10.1007/s40641-017-0058-x

Wild M (2020) The global energy balance as represented in cmip6 climate models. Clim Dyn 55(3–4):553–577. https://doi.org/10.1007/s00382-020-05282-7

Wild M, Folini D, Hakuba MZ, Schar C, Seneviratne SI, Kato S, Rutan D, Ammann C, Wood EF, Konig-Langlo G (2015) The energy balance over land and oceans: an assessment based on direct observations and cmip5 climate models. Clim Dyn 44(11–12):3393–3429. https://doi.org/10.1007/s00382-014-2430-z

Wild M, Folini D, Schar C, Loeb N, Dutton EG, Konig-Langlo G (2013) The global energy balance from a surface perspective. Clim Dyn 40(11–12):3107–3134. https://doi.org/10.1007/s00382-012-1569-8

Wild M, Hakuba MZ, Folini D, Dörig-Ott P, Schär C, Kato S, Long CN (2019) The cloud-free global energy balance and inferred cloud radiative effects: an assessment based on direct observations and climate models. Clim Dyn 52:4787–4812. https://doi.org/10.1007/s00382-018-4413-y

Wild M, Ohmura A, Gilgen H, Morcrette JJ, Slingo A (2001) Evaluation of downward longwave radiation in general circulation models. J Clim 14(15):3227–3239. https://doi.org/10.1175/1520-0442(2001)014%3c3227:Eodlri%3e2.0.Co;2

Wild M, Ohmura A, Gilgen H, Roeckner E (1995) Validation of general-circulation model radiative fluxes using surface observations. J Clim 8(5):1309–1324

Wild M, Ohmura A, Gilgen H, Roeckner E, Giorgetta M, Morcrette JJ (1998) The disposition of radiative energy in the global climate system: Gcm-calculated versus observational estimates. Clim Dyn 14(12):853–869

Yu LS (2019) Global air-sea fluxes of heat, fresh water, and momentum: Energy budget closure and unanswered questions. Annu Rev Mar Sci 11:227–248. https://doi.org/10.1146/annurev-marine-010816-060704

Publisher's Note Springer Nature remains neutral with regard to jurisdictional claims in published maps and institutional affiliations.

Assessment of Atmospheric and Surface Energy Budgets Using Observation-Based Data Products

Michael Mayer[1,2] · Seiji Kato[3] · Michael Bosilovich[4] · Peter Bechtold[1] · Johannes Mayer[2] · Marc Schröder[5] · Ali Behrangi[6] · Martin Wild[7] · Shinya Kobayashi[8] · Zhujun Li[3] · Tristan L'Ecuyer[9]

Received: 20 October 2023 / Accepted: 7 February 2024 / Published online: 17 April 2024
© The Author(s) 2024, corrected publication 2024

Abstract

Accurate diagnosis of regional atmospheric and surface energy budgets is critical for understanding the spatial distribution of heat uptake associated with the Earth's energy imbalance (EEI). This contribution discusses frameworks and methods for consistent evaluation of key quantities of those budgets using observationally constrained data sets. It thereby touches upon assumptions made in data products which have implications for these evaluations. We evaluate 2001–2020 average regional total (TE) and dry static energy (DSE) budgets using satellite-based and reanalysis data. For the first time, a consistent framework is applied to the ensemble of the 5th generation European Reanalysis (ERA5), version 2 of modern-era retrospective analysis for research and applications (MERRA-2), and the Japanese 55-year Reanalysis (JRA55). Uncertainties of the computed budgets are assessed through inter-product spread and evaluation of physical constraints. Furthermore, we use the TE budget to infer fields of net surface energy flux. Results indicate biases < 1 W/m^2 on the global, < 5 W/m^2 on the continental, and ~ 15 W/m^2 on the regional scale. Inferred net surface energy fluxes exhibit reduced large-scale biases compared to surface flux data based on remote sensing and models. We use the DSE budget to infer atmospheric diabatic heating from condensational processes. Comparison to observation-based precipitation data indicates larger uncertainties (10–15 Wm^{-2} globally) in the DSE budget compared to the TE budget, which is reflected by increased spread in reanalysis-based fields. Continued validation efforts of atmospheric energy budgets are needed to document progress in new and upcoming observational products, and to understand their limitations when performing EEI research.

Keywords Energy budget · Earth's energy imbalance · Surface energy flux · Reanalysis · Remote sensing

---

Extended author information available on the last page of the article

Article Highlights

- Consistent frameworks to evaluate atmospheric energy budgets from observationally constrained data sets are presented
- Total and dry static energy budgets are used to infer net surface energy flux and diabatic heating from precipitation, respectively
- Uncertainties of inferred net surface energy flux are demonstrably lower than those based on remote sensing products

1 Introduction

The atmosphere redistributes petawatts of energy globally, with annual mean zonally integrated poleward transports peaking over 4 petawatts in both hemispheres (e.g., Mayer et al. 2021). It thereby accomplishes a large fraction of planetary heat transport from the tropics to the poles (e.g., Trenberth et al. 2019), as required by latitudinal changes in net energy input at the top-of-atmosphere (TOA) associated with Earth's spherical geometry (Peixoto and Oort 1992). Furthermore, regional patterns of the divergence of atmospheric total energy transports are related to net surface energy fluxes into and out of the ocean: Climatological convergence (divergence) tends to occur over regions of enhanced ocean heat uptake (release), such as the equatorial cold tongues (Western Boundary Currents) (Trenberth and Stepaniak 2003). In the context of Earth's energy imbalance (EEI), divergent atmospheric energy transports determine how spatial variations and trends of net radiation at the TOA are redistributed, and thereby how heat input at the ocean surface is modulated. The atmospheric divergence is thus a critical quantity for understanding regional patterns of ocean heat uptake. Indeed, decadal changes in atmospheric circulation and/or gradients lead to pronounced decadal variations in divergent atmospheric energy transport and largely determine spatial patterns of decadal trends in net surface energy flux (Loeb et al. 2022).

As opposed to TOA, where high-quality observations exist, most notably products from the Clouds and the Earth's Radiant Energy System (CERES) program (Loeb et al. 2018), there do not exist in situ observations of atmospheric energy transports on a global scale, and direct measurements of air–sea fluxes are scarce. Atmospheric energy transports are typically diagnosed from reanalyses, which comes with several caveats regarding methods and data (Trenberth 1991; Mayer et al. 2017; Trenberth and Fasullo 2018; Kato et al. 2021). In the absence of direct measurements, e.g., through eddy covariance methods (which themselves exhibit large uncertainties, see Mauder et al. 2020 for a recent review), air–sea fluxes can be estimated using bulk formulae, using meteorological input data based on models or remotely sensed data, but such estimates exhibit considerable uncertainties, leading to large global imbalances of the obtained fluxes in this way (Yu 2019). An alternative approach infers net surface energy flux from the atmospheric energy budget, which has some advantages compared to the "direct" approach and has been followed in numerous studies (e.g., Trenberth and Fasullo 2017; Cheng et al. 2019; Mayer et al. 2022).

The focus of this paper is twofold. We first discuss in detail the practical evaluation of atmospheric total and dry static energy budgets from observationally constrained products. For this, we begin from a complete formulation of the total atmospheric energy budget, derivation of which has been presented elsewhere (e.g., Bannon 2002; Lauritzen et al. 2018; Kato et al 2021; Lauritzen et al. 2022). We then discuss simplifications typically

made in models underlying reanalysis products to provide a framework for practical evaluation of atmospheric energy budgets and discuss pragmatic choices made in the data products and the diagnostic framework. This includes discussion of technical aspects such as mass corrections. The formulation of diagnostic atmospheric energy budgets has received some attention in recent years, i.e., because data quality has become sufficient to reveal biases in previously employed frameworks (Mayer et al. 2017). In this context, we reconcile apparent discrepancies between budget formulations presented in recent works (Mayer et al. 2017; Trenberth and Fasullo 2018; Kato et al. 2021). The second aim of this paper is to provide an overview of uncertainty of the diagnosed quantities and flux estimates using state-of-the-art data products. To this end, we rely on (i) inter-comparison of data products, (ii) assessment of physical constraints, and (iii) employing the budget to infer a quantity which can be compared against an observational product (e.g., comparison of diabatic heating from precipitation against an observational precipitation product).

Section 2 provides an overview of employed data products and considerations regarding the presented diagnostics. Section 3 discusses atmospheric budgets, including the atmospheric budget and its relevance for energy budget diagnostics (3.1), the total energy budget (3.2), and the dry static energy budget (3.3). Section 4 covers the surface energy budget, including an assessment of physical constraints of the total energy budget. Conclusions, recommendations, and an outlook are provided in Sect. 5.

2 Data, Study Period, Diagnostics

We employ a vertically integrated framework to diagnose atmospheric budgets. The vertical coordinate is thus eliminated. Energy fluxes discussed in this study are, therefore, radiative flux at the TOA, horizontal transports and storage within the atmosphere, and fluxes at the surface.

We use TOA fluxes from the CERES Energy-Balanced and Filled product in Edition 4.2 (CERES-EBAF-TOA; Loeb et al. 2018). They are tuned (with a one-time global adjustment) to match the global net TOA flux averaged from July 2005 through June 2015 with an observational estimate of the sum of the rate of global ocean heating, ice heating and melt, and atmospheric and lithospheric heating averaged over this period.

Transports within the atmosphere (and their divergence) of total energy (precisely: moist static plus kinetic) and dry static energy as well as storage rates of total and dry static energy are obtained from three atmospheric reanalyses ERA5 (Hersbach et al. 2020), MERRA-2 (Gelaro et al. 2017), and JRA55 (Kobayashi et al. 2015). They differ in many aspects, including data assimilation system, resolution, boundary conditions, and treatment of observations. Please refer to the corresponding references for details. The three products are of different vintage, with ERA5 being the most recent of the group, but successors of the others are underway or currently being produced (e.g., JRA-3Q; Kosaka et al. 2024). The required quantities (winds, pressure, and thermodynamic fields) rely on analyzed state quantities from reanalyses, i.e., they are strongly constrained by observations. The evaluation of ERA5 and JRA55 data is performed using data on the model native vertical and horizontal grid at maximum available temporal resolution (1-hourly and 6-hourly, respectively) to avoid interpolation errors. The divergence fields are mass-adjusted as described in Sect. 3.2.4. Native grid 3D fields at high temporal resolution that are required for the presented budget diagnostics were not available from MERRA-2. Yet, the

MERRA-2 archive provides unadjusted vertically integrated monthly mean divergences, which are computed on the native grid at each model time step and temporally accumulated (Global Modeling and Assimilation Office (GMAO) 2015a). Methods for mass adjustments are discussed in Sect. 3.2.4.

Budgets derived from some reanalyses suffer from spectral noise of the divergence term, and typically useful resolution is limited to around 2.5° (e.g., Trenberth and Stepaniak 2003). However, progress has been made in recent years to reduce spectral noise in budgets derived from ERA5, but this requires advanced numerical methods (see Mayer et al. 2021 for computational details), allowing for an increase of useful resolution to about 1°. MERRA-2 budgets do not suffer from these problems thanks to the finite volume dynamical core of the underlying model and computation of the divergence at each model time step (see the above paragraph), and in principle, no truncation of the MERRA-2 results is necessary.

Diabatic heating from condensation processes is a crucial element of the dry static energy budget. In this context, we use the latest version (V3.2) of the Global Precipitation Climatology Project (GPCP) product (Huffman et al. 2023).

Surface energy fluxes are obtained from different types of data sets. We use satellite-based data sets of net surface radiation, namely CERES-EBAF surface in Edition 4.2 (CERES-EBAF-sfc; Kato et al. 2018) and CM SAF cLoud, Albedo and surface radiation data set from AVHRR data in version 3 (CLARA-A3; Karlsson et al. 2023), and turbulent fluxes derived from remote sensing data, namely OAflux version 3 (Yu and Weller 2007), Japanese Ocean Flux Data Sets with Use of Remote-Sensing Observations 3 (J-OFURO3) in version 1.1 (Tomita et al. 2019), IFREMER v4.1 (Bentamy et al. 2013), and SeaFlux v3 (Curry et al. 2004). CERES-EBAF-sfc fluxes are constrained by observed TOA fluxes by the method described in Kato et al. (2018).

Reanalyses also provide surface energy fluxes as output from short-term (typically 12-hourly) forecasts of the underlying models. By nature, these fields are less strongly constrained by observations (but through the initial conditions of the forecasts). We employ model-based fluxes from a range of reanalyses with varying degrees of observational data ingestion (see Table 1 for a complete list).

An alternative to satellite products and fluxes from model forecasts is to infer the net surface energy flux from the atmospheric energy budget (storage rates and divergences based on analyses) in combination with observed TOA fluxes, which will be elaborated on in Sect. 4. Long-term means of relevant budget terms are averaged over the standard period 2001–2020. This is the twenty-year period starting with the first full year of CERES data. This also coincides with a time when atmospheric reanalyses are relatively stable temporally. Some of the employed satellite products cover a shorter period, which dictates the averaging period for the respective diagnostics.

Given the comprehensive discussion of methods, we limit diagnostics to multi-year average fields (2001–2020 averages wherever data availability allows). While computations such as divergences are performed on the respective native grids (see above), all results and data products are interpolated to a common $1 \times 1°$ horizontal grid to facilitate inter-comparison. We consider long-term mean global averages of fields to check physical constraints. Unlike reanalyses, some of the employed satellite products have no full global coverage. For example, the satellite-based turbulent flux data have no data over land and in sea ice covered regions. For consistent inter-comparisons of these products with other estimates, we mask out grid points, where any of the involved products does not contain at least one valid value for each calendar month. Moreover, unbiased long-term means at grid points with temporally varying data coverage are ensured by first computing a mean annual

Table 1 Overview of employed data products, their type, reference papers, and the evaluated budget terms

Product	Type	References	Evaluated terms
CERES-EBAF-TOA ed4.2	Satellite, with global mean adjustment	Loeb et al. (2018)	Rad_{TOA}
ERA5	Reanalysis	Hersbach et al. (2020)	$\frac{\partial}{\partial t}AE$, TEDIV, DSEDIV, LH, SH, Rad_S, P_{snow}
JRA55	Reanalysis	Kobayashi et al. (2015)	$\frac{\partial}{\partial t}AE$, TEDIV, DSEDIV, LH, SH, Rad_S, P_{snow}
MERRA-2	Reanalysis	Gelaro et al. (2017), Global Modeling and Assimilation Office (GMAO) (2015a, b, c)	$\frac{\partial}{\partial t}AE$, TEDIV, DSEDIV, LH, SH, Rad_S, P_{snow}
GPCP V3.2	Satellite plus rain gauge	Huffman et al. (2023)	P
IFREMER	Satellite	Bentamy et al. (2013)	LH, SH
OAflux	Satellite	Yu and Weller (2007)	LH, SH
J-OFURO	Satellite	Tomita et al. (2019)	LH, SH
SeaFlux	Satellite	Curry et al. (2004)	LH, SH
CERES-EBAF-sfc ed4.2	Satellite plus radiative transfer model	Kato et al. (2018)	Rad_S
CLARA-A3	Satellite plus radiative transfer model	Karlsson et al. (2023)	Rad_S
MERRA	Reanalysis	Rienecker et al. (2011)	LH, SH, Rad_S, P_{snow}
MERRA-2-AMIP	Reanalysis (no data assimilation)	Collow et al. (2017)	LH, SH, Rad_S, P_{snow}
NCEP R2	Reanalysis	Kanamitsu et al. (2002)	LH, SH, Rad_S, P_{snow}
20CRv3	Reanalysis (surface data only)	Slivinski et al. (2019)	LH, SH, Rad_S, P_{snow}
ERA20C	Reanalysis (surface data only)	Poli et al. (2016)	LH, SH, Rad_S, P_{snow}
ERA20CM	Reanalysis (no data assimilation)	Hersbach et al. (2015)	LH, SH, Rad_S, P_{snow}

cycle (using the available data for each calendar month) and subsequently long-term means as an average of the monthly values. This makes sure that no calendar month is over- or underrepresented. The same spatiotemporal mask is applied to all products used for the respective diagnostics.

3 Atmospheric Energy Budgets

3.1 Atmospheric Mass Budget

We begin with a short discussion of the atmospheric moisture and mass budget and its relevance for the energy budget. The mass budget of the atmosphere including water in all states reads as follows:

$$P + E + \frac{1}{g}\frac{\partial}{\partial t}p_s = -\nabla \cdot \frac{1}{g}\int_0^{p_s} \bar{\mathbf{v}} dp \qquad (1)$$

Precipitation P and evaporation E represent surface fluxes of all species of water and are defined positive downward. Surface pressure p_s is proportional to the mass of the moist air column, scaled with gravitational acceleration g. The horizontal wind $\bar{\mathbf{v}}$ denotes barycentric velocity and represents the horizontal mass flux of moist air (including dry air and all species of water). Further degrees of freedom can be introduced by allowing for different velocities of dry air and water species (Kato et al. 2021). However, when the mass-weighted moist air velocity is considered it can be assumed to be the velocity of dry air because the mixing ratio is of the order of or less than 10^{-2} (Bannon 2002). In addition, observational products and reanalyses typically do not provide velocity of hydrometeors. The model and assimilation system underlying MERRA-2, satisfies Eq. (1) and thus conserves atmospheric dry mass (Takacs et al. 2016, see their Fig. 5).

An important approximation in the Integrated Forecasting System (IFS), the model underlying ERA5, is that surface precipitation and evaporation do not change the total mass of the atmosphere as a change in water mass is implicitly replaced by an equivalent change in dry air mass (Malardel et al. 2019). As a result, lateral convergence and divergence of moisture is balanced by unphysical divergence and convergence of dry air, respectively, instead of $P+E$. Consequently, surface pressure does not vary in response to moisture changes in the above column. The moist continuity equation in the IFS hence reads as follows (ECMWF 2021a):

$$\frac{1}{g}\frac{\partial}{\partial t}p_{s,\text{IFS}} = -\nabla \cdot \frac{1}{g}\int_0^{p_s} \bar{\mathbf{v}} dp \qquad (2)$$

The divergence term in Eq. (2) thus contains contributions from physical dry air divergence (which can change surface pressure) but no contributions from moisture flux divergence. Hence, the long-term mean of lateral mass flux divergence from ERA5 does not show the signature of $P+E$ (Mayer et al. 2021; their Fig. 2a), as one would expect from Eq. (1) and physical conception. The JMA spectral model makes similar approximations as the IFS and hence satisfies Eq. (2) as well (JMA 2007). This has implications for mass corrections typically applied for energy budget diagnostics, which will be discussed in Sect. 3.2.4.

3.2 Atmospheric Total Energy Budget

3.2.1 Complete Formulation

We use the total atmospheric energy budget equation using liquid water at 0 °C as a reference state from Lauritzen et al. (2022; their Eq. 12) as a starting point, but write it in vertical pressure coordinates, use specific moisture quantities instead of mixing ratios, and include lateral transports (i.e., a local instead of a globally integrated formulation):

$$\text{Rad}_{\text{TOA}} - \text{Rad}_{\text{S}} - \text{SH} - \text{LH} - L_{\text{f}}(T_{\text{p}})P_{\text{snow}} - P|h_{00} + c_{\text{l}}(T_{\text{p}} - T_{00}) + \phi_{\text{s}} + k_{\text{s}}| - E|h_{00} + c_{\text{l}}(T_{\text{s}} - T_{00}) + \phi_{\text{s}} + k_{\text{s}}|$$
$$= \frac{\partial}{\partial t} \frac{1}{g} \int_0^{p_s} ((1 - q_{\text{v}} - q_{\text{l}} - q_{\text{f}})c_{\text{a}}(T_a - T_{00}) + (q_{\text{v}} + q_{\text{l}} + q_{\text{f}})c_{\text{l}}(T_a - T_{00}) + L_{\text{v}}(T_a)q_{\text{v}} + L_{\text{f}}(T_a)q_{\text{f}} + (q_{\text{v}} + q_{\text{l}} + q_{\text{f}})h_{00} + \phi_{\text{s}} + k) \vec{v} \, dp$$
$$+ \nabla \cdot \frac{1}{g} \int_0^{p_s} ((1 - q_{\text{v}} - q_{\text{l}} - q_{\text{f}})c_{\text{a}}(T_a - T_{00}) + (q_{\text{v}} + q_{\text{l}} + q_{\text{f}})c_{\text{l}}(T_a - T_{00}) + L_{\text{v}}(T_a)q_{\text{v}} + L_{\text{f}}(T_a)q_{\text{f}} + (q_{\text{v}} + q_{\text{l}} + q_{\text{f}})h_{00} + \phi + k) \vec{v} \, dp$$

(3)

The left-hand side (lhs) of Eq. (3) contains the vertical fluxes: net TOA radiative flux Rad_{TOA}, net surface radiative flux Rad_{S}, sensible heat flux SH, latent heat flux LH (computed as $L_{\text{v}}(T_{00})E$, with T_{00} being the reference temperature), and latent heat flux associated with snowfall P_{snow}. The lhs also contains non-latent contributions of P (F_{P}) and E (F_{E}) to the energy budget, which include kinetic energy, surface geopotential ϕ_{s}, and enthalpy. Evaporation occurs at skin temperature T_{s}, and for temperature of precipitation (T_{P}) near-surface wet bulb temperature is deemed a good approximation (Gosnell et al. 1995). Figure 6 of Mayer et al. (2017) shows F_{P} and Fig. 3 of Kato et al. (2021) shows regional net enthalpy flux associated water mass exchanges at the surface, i.e., $F_{\text{E}} + F_{\text{P}}$. However, we will not evaluate these fluxes here as they are typically not provided by models or reanalyses and there are ambiguities related to the choice of reference temperature due to typically nonzero mass flux associated with P and E, which is reflected by the reference enthalpy term h_{00} (discussed below). However, as will be discussed later, F_{E} and F_{P} are real physical terms, and their energetic effect on other budget terms may be implicitly included in observational products.

The right-hand side (rhs) contains storage rate of total atmospheric energy ($\frac{\partial}{\partial t}\text{AE}$) and divergence of moist static plus kinetic energy (TEDIV), both including enthalpy, geopotential, kinetic energy, and latent heat. Note that water in all states (gaseous g, liquid l, solid s) is considered, as well as latent heat of vapor and snow/ice. The acronyms that are not explicitly mentioned have their standard meaning, and a list of used symbols and acronyms is provided in Table 2.

Equation (3) is almost identical to Mayer et al. (2017; their Eq. 19) but has surface geopotential and kinetic energy flux associated with precipitation and evaporation included. This equation is also consistent with equation D5 in (Kato et al. 2021) except that their formulation additionally allows for differing velocities of dry air and water particles (which we do not account for; see Sect. 3.1).

Equation (3) differs from the formulation of Trenberth and Fasullo (2018) as it only requires the energetic state of P and E at the surface, while theirs requires the full vertical profile of temperature at which condensation and evaporation occurs. This difference arises from the fact that Trenberth and Fasullo (2018) only include dry air and water vapor but not liquid water in their energy budget equations, i.e., water leaves the column where it condensates rather than when it hits the surface as rain (as is the case in Eq. 1 of this paper). The approach in Trenberth and Fasullo (2018) complicates the evaluation,

Table 2 List of acronyms

AE	Atmospheric total energy
C	Celsius
c_a	Isobaric specific heat capacity of dry air
c_l	Specific heat capacity of liquid water
$\dot{C}_{vl}$	Condensation rate from vapor to liquid cloud particles
$\dot{C}_{vi}$	Condensation rate from vapor to ice cloud particles
$\dot{C}_{li}$	Freezing rate from liquid to ice cloud particles
DSEDIV	Divergence of dry static plus kinetic energy transport
E	Evaporation rate at the surface (positive downward)
F_e	Surface enthalpy flux associated with evaporation
F_S	Net surface energy flux (latent plus sensible heat flux plus net radiation plus energetic effect of snowfall)
g	Gravitational acceleration
h_{00}	Reference enthalpy of water
K	Kelvin
k	Atmospheric kinetic energy
k_s	Atmospheric kinetic energy at the surface
LH	Latent heat flux
L_f	Latent heat of fusion
L_s	Latent heat of sublimation
L_v	Latent heat of vaporization
P	Total precipitation (sum of rain and snow; positive downward)
P_{rain}	Rain rate
P_{snow}	Snowfall rate
p	Atmospheric pressure
$\dot{P}_{vr}$	Column-integrated net conversion rate from vapor to rain
$\dot{P}_{vs}$	Column-integrated net conversion rate from vapor to snow
$\dot{P}_{ri}$	Column-integrated net conversion rate from rain to ice
$\dot{P}_{ls}$	Column-integrated net conversion rate from liquid particles to snow
$\dot{P}_{rs}$	Column-integrated net conversion rate from rain to snow
q_v	Specific vapor content
q_l	Specific liquid water content
q_f	Specific frozen water (snow, ice) content
Rad_S	Net radiation at the surface
Rad_{TOA}	Net radiation at the top-of-atmosphere
SH	Sensible heat flux
SST	Sea surface temperature
T	Temperature
T_a	Atmospheric temperature
T_p	Temperature of precipitation
T_S	Skin temperature
T_{00}	Reference temperature
TEDIV	Divergence of moist static plus kinetic energy transport
Φ	Geopotential
Φ_s	Surface geopotential
$\vec{v}$	Horizontal wind vector

especially given that there currently do not exist any observational products providing the height where precipitation is formed.

A diagnostic complication of Eq. (3) is that all terms involving mass exchanges or variations (P, E, storage, and divergence term) depend on the chosen reference enthalpy of water (h_{00}) and the chosen reference temperature (T_{00}). For the reference state of water, we make the typical choice to be liquid water at 0 °C and set this state to have zero enthalpy. Any other choice is valid (see Lauritzen et al. 2022 for variants of the total energy equation with different choices for h_{00} that are all equivalent), but for our application this seems convenient and is widely used. The effect of reference temperature on the evaluated terms can readily be seen for F_P and F_E. The terms will be excessively large when choosing K scale. The effect of T_{00} on the storage and divergence terms is similar. This is because, e.g., the divergence term can be decomposed into a gradient term (independent of T_{00}) and a mass divergence term that scales with energy and thus depends on T_{00} (in short form, $\nabla \cdot vT = v\nabla T + T\nabla \cdot v$; see extensive discussion in Mayer et al. 2017), only if the budget is mass-consistent (i.e., P, E, and lateral divergence of mass balance each other), and in the steady state, T_{00} drops out cleanly. Mass consistency can be achieved by mass corrections discussed in Sect. 3.2.4. However, the single terms remain dependent on T_{00}, which is most pronounced for terms representing pure mass exchanges, such as F_P and F_E. The effect of real mass variations arising from non-steady conditions on the magnitude of the diagnosed terms through T_{00} cannot be avoided but minimized by choosing C scale instead of K scale. However, as will be discussed next, simplifications typically made in atmospheric models help to avoid these complications.

3.2.2 Simplifications

Atmospheric reanalyses (as most atmospheric models, see Lauritzen et al. 2022) make several assumptions and simplifications in their energy and mass budgets. All water species are assumed to have the same heat capacity, namely that of dry air, and latent heats are taken as constants with the values at $T_{00}=0$ °C (see Sect. 3.3 for discussion of the introduced error). Hydrometeors can thus be advected but carry the same specific enthalpy as dry air. In addition, moist physics parameterizations in many models including the IFS and JMA model do not modify total mass, which in consequence leads to unphysical sources and sinks of dry air to balance moist mass changes associated with net condensation. Thus, precipitation in these models does not carry energy (the energy is "conserved" in the appearing dry air), and consistent with this, there is no exchange of sensible heat with the environment and no dissipation associated with falling precipitation. However, latent heat associated with falling snow is taken into account. With these assumptions, Eq. (3) simplifies considerably to

$$\text{Rad}_{\text{TOA}} - \text{Rad}_S - \text{LH} - \text{SH} - L_f(T_{00})P_{\text{snow}}$$
$$= \frac{\partial}{\partial t}\frac{1}{g}\int_0^{p_s}\left(c_a(T_a - T_{00}) + L_v(T_{00})q_v + L_f(T_{00})q_f + \phi_s + k\right)\vec{v}dp \quad (4)$$
$$+ \nabla \cdot \frac{1}{g}\int_0^{p_s}\left(c_a(T_a - T_{00}) + L_v(T_{00})q_v + L_f(T_{00})q_f + \phi + k\right)\vec{v}dp$$

This equation is consistent with the total energy budget equation as it is used in the IFS (sum of Eqs. 12.38 and 12.40 in ECMWF 2021b) and the JMA model. MERRA-2 follows a very similar total energy conservation equation as Eq. (4), but uses virtual temperature instead of T_a (Bosilovich et al. 2016). We dropped h_{00} in Eq. (4) as it is assumed zero, but a dependence on T_{00} in storage and divergence terms remains. As discussed above and will be shown below, the impact of the choice of T_{00} becomes small when mass consistency is ensured. It is also important to note that Eq. (4) is valid when all terms are evaluated from the same model-based data product. If the budget is evaluated using a mix of different products where some terms are evaluated using observational products which do not make the same simplifications, this will inevitably introduce inconsistencies.

3.2.3 Energy Conservation in Models and Reanalyses

After introducing the conservation equations of the physical and dynamical models underlying reanalyses, it is of interest to assess how well those are satisfied. Moist physics parametrizations in the IFS conserve total mass and enthalpy with the exception of a globally small $< = 0.1$ W/m^2 enthalpy error due to the handling of the mixed phase in convection. Importantly, the semi-Lagrangian advection scheme is non-conserving and introduces, depending on horizontal resolution, a spurious moisture and enthalpy source. As a result, the ERA5 energy budget exhibits a global mean imbalance of -2.4 W/m^2 (based on 2001–2020 mean global averages of 12-hourly short-term forecasts of vertical fluxes and atmospheric tendencies, where the latter is a proxy for the energetic effect of analysis increments), where the largest contribution (-2.1 W/m^2) stems from the non-closure of the moisture budget. This value is higher than those provided for two resolutions of the IFS (50 and 100 km versus ~31 km of ERA5) in Roberts et al. (2018), which indicates dependence on model resolution and the version of the IFS (Cy43R1 versus Cy41R2 in ERA5).

For consistency with the ERA5 estimate, we estimate the degree of non-closure of the energy budget in MERRA-2 by combining global mean net TOA and surface energy fluxes and total energy increments. The 2001–2020 average is -1.0 W/m^2, which represents the balance between physical tendencies and the tendencies introduces by the data assimilation. However, we note that the energy budget in MERRA-2 has a feature writing terms that exactly balance (Bosilovich et al. 2016). This is accomplished through the use of the Analysis Increment Update data assimilation scheme (Bloom et al. 1996) which acts as a numerical tendency alongside the physical tendencies in the forecast model's budget equations together with an "energy fixer" for numerical dissipation.

We assessed energy conservation of the JMA model by looking at an AMIP-type run (JRA55-AMIP; Kobayashi et al. 2015), where no diagnostic complication from analysis increments arises. Global average of TOA minus net surface fluxes (i.e., the left side of Eq. 4) is 0.4 W/m^2 over 2001–2012, which can be considered the model imbalance when neglecting atmospheric storage. This value is consistent with tests using JRA55 (taking account of the increments). We note that no corrections to improve budget closure have been applied during the production of JRA55-AMIP.

3.2.4 Mass Adjustment

As pointed out in numerous studies (e.g., Savijärvi 1982; Trenberth 1991, 1997; Chiodo and Haimberger 2010), analyzed winds from atmospheric (re)analyses do not satisfy the continuity equation discussed in Sect. 3.1, which has detrimental effects on the diagnosed energy budgets (Fig. 2 will illustrate this). The standard approach is a barotropic mass correction that is applied to the wind field at each time step TEDIV is computed and enforces satisfaction of the moist atmosphere's mass budget (Eq. (1), see, e.g., Trenberth 1991). It is important to stress that the mass correction is applied to make the employed data self-consistent and not to adjust the mass budget towards some best estimate, e.g., of

precipitation. Mayer et al. (2017) have laid out that application of this correction (and thus implicit inclusion of lateral energy fluxes associated with moisture) requires quantification of enthalpy fluxes associated with P and E. They also showed that neglect of those fluxes in conjunction with the typical choice for K temperature scale introduces a bias to the diagnosed fields on the order of 20–30 W/m^2.

However, this approach for mass adjustment contrasts with the implementation of the mass budget in many models including the IFS, where lateral moisture transports do not accomplish a net energy transport except for that of latent heat, and, similarly, P and E do not carry any enthalpy except for latent heat (see Sect. 3.1). From that perspective, it is more in line with the underlying models to adjust the winds to satisfy Eq. (2), as done, e.g., by Chiodo and Haimberger (2010) who neglected P and E during their mass adjustment. We argue that this discrepancy can be reconciled by use of the simplified diagnostic equations proposed by Mayer et al. (2017) which consistently remove moisture enthalpies from lateral and vertical fluxes. This simplification assumes that moisture does not change temperature on its passage through the column and as a result moisture transports do not have a net energetic effect on the column (again, except for the release of latent heat). Hence, it is approximately equivalent to employ energy budget Eq. (4) in conjunction with winds adjusted to satisfy Eq. (2) (i.e., winds balance only surface pressure changes) or a variant of Eq. (4) with moisture enthalpy removed [similar to Eq. (24) in Mayer et al. (2017)] but with winds adjusted to satisfy Eq. (1) (i.e., winds additionally balance $P + E$) of this paper. Indeed, internal consistency of the ERA5 energy budget, i.e., satisfaction of Eq. (4) when computing all terms from ERA5 (not shown), is actually slightly better when using the latter variant. Hence, we will use that approach for ERA5 and JRA55 data for evaluations of TEDIV in the present paper as well. The ERA5-based TEDIV fields shown in this paper are thus the same as available from the Copernicus Data Store (CDS 2021) except that the latter do not account for atmospheric snow and ice.

Given that the MERRA-2 archive does not provide sub-monthly native grid 3D fields (see Sect. 2) needed for the mass adjustment discussed above, we resorted to a simplified mass correction that can be applied to vertically integrated divergences as described in Chiodo and Haimberger (2010). This method only takes into account the effect of the corrected mass flux divergence on TEDIV, but not the effect of modified advection on TEDIV. The latter effect is however small.

3.2.5 Evaluation

Figure 1a shows 2001–2020 average TEDIV from ERA5 following Eq. (4). As discussed in Sect. 3.2.4, winds have been adjusted such that they satisfy Eq. (4), moisture enthalpy has been removed from the transports, and computations were performed in Celsius scale. Main areas of divergence are warm tropical oceans and equatorial land masses, whereas convergence occurs over the equatorial cold tongues, extra tropical continents, and generally at high latitudes. The field in Fig. 1a has been evaluated using numerical methods as used in the IFS (as described in Mayer et al. 2021), which helps to reduce spectral noise and allows to truncate at relatively high wave numbers (here T179), i.e., the field effectively retains features at 1 degree resolution, which is much improved compared to earlier evaluations based on reanalyses that are based on spectral models (e.g., Trenberth and Stepaniak 2003; Edwards 2007; Mayer et al. 2017). Figure 1b presents an estimate of structural uncertainty in the divergence term, as estimated from the standard deviations (at each grid point) across the long-term average fields diagnosed from ERA5, MERRA-2, and JRA55.

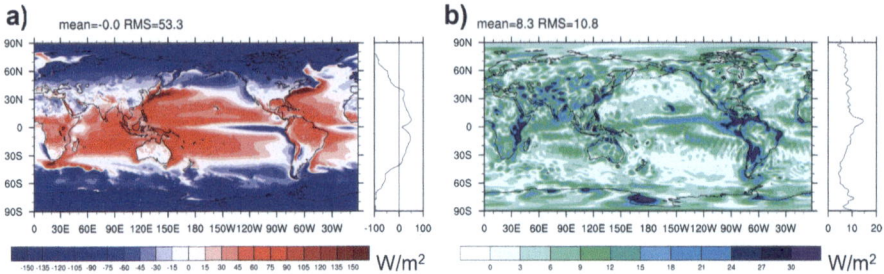

Fig. 1 **a** Divergence of atmospheric moist static plus kinetic energy transports based on ERA5 averaged over 2001–2020 (truncated at T179); **b** spread in 2001–2020 mean TEDIV from ERA5, MERRA-2, and JRA55. Field in panel **b** is truncated at T63 to emphasize the larger scales

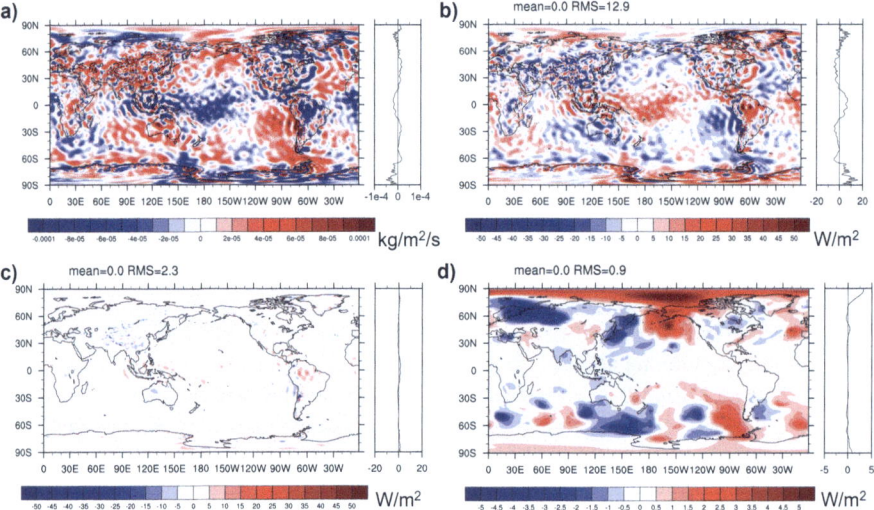

Fig. 2 Impact of mass adjustment and different choices for reference temperature. **a** Vertically integrated mass flux divergence from ERA5 for 2015, **b** impact of mass-adjusting winds on TEDIV from ERA5 2015 (using Kelvin scale), **c** impact of mass-adjusting winds on TEDIV from ERA5 2015 (using Celsius scale), **d** difference of mass-adjusted TEDIV in 2015 using Kelvin or Celsius scale

Large spread is found in the Intertropical Convergence Zone (ITCZ), where systematic differences in the divergent circulation between reanalyses likely affect the divergent energy transport. Spread over extratropical oceans, including western boundary currents with their strong divergences, is remarkably small. Spread over land is relatively large (especially in the tropics), and this is related to biases in the reanalyses and spectral artifacts projecting on larger scales.

The effect of mass adjustment and choice of T_{00} is illustrated in Fig. 2. Panel (a) shows the 2015 average mass flux divergence from ERA5. The shown pattern is unrealistic, as according to Eq. (2) the mass flux divergence should balance surface pressure variations, but the illustrated mass divergence patterns of order 10^{-4} kg/m²/s would imply annual surface pressure changes around 315 hPa/a. Panels (b) and (c) show the impact of a barotropic mass adjustment on TEDIV using either K or C temperature scale, respectively. The impact is larger in K scale, as expected from considerations in Sect. 3.2.1. The ambiguity of TEDIV arising from the choice of temperature scale becomes small after mass correction, as can be seen from panel (d). Differences only remain in the mid-latitudes, where real surface pressure variations within one year are non-negligible.

Figure 3 shows the 2015 average difference in the energy flux divergence from ERA5 with the mass adjustment applied to satisfy Eq. (2) or to satisfy Eq. (1) but with 3D moisture enthalpy fluxes removed consistently (as proposed by Mayer et al. 2017). As expected from the discussion in 3.2.4, the difference is small, with a low-amplitude P–E pattern, which likely arises from the implicit consideration of potential energy of moisture (and its divergence) in the latter formulation (which is not taken into account in the IFS because of the discussed replacement of moisture with dry air). Results are thus very similar, if the formulations are self-consistent and mass consistency is ensured.

Figure 4 presents two fields that have often been neglected but should be included in evaluations of the total energy budget according to Eq. (4). Panel 4a shows the divergence of latent heat transports associated with snow and ice. Values are generally small except where the average atmospheric flow crosses mountain ranges, with convergence (divergence) on the windward (lee) side due to uplift and descent, respectively. Note the reverted sign compared to the divergence of atmospheric ice transport because the latent heat of snow and ice is relative to liquid water, and hence the contribution of L_f is negative. Panel 4b shows latent heat flux associated with snowfall, which represents a cooling of the surface [either by cooling the ocean or by depositing snow (because $L_f<0$) on the land surface]. The global average effect is -0.9 W/m^2. As noted by Mayer et al. (2017), the typically considered surface fluxes $Rad_S + SH + LH$ must be increased by this value to balance observed ocean warming.

Direct validation of diagnosed atmospheric energy transports is impossible because no in situ observations of this quantity exist. The inter-product spread (Fig. 1b) is one indicator of uncertainty, but this could overestimate uncertainty if one product is an outlier but could also underestimate uncertainty in the case of similar structural biases. Since ERA5 is an ensemble product, we can also assess internal uncertainty from the

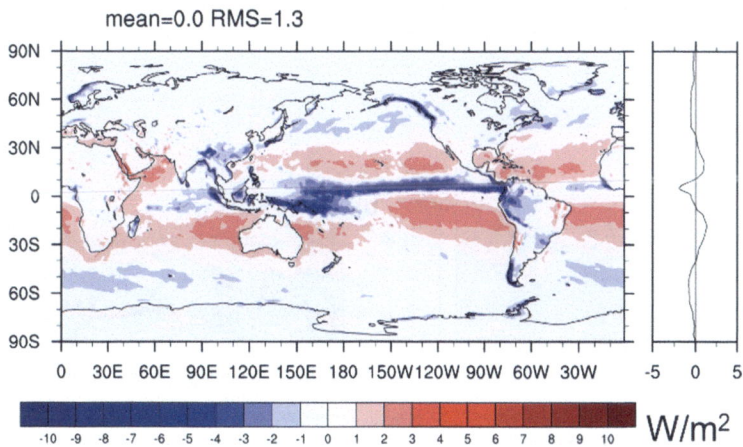

Fig. 3 Difference of TEDIV computed following Mayer et al. (2017), i.e., with winds adjusted to satisfy Eq. (1) but moisture enthalpy taken out from Eq. (4), or following the IFS formulation, i.e., with winds adjusted to satisfy Eq. (2) and strictly following Eq. (4). Based on ERA5 data for 2015

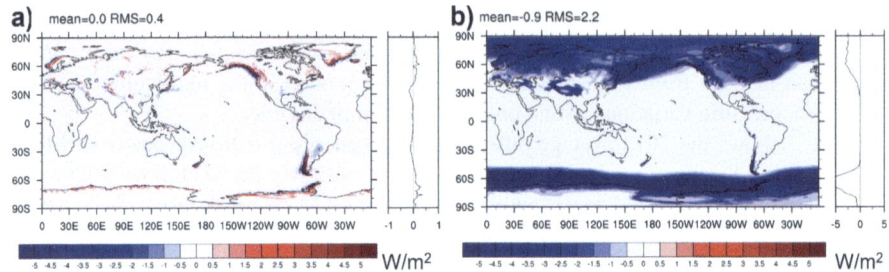

Fig. 4 a Divergence of snow and ice-related latent heat transports, **b** $L_f(T_{00})P_{snow}$ averaged over 2001–2020 based on ERA5 data

ensemble spread, which is shown for the year 2015 in Fig. 5a. For a fair comparison, Fig. 5b shows the inter-product spread as presented in Fig. 1b but only for 2015. The magnitude of the ensemble spread of TEDIV from ERA5 is generally lower compared to the inter-product spread and there are regional differences, e.g., in the mid-latitudes. The latter is likely related to relatively high uncertainties in ERA5 associated with synoptic-scale activity which still stand out in a one-year average. The generally lower magnitude of the ERA5 spread can be explained by the largely random nature of the perturbations used for ensemble generation (Hersbach et al. 2020), a large fraction of which is averaged out when considering one-year means (the spread is considerably larger on monthly timescales; not shown). Figure 5b is very similar to Fig. 1b, with modestly enhanced magnitude (RMS increased by ~ 15%), which indicates that most of the inter-model spread on annual timescales arises from systematic differences and has relatively small temporal variability.

Another option for validation is to compare convergence of atmospheric energy over land to observed net TOA radiation, as there these quantities should balance locally in the long term if land heat uptake is neglected. Similarly, TOA radiation and atmospheric divergence can be combined to yield net surface energy flux, which in the long-term mean should be small everywhere over land. This validation approach will be adopted in Sect. 4, where the net surface energy flux will be discussed.

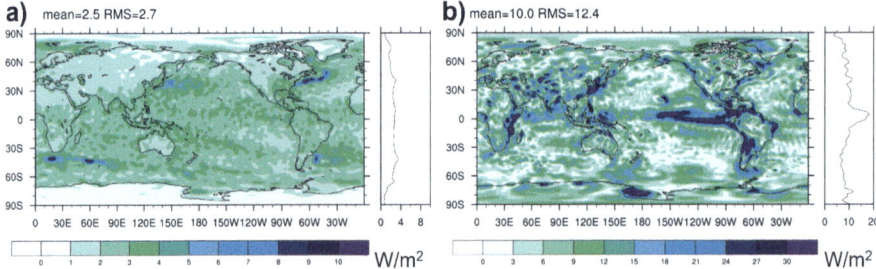

Fig. 5 a Spread of TEDIV from the 10-member ERA5 ensemble during 2015, computed as spread of the 10 annual means; **b** inter-product spread in 2015 mean TEDIV from ERA5, MERRA-2, and JRA55

3.3 Atmospheric Dry Static Energy Budget

The dry static energy (DSE) budget equation integrated over the atmosphere expresses the energy balance of the atmospheric column after water mass multiplied by latent heat has been subtracted from the reference state from Eq. (4) (Kato et al. 2021). The difference of the DSE equation from the total energy equation discussed in Sect. 3.2 is that water contribution appears as a diabatic heating term because lateral divergence and convergence of water mass is balanced by condensation and (re-)evaporation. In earlier studies, the DSE equation has been used to assess regions with a large energy balance residual when satellite derived data products are combined with reanalysis data (e.g., Kato et al. 2016, 2021).

To highlight fluxes that are not included in energy flux data products but needed for the energy balance in this section, we set up the DSE equation with assumptions that are mentioned in Sect. 3.2.2. In addition, because we assume that velocities of all water species are the same as the dry air velocity, the assumption yields a vertically integrated dry static energy equation which is simplified compared to Kato et al. (2021)

$$\frac{1}{g}\frac{\partial}{\partial t}\int_0^{p_s}\left[\overline{c_a}T+\Phi_s+k\right]dp+\frac{1}{g}\nabla\cdot\int_0^{p_s}\vec{v}\left[\overline{c_a}T+\Phi+k\right]dp=\left(\text{Rad}_{\text{TOA}}-\text{Rad}_s\right)$$
$$+L_v\left(\dot{C}_{vl}+\dot{P}_{vr}\right)+L_s\left(\dot{C}_{vi}+\dot{P}_{vs}\right)+L_f\left(\dot{C}_{li}+\dot{P}_{ri}+\dot{P}_{ls}+\dot{P}_{rs}\right)+\text{SH}+F_E+F_P \quad (5)$$

where the sum of enthalpy (with $\overline{c_a}$ representing mass-weighted specific heat as a function of atmospheric moisture) and surface geopotential is the dry static energy. Since the two terms on the lhs are typically evaluated using reanalysis data, we follow the simplification made in most models and use heat capacity of dry air (see Sect. 3.2.2). L_v, L_s, and L_f are, respectively, the enthalpy of vaporization, sublimation, and fusion and $\dot{P}_{vr}$ is the rate of vapor condensed to raindrops that precipitate in the column and $\dot{C}_{vl}$ is the rate of vapor condensed to form cloud droplets. Similarly, P and C in the diabatic heating term indicate, respectively, precipitating and non-precipitating (i.e., clouds) hydrometeors. Subscripts v, l, i, r, and s indicate, respectively, water vapor, liquid, ice, rain, and snow. In reality, latent heats depend on temperature, but most models use constant latent heats (see Sect. 3.2.2). Kato et al. (2021) estimate the bias in diabatic heating rate when a constant latent heat of vaporization at 0 °C is used. The bias averaged over a year can be <-5 Wm^{-2} in the tropics (because there typically $L_v(T_a) < L_v(T_{00})$) and $>+5$ Wm^{-2} in mid-latitude and polar regions (because there typically $L_v(T_a) > L_v(T_{00})$). Note that additional underestimation of diabatic heating in mid-latitude and polar regions arises if the enhancement of latent heat release associated with snowfall is neglected ($L_s(T_{00}) > L_v(T_{00})$) (see Fig. 4b). Since in this section we derive diabatic heating rates based on an observational precipitation product, we use latent heat of vaporization at 0 °C for conversion.

Some earlier studies assume that hydrometeors formed in the column are precipitated out in the same column. Therefore, they neglect $\dot{C}_{vl}$, $\dot{C}_{vi}$, and $\dot{C}_{li}$ in Eq. (5). We used MERRA-2 data to estimate the energetic effect of $\dot{C}_{vl}$ (liquid water tendency due to dynamics multiplied by L_v), and long-term mean values range in ± 2 W/m^2 regionally (not shown). The energetic effect of $\dot{C}_{vi}$ and $\dot{C}_{li}$ is difficult to estimate separately, as the freezing and sublimation rates are typically not output by reanalyses. However, Fig. 4a provides an approximate estimation of the combined effect of the two terms, and it is concluded to be sizeable only close to high orography. The uncertainty in these estimates is expected to be large especially at a regional scale because of little observational constraint. Nevertheless, the result indicates that diabatic heating due to cloud particles may not be negligible for some regions.

F_E and F_P are generally not included in energy fluxes flux data products. These fluxes are needed to balance energy for the atmospheric column but add complexities in balancing regional energy fluxes using dry static energy equations. Mayer et al. (2017; their Fig. 7) showed that most of $F_E + F_P$ is balanced by lateral moisture enthalpy transports on the lhs of Eq. (5). Thus, removing those from the divergence (as we do here, see Sect. 3.2.4) is assumed to remove most of the $F_E + F_P$ pattern from the inferred diabatic heating. However, the imprint of F_E and F_P may implicitly be included when evaluating some terms of Eq. (5) using observational products, which do not make the same simplifications as made in reanalyses (see discussion in Sect. 3.2.2). With these considerations in mind, we drop F_E and F_P to write the DSE budget as

$$\frac{1}{g}\frac{\partial}{\partial t}\int_0^{p_s}\left[c_a T(1-q_v-q_l-q_f)+\Phi_s+k\right]dp + \frac{1}{g}\nabla\cdot\int_0^{p_s}\vec{v}\left[c_a T(1-q_v-q_l-q_f)+\Phi+k\right]dp - (\text{Rad}_{\text{TOA}} \quad (6)$$
$$-\text{Rad}_s) - \text{SH} = L_v(\dot{C}_{vl}+\dot{P}_{vr}) + L_s(\dot{C}_{vi}+\dot{P}_{vs}) + L_f(\dot{C}_{li}+\dot{P}_{ri}+\dot{P}_{ls}+\dot{P}_{rs})$$

Figure 6a shows the 2001–2020 mean divergence of DSE (DSEDIV) as written in Eq. (6) based on ERA5 data. Compared to the divergence of total energy transport (Fig. 1a), the field shows maxima where precipitation is maximal, i.e., in the ITCZ. Figure 6b shows the spread across the three DSE divergence estimates from ERA5, MERRA-2, and JRA55, as a measure of structural uncertainty. Uncertainties are generally larger compared to total energy (global RMS 23.3 W/m² compared to 10.8 W/m²) and the spread peaks in the regions of deep convection in the tropics, which reflects the large uncertainty in DSEDIV. The spread of the total energy divergence is lower because total energy is conserved during latent heat release, i.e., in TEDIV, differences in DSE divergence are balanced by opposing differences in the divergence of latent heat transport (both terms are included in TEDIV) in the different products.

Figure 7 shows the DSE budget residual (defined by bringing all terms of Eq. (6) to the rhs) averaged March 2000 through December 2018. Data products used here are CERES-EBAF (radiation), GPCP V3.2 (precipitation), SeaFlux v3 (Sensible heat flux over ocean), ERA5 (sensible heat flux over land), as well as ERA5, MERRA-2, and JRA55 (Dry static energy divergence and tendency). All fluxes in Eq. (6) are included. The DSE budget residual is considerable, with RMS values between 21.5 and 37.4 W/m² (a considerable fraction of the RMS magnitude of the DSEDIV itself—compare to Fig. 6a), depending on the employed reanalysis product. All solutions for the residual exhibit a negative global mean, indicating inconsistencies between radiative fluxes, precipitation, and sensible heat flux

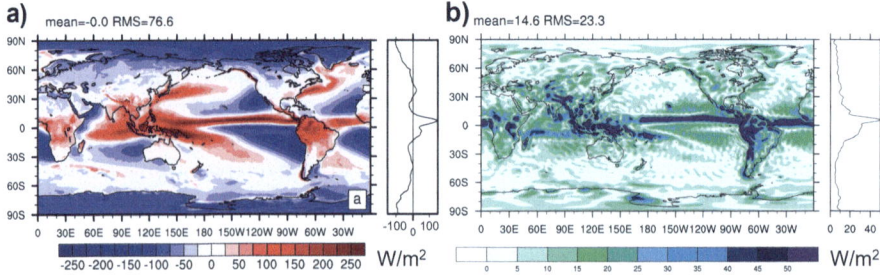

Fig. 6 **a** Divergence of atmospheric dry static energy transports based on ERA5 averaged over 2001–2020 (truncated at T179); **b** spread in 2001–2020 mean DSEDIV from ERA5, MERRA-2, and JRA55. Field in panel **b** is truncated at T63 to emphasize the larger scales

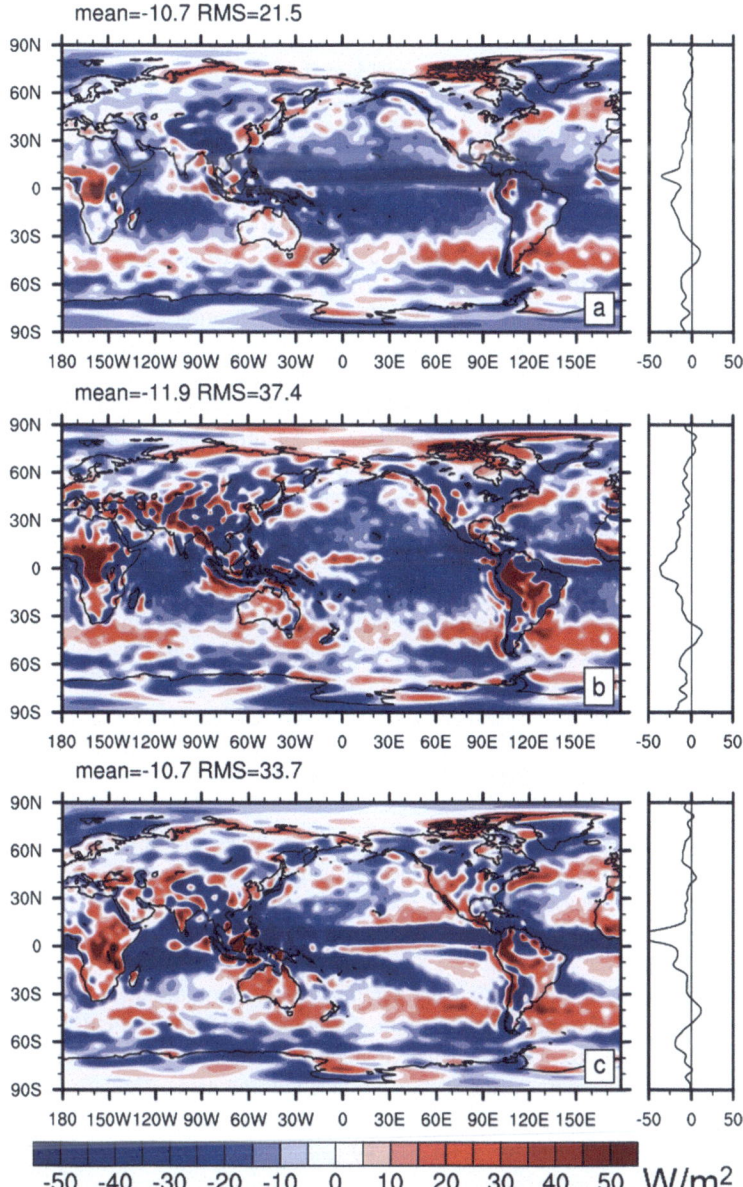

Fig. 7 a Spatial distribution of atmospheric energy balance residual averaged from March 2000 through December 2018 computed with Eq. (6). Energy flux products used are Edition 4.2 EBAF (radiation), GPCP V3.2 (precipitation), SeaFlux (Sensible heat flux over ocean), ERA5 (sensible heat flux over land). **b** Same as (**a**) except that it uses MERRA-2 dry static and kinetic energy divergence and tendency and MERRA-2 sensible heat flux. **c** As (**a**) but uses JRA55 dry static and kinetic energy divergence and tendency. All fields are truncated at T63 for a fair comparison

data (note that global mean DSEDIV is 0 by construction and the DSE tendency is negligible). The residual patterns are not straightforward to interpret, but an imprint from tropical precipitation is clearly visible for all. This indicates that tropical DSEDIV from reanalyses is stronger than the diabatic heating suggested by precipitation from GPCP V3.2 and the radiative flux convergence from CERES-EBAF.

While the atmospheric energy balance residual shown in Fig. 7 remains, we can compute diabatic heating rate due to water phase changes by evaluating the lhs of Eq. (6) using a combination of reanalyses and observational products. Figure 8 (left) shows the diabatic heating rate derived from the left side of Eq. (6) using ERA5 reanalysis data (tendency and divergence terms), CERES-EBAF (TOA and surface irradiance), as well as Seaflux v3 and ERA5 sensible heat fluxes over ocean and land, respectively. The equivalent diabatic heating by precipitation derived from the right side of Eq. (6) is shown in the right panel of Fig. 8. In computing the right side of Eq. (6), the diabatic heating by non-precipitating hydrometeors and fusion terms are ignored. The diabatic heating rate estimated from the DSE budget and from GPCP averaged over different regions are shown in Table 3. Diabatic heating rates estimated from the DSE budget are generally higher than those based on GPCP (~14% in the global ocean mean and ~21% in the tropics). This is a substantial difference, still larger than can be expected through potential increase in GPCP guided by merging the currently best precipitation sensors over the oceans and considering the latest version of the precipitation products (Behrangi and Song 2020; Behrangi et al. 2023) and suggests a positive bias of the DSE-based estimates of diabatic heating.

Because the diabatic heating is derived from the sum of terms on the left side of Eq. (6), biases in dry static energy divergence and tendency, atmospheric net irradiance, and sensible heat flux affect the diabatic heating. Although the regional bias in each term is difficult to quantify, the comparison of diabatic heating derived from Eq. (6) and precipitation products can be used to infer the uncertainty in the estimate of the diabatic heating by precipitation. A thorough investigation is left for the future.

4 Net Surface Energy Budget

The net surface energy flux (F_S) is a key quantity, as it drives atmospheric circulation and energy transports, and similarly oceanic heat redistribution. Moreover, regional changes in F_S are a key contributor to regional energy imbalance of the climate system. Direct measurements of F_S via eddy covariance methods are scarce, but there exist several alternative approaches.

F_S can be diagnosed "directly" using satellite products of turbulent fluxes obtained via remotely sensed quantities input to bulk formulae (OAflux, J-OFURO, IFREMER, SeaFlux) and radiative fluxes (CERES-EBAF-sfc, CLARA-A3). Reanalyses output surface fluxes "directly" during short-term forecasts produced during data assimilation.

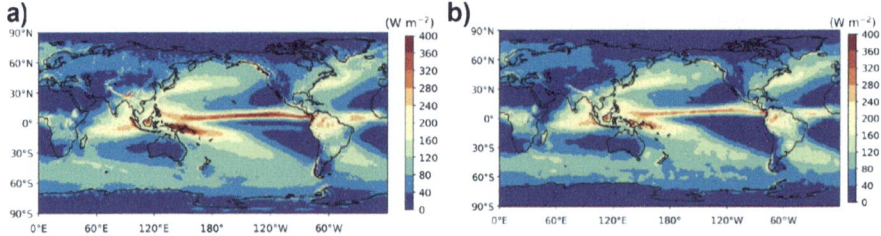

Fig. 8 a Diabatic heating rate associated with water phase change (right side of Eq. 6) inferred from the sum of tendency and divergence terms, TOA and surface net irradiance, and sensible heat fluxes (i.e., left-hand side terms of Eq. 6) averaged from January 2001 through December 2018. b diabatic heating computed with GPCP V 3.2 precipitation for $L_v \dot{P}_{vr} + L_s \dot{P}_{vs}$ (i.e., the right-hand side of Eq. 6)

Table 3 Global and regional mean diabatic heating rates averaged over the time period from January 2001 through December 2018

	Evaluate lhs of Eq. (6)		Evaluate rhs of Eq. (6)	
	Data product			
$\frac{1}{g}\nabla \cdot \int_0^{p_s} \vec{v}[c_a T + \Phi + k] dp$	ERA5	MERRA-2		
$\mathrm{Rad}_{\mathrm{TOA}} - \mathrm{Rad}_s$	EBAF Ed4.2	EBAF Ed4.2		
SH	SeaFlux 3.0 (Ocean), ERA5 (Land)	MERRA-2		
$L_v(\dot{C}_{vl} + \dot{P}_{vr}) +$ $L_s(\dot{C}_{vi} + \dot{P}_{vs}) +$ $L_f(\dot{C}_{li} + \dot{P}_{ri} + \dot{P}_{ls} + \dot{P}_{rs})$			GPCP+MERRA-2	GPCP+ERA5
Global (Wm^{-2})	93.3	91.5	81.7	81.6
Ocean (Wm^{-2})	105.2	101.4	92.2	92.0
Land (Wm^{-2})	72.6	74.1	63.4	63.3
Northern hemisphere (Wm^{-2})	96.0	94.5	84.4	83.3
Southern hemisphere (Wm^{-2})	90.7	88.5	80.0	79.9
30°N–30°S (Wm^{-2})	110.4	108.3	91.2	91.0
30°N–60°N plus 30°S–60°S (Wm^{-2})	87.0	84.1	83.5	83.5
60°N–90°N plus 60°S–90°S (Wm^{-2})	46.9	48.8	41.2	41.2

Alternatively, the total energy budget can be used to infer net surface energy flux F_S as a residual, i.e., by rearranging Eq. (4) and evaluating the lhs of the following equation:

$$\mathrm{Rad}_{\mathrm{TOA}} - \mathrm{TEDIV} - \frac{\partial}{\partial t}\mathrm{AE} = \mathrm{Rad}_S + \mathrm{LH} + \mathrm{SH} + L_f(T_{00})P_{\mathrm{snow}} \quad (7)$$

We define F_S as the sum of net surface radiation, turbulent surface fluxes, and the energetic effect of snowfall (i.e., the rhs of Eq. 7) The lhs of Eq. (7) is typically evaluated by combining net TOA radiation from a satellite product such as CERES-EBAF with atmospheric divergence and tendencies obtained from reanalyses. The advantage of this approach is that reanalysis-based analysis fields are strongly constrained by observations and thus are deemed more accurate than vertical fluxes output during the short-term forecasts. Another key feature of the residual approach is that, since global mean divergence vanishes and global mean $\frac{\partial}{\partial t}\mathrm{AE}$ is very well constrained by observations (Johnson et al. 2023; Mayer et al. 2019; Von Schuckmann et al. 2022), the global mean bias of the inferred flux roughly equals that of the employed TOA flux product (i.e., < 1 W/m^2).

We begin the evaluation with an inter-comparison of inferred F_S averaged 2001–2020 based on the atmospheric energy budgets from ERA5, MERRA-2, JRA55, as shown in Fig. 9a–c, respectively. The spread (Fig. 9d) is almost identical to that of the divergence (Fig. 1b), indicating that atmospheric storage and especially its inter-product spread is very small (not shown). F_S over land is expected to be small (< 1 Wm^{-2}) locally on 20-year timescales, and hence patterns of inferred F_S over land inform about uncertainties. The spatial RMS of the long-term means over land ranges in ~ 13–19 W/m^2 (Table 4), depending on the product. This represents an uncertainty estimate on the local scale, but we note

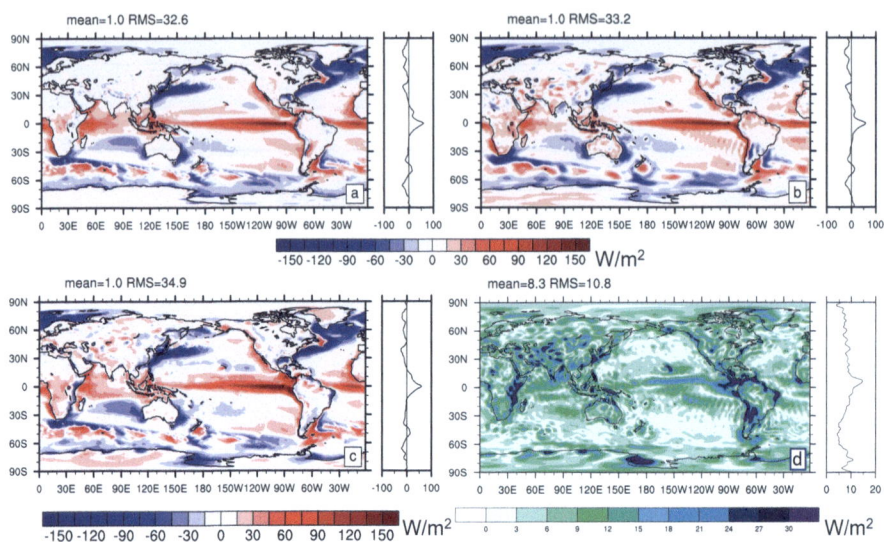

Fig. 9 Inferred net surface energy flux over ocean based on CERES-EBAF-TOA and **a** ERA5 (truncated at T179), **b** JRA55 (truncated at T63), and **c** MERRA-2 (untruncated) data averaged over 2001–2020. Panel **d** at bottom right shows the spread (measured as standard deviation) across the three fields shown in (**a**)–(**c**)

that spectral noise is maximal over high topography as visible in fields based on ERA5 and JRA55 (both are based on spectral models) and thus contributes to uncertainty mostly over land. We thus we expect local errors over the ocean to be generally smaller. Comparison of Fig. 9a–c also reveals common errors such as the positive bias over central Africa present

Table 4 Satisfaction of physical constraints on a regional scale

Region	$FS_{inferred}$ using CERES-EBAF-TOA combined with atmospheric budgets from		
	ERA5	MERRA-2	JRA55
South America	− 0.8	− 2.3	6.2
North America	2.4	− 2.2	3.0
Australia	2.7	0.8	12.4
Maritime Continent	− 8.8	20.8	32.7
Africa	1.4	5.1	9.3
Eurasia	− 3.1	0.9	3.1
Global land	− 0.6	1.1	5.4
Global Ocean	1.5	0.9	− 0.5
Global Ocean + Land	1.0	1.0	1.0
Global land spatial RMS value	13.2	18.8	19.0

Results are shown for inferred net surface energy flux [W/m^2] combining CERES-EBAF-TOA fluxes and atmospheric budgets (divergence and tendency) from ERA5, MERRA-2, and JRA55, averaged over 2001–2020

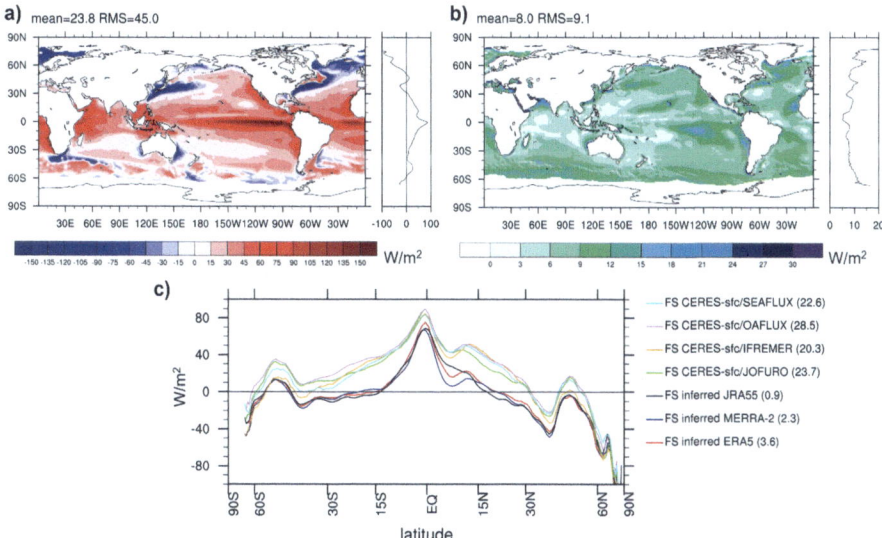

Fig. 10 Net surface energy flux derived from satellite-based products (energetic effect of snowfall taken from ERA5); **a** ensemble mean of four estimates combining CERES-EBAF-sfc and SeaFlux, OAflux, IFREMER, J-OFURO, respectively; **b** ensemble spread (standard deviation) across the four estimates; **c** zonally averaged net surface energy flux over the ocean from satellite-based estimates and inferred flux estimates. Global averages over all grid points with valid data are provided in parentheses (units are W/m²). All estimates shown in (**c**) have the same spatiotemporal mask applied and are averaged over 2001–2017

in all three estimates, which are not revealed by the inter-product spread. On a larger (continental) scale, the bias of inferred F_S becomes significantly smaller, ranging in ±3 W/m² except for the Maritime Continent, where all estimates have significant biases, likely due to the dominance of coastal areas with strong gradients (see Table 4). Global land and ocean averages of the three estimates exhibit fairly small biases, suggesting only small biases in global-scale ocean-to-land energy transport in all three reanalyses. The global mean F_S is 1 W/m², as dictated by the global mean of CERES-EBAF-TOA fluxes. Note that the implied fluxes include the effect of snowfall (see rhs of Eq. 7). We can obtain the global average of surface net radiation plus turbulent heat fluxes (often used to approximate net surface energy flux) using the quantification of the snowfall term based on ERA5 data (global average is − 0.9 W/m², see Fig. 4b), which is 1.9 W/m² for 2001–2020.

Figure 10a presents the multi-product mean F_S obtained from four satellite products averaged over 2001–2017. For a clean comparison to the inferred fluxes, we added the energetic effect of snowfall based on ERA5 data (as shown in Fig. 4b) to the satellite-based radiative and turbulent fluxes. From comparison with Fig. 9a–c it is evident that the satellite-based estimates exhibit positive values (i.e., net ocean heat uptake) over much larger regions. Inter-product spread is relatively large over subtropical basins and the mid-latitudes and small in the Warm Pool regions (Fig. 10b). Figure 10c presents an inter-comparison of zonally averaged inferred and satellite-based F_S estimates. It reveals substantial differences between estimates from the two approaches, with the satellite-based fluxes being higher (i.e., more flux into the ocean) across almost all latitudes, with smaller differences in high latitudes. The satellite-based estimates exhibit strongly positive global mean values > 20 W/m², which is a much higher value than can be expected from the removal of the grid points in the vicinity of sea ice with typically negative F_S values. This effect is estimated to be order 4–5 W/m^2, as can be seen from a comparison of the quasi-global ocean means of inferred F_S in Fig. 10c to true ocean averages provided in Table 4. Thus, satellite-based F_S estimates exhibit a long-term global ocean mean bias order ~ 15 W/m^2. Comparison of Rad$_S$ from CERES-EBAF-sfc and CLARA-A3 shows very good agreement (global long-term mean difference − 2.2 W/m^2 and spatial RMS of long-term mean difference is 5.3 W/m^2) of the two products. It is noted that both products utilize similar approaches to estimate surface radiation, i.e., via radiative transfer simulations and the use of simultaneous AVHRR and CERES data for the estimation of broadband albedo in CLARA-A3. Apart from this, they are independent as they utilize different satellite input, auxiliary data and radiative transfer models. Thus, the agreement between both radiation records suggests that the main source of uncertainty are the turbulent fluxes.

Lastly, in Table 5 we present long-term global, land, and ocean mean F_s values from forecasts of different reanalysis products, which are listed in Table 1. Values range in − 11.2 to + 11.4 W/m^2, with the largest values obtained from reanalyses with full data assimilation and relatively small values from the AMIP-type runs. All reanalyses except for NCEP R2 have the largest biases over ocean, where prescribed SSTs imply infinite heat capacity of the ocean and as a result lead to unrealistic fluxes. This indicates that assimilation of observational data into reanalyses arguably yields a more realistic atmospheric state but introduces discrepancies between observed and model climates that are reflected by unrealistically large vertical energy flux during the short-term forecasts.

5 Summary, Conclusions, and Challenges Ahead

This paper discussed frameworks to diagnose the atmospheric mass and energy budgets, including storage terms, vertical fluxes, and lateral transports. We detailed assumptions typically made in atmospheric models that underlie reanalysis products, which are essential data sets for some terms of the energy budget, most notably the lateral transports and their divergence. These assumptions simplify the diagnostic equations, and optimal use of the data can only be made if diagnostics take into account the models' conservation equations in a consistent manner. For example, models typically use constant latent heats rather than more realistic temperature-dependent formulations, which should be then followed when using the data.

We evaluated the total and dry static energy budget using observational data and three atmospheric reanalyses ERA5, JRA55, and MERRA-2. Thereby, we illustrated the impact of mass corrections which are still needed despite the progress made in data quality over past decades. Specific attention has been paid to the sensitivity of results to the employed temperature scale, and it has been shown that the Celsius scale is preferable since it helps to minimize ambiguities that remain in diagnosed budgets after mass correction due to real mass variations.

All these assumptions and errors contribute to the regional energy balance residual, and a complication arises from the fact that different products, most notably reanalyses compared to observations, use different assumptions. As a consequence, when we compare individual estimates of the errors in different terms (e.g., Fig. 6) with the regional energy balance residual shown in Fig. 7, individual maps do not fully resemble the map of the regional energy balance residual. This means that errors are present in divergence, radiation, and turbulent flux products. While balancing regional energy using observation-based products is difficult, we believe two approaches are useful to move forward toward balancing regional energy budgets. One approach is, as demonstrated by L'Ecuyer et al. (2015) and similarly Mayer

Table 5 2001–2010 average net surface energy flux in W/m^2 obtained from reanalyses with varying amounts of data assimilated (based on short-term forecasts) and free AMIP-style runs of atmospheric models underlying some of the employed reanalyses (MERRA-2 AMIP and ERA20CM)

Region	ERA5	MERRA-2	JRA55	MERRA-2 AMIP	NCEP R2	20CRv3	ERA20C	ERA20CM	MERRA
Global land	1.0	0.8	0.5	1.1	8.1	1.2	0.8	0.9	1.7
Global Ocean	5.1	− 6.6	− 15.7	− 1.8	0.9	9.1	1.8	2.3	15.2
Global Ocean + Land	3.9	− 4.6	− 11.2	− 1.0	2.8	6.9	1.5	1.9	11.5

et al. (2019), to adjust each component of regional energy fluxes within their uncertainty. These studies demonstrate that regional energy fluxes can be balanced at a relatively large scale. Closing at smaller scales (e.g., $1° \times 1°$) is, however, difficult because of the increasing influence of energy and moisture transport in the local energy balance. Furthermore, estimating time- and regionally dependent errors for all products at a smaller temporal and spatial scales is difficult. Another challenge for the provision of realistic uncertainty estimates is that inter-product spread for the budget term in question can underestimate (due to common biases in different products) or overestimate (due to inferior quality of one or more employed products) true uncertainty. Nevertheless, the inter-product spread appears to provide a more realistic uncertainty estimate than the spread of current ensemble products such as ERA5, since ensemble perturbations do not take systematic errors (e.g., of the assimilating model) into account which seem to be dominant on the timescales considered here.

An alternative approach that we showed in this study is to use the energy balance equations to estimate one flux component from the sum of the rest of flux components as residual. While this approach guarantees to balance regional energy, all errors (arising from both shortcomings of the individual data sets but also inconsistencies arising from different assumptions made in different products) are in the residual. We showed the regional net surface energy flux estimated with net TOA flux and total energy divergence. Various metrics demonstrated a low bias of inferred F_S (across spatial scales, ranging from close to zero on global scale, over < 5 W/m^2 on continental scale for ERA5 and MERRA-2 and ~ 15 W/m^2 on the local scale of a 1×1 degree grid) compared to other approaches to estimate net surface energy flux, such as direct estimation from model output or satellite-based data. For the latter, the main source of uncertainty lies in the turbulent fluxes. Satellite-based estimates have been tuned to minimize differences with flux estimates from buoy-based measurements, whereby fluxes are obtained using bulk formulae. The large quasi-global biases of the satellite-based F_S estimates suggest that they are either overfitted to conditions at the buoy locations (which are largely in tropical seas) or the employed bulk formulae, which are known to have large uncertainties (Yu 2019), yield biased fluxes. Thus, in terms of long-term means inferred F_S fields appear to be of relatively high accuracy, and Mayer et al. (2022) have demonstrated that they are also superior in terms of temporal consistency. A major difficulty for evaluating surface fluxes is the limited availability of ground-truth data, such as measurements using eddy covariance methods over the ocean. As a consequence, we have to resort to the evaluation of physical constraints, which only exist for the long-term mean and only for the large scale over the ocean.

We also showed the regional diabatic heating associated with water estimated indirectly using radiative fluxes, dry static and kinetic energy divergence, and surface sensible heat flux. Both examples of inferring quantities as a residual take advantage of TOA flux constrained by ocean temperature measurements. The second example further extends the TOA flux constraint to surface radiative flux. Divergence terms in both examples use the same wind and temperature fields. The advantage of this residual approaches is that once assumptions and simplifications made in the employed data are accounted for in the diagnostic framework, inferred regional fluxes can be evaluated with measurements. For example, Trenberth and Fasullo (2017), Liu et al. (2020), or Mayer et al. (2022) used ocean heat transport measured at the RAPID section (Johns et al. 2011) and ocean temperature data in the North Atlantic to evaluate regional net surface fluxes derived as residual. The diabatic heating rate derived as residual can be evaluated by precipitation measurements. Results presented here indicate a positive bias of diabatic heating rates inferred from the dry static energy budget when compared to GPCP-based estimates, also when accounting for documented low biases of the latter based on the merger of the currently best and latest version of the precipitation products (Behrangi and Song 2020; Behrangi et al. 2023). However, the comparison in this paper is merely a demonstration of this application of the atmospheric energy budget and more work is needed to investigate the discrepancies found in more detail.

The evaluations presented here focused on long-term means. Assessing trends of inferred F_S, an important application in the context of EEI research, still pushes the limits of the data, and only after ~ 2000 the involved products appear to be sufficiently stable (Loeb et al. 2022). Stable TOA fluxes such as those from CERES-EBAF are an essential ingredient for this, and their continuity into the future should have high priority. A remaining limiting factor of trend diagnostics is temporal stability of reanalyses, although progress has demonstrably been made over the years (e.g., Mayer et al. 2021). Nevertheless, for example TEDIV from ERA5 exhibits discontinuities in the late 1990s, which demonstrates the need for continued efforts in the areas of data rescue, homogenization, satellite data reprocessing activities, and bias estimation in future reanalyses, as advocated by Buizza et al. (2018).

It is expected that future reanalyses will increasingly adopt an Earth system approach, with atmosphere, ocean, sea ice, and land sub-systems being coupled. This will open new possibilities for coupled budget diagnostics, but the challenge of setting up a diagnostic framework that consistently (e.g., in terms of reference temperatures) tracks all relevant exchanges between the different compartments will remain. Generally, diagnostic frameworks need to take account of further development of data sets in the future and relaxation of simplification and assumptions in models. Evaluation efforts such as those presented here should be continued to document progress with updated data sets, and to inform about remaining challenges.

Acknowledgements This paper is an outcome of the Workshop "Challenges in Understanding the Global Water Energy Cycle and its Changes in Response to Greenhouse Gas Emissions" held at the International Space Science Institute (ISSI) in Bern, Switzerland (26–30 September 2023). M. Mayer and J. Mayer received funding from Austrian Science Fund (FWF) P33177. S. Kato was supported by the NASA CERES project and Z. Li was supported by the NASA Energy and Water Cycle Study (NNH18ZDA001N-NEWS). M. Schröder acknowledges the financial support by the EUMETSAT (European Organisation for the Exploitation of Meteorological Satellites) member states through CM SAF (Satellite Application Facility on Climate Monitoring). Financial support for A. Behrangi was made available from NASA MEaSUREs (NNH17ZDA001N-MEASURES; NNH22ZDA001N-MEASURES). The authors thank two anonymous reviewers for insightful and constructive comments. The authors also thank Brent Roberts (NASA) for useful discussions on MERRA-2 data.

Funding Open access funding provided by University of Vienna.

Declarations

Conflict of interest The authors have no relevant financial or non-financial interests to disclose.

Open Access This article is licensed under a Creative Commons Attribution 4.0 International License, which permits use, sharing, adaptation, distribution and reproduction in any medium or format, as long as you give appropriate credit to the original author(s) and the source, provide a link to the Creative Commons licence, and indicate if changes were made. The images or other third party material in this article are included in the article's Creative Commons licence, unless indicated otherwise in a credit line to the material. If material is not included in the article's Creative Commons licence and your intended use is not permitted by statutory regulation or exceeds the permitted use, you will need to obtain permission directly from the copyright holder. To view a copy of this licence, visit http://creativecommons.org/licenses/by/4.0/.

References

Bannon PR (2002) Theoretical foundations for models of moist convection. J Atmospheric Sci 59:1967–1982. https://doi.org/10.1175/1520-0469(2002)059%3c1967:TFFMOM%3e2.0.CO;2

Behrangi A, Song Y (2020) A new estimate for oceanic precipitation amount and distribution using complementary precipitation observations from space and comparison with GPCP. Environ Res Lett 15:124042. https://doi.org/10.1088/1748-9326/abc6d1

Behrangi A, Song Y, Huffmann GJ, Adler R (2023) Comparative analysis of the latest global oceanic precipitation estimates from GPM V07 and GPCP V3.2 products. J Hydrometeorol 25:293–309. https://doi.org/10.1175/JHM-D-23-0082.1

Bentamy A, Grodsky SA, Katsaros K, Mestas-Nuñez AM, Blanke B, Desbiolles F (2013) Improvement in air–sea flux estimates derived from satellite observations. Int J Remote Sens 34:5243–5261. https://doi.org/10.1080/01431161.2013.787502

Bloom S, Takacs L, Da Silva A, Ledvina D (1996) Data assimilation using incremental analysis updates. Mon Weather Rev 124:1256–1271. https://doi.org/10.1175/1520-0493(1996)124%3c1256:DAUIAU%3e2.0.CO;2

Bosilovich MG, Lucchesi R, Suarez M (2016) MERRA-2: file specification. GMAO Off. Note No 9 Version 11, 73. http://gmao.gsfc.nasa.gov/pubs/office_notes

Buizza R et al (2018) The EU-FP7 ERA-CLIM2 project contribution to advancing science and production of earth system climate reanalyses. Bull Am Meteorol Soc 99:1003–1014

CDS (2021) Mass-consistent atmospheric energy and moisture budget monthly data from 1979 to present derived from ERA5 reanalysis. https://doi.org/10.24381/cds.c2451f6b

Cheng L, Trenberth KE, Fasullo JT, Mayer M, Balmaseda M, Zhu J (2019) Evolution of ocean heat content related to ENSO. J Clim 32:3529–3556. https://doi.org/10.1175/JCLI-D-18-0607.1

Chiodo G, Haimberger L (2010) Interannual changes in mass consistent energy budgets from ERA-Interim and satellite data. J Geophys Res. https://doi.org/10.1029/2009JD012049

Collow ABM, Mahanama SP, Bosilovich MG, Koster RD, Schubert SD (2017) An evaluation of teleconnections over the United States in an ensemble of AMIP simulations with the MERRA-2 configuration of the GEOS atmospheric model. NASATM-2017-104606 47:78

Curry JA et al (2004) Seaflux. Bull Am Meteorol Soc 85:409–424. https://doi.org/10.1175/BAMS-85-3-409

ECMWF (2021a) IFS documentation CY47R3—part III Dynamics and numerical procedures. IFS Documentation CY47R3, IFS Documentation, ECMWF

ECMWF (2021b) IFS documentation CY47R3—part IV Physical processes. IFS Documentation CY47R3, IFS Documentation, ECMWF

Edwards JM (2007) Oceanic latent heat fluxes: consistency with the atmospheric hydrological and energy cycles and general circulation modeling. J Geophys Res Atmos. https://doi.org/10.1029/2006JD007324

Gelaro R et al (2017) The modern-era retrospective analysis for research and applications, version 2 (MERRA-2). J Clim 30:5419–5454. https://doi.org/10.1175/JCLI-D-16-0758.1

Global Modeling and Assimilation Office (GMAO) (2015a) MERRA-2 tavgM_2d_int_Nx: 2d, monthly mean, time-averaged, single-level, assimilation, vertically integrated diagnostics V5.12.4. https://doi.org/10.5067/FQPTQ4OJ22TL

Global Modeling and Assimilation Office (GMAO) (2015b) MERRA-2 tavgM_2d_flx_Nx: 2d, monthly mean, time-averaged, single-level, assimilation, surface flux diagnostics V5.12.4. https://doi.org/10.5067/0JRLVL8YV2Y4

Global Modeling and Assimilation Office (GMAO) (2015c) MERRA-2 tavgM_2d_rad_Nx: 2d, monthly mean, time-averaged, single-level, assimilation, radiation diagnostics V5.12.4. https://doi.org/10.5067/OU3HJDS97300

Gosnell R, Fairall CW, Webster PJ (1995) The sensible heat of rainfall in the tropical ocean. J Geophys Res Oceans 100:18437–18442. https://doi.org/10.1029/95JC01833

Hersbach H, Peubey C, Simmons A, Berrisford P, Poli P, Dee D (2015) ERA-20CM: a twentieth-century atmospheric model ensemble. Q J R Meteorol Soc 141:2350–2375. https://doi.org/10.1002/qj.2528

Hersbach H et al (2020) The ERA5 global reanalysis. Q J R Meteorol Soc 146:1999–2049. https://doi.org/10.1002/qj.3803

Huffman GJ, Adler RF, Behrangi A, Bolvin DT, Nelkin EJ, Gu G, Ehsani MR (2023) The new version 3.2 global precipitation climatology project (GPCP) monthly and daily precipitation products. J Clim 36:7635–7655. https://doi.org/10.1175/JCLI-D-23-0123.1

JMA (2007) Outline of the operational numerical weather prediction at the Japan Meteorological Agency. Appendix to WMO Technical Progress Report on the Global Data-Processing and Forecasting System and Numerical Weather Prediction. JMA

Johns WE et al (2011) Continuous, array-based estimates of Atlantic Ocean heat transport at 26.5 N. J Clim 24:2429–2449. https://doi.org/10.1175/2010JCLI3997.1

Johnson GC, Landerer FW, Loeb NG, Lyman JM, Mayer M, Swann AL, Zhang J (2023) Closure of earth's global seasonal cycle of energy storage. Surv Geophys. https://doi.org/10.1007/s10712-023-09797-6

Kanamitsu M, Ebisuzaki W, Woollen J, Yang S-K, Hnilo JJ, Fiorino M, Potter GL (2002) NCEP–DOE AMIP-II reanalysis (R-2). Bull Am Meteorol Soc 83:1631–1644. https://doi.org/10.1175/BAMS-83-11-1631

Karlsson K-G et al. (2023) CLARA-A3: the third edition of the AVHRR-based CM SAF climate data record on clouds, radiation and surface albedo covering the period 1979 to 2023. Earth Syst Sci Data. https://doi.org/10.5194/essd-15-4901-2023

Kato S, Xu K-M, Wong T, Loeb NG, Rose FG, Trenberth KE, Thorsen TJ (2016) Investigation of the residual in column-integrated atmospheric energy balance using cloud objects. J Clim 29:7435–7452. https://doi.org/10.1175/JCLI-D-15-0782.1

Kato S et al. (2018) Surface irradiances of edition 4.0 clouds and the earth's radiant energy system (CERES) energy balanced and filled (EBAF) data product. J Clim 31:4501–4527. https://doi.org/10.1175/JCLI-D-17-0523.1

Kato S, Loeb NG, Fasullo JT, Trenberth KE, Lauritzen PH, Rose FG, Rutan DA, Satoh M (2021) Regional energy and water budget of a precipitating atmosphere over ocean. J Clim 34:4189–4205. https://doi.org/10.1175/JCLI-D-20-0175.1

Kobayashi S et al (2015) The JRA-55 reanalysis: general specifications and basic characteristics. J Meteorol Soc Jpn Ser II 93:5–48. https://doi.org/10.2151/jmsj.2015-001

Kosaka Y et al (2024) The JRA-3Q reanalysis. J Meteorol Soc Jpn Ser II 102:49–109. https://doi.org/10.2151/jmsj.2024-004

Lauritzen, PH et al (2018) NCAR release of CAM-SE in CESM2.0: A reformulation of the spectral element dynamical core in dry-mass vertical coordinates with comprehensive treatment of condensates and energy. J Adv Model Earth Syst 10:1537–1570. https://doi.org/10.1029/2017MS001257

Lauritzen PH et al (2022) Reconciling and improving formulations for thermodynamics and conservation principles in Earth System Models (ESMs). J Adv Model Earth Syst 1(4):e2022MS003117

L'Ecuyer TS et al (2015) The observed state of the energy budget in the early twenty-first century. J Clim 28:8319–8346. https://doi.org/10.1175/JCLI-D-14-00556.1

Liu C et al (2020) Variability in the global energy budget and transports 1985–2017. Clim Dyn 55:3381–3396. https://doi.org/10.1007/s00382-020-05451-8

Loeb NG et al (2018) Clouds and the earth's radiant energy system (CERES) energy balanced and filled (EBAF) top-of-atmosphere (TOA) edition-4.0 data product. J Clim 31:895–918. https://doi.org/10.1175/JCLI-D-17-0208.1

Loeb NG et al (2022) Evaluating twenty-year trends in Earth's energy flows from observations and reanalyses. J Geophys Res Atmos 127:e2022JD036686. https://doi.org/10.1029/2022JD036686

Malardel S, Diamantakis M, Agusti-Panareda A, Flemming J (2019) Dry mass versus total mass conservation in the IFS. ECMWF Technical Memorandum 849

Mauder M, Foken T, Cuxart J (2020) Surface-energy-balance closure over land: a review. Bound-Layer Meteorol 177:395–426. https://doi.org/10.1007/s10546-020-00529-6

Mayer J, Mayer M, Haimberger L (2021) Consistency and homogeneity of atmospheric energy, moisture, and mass budgets in ERA5. J Clim 34:3955–3974. https://doi.org/10.1175/JCLI-D-20-0676.1

Mayer J, Mayer M, Haimberger L, Liu C (2022) Comparison of surface energy fluxes from global to local scale. J Clim 35:4551–4569. https://doi.org/10.1175/JCLI-D-21-0598.1

Mayer M, Haimberger L, Edwards JM, Hyder P (2017) Toward consistent diagnostics of the coupled atmosphere and ocean energy budgets. J Clim 30:9225–9246. https://doi.org/10.1175/JCLI-D-17-0137.1

Mayer M, Tietsche S, Haimberger L, Tsubouchi T, Mayer J, Zuo H (2019) An improved estimate of the coupled Arctic energy budget. J Clim 32:7915–7934. https://doi.org/10.1175/JCLI-D-19-0233.1

Peixoto JP, Oort AH (1992) Physics of climate. American Institute of Physics Melville, New York, p 520

Poli P et al (2016) ERA-20C: an atmospheric reanalysis of the twentieth century. J Clim 29:4083–4097. https://doi.org/10.1175/JCLI-D-15-0556.1

Rienecker MM et al (2011) MERRA: NASA's modern-era retrospective analysis for research and applications. J Clim 24:3624–3648. https://doi.org/10.1175/JCLI-D-11-00015.1

Roberts CD, Senan R, Molteni F, Boussetta S, Mayer M, Keeley SP (2018) Climate model configurations of the ECMWF integrated forecasting system (ECMWF-IFS cycle 43r1) for HighResMIP. Geosci Model Dev 11:3681–3712. https://doi.org/10.5194/gmd-11-3681-2018

Savijärvi H (1982) The mass balance in diagnostic studies: an example of analysed and forecast data calculations. Tellus 34:540–544. https://doi.org/10.3402/tellusa.v34i6.10839

Slivinski LC et al (2019) Towards a more reliable historical reanalysis: Improvements for version 3 of the Twentieth Century Reanalysis system. Q J R Meteorol Soc 145:2876–2908

Takacs LL, Suárez MJ, Todling R (2016) Maintaining atmospheric mass and water balance in reanalyses. Q J R Meteorol Soc 142:1565–1573. https://doi.org/10.1002/qj.2763

Tomita H, Hihara T, Kako S, Kubota M, Kutsuwada K (2019) An introduction to J-OFURO3, a third-generation Japanese ocean flux data set using remote-sensing observations. J Oceanogr 75:171–194. https://doi.org/10.1007/s10872-018-0493-x

Trenberth K (1991) Climate diagnosics from global analyses: conservation of mass in ECMWF analyses. J Clim 4:707–722. https://doi.org/10.1175/1520-0442(1991)004%3c0707:CDFGAC%3e2.0.CO;2

Trenberth KE (1997) Using atmospheric budgets as a constraint on surface fluxes. J Clim 10:2796–2809. https://doi.org/10.1175/1520-0442(1997)010%3c2796:UABAAC%3e2.0.CO;2

Trenberth KE, Stepaniak DP (2003) Seamless poleward atmospheric energy transports and implications for the hadley circulation. J Clim 16:3706–3722. https://doi.org/10.1175/1520-0442(2003)016%3c3706:SPAETA%3e2.0.CO;2

Trenberth KE, Fasullo JT (2017) Atlantic meridional heat transports computed from balancing Earth's energy locally. Geophys Res Lett 44:1919–1927. https://doi.org/10.1002/2016GL072475

Trenberth KE, Fasullo JT (2018) Applications of an updated atmospheric energetics formulation. J Clim 31:6263–6279. https://doi.org/10.1175/JCLI-D-17-0838.1

Trenberth KE, Zhang Y, Fasullo JT, Cheng L (2019) Observation-based estimates of global and basin ocean meridional heat transport time series. J Clim 32:4567–4583. https://doi.org/10.1175/JCLI-D-18-0872.1

Von Schuckmann K et al (2022) Heat stored in the Earth system 1960–2020: where does the energy go? Earth Syst Sci Data 2022:1675–1709. https://doi.org/10.5194/essd-15-1675-2023

Yu L (2019) Global air–sea fluxes of heat, fresh water, and momentum: energy budget closure and unanswered questions. Annu Rev Mar Sci 11:227–248. https://doi.org/10.1146/annurev-marine-010816-060704

Yu L, Weller RA (2007) Objectively analyzed air–sea heat fluxes for the global ice-free oceans (1981–2005). Bull Am Meteorol Soc 88:527–540. https://doi.org/10.1175/BAMS-88-4-527

Publisher's Note Springer Nature remains neutral with regard to jurisdictional claims in published maps and institutional affiliations.

Authors and Affiliations

Michael Mayer[1,2] · Seiji Kato[3] · Michael Bosilovich[4] · Peter Bechtold[1] · Johannes Mayer[2] · Marc Schröder[5] · Ali Behrangi[6] · Martin Wild[7] · Shinya Kobayashi[8] · Zhujun Li[3] · Tristan L'Ecuyer[9]

✉ Michael Mayer
  michael.mayer@ecmwf.int

[1] Research Department, European Centre for Medium-Range Weather Forecasts, Reading RG2 9AX, UK

[2] Department of Meteorology and Geophysics, University of Vienna, 1090 Vienna, Austria

[3] NASA Langley Research Center, Hampton, VA 23681-2199, USA

[4] NASA Global Modeling and Assimilation Office, Goddard Space Flight Center, Greenbelt, MD 20771, USA

[5] Satellite-Based Climate Monitoring, Deutscher Wetterdienst, 63067 Offenbach, Germany

[6] Department of Hydrology and Atmospheric Sciences, University of Arizona, Tucson, AZ 85721, USA

[7] Institute for Atmospheric and Climate Science, ETH Zurich, 8092 Zurich, Switzerland

[8] Numerical Prediction Development Center, Japan Meteorological Agency, Tsukuba City 305-0052, Japan

[9] Department of Atmospheric and Oceanic Science, University of Wisconsin, Madison, WI 53706, USA

North Atlantic Heat Transport Convergence Derived from a Regional Energy Budget Using Different Ocean Heat Content Estimates

B. Meyssignac[1] · S. Fourest[1] · Michael Mayer[2,3] · G. C. Johnson[4] · F. M. Calafat[5] · M. Ablain[6] · T. Boyer[7] · L. Cheng[8] · D. Desbruyères[9] · G. Forget[10] · D. Giglio[11] · M. Kuusela[12] · R. Locarnini[7] · J. M. Lyman[4,13] · W. Llovel[9] · A. Mishonov[7,14] · J. Reagan[7] · V. Rousseau[6] · J. Benveniste[15]

Received: 11 December 2023 / Accepted: 7 September 2024 / Published online: 24 October 2024
© The Author(s) 2024

Abstract

This study uses an oceanic energy budget to estimate the ocean heat transport convergence in the North Atlantic during 2005–2018. The horizontal convergence of the ocean heat transport is estimated using ocean heat content tendency primarily derived from satellite altimetry combined with space gravimetry. The net surface energy fluxes are inferred from mass-corrected divergence of atmospheric energy transport and tendency of the ECMWF ERA5 reanalysis combined with top-of-the-atmosphere radiative fluxes from the clouds and the Earth's radiant energy system project. The indirectly estimated horizontal convergence of the ocean heat transport is integrated between the rapid climate change-meridional overturning circulation and heatflux array (RAPID) section at 26.5°N (operating since 2004) and the overturning in the subpolar north atlantic program (OSNAP) section, situated at 53°–60°N (operating since 2014). This is to validate the ocean heat transport convergence estimate against an independent estimate derived from RAPID and OSNAP in-situ measurements. The mean ocean energy budget of the North Atlantic is closed to within ±0.25 PW between RAPID and OSNAP sections. The mean oceanic heat transport convergence between these sections is 0.58±0.25 PW, which agrees well with observed section transports. Interannual variability of the inferred oceanic heat transport convergence is also in reasonable agreement with the interannual variability observed at RAPID and OSNAP, with a correlation of 0.54 between annual time series. The correlation increases to 0.67 for biannual time series. Other estimates of the ocean energy budget based on ocean heat content tendency derived from various methods give similar results. Despite a large spread, the correlation is always significant meaning the results are robust against the method to estimate the ocean heat content tendency.

Keywords North Atlantic heat transport · Regional energy budget · Energy transport · Climate variability · Energy budget/balance · Heat budgets/fluxes · Surface fluxes · In situ observations · Satellite observations · Ocean heat content

Extended author information available on the last page of the article

Article Highlights

- The mean ocean energy budget of the North Atlantic is closed, within ±0.25 PW, between RAPID and OSNAP sections
- The inferred mean oceanic heat transport convergence between RAPID and OSNAP sections is 0.58 ±0.26 PW, which agrees well with observed section transports
- Interannual variability of the North Atlantic oceanic heat transport convergence is in reasonable agreement with the interannual variability observed at RAPID and OSNAP, with a correlation of 0.54 between annual time series and 0.67 for biannual time series
- The results are robust against those products used to estimate the ocean heat content (OHC) tendency, whether these products are based on in situ measurements, satellite altimetry, and space gravimetry or a combination of them

1 Introduction

The Atlantic meridional overturning circulation (AMOC) is often represented as stream function in latitude-depth or sometimes latitude-density space derived from the zonally-integrated meridional velocities in the Atlantic Ocean (Frajka-Williams et al. 2019; Jackson et al. 2019; Rousselet et al. 2021). It is characterized by a northward flow of warm, salty upper ocean water, and a southward flow of colder, denser, deep waters that sink in the subpolar North Atlantic and Greenland-Iceland-Norwegian Sea after transferring heat to the atmosphere. The AMOC affects a meridional transport of heat (MHT) in the North Atlantic basin that is positive northward. It is responsible for most of the meridional transport of heat by the midlatitude northern hemisphere ocean (and up to 25% of the northward global atmosphere–ocean heat transport in the northern hemisphere -e.g., Bryden et al. 2001-). This North Atlantic MHT plays a crucial role in the climate variability of the entire Northern hemisphere, particularly affecting ocean temperature and circulation, and the Atlantic storm track (e.g., Rhines et al. 2008). Additionally, the Atlantic MHT impacts the regional climate by warming the atmosphere over Europe, thereby influencing air temperature and precipitation in this densely populated region (e.g., Palter 2015). Future projections with global warming suggest a weakening of the North Atlantic MHT with potential profound local climate effects. However, there are substantial disagreements among models both in terms of how well they simulate the MHT and the magnitude of future changes (e.g. Collins et al. 2019; Mecking and Drijfhout 2023). Accordingly, there is considerable interest in quantifying the North Atlantic variability to better validate and constrain climate simulations and determine whether the projected trends are a response to the forced climate or internal variability.

In this work, we evaluate the zonal mean MHT in the North Atlantic as a residual of the ocean energy budget (see Sect. 2). An advantage of this approach is that it provides estimates of the zonal mean MHT at any latitude. For validation, we compare our estimate of the zonal mean MHT with estimates derived from in situ measurements of the RAPID array, at the RAPID section at 26.5°N, and of the OSNAP array, along OSNAP sections in the subpolar North Atlantic (see Fig. 1 and Sect. 4). We further analyze the spatio-temporal variability of the North Atlantic MHT and its causes (see Sect. 4). As the MHT is derived from an energy budget, we evaluate the role of the different terms of the oceanic energy budget in the North Atlantic MHT spatiotemporal variability (see Sect. 4 and 5).

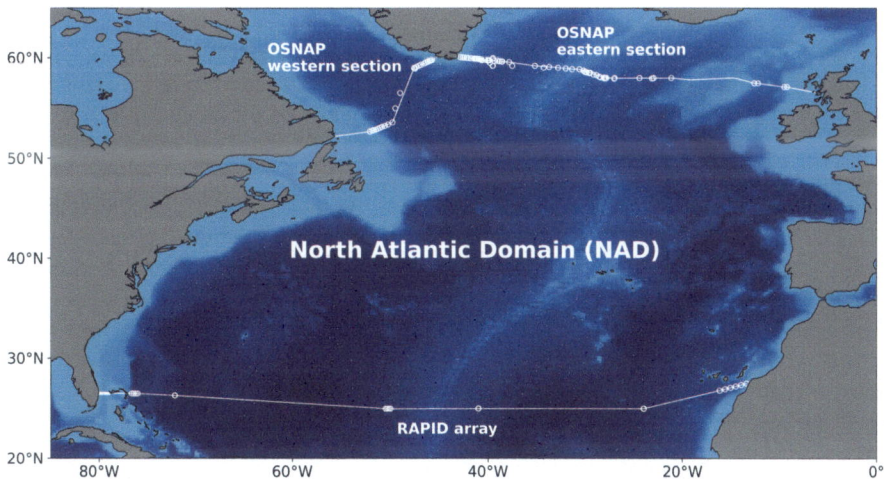

Fig. 1 Map of the North Atlantic domain (NAD) in which the ocean energy budget is evaluated. The southern white line indicates the 26.5°N latitude where the RAPID array is located. The white dots along the line indicate the position of the RAPID array moorings. The Northern white lines indicate the two legs of OSNAP and the dots indicate the location of the OSNAP moorings

The indirect method that we use here to evaluate the zonal mean MHT in the North Atlantic (which relies on an ocean energy budget) has already been developed in the literature (Mayer et al. 2022; Liu et al. 2020, Trenberth and Fasullo 2017). Previous studies use atmospheric reanalysis (ERA-interim -Dee et al. 2011 and ERA5—Hersbach et al. 2020) combined with top-of-the-atmosphere (TOA) radiative fluxes (from the clouds and the earth's radiant energy system project-CERES) to estimate the net surface fluxes and ocean reanalyses (ORAS5 -Zuo et al. 2019 and ORAP6 -Zuo et al. 2021) to estimate the ocean heat content tendency (OHCT). They show that the ocean heat content tendency term of the energy budget explains most of the time variability in the Atlantic MHT at the RAPID section on interannual and longer time scales. However, it has never been explored the extent to which the estimate of the Atlantic MHT time variability is dependent on the North Atlantic OHCT estimate derived from ocean reanalysis.

In this study, we use a recent estimate of the ocean heat content changes over the North Atlantic derived from a geodetic approach which combines satellite altimetry data with space gravimetry data (update of Marti et al. 2022, 2024) to estimate the OHCT. We also use other estimates of the OHCT either derived from in situ measurements of the ocean temperature by the Argo network of profiling floats augmented by other ocean observing systems, or derived from in situ data combined with satellite altimetry data, or derived from an ocean reanalysis. With this ensemble of OHCT estimates, we assess the closure of the ocean energy budget and evaluate the spread in MHT that is induced by the differences across different OHCT estimates. This approach has three important benefits: First, it enables exploration of the sensitivity of the North Atlantic MHT estimate on the OHC product that is used and thus provides insights on the uncertainty in the MHT variability that is due to the uncertainty in the OHC estimate (which is likely the largest source of uncertainty in the estimated MHT variability). Second, it enables testing for closure of the North Atlantic Ocean energy budget. This important piece of information reveals whether the important causes of the North Atlantic MHT are identified and how closely their combination matches the MHT estimated using mooring data at RAPID and OSNAP sections. It provides a quantitative understanding of why and how the North Atlantic MHT is changing with time. It also cross-validates different products based on different observations such as the OHCT based on satellite altimetry and space gravimetry or argo observations, the atmospheric lateral energy transport divergence and tendency of ERA5, and the TOA radiative fluxes from CERES. This cross-validation is an efficient approach to evaluate the consistency of different observed North Atlantic variables including OHCT, TOA radiation fluxes, and MHT with regard to the energy conservation law. This aspect is particularly important when we want to use OHCT products, TOA radiation products, and MHT products in globally consistent objective estimates of the global energy budget such as the one developed in the world climate research program global energy and water exchanges (GEWEX) project (Stephens et al. 2023) and in the NASA NEWS project (e.g., L'Ecuyer et al. 2015 and Roberts et al. this issue). We evaluate and discuss these three benefits in the discussion section.

2 Method

The oceanic energy budget is evaluated over a closed domain of the North Atlantic region, between the RAPID section and the OSNAP sections, for the period of 2005–2018. The OSNAP sections comprise two legs: one leg from Southern Labrador to South-west Greenland and one leg from South-east Greenland to the coast of Scotland (see Fig. 1). The Mediterranean sea is excluded from the domain. This closed North Atlantic domain is called NAD hereafter.

We write the vertical integral of the oceanic energy budget for any column of water located in NAD as

$$\nabla.\text{OHF} = \text{SHF} - \text{OHCT} \qquad (1)$$

where $\nabla.\text{OHF}$ is the vertical integral of the horizontal divergence (i.e., the opposite of the horizontal convergence) of the ocean heat transport in the water column, SHF is the net surface heat flux at the top of the water column, and OHCT is the temporal ocean heat content tendency of the water column. Since 2000s, Sea ice is only presented in the NAD in a narrow band along southern Labrador in winter, so the tendency in the sea ice melt energy is neglected in Eq. (1).

When ∇OHF is integrated horizontally over any large domain such NAD, it corresponds to the heat transport convergence or to the difference in MHT through the frontiers of the domain (Gauss theorem). In the case of the NAD, the horizontal integration of ∇OHF represents the difference between the MHT through the RAPID section and the MHT through the OSNAP sections.

In this study, the vertical integral of the convergence of the ocean heat transport (i.e., the integral of $-\nabla$OHF) is estimated as a residual of Eq. (1). The surface heat flux is derived as the residual of the mass-corrected atmospheric energy budget as in Mayer et al. (2017) and discussed also in this issue (Mayer et al. 2024). That is, the surface heat flux is estimated as the difference of the TOA net radiative fluxes with the atmospheric tendency of energy and the vertically integrated divergence of atmospheric moist static plus kinetic energy fluxes. The OHCT is estimated by derivation of the ocean heat content which is derived either from optimal interpolation of in situ measurements of the ocean temperature and salinity (with a vertical integration of the specific heat of sea water multiplied by the local density of seawater and the oceanic temperature as in Melet and Meyssignac 2015), or from the observed thermal expansion of the sea water (with the local thermosteric sea level rise multiplied by the local integrated expansion efficiency of heat, e.g., Marti et al. 2022), or from the reconstruction of the ocean state by an ocean reanalysis.

Small enthalpy fluxes, associated to evaporation, precipitation, river run-off, and the inflow from the Mediterranean, are entering or leaving the NAD. In total, these enthalpy fluxes are of the order of 10TW (the dominating enthalpy flux is due to the Mediterranean inflow and amounts less than 15TW on average over decadal time scales, Macdonald et al. 1994; Wu and Haines 1998). 10TW is about two orders of magnitude smaller than the different terms of the budget represented in Eq. (1) so we neglect these enthalpy fluxes in the ocean energy budget. Small amounts of mass may enter or leave the NAD without going through the RAPID and OSNAP sections. It can be sea ice floating in and out through the northern boundary, evaporation and precipitation through the surface, river run-off and river discharge through the coast, or the inflow from the Mediterranean. It is also likely that RAPID and OSNAP do not fully sample all the net mass flux going through their respective sections. This small amount of mass generates a bias in the NAD oceanic energy budget (see Mayer et al. 2022 for more details). A mass correction should be applied to the ocean energy budget to balance RAPID, OSNAP, and other term volume fluxes. However, in this study, we use heat transports at both RAPID and OSNAP which are computed with a flow field that is constrained to have zero net mass transport through both sections. This constraint on the flow field implicitly accounts for precipitation minus evaporation plus other small input of water in the NAD region. So, the mass correction is not needed here.

3 Data

3.1 Data Used to Estimate the Net Surface Heat Flux SHF

The net surface heat flux SHF is inferred over the period of 2005–2018 from the mass corrected vertically integrated total atmospheric energy budget in which, the mass-balanced atmospheric horizontal energy transport divergence and the atmospheric energy tendency are derived from ECMWF's latest reanalysis dataset ERA5 as in Mayer et al. (2021a). ERA5 provides a four-dimensional estimate of the atmospheric state at ~31 km spatial and hourly temporal resolution, generated using a 4-dimensional variational data assimilation method that ingests a wealth of remotely sensed and in situ-based observational information (Hersbach et al. 2020). The atmospheric budget data are available from the Copernicus Climate Data Store (CDS; Mayer et al. 2021b). The net TOA radiation fluxes are derived from the CERES–Energy Balanced and Filled (CERES-EBAF) product in version 4.1 (Loeb et al. 2018). The uncertainty on the net surface heat flux is evaluated using two other atmospheric reanalysis (namely JRA55 and MERRA2) to estimate the net surface heat flux SHF. For each month, we consider the maximum difference across the three reanalysis estimates of the SHF as the standard deviation of the uncertainty for this monthly SHF estimate. Note that we could have considered the ensemble mean of the three reanalysis estimates of the SHF as the best estimate of the SHF, but the literature suggests that the ERA5 reanalysis performs best for ocean energy budgets (see Mayer et al. 2022) so we use ERA5 estimate as best estimate.

3.2 Data Used to Estimate the Ocean Heat Content Tendency OHCT

OHCT is estimated from eight different products in total, comprising ocean in situ temperature products from the National Oceanographic and Atmospheric Administration (NCEI from Levitus et al. 2012), from Ifremer (ISAS21 Kolodziejczyk et al. 2023), from the National Oceanic Center (NOC, King 2023), by Giglio et al. 2023 (LocalGPspace), and from the Institute of Atmospheric Physics (IAP, Cheng et al. 2017), plus Ocean heat content products from the European Space Agency (Magellium/LEGOS, 2022; Marti et al. 2024 in revision), NOAA/PMEL (RFROM, Lyman and Johnson 2023), and an ocean state reanalysis (ECCO, Fukumori et al 2021). The ECCO and Magellium/LEGOS estimates are full depth, but the others are from the surface to 2000 m depth.

The ocean in situ products consist of statistical optimal interpolation of the observed in situ temperature profiles mainly from the argo profiling float network over the period of 2005–2018. The argo data are augmented by ocean temperature observations from ship-based observations (research ships and ship of opportunity merchant ships), moored buoys (mainly the tropical moored buoy array with additional OceanSITES buoys) and sometimes gliders, and pinniped mounted sensors.

The NCEI product is a gridded product generated by objective analysis of binned one-degree latitude/longitude means of monthly temperature anomalies calculated from all available ocean profile temperature data subtracted from the corresponding one-degree climatological mean temperature at standard depth levels for years of 1955–2006 (World Ocean Atlas 2009, Locarnini et al. 2010) as described in Levitus et al. 2012. Analyzed temperature anomalies are added back to the climatological mean field to obtain temperature and salinity fields for the time period in question (each month for years 2005–2022) at 26 vertical levels between the surface and 2000 m.

The ISAS21 product is a gridded product of temperature and salinity data derived from the in situ analysis system (ISAS, Gaillard et al. 2016). This version is an update of ISAS21_ARGO (https://www.seanoe.org/data/00412/52367/data/86436.pdf). This product merges all the available delayed-mode quality-controlled (core and deep) argo profiles in the Atlantic domain along with oceanographic campaigns. ISAS uses an optimal interpolation scheme that preserves as much as possible the time and space sampling capabilities of the in situ profiles. The ISAS procedure and products are described in Gaillard et al. (2016).and in Kolodziejczyk et al. (2023).

The NOC_OI product provides gridded fields of in situ temperature and practical salinity encompassing the period from 2004 to 2022. Such fields have been generated through objective mapping of argo profiling float data and quality-controlled mooring data at 26.5°N from RAPID. The data are provided on a $1° \times 1°$ grid with a vertical spacing of 20 decibars and a temporal resolution of 10 days. Quality control procedures were conducted by the source data programs, and no additional quality control was applied before the mapping process.

The LocalGPspace product (Giglio et al. 2023) provides OHC fields that are mapped using locally stationary Gaussian processes with data-driven decorrelation scales (Kuusela and Stein 2018). A linear time trend is included in the estimate of the mean field, along with spatial terms and harmonics for the annual cycle. Mapping is done separately for different vertical sections, which are then combined to estimate global OHC timeseries in the upper 2000 dbar of the ocean. Regions of the ocean that are not sufficiently well sampled by the Argo array are not included.

The IAP product is a gridded product generated by the Institute of Atmospheric Physics (IAP hereinafter, Cheng and Zhu 2016; Cheng et al. 2017). It has advantages in both instrumental error reduction and its gap-filling method. IAP mapping technique used spatial covariance from model simulations to help provide spatial interpolation. This product merges all the available bias-corrected in situ ocean temperature observations from a variety of instruments held in the World Ocean Database. The spatial resolution of IAP data is 1° by 1° mesh grid from 1 to 2000 m (41 levels), and the temporal resolution is monthly.

The ocean heat content products consist of a combination of ocean in situ temperature profiles with satellite data derived from satellite altimetry and space gravimetry.

The Magellium/LEGOS product is a gridded OHC product available from April 2002 to December 2020 over the Atlantic Ocean (Magellium/LEGOS, 2022). The spatial resolution is 1° in latitude and longitude with a monthly temporal resolution. The product is the combination of satellite altimetry-based data (C3S, Legeais et al. 2021) and satellite gravimetry-based data (update of Blazquez et al. 2018) to estimate local expansion of sea water in addition to in situ data (EN4.2.2.l09, Levitus et al. 2009, and ISAS21, Kolodziejczyk et al. 2021) to correct for the salinity effect and convert the resulting thermal expansion into OHC with the integrated expansion efficiency of heat. Uncertainties are provided regionally at a yearly timescale in a variance–covariance matrix and are estimated by propagating uncertainties from the input satellite data until the OHC change. The regional uncertainties associated with altimetry data are derived from the error budget of Prandi et al. (2021) while uncertainties associated with gravimetry data are estimated with an ensemble approach derived from Blazquez et al. (2018). The Magellium/LEGOS product and its uncertainty are described in Marti et al. (2024) and references therein. From the local variance–covariance matrix of the uncertainties of the Magellium/LEGOS product, we compute the variance of the uncertainty in OHCT at each time step for each location in the NAD. The uncertainty variance–covariance matrix of the Magellium/LEGOS product only includes the temporal correlation of the errors and does not include the spatial correlation in errors. We adopt a conservative approach and consider that the Magellium/LEGOS product errors are fully correlated spatially. So, we add linearly the uncertainties of the OHCT of each location when the OHCT is aggregated over the NAD to estimate the uncertainty on the OHCT spatial mean. Among all OHC products, the LEGOS/Magellium product is the most comprehensive one as it covers the global ocean down to the bottom of the ocean and it provides an uncertainty estimates which accounts for all sources of known errors including the time correlation in errors. For this reason, we use this product to test wether the NAD energy budget (on Fig. 5) is closed within uncertainties.

RFROM (random forest regression ocean maps) is a gridded product of ocean heat content anomaly maps produced with a machine learning algorithm, random forest regression. In situ ocean temperature profile data are used as training data. Geographic location, time, and gridded satellite sea-surface height (SSH) and sea-surface temperature (SST) maps are used as predictors. The end result is ocean heat content anomaly maps for 10 different pressure levels between the ocean surface and 2000 dbar at 7-day × ¼° × ¼° resolution, starting in January 1993 (Lyman and Johnson, 2023).

The ocean reanalysis is the ECCO4 reanalysis which estimates the ocean state from 1992 to 2017 from a collection of global datasets (Forget et al 2015a; Fukumori et al 2021). ECCO4 notably includes argo, altimetry, gravimetry, and atmospheric data as constraints. The fit of model to data is achieved through optimization of forcing fields and parameters that control turbulent transport rates in the ocean interior (Forget et al 2015a, b; Forget and Ponte 2015). Through this technique, ECCO4 provides a dynamically consistent estimate of ocean transports, OHU, and OHC variability that form a closed heat budget unlike other data assimilative products that involve state variable increments of unknown nature (Storto et al 2019). A detailed analysis of ECCO4's heat budget in terms of meridional transports is available in Forget and Ferreira 2019.

3.3 Data Used to Validate the Vertical Integral of the Oceanic Energy Budget ∇OHF

The vertical integral of the horizontal divergence of the ocean heat transport ∇OHF is inferred from the net surface heat flux (SHF) and the ocean heat content tendency (OHCT) (following the equation of the oceanic energy budget, Eq. (1)). Then, it is integrated horizontally over the NAD and compared with the difference of the MHT between the RAPID section and the OSNAP sections for validation.

The MHT across the RAPID (26.5°N) and OSNAP (between 50 and 60°N) trans-basin sections is calculated by integrating the product of the cross-sectional velocity, potential temperature, specific heat and density along each section based on data from mooring arrays, argo floats, and (at OSNAP) gliders. The MHT due to Ekman transport is derived from the wind fields of the ERA5 reanalysis (RAPID) and the ERA-Interim reanalysis (OSNAP). At the RAPID section, MHT through the Florida Straits is estimated based on measurements from a submarine cable. For RAPID, the data are available as 12-hourly estimates of MHT spanning the period from April 2004 to August 2018, whereas for OSNAP, the data are provided as 30-day mean estimates for the period from August 2014 to May 2018. We note that these transports represent true heat transports, at both RAPID and OSNAP, since the flow field used in the calculation is constrained to have zero net mass transport through both sections. The approaches to calculate MHT at the RAPID and OSNAP sections are described, respectively, by Johns et al. (2011) and Lozier et al. (2019), and we defer to those studies for full details. For the MHT uncertainties at the RAPID section, we consider an uncertainty of ± 0.22 PW (at 1 sigma) on the monthly time series that is derived from the ± 0.21 PW uncertainty on daily times series from Johns et al. (2011) onto which we added a possible measurement bias error estimated to be ± 0.07 PW (W. Johns personal communication). Since only one degree of freedom per 40 days of observations were considered in the Johns et al. (2011) uncertainty estimate, the error bar on the RAPID MHT monthly time series is estimated to be same as the daily error bar. For the MHT uncertainties at the OSNAP sections, we use the uncertainties given in the product for the period of 2015–2018. For the period of 2005–2014, we use the same uncertainty at each month calculated as the mean monthly uncertainty over the period of 2015–2018 (i.e., ± 0.1 PW at 1 sigma). We consider the uncertainty in the OSNAP MHT and the RAPID MHT as uncorrelated because they are derived from independent data. So, we add quadratically their uncertainty to estimate the uncertainty on the NAD heat transport convergence.

All calculations in this work are done with monthly time series of the different datasets on their native grid. The datasets are then resampled on the same $1° \times 1°$ grid at the last step and then combined together to infer the vertical integral of the oceanic energy budget. The climatology is then removed to get anomalies. When time series are filtered in time, we use the same Lanczos filter to filter them all in a consistent way. All correlations are Pearson correlations and the associated confidence intervals is estimated with a two-tailed p-value test.

4 Results

4.1 The Mean Ocean Energy Budget over 2005–2018

The mean surface heat flux is the dominant term of the mean NAD energy budget from 2005 to 2018 in both amplitude and spatial variability (Fig. 2b). Over 2005–2018, the surface heat flux shows a mean heat loss from the ocean to the atmosphere of 0.54 PW over the NAD. Ocean heat uptake in the tropics is discharged from the ocean to the atmosphere at higher latitudes and in particular along the Gulf stream (Fig. 2b). This picture is consistent with the general poleward redistribution of heat in the upper ocean. This air-sea heat flux is about three times as large as the mean OHCT in the region (Fig. 2a).

Over 2005–2018, the time-mean OHCT is small on average over the NAD (0.062 PW), with warming generally in the subtropics that is strongest in the north portions of the Gulf stream and cooling through much of the subpolar region (Fig. 2a). This spatial variability is consistent across all estimates of the mean OHCT whether they are based on satellite data, in situ observations, or reanalysis (Fig. 3). The OHCT estimates using satellite altimetry data tend to show finer spatial structures coming from the mesoscale activity partially resolved by satellite radar altimeters, in particular along the Gulf stream and North Atlantic current (Fig. 3a,b). When combined together, the different terms of the mean ocean energy budget lead to an estimate of the time-mean vertical integral of the horizontal heat flux that shows a maximum along the Gulf stream and at high latitude (Fig. 2d). This pattern is owing to the role of the mean surface heat flux, which dominates the mean energy budget. The spatial variability of the OHCT is small and the spatial variability of the mean vertical integral of the horizontal heat flux convergence is actually almost all compensated for by the spatial variability of the mean surface heat flux.

4.2 Meridional Heat Transport Between the RAPID and OSNAP Sections over 2005–2018

Poleward heat transport of 0.50 ± 0.1 PW is estimated to cross the OSNAP sections from August 2014 to May 2018 (Fig. 4). The overturning circulation is largely responsible for setting this heat transport (Li et al. 2021). In contrast, the RAPID array indicates a much larger mean heat transport across 26°N of 1.18 ± 0.22 PW over August 2014 to May 2018 and of 1.17 ± 0.22 PW over 2005–2018 (Fig. 4b). These results are expected as the Atlantic poleward heat transport is known to be stronger in the subtropical North Atlantic than in the subpolar North Atlantic (Ganachaud and Wunsch 2000, Trenberth et al., 2001). At the RAPID section, the overturning circulation is also largely responsible for the MHT, even more than at OSNAP sections (McCarthy et al., 2015). The comparison of the RAPID and OSNAP MHT estimates shows another difference: the subtropical MHT is markedly more variable over monthly to interannual time scales than the subpolar MHT. Indeed, the RAPID MHT monthly standard deviation (0.28 PW) is about five times larger than that at OSNAP (0.051 PW).

Combining the OSNAP and RAPID MHT estimates reveals a convergence of heat transport in the NAD of 0.67 ± 0.24 PW during the overlapping time period of 2014–2018. The temporal variability of this heat transport convergence is largely dominated by the RAPID heat transport variability in the subtropics (see Fig. 4). The same situation probably holds before 2014, up to 2005, because the decadal changes in the

Fig. 2 Maps of North Atlantic mean ocean heat content tendency (mean OHCT) (**a**), mean surface heat flux derived from ERA5 and CERES (mean SHF) (**b**), and mean vertical integral of the horizontal heat flux convergence inferred from the energy budget residual (mean ∇OHF) (**c**), for the period of 2005–2018. In **a** and **c**, the OHCT is derived with the geodetic method from the Magellium/LEGOS product (see Fig. 3 For the OHCT over the same period derived from other OHC products)

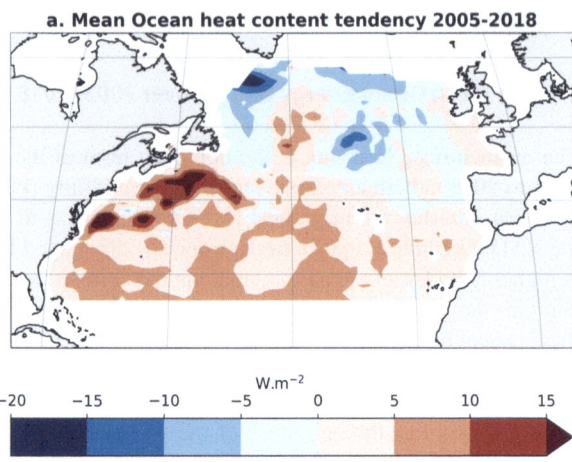

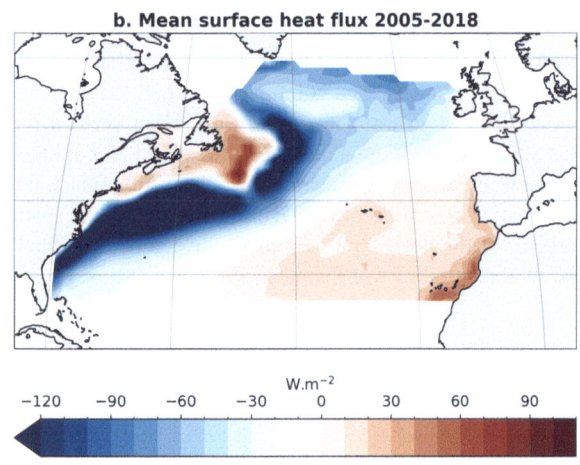

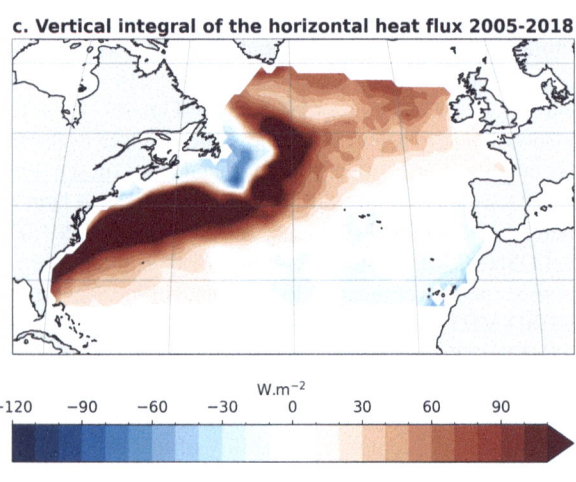

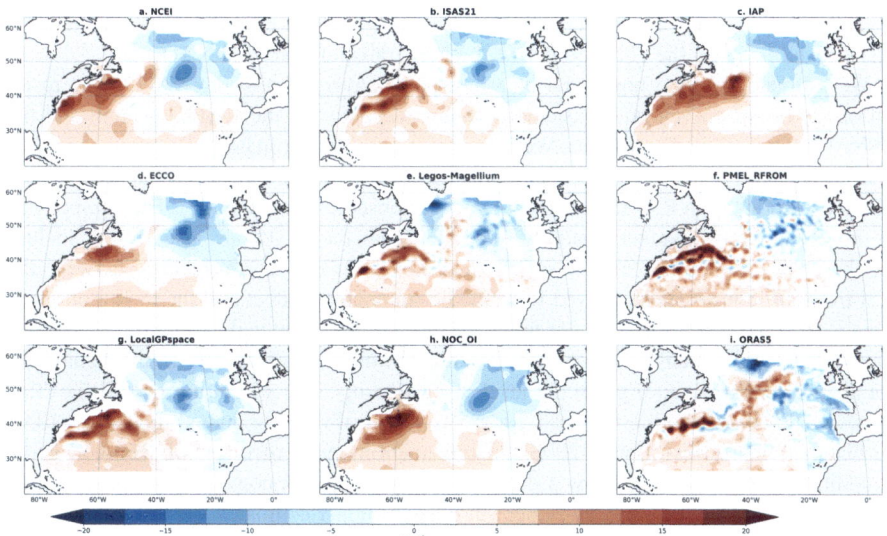

Fig. 3 OHCT averaged over 2005–2018 derived from the different OHC products considered in this study. The middle panel corresponds to Fig. 2 panel a

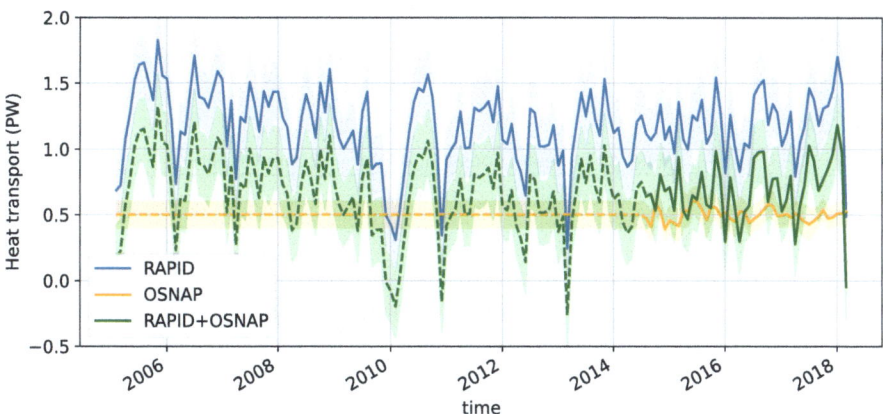

Fig. 4 OSNAP (yellow plain line) and RAPID (blue line) meridional heat transport estimated from, respectively, OSNAP and RAPID arrays in situ observations. Approximation of OSNAP meridional heat transport with a constant of 0.5 ± 0.05 PW over 2005–2014 (yellow dashed line). NAD heat transport convergence from the difference of RAPID and OSNAP meridional heat transport (green line)

subpolar heat transports are known to be weak during the past few decades (around a few tenth of PW, Li et al. 2021).

To extend our ocean heat budget analysis over the whole period of 2005–2018, we make the hypothesis that the mean MHT at OSNAP has not changed compared to August 2014 to May 2018 and is 0.50 ± 0.1 PW over 2005–2018. We also neglect the variability in OSNAP MHT before 2014 under the assumption that it has remained weak over the past decades and it plays only a marginal role (in comparison to the RAPID MHT variability) on the convergence of the heat transport over the NAD.

We estimate the total horizontal heat transport convergence in the NAD as the residual of the ocean energy budget spatially integrated between the RAPID section and the OSNAP section (Fig. 5, red line). We compare it with the estimate of the convergence of heat transport in the NAD estimated as the difference in MHT between the RAPID and OSNAP sections with RAPID and OSNAP in situ observations (red line). The components of the ocean energy budget, spatially integrated over the NAD, are also shown, including the OHCT (blue line) and the surface heat flux (orange line). Agreement between the heat transport convergence derived from the ocean energy budget and the in situ observations is reasonable both in terms of temporal mean and interannual variability. The difference in long-term means is 0.06 ± 0.4 PW, which is small (less than 15% of the signal) and not distinguishable from 0 given the level of uncertainty. The variability over 2008–2018 agrees well between both estimates with a correlation of 0.73 for annual time series (0.91 for biannual timeseries). Before 2008, the agreement in terms of variability is not as good. In addition, the in situ data show a decrease in heat transport convergence between 2005 and 2009 that is not captured by the ocean energy budget estimate. The decrease of heat transport convergence in 2009–2010, primarily wind-driven (McCarthy et al. 2012), lasts longer in the energy budget, until 2011. Over the whole record, the correlation between the in situ estimate and the energy budget estimate of the heat transport convergence is 0.54 and remains significant at the 95% confidence level (the confidence level is estimated with a two-tailed p-value test). On 2 year and longer time scales, agreement is better between both estimates of the NAD heat transport convergence with a correlation of 0.67 over the period of 2005–2018 (Fig. 6).

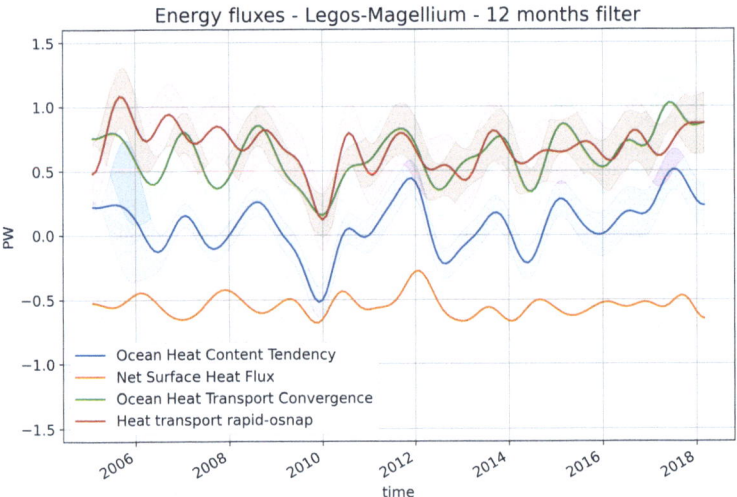

Fig. 5 Monthly time series filtered with a Lanczos low-pass with a cutoff period at 1 year of the net surface heat flux (orange line) and OHCT estimated from Magellium/LEGOS (blue line) over the NAD (see Fig. 6 for OHCT time series derived from other OHCT products). Monthly time series series low-pass filtered at 1 year of the convergence of heat transport between the RAPID and the OSNAP sections estimated from in situ measurement of the RAPID array and the OSNAP array (red line) and from the residual of the oceanic energy budget (green line)

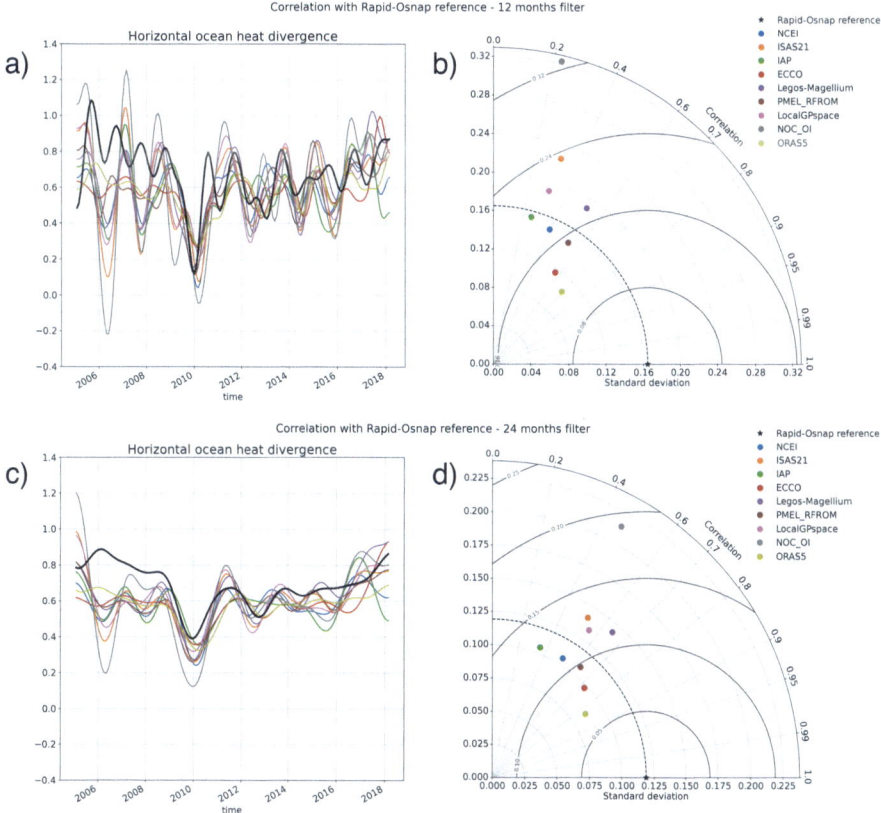

Fig. 6 a Monthly time series low-pass filtered at 1 year of the meridional heat transport convergence between the RAPID and the OSNAP arrays estimated from in situ observations of the RAPID and the OSNAP arrays (black curve) and from the NAD energy budget residual with the OHCT derived from Magellium/LEGOS (purple line, same time series as the blue line in Fig. 5), RFROM (brown), NCEI (blue) ISAS21 (orange), IAP (green), ECCO (red), NOC_OI (gray), and LocalGPspace (pink). **b** Taylor diagram of the ocean heat transport convergence inferred from NAD energy budget residuals against the in situ estimate from the RAPID and the OSNAP arrays. **c** and **d** are, respectively, equivalent to **a** and **b** but with monthly time series low-pass filtered at 2 years

The energy budget estimate of the NAD allows us to analyze the causes of the temporal variability in the NAD OHCT and evaluate the role of heat transport convergence in NAD OHCT. We find that, on average over 2005–2018, the heat transport convergence in NAD is almost compensated by the NAD surface heat loss to the atmosphere leading to a 2005–2018 mean NAD OHCT of 0.062 ± 0.17 PW which is close to zero. On interannual time scales, the picture is different. The role of surface heat fluxes is sizable on the OHCT interannual variability but remains small except for some periods such as at the end of 2009 and at the beginning of 2012. The NAD heat convergence shows an increase from 2014 to 2018 that is captured by both the in situ estimate and the ocean energy budget. This increase results in a warming of the ocean in the NAD (OHCT positive) that is concomitant with a decrease of heat loss from the ocean to the atmosphere (decrease in surface heat flux).

All conclusions presented above hold for estimates of OHCT derived from any OHC product considered in this study. But, they show different levels of correlation, variance, and rms differences. Part of these differences in terms of correlation variance and rms differences is likely explained by the limited depth of several OHC products (indeed products based on in situ data except ISAS21, only cover the upper ocean down to 2000 m depth and probably miss important parts of the deeper branch of the AMOC) and their limited resolution. On annual time scales, products that use satellite altimetry (Magellium/LEGOS, RFROM, ECCO4) show better correlation between the energy budget estimate and the in situ estimate of the NAD heat transport convergence (Fig. 6a, b). They also show better or equivalent rms difference, but RFROM standard deviations are lower than that of the NAD heat transport convergence and the Magellium/LEGOS are higher (as are the NOC_OI standard deviations). On 2 year and longer time scales, the correlation increases for all product and the difference in correlation among them is reduced (Fig. 6a). For that time scale, the Magellium/LEGOS and NOC_OI standard deviations are still noticeably higher than the others. While the lateral heat transport convergence is cast as the reference in the Taylor diagrams, all of the terms in the heat budget have uncertainties and some errors can also be hidden in the estimate of the lateral heat transport convergence.

Generally, on 2 year and longer time scales, the energy budget estimates of the NAD heat transport convergence show similar temporal variability regardless of the product used for OHCT (Fig. 6b). Despite this similar temporal variability across energy budget estimates, some systematic differences remain with the in situ estimate of the NAD heat transport convergence, in particular before 2009 (see Fig. 6). These discrepancies could be due to some systematic errors common to all OHC content products. But, some of the OHC products (e.g., Magellium/LEGOS) are based on largely independent data. It rather suggests some errors in other terms of the energy budget such as the surface heat flux component, the assumption that the OSNAP MHT has not changed over 2005–2014, or the approximation associated with the net zero mass transport constraint applied to RAPID and OSNAP heat transport calculation. The latter option is unlikely given the magnitude of the difference between the energy-based and the in situ-based estimate of the NAD heat transport convergence, in particular before 2009. Previous studies also suspected some issue in the in situ estimate of the RAPID MHT before 2009 (Trenberth and Fasullo 2017).

On annual time scales, the situation is slightly different as we find some significant differences in the temporal variability across the energy budget-based estimates of the NAD heat transport convergence. The differences are particularly large during 2009 and 2011–2014. They may be due to several issues including mapping errors, differences in the climatology, differences in the data quality criteria, and others.

5 Discussion and Conclusions

In this study, we evaluate the heat transport convergence over the North Atlantic with an energy budget approach using different ocean heat content products derived from in situ temperature measurements, satellite altimetry, satellite gravimetry data, and reanalysis. We compare the energy-based estimates of the North Atlantic heat transport convergence with an estimate derived from the RAPID and OSNAP arrays' in situ data over the period of 2005–2018. Confirming previous studies (Trenberth and Fasullo 2017, Forget and Ferreira 2019, Liu et al. 2020, Mayer et al. 2022), we find a good agreement in terms of mean (agreement at precision of ±0.24 PW) and a reasonable agreement in terms of annual and interannual temporal variability between the energy-based estimates and the in situ-based estimate of the North Atlantic heat transport convergence (with significant correlations between 0.22 and 0.7 for time series low-pass filtered at 1 year). The agreement in temporal variability increases at 2-year and longer time scales (with significant correlations between 0.35 and 0.83 for time series low-pass filtered at 2 years). We close the oceanic energy budget in the North Atlantic Ocean to within ± 0.2 PW (residual STD) for the closed ocean domain between RAPID array and the OSNAP array with an average budget residual of 0.06 PW (residual mean).

We find that the temporal mean spatial variability of the local North Atlantic heat transport convergence is compensated for by the local time-mean surface heat fluxes leading to a time-mean OHCT that is closed to uniform over 2005–2018 in the NAD. However, it is the North Atlantic heat transport convergence that explains most of the time variability of OHCT averaged over the NAD. So, the agreement in terms of annual and interannual temporal variability between the energy-based and the in situ-based estimate of the North Atlantic heat transport convergence means that the OHC products contain part of the information associated to the MHT. This is true at 2 year and longer time scales for all OHC products tested but less so at annual time scales where OHC products using satellite altimetry tend to perform better. The comparisons made here provide an independent framework for assessing regional ocean heat content and their capacity to represent the ocean heat transport means and variability, in particular, the heat transport associated with the main branches of the overturning circulation (Forget and Ferreira 2019; Rousselet et al 2021). Such validation is useful if ocean heat content products are to be used in objective regionalized Earth energy budget to evaluate the changes in regional energy fluxes as in the GEWEX effort from Stephens et al. (2023). In this respect, continuous monitoring of the heat transport at ocean cross-sections such as RAPID and OSNAP is essential to test the closure of ocean energy budget and validate ocean heat content estimates at regional scale.

The OHC product that leads to the highest correlation between the energy-based and the in situ-based estimates of the NAD heat transport convergence on annual time scales is based on satellite altimetry and space gravimetry data. It shows a correlation of 0.54 for annual time series over 2005–2018 (0.67 for biannual time series). This is comparable to Trenberth and Fasullo (2017) but smaller than Liu et al. (2020) who obtained a correlation of 0.66 and Mayer et al. (2022) who obtained a correlation of 0.72. However, our approach is different from Trenberth and Fasullo (2017) and Liu et al. (2020) attempts, which inferred OHT between the RAPID section and the Bering Strait, and from Mayer et al. (2022) attempt which inferred OHT between the RAPID section and the Greenland–Scotland Ridge and Davis Strait. It is unlikely though that the difference in the domain size explain the difference in correlation as the time variability in heat transport convergence over the North Atlantic is generally dominated by the RAPID MHT variability. Liu et al. (2020) and Mayer et al (2022) have in common that they use the ocean reanalysis ORAS5 to infer the OHCT. We tested ORAS5 to derive the OHCT, and we found indeed a better correlation of 0.7 (for annual time series and 0.82 for time series filtered at 2 years). A simple interpretation of this better performance of ORAS5 is that ORAS5 better resolves the Atlantic MHT in its OHC reconstruction. However, ORAS5 is an ocean reanalysis which uses ERA-interim atmospheric reanalysis for the surface forcing. Over the common period 2005–2014, ERA-interim surface heat fluxes averaged over the NAD are close to ERA5 surface fluxes (not shown). We suspect that using a similar surface heat flux to force ORAS5 reanalysis and to infer the ocean energy budget, leads to compensation of errors in Liu et al. (2020) and Mayer et al. (2022). In our study, these errors cannot compensate and may lead to a poorer correlation with in situ-based estimate of the heat transport convergence. More analyses are needed to test this hypothesis.

The drop from 2005 to 2009 in NAD heat transport convergence observed by the in situ data of RAPID and OSNAP is not captured by any energy budget estimate of this study, regardless of the OHC product used. It is not captured by previous energy budget studies either (Trenberth and Fasullo 2017, Liu et al. 2020, Mayer et al. 2022). In this study, we find that this discrepancy is unlikely related to the OHC data as we tested different estimates of OHC based on independent data. It raises questions on the cause for this discrepancy. Is it an issue in surface heat fluxes? Or in the in situ data? How substantial is the vertical heat flux through the 2000 m (the bottom depth of most of the ocean heat content estimates) in the NAD? Can we really assume no changes in the MHT variability across the OSNAP section before 2014?

Although we find the North Atlantic Ocean energy budget is closed on annual time scales, there is room for improvement. It would be interesting to get to monthly estimates but this objective remains challenging for the argo observing system and for the space gravimetry observing system, but also the in situ-measurements in the ocean. In addition, the processing of the convergence (interpolation and truncation) and the processing of the tendency (derivation) both amplify the noise and introduce uncertainties that are larger for higher temporal and spatial resolution. These issues pose a challenge as well.

Acknowledgements This paper is an outcome of the Workshop "Challenges in Understanding the Global Water Energy Cycle and its Changes in Response to Greenhouse Gas Emissions" co-sponsored by the GEWEX Data Analysis Panel and held at the International Space Science Institute (ISSI) in Bern, Switzerland (26–30 September 2022). This work was supported by the European Space Agency (ESA) 4DATLANTIC-OHC project study contract no. 4000134928/21/I-NB. It was also supported by the Centre National d'Etudes Spatiales, with a focus on Sentinel 6/MF. GCJ and JML are supported by NOAA Research and the NOAA Global Ocean Monitoring and Observation Program. PMEL Contribution Number 5573. MM was supported by Austrian Science Fund P33177. Donata Giglio and Mikael Kuusela were supported by NOAA (awards NA21OAR4310261 and NA21OAR4310258). Gael Forget acknowledges support from NASA awards 80NSSC20K0796, 80NSSC23K0355, 80NSSC22K1697, 1676067, and 1686358. ISAS temperature and salinity monthly gridded field products are made freely available by SNO Argo France at LOPS Laboratory (supported by UBO/CNRS/Ifremer/IRD) and IUEM Observatory (OSU IUEM/CNRS/INSU) at doi: https://doi.org/https://doi.org/10.17882/52367. We thank Annaig Prigent (LOPS/IFREMER) for the computation of ISAS21 gridded product.

Funding Global Ocean Monitoring and Observing Program,5573,Gregory C Johnson,5573, John M Lyman,NA21OAR4310261,Donata Giglio,NA21OAR4310258,Mikael Kuusela,Austrian Science Fund,P33177, Michael Mayer.

Material Availability The authors have no financial or proprietary interests in any material discussed in this article.

Declarations

Conflict of interest The authors have no relevant financial or non-financial interests to disclose. The authors have no competing interests to declare that are relevant to the content of this article. All authors certify that they have no affiliations with or involvement in any organization or entity with any financial interest or non-financial interest in the subject matter or materials discussed in this manuscript.

Open Access This article is licensed under a Creative Commons Attribution 4.0 International License, which permits use, sharing, adaptation, distribution and reproduction in any medium or format, as long as you give appropriate credit to the original author(s) and the source, provide a link to the Creative Commons licence, and indicate if changes were made. The images or other third party material in this article are included in the article's Creative Commons licence, unless indicated otherwise in a credit line to the material. If material is not included in the article's Creative Commons licence and your intended use is not permitted by statutory regulation or exceeds the permitted use, you will need to obtain permission directly from the copyright holder. To view a copy of this licence, visit http://creativecommons.org/licenses/by/4.0/.

References

Blazquez A, Meyssignac B, Lemoine J, Berthier E, Ribes A, Cazenave A (2018) Exploring the uncertainty in GRACE estimates of the mass redistributions at the Earth surface: implications for the global water and sea level budgets. Geophys J Int 215:415–430. https://doi.org/10.1093/gji/ggy293

Bryden HL, Imawaki S (2001) Ocean heat transport. Int Geophys. Acad Press 77:455–474

Cheng L, Zhu J (2016) Benefits of CMIP5 multimodel ensemble in reconstructing historical ocean subsurface temperature variation. J Climate 29:5393–5416. https://doi.org/10.1175/JCLI-D-15-0730.1

Cheng L, Trenberth KE, Fasullo J, Boyer T, Abraham J, Zhu J (2017) Improved estimates of ocean heat content from 1960–2015. Sci Adv 3:e1601545. https://doi.org/10.1126/sciadv.1601545

Collins M, Sutherland M, Bouwer L, Cheong S-M, Frölicher, Jacot Des Combes H, Koll Roxy M, Losada I, McInnes K, Ratter B, Rivera-Arriaga E, Susanto RD, Swingedouw D, Tibig L (2019) Extremes, abrupt changes and managing risk. In: IPCC special report on the ocean and cryosphere in a changing climate [Pörtner H-O, Roberts DC, Masson-Delmotte V, Zhai P, Tignor M, Poloczanska E, Mintenbeck K, Alegría A, Nicolai M, Okem A, Petzold J, Rama B, Weyer NM (eds)]. In press.

Dee DP et al (2011) The ERA-Interim reanalysis: configuration and performance of the data assimilation system. Q J R Meteorol Soc 137:553–597

Forget G, Ferreira D (2019) Global ocean heat transport dominated by heat export from the tropical Pacific. Nat Geosci 12:351–354. https://doi.org/10.1038/s41561-019-0333-7

Forget G, Ponte RM (2015) The partition of regional sea level variability. Prog Oceanogr 137:173–195

Forget G, Campin J-M, Heimbach P, Hill CN, Ponte RM, Wunsch C (2015a) ECCO version 4: an integrated framework for non-linear inverse modeling and global ocean state estimation. Geosci Model Dev 8:3071–3104. https://doi.org/10.5194/gmd-8-3071-2015

Forget G, Ferreira D, Liang X (2015b) On the observability of turbulent transport rates by Argo: supporting evidence from an inversion experiment. Ocean Sci 11(5):839–853

Frajka-Williams E et al (2019) Atlantic meridional overturning circulation: observed transport and variability. Front Mar Sci 6:260. https://doi.org/10.3389/fmars.2019.00260

Fukumori I, Wang O, Fenty I, Forget G, Heimbach P, Ponte RM (2021) Synopsis of the ECCO central production global ocean and sea-ice state estimate, version 4 release 4 (4 release 4). Zenodo. https://doi.org/10.5281/zenodo.4533349

Gaillard F, Reynaud T, Thierry V, Kolodziejczyk N, von Schuckmann K (2016) In situ-based reanalysis of the global ocean temperature and salinity with ISAS: variability of the heat content and steric height. J Clim 29:1305–1323. https://doi.org/10.1175/JCLI-D-15-0028.1

Ganachaud A, Wunsch C (2000) Improved estimates of global ocean circulation, heat transport and mixing from hydrographic data. Nature 408:453–457

Giglio D., Sukianto T., Kuusela M. (2023). Ocean heat content anomalies in the North Atlantic based on mapping Argo data using local Gaussian processes defined over space (1.0.0). enodo. https://doi.org/10.5281/zenodo.10183869

Hersbach H, Bell B, Berrisford P, Hirahara S, Horányi A, Muñoz-Sabater J, Thépaut JN (2020) The ERA5 global reanalysis. Quart J Roy Meteor Soc 146:1999–2049. https://doi.org/10.1002/qj.3803

Jackson LC, Dubois C, Forget G, Haines K, Harrison M, Iovino D et al (2019) The mean state and variability of the North Atlantic circulation: A perspective from ocean reanalyses. J Geophys Res: Oceans 124:9141–9170. https://doi.org/10.1029/2019JC015210

Johns WE et al (2011) Continuous, array-based estimates of Atlantic Ocean heat transport at 26.5°N. J Clim 24:2429–2449

King BA (2023) Objectively mapped Argo profiling float data and RAPID moored microcat data from the North Atlantic Ocean, 2004–2022. NERC EDS Br Oceanogr Data Centre NOC. https://doi.org/10.5285/fe8e524d-7f04-41f3-e053-6c86abc04d51

Kolodziejczyk N, Prigent-Mazella A, Gaillard F (2021). ISAS temperature and salinity gridded fields. SEANOE. https://doi.org/10.17882/52367

Kolodziejczyk Nicolas, Prigent-Mazella Annaig, Gaillard Fabienne (2023). ISAS temperature, salinity, dissolved oxygen gridded fields. SEANOE. https://doi.org/10.17882/52367

Kuusela M., Stein M.L. (2018) Locally stationary spatio-temporal interpolation of Argo profiling float data. In: Proc. R. Soc. A, 474, pp 20180400. https://doi.org/10.1098/rspa.2018.0400

L'Ecuyer TS, Beaudoing HK, Rodell M, Olson W, Lin B, Kato S, Clayson CA et al (2015) The observed state of the energy budget in the early twenty-first century. J Clim 21:8319–8346

Legeais J-F, Meyssignac B, Faugère Y, Guerou A, Ablain M, Pujol M-I, Dufau S, Dibarboure G (2021) Copernicus sea level space observations: a basis for assessing mitigation and developing adaptation strategies to sea level rise. Front Mar Sci. https://doi.org/10.3389/fmars.2021.704721

Levitus S, Antonov JI, Boyer TP, Locarnini RA, Garcia HE, Mishonov AV (2009) Global ocean heat content 1955–2008 in light of recently revealed instrumentation problems. Geophys Res Lett. https://doi.org/10.1029/2008GL037155

Levitus S et al (2012) World ocean heat content and thermosteric sea level change (0–2000 m), 1955–2010. Geophys Res Lett 39:L10603. https://doi.org/10.1029/2012GL051106

Li F, Lozier M, Holliday N, Johns W, Le Bras I, Moat B, Cunningham S, de Jong M (2021) Observation-based estimates of heat and freshwater exchanges from the subtropical North Atlantic to the Arctic. Prog Oceanogr 197:102640

Liu C et al (2020) Variability in the global energy budget and transports 1985-2017. Climate Dyn 55:3381–3396. https://doi.org/10.1007/s00382-020-05451-8

Locarnini RA, Mishonov AV, Antonov JI, Boyer TP, Garcia HE (2010) World Ocean Atlas 2009, Volume1: Temperature. In: S. Levitus, Ed., NOAA Atlas NESDIS 68, U.S. Gov. Printing Office, Washington, D.C., pp 184

Loeb NG, Thorsen TJ, Norris JR, Hailan W, Wenying S (2018) Changes in Earth's energy budget during and after the "pause" in global warming: an observational perspective. Climate 6(3):62. https://doi.org/10.3390/cli6030062

Lozier MS, Li F, Bacon S, Bahr F, Bower AS, Cunningham SA, de Jong MF, de Steur L, de Young B, Fischer J, Gary SF, Greenan BJW, Holliday NP, Houk A, Houpert L, Inall ME, Johns WE, Johnson HL, Johnson C, Karstensen J, Koman G, Le Bras IA, Lin X, Mackay N, Marshall DP, Mercier H, Oltmanns M, Pickart RS, Ramsey AL, Rayner D, Straneo VF, Thierry Torres DJ, Williams RG, Wilson C, Yang J, Yashayaev I, Zhao I (2019) A sea change in our view of overturning in the subpolar North Atlantic. Science 363(6426):516–521

Lyman JM, Johnson GC (2023) Global high-resolution random forest regression maps of ocean heat content anomalies using in situ and satellite data. J Atmos Ocean Technol 40(5):575–586. https://doi.org/10.1175/JTECH-D-22-0058.1

Macdonald AM, Candela J, Bryden HL (1994) An estimate of the net heat transport through the Strait of Gibraltar. In: Seasonal and Interannual Variability of the Western Mediterranean Sea, LaViolette PE, (ed) Amer Geophys. Union, pp 13–32

Magellium/LEGOS: Atlantic OHC from space: Heat content change over the Atlantic Ocean by space geodetic approach, https://doi.org/10.24400/527896/A01-2022.012, 2022.

Marti F, Blazquez A, Meyssignac B, Ablain M, Barnoud A, Fraudeau R, Jugier R, Chenal J, Larnicol G, Pfeffer J, Restano M, Benveniste J (2022) Monitoring the ocean heat content change and the Earth energy imbalance from space altimetry and space gravimetry. Earth Syst Sci Data 14:229–249. https://doi.org/10.5194/essd-14-229-2022

Marti F, Meyssignac B, Rousseau V, Ablain M, Fraudeau R, Blazquez A, Fourest S (2024) Monitoring global ocean heat content from space geodetic observations to estimate the Earth energy imbalance. In: von Schuckmann K, Moreira L, Grégoire M, Marcos M, Staneva J, Brasseur P, Garric G, Lionello P, Karstensen J, Neukermans G (eds) 8th edition of the Copernicus Ocean State Report (OSR8). Copernicus Publications, State Planet, 4-osr8, 3. https://doi.org/10.5194/sp-4-osr8-3-2024

Mayer J, Haimberger MML (2021) Consistency and homogeneity of atmospheric energy, moisture, and mass budgets in ERA5. J Climate 34:3955–3974. https://doi.org/10.1175/JCLI-D-20-0676.1

Mayer M, Haimberger L, Edwards JM, Hyder P (2017) Toward consistent diagnostics of the coupled atmosphere and ocean energy budgets. J Climate 30:9225–9246. https://doi.org/10.1175/JCLI-D-17-0137.1

Mayer J, Mayer M, Haimberger L (2021) Mass-consistent atmospheric energy and moisture budget monthly data from 1979 to present derived from ERA5 reanalysis. Copernic Clim Change Serv (C3S) Clim Data Store (CDS). https://doi.org/10.24381/cds.c2451f6b

Mayer J, Mayer M, Haimberger L, Liu C (2022) Comparison of surface energy fluxes from global to local scale. J Clim. https://doi.org/10.1175/JCLI-D-21-0598.1

Mayer M, Kato S, Bosilovich M et al (2024) Assessment of atmospheric and surface energy budgets using observation-based data products. Surv Geophys. https://doi-org.insu.bib.cnrs.fr/10.1007/s10712-024-09827-x

McCarthy G, Frajka-Williams E, Johns WE, Baringer MO, Meinen CS, Bryden HL, Rayner D, Duchez A, Roberts CD, Cunningham SA (2012) Observed interannual variability of the Atlantic meridional overturning circulation at 26.5°N. Geophys Res Lett 39:L19609. https://doi.org/10.1029/2012GL052933

McCarthy GD, Smeed DA, Johns WE, Frajka-Williams E, Moat BI, Rayner D, Baringer MO, Meinen CS, Collins J, Bryden HL (2015) Measuring the Atlantic meridional overturning circulation at 26°N. Prog Oceanogr 130:91–111

Mecking JV, Drijfhout SS (2023) The decrease in ocean heat transport in response to global warming. Nat Clim Change. https://doi.org/10.1038/s41558-023-01829-8

Melet A, Meyssignac B (2015) Explaining the spread in global mean thermosteric sea level rise in CMIP5 climate models. J Clim 28:9918–9940. https://doi.org/10.1175/JCLI-D-15-0200.1

Palter JB (2015) The role of the Gulf stream in European climate. Ann Rev Mar Sci 7(1):113–137

Prandi P, Meyssignac B, Ablain M, Spada G, Ribes A, Benveniste J (2021) Local sea level trends, accelerations and uncertainties over 1993–2019. Sci Data 8:1. https://doi.org/10.1038/s41597-020-00786-7

Rhines P, Häkkinen S, Josey SA (2008) Is oceanic heat transport significant in the climate system? In: Dickson RR, Meincke J, Rhines P (eds) Arctic-Subarctic Ocean Fluxes. Springer, Dordrecht

Rousselet L, Cessi P, Forget G (2021) Coupling of the mid-depth and abyssal components of the global overturning circulation according to a state estimate. Sci Adv 7:eabf5478. https://doi.org/10.1126/sciadv.abf5478

Stephens G, Polcher J, Zeng X, Van Oevelen P, Poveda G, Bosilovich M, Ahn MH, Balsamo G, Duan Q, Hegerl G, Jakob C (2023) The first 30 years of GEWEX. Bull Amer Meteor Soc 104:E126–E157. https://doi.org/10.1175/BAMS-D-22-0061.1

Storto A et al (2019) Ocean reanalyses: recent advances and unsolved challenges. Front Mar Sci 6:418

Trenberth KE, Caron JM (2001) Estimates of meridional atmosphere and ocean heat transports. J Clim 14:3433–3443

Trenberth KE, Fasullo JT (2017) Atlantic meridional heat transports computed from balancing Earth's energy locally. Geophys Res Lett 44:1919–1927. https://doi.org/10.1002/2016GL072475

Wu P, Haines K (1998) The general circulation of the Mediterranean Sea from a 100-year simulation. J Geophys Res 103:1121–1135. https://doi.org/10.1029/97JC02720

Zuo H, Balmaseda MA, Tietsche S, Mogensen K, Mayer M (2019) The ECMWF operational ensemble reanalysis–analysis system for ocean and sea ice: a description of the system and assessment. Ocean Sci 15(3):779–808. https://doi.org/10.5194/os-15-779-2019

Zuo H, Balmaseda MA, de Boisseson E, Tietsche S, Mayer M, de Rosnay P, (2021) The ORAP6 ocean and sea-ice reanalysis: Description and evaluation. EGU General Assembly Conference Abstracts, EGU21-9997, EGU General Assembly Conference Abstracts.

Publisher's Note Springer Nature remains neutral with regard to jurisdictional claims in published maps and institutional affiliations.

Authors and Affiliations

B. Meyssignac[1] · S. Fourest[1] · Michael Mayer[2,3] · G. C. Johnson[4] · F. M. Calafat[5] · M. Ablain[6] · T. Boyer[7] · L. Cheng[8] · D. Desbruyères[9] · G. Forget[10] · D. Giglio[11] · M. Kuusela[12] · R. Locarnini[7] · J. M. Lyman[4,13] · W. Llovel[9] · A. Mishonov[7,14] · J. Reagan[7] · V. Rousseau[6] · J. Benveniste[15]

✉ Michael Mayer
michael.mayer@univie.ac.at

[1] Université de Toulouse, LEGOS (CNES/CNRS/IRD/UT3), 31400 Toulouse, France

[2] Research Department, European Centre for Medium-Range Weather Forecasts, 53175 Bonn, Germany

[3] Department for Meteorology and Geophysics, University of Vienna, 1090 Vienna, Austria

[4] NOAA/Pacific Marine Environmental Laboratory, Seattle, Washington 98115, USA

[5] National Oceanography Centre, Liverpool L3 5DA, UK

[6] MAGELLIUM, 31250 Ramonville Saint-Agne, France

[7] NOAA National Centers for Environmental Information, Silver Spring, MD 20910, USA

[8] Institute of Atmospheric Physics, Chinese Academy of Sciences, Beijing 100029, China

[9] CNRS, Ifremer, IRD, Laboratoire d'Océanographie Physique et Spatiale (LOPS), IUEM,

University of Brest, 29280 Plouzané, France

[10] Department of Earth, Atmospheric and Planetary Sciences, Massachusetts Institute of Technology, Cambridge, MA 02139-4307, USA

[11] Department of Atmospheric and Oceanic Sciences, University of Colorado Boulder, Boulder, CO 80309-0311, USA

[12] Department of Statistics and Data Science, Carnegie Mellon University, Pittsburgh, PA 15213, USA

[13] CIMAR, University of Hawaii, Honolulu, HI 96822, USA

[14] Cooperative Institute for Satellite Earth Systems Studies, Earth System Science Interdisciplinary Center, University of Maryland, College Park, MD 20742, USA

[15] European Space Agency (ESA-ESRIN), 00044 Frascati, Italy

Surveys in Geophysics (2024) 45:1875–1902
https://doi.org/10.1007/s10712-024-09860-w

An Abrupt Decline in Global Terrestrial Water Storage and Its Relationship with Sea Level Change

Matthew Rodell[1] · Anne Barnoud[2] · Franklin R. Robertson[3] · Richard P. Allan[4] · Ashley Bellas-Manley[5] · Michael G. Bosilovich[1] · Don Chambers[6] · Felix Landerer[7] · Bryant Loomis[1] · R. Steven Nerem[5] · Mary Michael O'Neill[1,8] · David Wiese[7] · Sonia I. Seneviratne[9]

Received: 21 December 2023 / Accepted: 2 September 2024 / Published online: 4 November 2024
© The Author(s) 2024

Abstract

As observed by the Gravity Recovery and Climate Experiment (GRACE) and GRACE Follow On (GRACE-FO) missions, global terrestrial water storage (TWS), excluding ice sheets and glaciers, declined rapidly between May 2014 and March 2016. By 2023, it had not yet recovered, with the upper end of its range remaining 1 cm equivalent height of water below the upper end of the earlier range. Beginning with a record-setting drought in northeastern South America, a series of droughts on five continents helped to prevent global TWS from rebounding. While back-to-back El Niño events are largely responsible for the South American drought and others in the 2014–2016 timeframe, the possibility exists that global warming has contributed to a net drying of the land since then, through enhanced evapotranspiration and increasing frequency and intensity of drought. Corollary to the decline in global TWS since 2015 has been a rise in barystatic sea level (i.e., global mean ocean mass). However, we find no evidence that it is anything other than a coincidence that, also in 2015, two estimates of barystatic sea level change, one from GRACE/FO and the other from a combination of satellite altimetry and Argo float ocean temperature measurements, began to diverge. Herein, we discuss both the mechanisms that account for the abrupt decline in terrestrial water storage and the possible explanations for the divergence of the barystatic sea level change estimates.

Keywords Climate change · Terrestrial water storage · Sea level · GRACE

Article Highlights

- Global terrestrial water storage, excluding glaciers and ice sheets, declined abruptly between May 2014 and March 2016, with a corollary increase in sea level
- A series of droughts, possibly linked to global warming, has since helped to prevent global terrestrial water storage from recovering
- Also around 2015, two independent estimates of barystatic sea level began to diverge, but we find no evidence of a connection with the terrestrial water storage decline

---

Extended author information available on the last page of the article

1 Introduction

Terrestrial water storage (TWS; i.e., the sum of groundwater, soil moisture, surface waters, snow water equivalent, and ice) is an Essential Climate Variable (https://gcos.wmo.int/en/essential-climate-variables/tws) and a natural resource vital to ecosystems and societies. It exhibits substantial variability seasonally and over longer periods due to climate change and human water usage (Rodell et al. 2018; Intergovernmental Panel On Climate Change 2023). Some TWS variations are better understood in terms of physical processes than others, and understanding is limited by a satellite observational record (2002–present) that is short relative to most in situ data records and model analyses. In this study, we investigate an apparent abrupt decline in global, unfrozen TWS during 2014–2016 and a simultaneous divergence of independent estimates of changes in barystatic sea level (BSL, sometimes called global mean ocean mass), based on a combination of satellite data and observations-forced model and reanalysis output.

Observed changes in water stored on and in the land surface are balanced almost perfectly by changes in water stored in the ocean and atmosphere. There is a net loss of water from the ocean to the mantle, which is on the order of 0.4 to 1.3 GT/yr (Bounama et al. 2001). To put that into context, eight estimates of the linear trend in BSL based on Gravity Recovery and Climate Experiment (GRACE) (Tapley et al. 2004) and GRACE Follow On (GRACE-FO) (Landerer et al. 2020) satellite gravimetry measurements, an ensemble compiled by Chen et al. (2020), had a standard deviation of 0.20 mm/yr. Using the conversion rate of 360 GT water (~360 km^3 water) per 1.0 mm sea level change, that equates to 72 GT/yr uncertainty in BSL change. The net loss of water to the mantle is at least an order of magnitude smaller. There is a tiny net loss of water (as hydrogen) from the stratosphere to space, less than 1 GT per 1000 years (Bounama et al. 2001). Ignoring rare impacts of > 1000 kg meteors, Earth's current accretion rate of extraterrestrial material is smaller still, 0.049–0.056 GT per 1000 years (Esser and Turekian 1988; Love and Brownlee 1993), of which only a fraction may be water. Ergo, on timescales of seconds to centuries, it is appropriate to assume that the global water cycle is a closed system in which the law of conservation of mass applies. Multi-annual changes in global atmospheric water storage are small compared with those in ocean or land water mass, though not necessarily negligible. Trent et al. (2023) compared multiple sources of atmospheric water storage variations and trends, finding that recent trends are less than 0.5 mm/decade (excluding one outlier estimate). Regional variations can be larger owing to significant modes of variability (e.g., ENSO) and human-induced climate change (Intergovernmental Panel On Climate Change 2023), with local variations of ±3 mm. Changes in atmospheric water vapor are substantially constrained by the Clausius-Clapeyron relationship (Trenberth et al. 2005; Allan et al. 2022; Intergovernmental Panel On Climate Change 2023) to small long-term trends at the global scale. Though small, global atmospheric moisture storage changes may be non-negligible when balancing TWS against ocean mass changes over decadal timescales and a source of error if ignored. Polar ice sheet and major glacier system mass changes are often separated from the remaining TWS in sea level budget accounting. Hereafter, unless otherwise noted, we define TWS as the aggregate of groundwater, soil moisture, surface waters, and ephemeral snow and ice (i.e., excluding ice sheets and major glacier systems).

Changes in global mean sea level (GMSL) comprise both the barystatic component due to ocean mass changes and the thermosteric component related to the temperature and salinity of the ocean water (Gregory et al. 2019). The rate of sea level rise during 1993–2017 was about 3.0 mm/yr—roughly 55% thermosteric and 45% barystatic gains

(Nerem et al. 2018). Most of the barystatic gain is attributed to ablation of polar ice sheets and glaciers. GRACE and GRACE-FO (collectively, GRACE/FO) data indicate that the Greenland and Antarctic ice sheets have contributed 261 ± 45 GT/yr and 104 ± 57 GT/yr to BSL since 2002 (Velicogna et al. 2020). The rest of the world's glacier systems contribute another 199 ± 32 GT/yr to BSL, including 53 ± 14 GT/yr from the glaciers along the gulf coast of Alaska (Wouters et al. 2019). While ice loss is the dominant driver of ocean mass change, Cáceres et al. (2020) showed that non-ice TWS is the primary control on seasonal and interannual variations of the land–ocean water balance excluding Greenland and Antarctica.

Shortly after launching in 2002, GRACE's measurements of time variable gravity changes proved valuable for elucidating characteristics of the mass balance between ocean and land, including changes in TWS (Chambers et al. 2004; Chen et al. 2005; Ramillien et al. 2008; Cazenave and Llovel 2010; Riva et al. 2010). More recently, Wada et al. (2016) focused on the contribution of groundwater pumping and depletion to sea level rise. Chandanpurkar et al. (2021) investigated the amplitudes of seasonal exchanges of water between the global land and oceans during the GRACE/FO era, which average about 17 mm equivalent sea level with significant interannual variability. GRACE allowed scientists to explain that the unusual decline in global mean sea level during 2010–2011 was caused by a massive increase in TWS, largely in Australia (Boening et al. 2012; Fasullo et al. 2013). Rodell et al. (2015) and L'Ecuyer et al. (2015) constrained estimates of mean monthly water and energy cycle fluxes at continental and global scales by enforcing water and energy budget closure, with GRACE-based estimates of seasonal changes in land and ocean water storage playing a key role. Of particular relevance, Reager et al. (2016) calculated that regional increases in precipitation during 2002–2014 had raised global TWS to the extent that the rate of sea level rise was reduced by about 15% after removing the effects of irrigation-enhanced groundwater depletion. Rietbroek et al. (2016) obtained similar results. However, improvements in two key sources of auxiliary information used in GRACE/FO data processing (a glacial isostatic adjustment (GIA) model and the determination of the C_{20} spherical harmonic coefficient based on satellite laser ranging data) flipped what had been a 0.32 mm/yr sea level equivalent increase in TWS to a 0.09 mm/yr decrease over the period of April 2002 to November 2014, noting that this magnitude of change is within a standard two-sigma uncertainty range for systematic errors (Chambers et al. 2017).

The timeframe of the latter two studies, 2002–2014, immediately preceded a rapid decrease and two decade minimum in TWS that is the focus of the present study. As shown in Fig. 1, global mean TWS anomalies (deviations from the 2003–2020 average) remained within the range -8 to +16 mm from the onset of observations until the end of 2014. By the end of 2015, a new record minimum of −15 mm had been set, and the upper end of the range in the following years was about +6 mm. The April 2002 to December 2014 average anomaly was +33 mm, and the January 2015 to May 2023 average was −51 mm. As detailed in the Results, the two independently derived estimates of BSL, one from GRACE/FO and the other from satellite altimetry minus Argo float based thermosteric sea level change, began to diverge, also around 2015, after being generally consistent beforehand (Barnoud et al. 2021). This led to speculation in the climate change community about errors or drift in one or more of the observational time series. The main purpose of the present study is to investigate the abrupt decline in TWS. We also discuss uncertainties in the various contributing datasets that may have led to the divergence between the direct (from GRACE/FO) and indirect (altimeter minus Argo) BSL estimates of global mean thermosteric sea level change, but we do not offer a definitive conclusion.

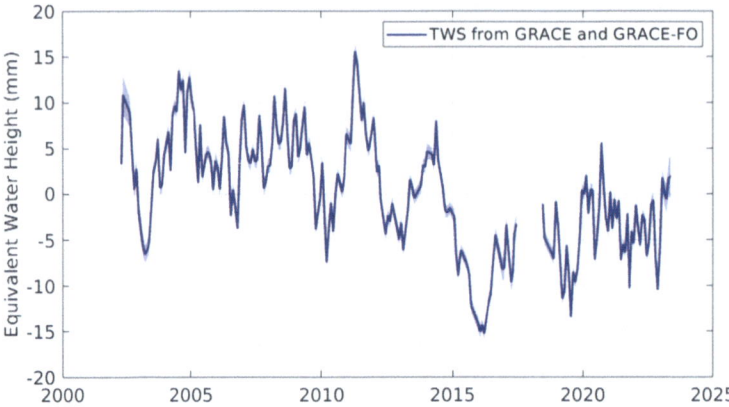

Fig. 1 Anomalies (relative to the 2003–2020 mean) of global mean terrestrial water storage (equivalent height of water over land) from GRACE and GRACE-FO after removing annual and semiannual components and the S2 and K2 tidal alias terms. Ice covered land regions excluded from the global mean are shown in Figure S1. The shading conveys the formal error estimates, which average about ±0.9 mm

2 Data and Methods

2.1 GRACE and GRACE-FO

GRACE (2002–2017) was a twin satellite mission that produced global, monthly fields of surface gravitational and (equivalently) mass anomalies. Micron-scale microwave-ranging measurements of changes in the distance between the two satellites (nominally around 200 km) as they orbited the Earth were combined with precise location and onboard accelerometer data to infer surface mass effects on the orbits and hence to derive the global anomaly fields (Wahr et al. 1998; Tapley et al. 2004). GRACE-FO (2018-present) is nearly identical to GRACE, with the addition of a laser ranging instrument (Landerer et al. 2020). The mass anomaly data provided by both missions have proven valuable for climatology, oceanography, cryoscience, carbon cycle science, and hydrology (Humphrey et al. 2018, 2023; Tapley et al. 2019; Flechtner et al. 2021; Rodell and Reager 2023).

We used the Jet Propulsion Laboratory RL06.1Mv03 GRACE/GRACE-FO mascon solution for our primary estimates of TWS anomalies (Wiese et al. 2016). The solution defines 4,551 equal-area 3-degree spherical cap mascon (mass concentration) elements directly from the intersatellite range-rate measurements within a Bayesian framework to remove correlated errors (Watkins et al. 2015). A coastal resolution improvement filter is subsequently applied to separate ocean and land mass from mascons that span coastlines (Wiese et al. 2016). Tidal alias terms (Ray and Luthcke 2006) were co-estimated along with the annual and semiannual components from the TWS time series. During the estimation process, a phase shift of 100 days was applied to each GRACE-FO alias term relative to those for GRACE in order to account for nodal plane differences. We used the formal errors that are distributed with the JPL mascon product, which represent the GRACE/FO measurement system error. We combined these in quadrature with leakage error associated with imperfections in the coastal resolution improvement filter by conservatively assuming the separation of land and ocean mass is in error by 50% (Wiese et al. 2016). Additional GRACE/FO data products (Save et al. 2016; Loomis et al. 2019) were used as noted below. Mascons in which ice cover exceeded 1% of the area were classified as "ice mascons" and were excluded from the TWS calculations (see Figure S2). Tests indicate that the ice coverage threshold (1%, 10%, etc.) does not affect the interannual variability as shown in Fig. 1; however, it does affect the trend. Thus, in an effort to be conservative in removing ice mass signal from the analysis, we used a 1% area cutoff to remove linear trends associated with ice mass loss.

In discussing the major droughts that have contributed to the step decrease in TWS, we refer to droughts and their durations and intensities as reported by Rodell and Li (2023). They applied a data clustering algorithm to the global, 2002–2021 GRACE/FO data record (Save et al. 2016; Save 2020) in order to identify spatially contiguous regions where the TWS anomaly exceeded one standard deviation from the mean seasonal cycle. The clusters of such data pixels were grouped temporally as well. This enabled automated, objective identification of droughts and pluvials (large scale and extreme wet events). These were then quantified using an intensity metric (Thomas et al. 2014) that represents the integral under the curve of non-seasonal TWS anomaly versus time, in units of km^3 months (equivalently, GT•months).

2.2 Ocean Altimetry and Temperature

Global mean sea level changes were computed using the daily gridded product provided by the Copernicus Climate Change Service (C3S), version vDT2021 (Legeais et al. 2021). The effects of GIA and of subsidence due to present-day ice melting were removed from the GMSL time series with respective values of −0.3 mm/yr (Peltier 2004) and −0.13 mm/yr (Frederikse et al. 2017; Lickley et al. 2018). The drift of the wet troposphere correction of Jason-3 altimetry was corrected using the empirical estimate of Barnoud et al. (2023). The standard uncertainties associated with the GMSL time series were derived from the method and uncertainty budget detailed by Guérou et al. (2023). Note that the GIA model is different from the one used for the processing of the gravimetric data, but the uncertainties associated with the GIA correction are incorporated in the GMSL uncertainty budget.

Global mean thermosteric contributions to GMSL change can be estimated by subtracting satellite gravimetry measurements of BSL change (global mean ocean mass change converted to an equivalent height of water) from altimetry measurements of GMSL change or can be computed directly from temperature and salinity observations in the water column (e.g., Jayne et al. 2003). As the global mean halosteric sea level change due to salinity changes is supposed to be negligible (Gregory and Lowe 2000; Llovel et al. 2019), we only accounted for the global mean thermosteric sea level component. Only the local time average over 2005–2015 of the salinity measurements was used in the computation of the thermosteric component, avoiding any problem linked to salinity measurement drift observed after 2015 (Wong et al. 2020; Barnoud et al. 2021). Despite the fact that Argo floats only sample the top 2000 m of the ocean, multiple studies have demonstrated that the two estimates match reasonably well over the period between 2005 and roughly 2016 (Chambers et al. 2017; World Climate Research Programme Global Sea Level Budget Group 2018; Barnoud et al. 2021). Herein, the thermosteric component was computed as the mean of an ensemble of 10 in situ datasets based on temperature and salinity measurements from the Argo network (Argo 2023). The 10 datasets included four EN4.2.2 datasets provided by the Met Office Hadley Center (Good et al. 2013) with four different combinations of corrections applied for the expendable bathythermograph (XBT) and mechanical bathythermograph (MBT) data—XBT correction from Gouretski and Reseghetti (2010) and MBT correction from Gouretski and Cheng (2020), XBT and MBT corrections from Levitus et al. (2009), XBT correction from Cowley et al. (2013) and MBT correction from Levitus et al. (2009), and XBT correction from Cheng et al. (2014) and MBT correction from Gouretski and Cheng (2020), a dataset from the Institute of Atmospheric Physics from the Chinese Academy of Sciences (Cheng et al. 2017, 2020), the In Situ Analyse System (ISAS) 20 dataset (Gaillard et al. 2016), data from Ishii et al. (2017), the Grid Point Value of the Monthly Objective Analysis using the Argo data (MOAA GPV) version 2021 data (Hosoda et al. 2010), a dataset from the National Oceanic and Atmospheric Administration (Levitus et al. 2012; Garcia et al. 2019), and the Roemmich and Gilson (2009) data from the Scripps Institute of Oceanography. From these 10 datasets, the thermosteric sea level change of the ocean was computed up to 2000 m depth. The contribution of the deep ocean (below 2000 m depth) was added with a linear trend of 0.12 mm/yr (Chang et al. 2019). The standard deviation of the 10 members of the ensemble was used as a measure of uncertainty in the thermosteric sea level change data.

2.3 Ocean Mass and Sea Level Budgets

The BSL was estimated over the ocean using the mean of three GRACE and GRACE-FO mascon solutions, from the Jet Propulsion Laboratory, the Center of Space Research (Save et al. 2016; Save 2020) and the Goddard Space Flight Center (Loomis et al. 2019). The ocean mass change was computed from the provided ocean bottom pressure data by removing the spatial mean of the so-called GAD product (Dobslaw et al. 2017) which accounts for the static atmospheric surface pressure (Chen et al. 2019). The effect of GIA was already removed from the mascon data using the ICE6G-D model (Peltier et al. 2018). The standard uncertainty of the BSL was conservatively estimated from the difference between the maximum and minimum values among the three solutions at each time.

Given BSL either from GRACE/FO or from altimetry less global mean thermosteric sea level, the contribution of TWS change to GMSL change can be computed by removing estimates of Greenland and Antarctica ice sheet mass losses, of other land glacier and ice cap ablation, and of atmosphere water vapor content variations. We corrected for the contribution of land ice mass changes using GRACE/FO data over glaciated regions (white areas in Figure S2). The monthly water vapor content variations of the atmosphere were computed from the European Centre for Medium-Range Forecasts atmospheric reanalysis version 5 (ERA5; Hersbach et al. 2020). A common mask was applied to altimetry, gravimetry, and Argo data to enable comparison. This mask excluded areas beyond $\pm 60°$ (lack of Argo data beyond $\pm 60°$ and of altimetry data beyond $\pm 66°$), closed seas (lack of Argo data) and Indonesian seas, and coastal areas up to 200 km away from the coasts (lack of Argo data and issues of gravimetric signal leakage near the coasts). The common mask ensures comparison of the altimetry, gravimetry, and Argo data over areas covered by all three observing systems with good quality data.

2.4 Satellite Laser Ranging

Prior to GRACE, space-based gravimetry was accomplished by satellite laser ranging (SLR) to mirrored, passive geodetic satellites, beginning with the launch of STARLETTE (*Satellite de taille adaptée avec réflecteurs laser pour les études de la terre*) in 1975 (Flechtner et al. 2021). After Stella launched in 1993, there were enough laser ranging satellites in orbit to derive mass change time series using the mascon approach (Rowlands et al.

2005) with sub-continental scale resolution. Herein, we use TWS mascon data (Figure S1) derived from SLR observations from 5–7 geodetic satellites; Laser Geodynamic Satellites 1 and 2 (LAGEOS-1, LAGEOS-2), STARLETTE, Stella, and Ajisai were available for the full span, and Larets and the Laser Relativity Satellite (LARES) were added to the solution when they became available in 2003 and 2012, respectively. We generated partial derivatives to spherical harmonic degree and order 10 and applied regularization following the same general procedures described by Rowlands et al. (2005). The solution time series was adaptively deseasonalized using the complete ensemble empirical mode decomposition with adaptive noise (Torres et al. 2011; Loomis and Luthcke 2014). Mass trends were calibrated to match GRACE/FO over a common span to mitigate impacts of parameter correlations (Loomis et al. 2019). This was further justified by results of our tests which demonstrated that the recovered SLR interannual signals are not dependent on the background gravity model, but that the recovered trends are.

2.5 Model, Reanalysis, and Ancillary Data

TWS from hydrological models forced with atmospheric analysis and observation-based near-surface meteorology was used for comparison with and interpretation of the observed time series. The ISBA-CTRIP (Interaction Soil-Biosphere–Atmosphere, Total Runoff Integrating Pathways from the Centre National de Recherches Météorologiques) hydrological model provides an estimate of the climate-driven TWS variations until 2018 (Decharme et al. 2019). However, human-induced contributions to TWS changes (e.g., irrigation and other consumptive uses of surface and groundwater) have become significant over the last two decades (Rodell et al. 2018). Based on the WaterGAP Hydrological Model (Müller Schmied et al. 2021), Cáceres et al. (2020) estimated a human contribution of 0.37 (0.30 to 0.45) mm/yr to GMSL change over 2003–2016. To account for the TWS contribution to GMSL change, we summed the climate-driven contribution from the ISBA-CTRIP model and the 0.37 mm/yr sea level equivalent anthropogenic trend from WaterGAP.

The ERA5-Land reanalysis (Muñoz-Sabater et al. 2021) is a 9-km resolution simulation of the ERA5 land model with four soil layers totaling 289 cm depth, forced with hourly mean ERA5 10 m near-surface meteorology, incident long- and short-wave radiation, and precipitation (P). Monthly mean fields of resulting total evapotranspiration (ET), runoff (RO), P, and TWS were sourced through the Copernicus Climate Change Service (C3S), as were ERA5 (Muñoz-Sabater et al. 2021) sea-surface temperature (SST) and vertically integrated moisture flux divergence. ERA5-Land TWS was found by aggregating the volumetric water content of four soil layers and further adding snow water equivalent and vegetative water. Ancillary monthly mean data used to help interpret GRACE/FO TWS changes include the Global Precipitation Climatology Project, GPCP v3.2 P (Huffman et al. 2021); Global Land Evaporation Amsterdam Model, GLEAM v3.6a ET (Martens et al. 2017; https://www.gleam.eu/); the Multi-Forcing Observation-Based Global Runoff Reanalysis, G-RUN RO (Ghiggi et al. 2021; https://doi.org/https://doi.org/10.6084/m9.figshare.12794 075); and TWS from the WaterGAP 2.2d model (Müller Schmied et al. 2021) which was forced with observationally bias adjusted ERA5 data (Lange 2019; Cucchi et al. 2020). All data were interpolated to a 1.0° grid using the Grid Analysis and Display System (GrADS) analysis system.

3 Results

3.1 GRACE/FO TWS Variability

As shown in Fig. 1, there was a large, abrupt decline in TWS between May 2014 and March 2016, when the GRACE/FO era minimum occurred. During this period, the mean (deseasonalized) TWS decrease over non-ice land mass was approximately 22 mm. To determine if this decline represents a statistically significant structural change in the global land TWS time series, we used a Bayesian ensemble algorithm for changepoint detection (Zhao et al. 2019). This method quantifies the likelihood of detected shifts in the mean and trend by randomizing parameters and structure of time-series decomposition models, creating a posterior probability distribution from the ensemble, and arriving at a weighted average model using Bayesian model averaging. Stochastic sampling of the model space was implemented via Markov Chain Monte Carlo. Figure S3 shows the raw and deseasonalized mean global TWS GRACE/FO anomalies, along with examples of individual trend components from random model samples. From a complex model space of 60,000 Monte Carlo iterations, an abrupt (single month) change of −3.2 mm was detected in January, 2015, with greater than 99.9% probability. Other possible changepoints were detected in April 2012 (−1.9 mm) and December 2019 (+1.9 mm), but the 2015 decline represents the largest (mostly likely single) shift in global TWS in the GRACE/FO record. By repeating this test for all land mascons and mapping the probability of changepoint occurrence in 2015 (Figure S3b), we can determine the spatial origin of the abrupt decline. The proximal source was a drought in northeastern South America (delineated in Figs. 3 and S3b) that was the most intense dry event in the GRACE/FO data record (Rodell and Li 2023). Figure 2 plots a time series of TWS from that region. Concurrent droughts elsewhere in the world also contributed (Intergovernmental Panel On Climate Change 2023).

That global TWS has remained in a lower range since the initial decline can be attributed in large part to dry events in other areas of the world that followed the drought in South America. To illustrate, Fig. 3 indicates locations where the GRACE/FO era minimum TWS was recorded between January 2015 and May 2023. These locations encompass 52% of the global land excluding Greenland and Antarctica. All else being equal, the expected area percentage would be equivalent to the ratio of the number of 2015–2023 to 2002–2023 GRACE/FO monthly solutions, which is 37%. To determine the field significance of the 52% statistic, we applied a block bootstrap technique (Douglas et al. 2000), which maintains both temporal and spatial autocorrelations. This method involves permuting blocks of the available GRACE/FO months, retrieving mascon-level TWS time series at the corresponding sample times, and repeating the global test statistic for each iteration. Given a block length of 24 months (to capture autocorrelation from seasonal dependence) and a distribution of 500 bootstraps, the percent of non-ice land area reaching a post-2015 minimum is 52% or higher in only 3.6% of the bootstrapped samples. These tests strongly suggest that the 2014–2016 decline represents a statistically unusual, abrupt shift in global land TWS, while Figs. 2 and S3 implicate central Brazil as the primary source of the initial decline.

3.2 Relationship between TWS and Sea Level

The effect of the decline in TWS on sea level is shown in Fig. 4, which compares the contributions of TWS to BSL as estimated using three different approaches. First, BSL was

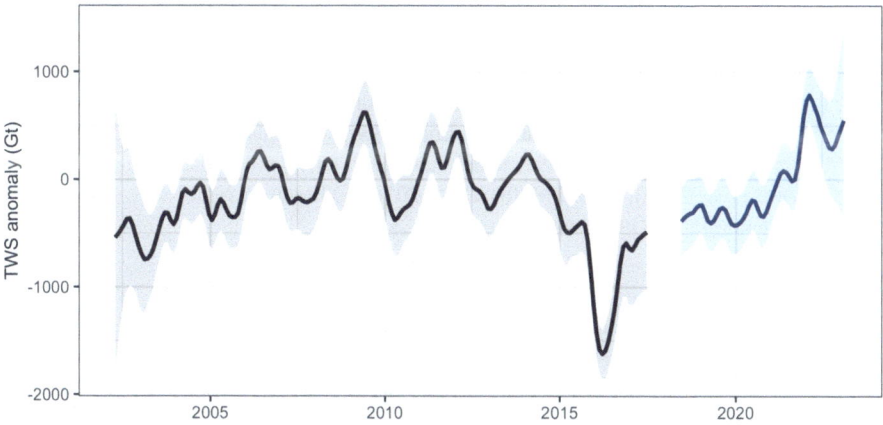

Fig. 2 Anomalies (relative to the 2004–2010 mean) of terrestrial water storage (GT) from GRACE (black) and GRACE-FO (blue) averaged over the region of South America delineated in Fig. 3. The time series was deseasonalized and smoothed (7-month moving window with seasonal and trend decomposition using Loess). The shading indicates the formal errors

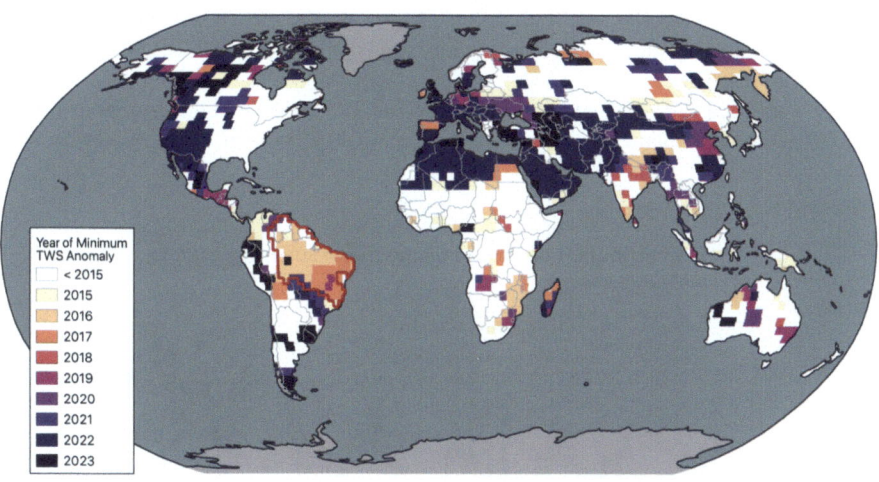

Fig. 3 Map of mascons in which the GRACE/FO era minimum terrestrial water storage occurred after the start of 2015, color coded for the year of that minimum. The time series plotted in Fig. 2 is averaged over the South American region delineated in red

estimated from GRACE/FO observations over the ocean. Second, it was estimated from satellite altimetry measurements of GMSL after correcting for thermosteric effects using Argo float data. Land ice and atmospheric water vapor contributions were removed from both of these BSL time series. Third is a time series based on hydrological model output. While the two observation-based time series are in very good agreement during the first half of the period, typically remaining within 7 mm of each other and crossing frequently, they diverge after 2015 and end the study period about 10 mm apart, signifying an overestimate of the total sea level change from altimetry and/or an underestimate of its components from GRACE/FO or Argo. The hydrological model agrees well with GRACE/FO even beyond 2015, but with subdued extremes. Despite the discrepancies over the later years, both the GRACE/FO and altimetry-based estimates indicate the large rise in BSL between 2014 and 2016. The associated decline in TWS manifested as droughts in northeastern South America and elsewhere, which have been attributed to back-to-back El Niño events (Llovel et al. 2023), including the 2015–2016 "Extreme El Niño" (see Box 11.4 in Intergovernmental Panel On Climate Change (2023)).

3.3 Historical Context

An important question is whether abrupt declines in TWS such as that during 2014–2016 (and perhaps abrupt gains) are unusual over the course of many decades. This question is difficult to answer using observations because global in situ measurements of the TWS components are woefully inadequate (Rodell and Reager 2023) while GRACE/FO has provided only two decades of observations to date. To provide historical context for the 2015 TWS decline, Fig. 5 compares time series from GRACE/FO (2002–2023), SLR (1994–2023), ERA5-Land (1980–2023), and WaterGAP (1980–2020). Reanalyses and global hydrologic models synthesize observational data with physically based constraints; thus, they may be instructive when investigating natural processes contributing to an abrupt decline in global TWS. During the concurrent period, 2002–2020, the four time series agree well, with correlation coefficients ranging from 0.70 (ERA5 vs. SLR) to 0.86 (WaterGAP vs. GRACE/FO).

TWS's abrupt decline around 2015 and its persistence in a lower range appears in all four time series. ERA5 displays a similar step decrease in 2002; however, it is not corroborated

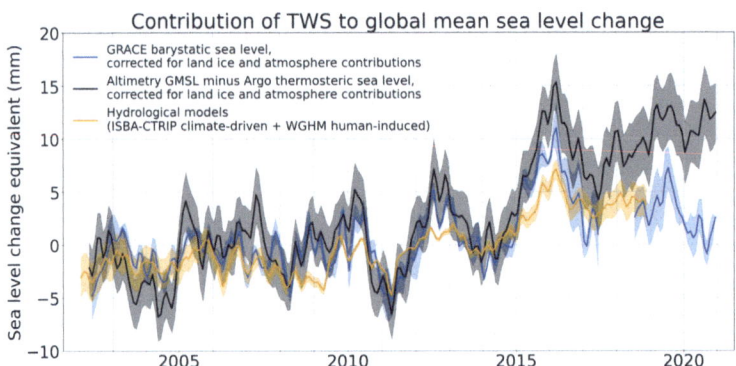

Fig. 4 Time series of three estimates of the contribution of TWS to BSL (as sea level change equivalent in mm): (1) GRACE/FO observations of BSL after subtracting the contributions of glacier and ice sheet melt water inputs (based on Jet Propulsion Laboratory GRACE/FO data) and of atmospheric water vapor content variations (using the ERA5 reanalysis); (2) satellite altimetry measurements of GMSL after subtracting the contributions of melt water inputs, atmospheric water vapor content, and thermal expansion (based on Argo float data); and (3) hydrological model output (sum of the climate-driven contribution from ISBA-CTRIP model and human-induced linear contribution based on WaterGAP Hydrological Model, WGHM). Annual and semiannual signals have been removed. A mask is applied to gravimetry, altimetry, and Argo data over the ocean, excluding latitudes beyond $\pm 60°$, closed seas, and Indonesian seas and coastal areas up to 200 km from the coastline

by the SLR or WaterGAP time series, which exhibit more gentle declines (with similar total magnitudes) between 1994 and 2005. To assess how unusual these events are in the context of interannual variability (distinct from long-term change), we first removed the linear trend and seasonal signals from each of the datasets and standardized the TWS anomalies to time series of Z-score statistics. We then identified TWS drying events with a Z-score < −1 for at least 2 consecutive months. These events are ranked by duration and intensity in Figure S4. In all four datasets, the 2015 TWS decline was the strongest in terms of minimum Z-score. In all but ERA5, it also had the longest duration. Clearly, the 2014–2016 decline was unusual. The regional expression of the overall GRACE/FO TWS global trend and the ability of ERA5-Land to replicate this variability are displayed in Fig. 6. The confidence we have in the former is rooted in it being an observational product that has been evaluated and trusted for more than two decades (Humphrey et al. 2023; Rodell and Reager 2023). There is broad agreement between ERA5-Land and GRACE/FO in terms of monthly, regional TWS trend patterns over much of the globe (Fig. 6c). However, the pattern correlation between GRACE/FO and ERA5-Land trends is only 0.24 and ERA5 trends are roughly half those of GRACE/FO. Widespread losses across semi-arid to arid climates over Eurasia, the Middle East, and western North America constitute the major driver of the global downward GRACE/FO TWS trend. Though pattern agreement over Asia is fair, significant extraction of groundwater, river damming, and other human influences spanning the Middle East (Joodaki et al. 2014; Chao et al. 2018; Nikraftar et al. 2024) to north-west India (Rodell et al. 2009; Bhanja et al. 2020; Swain et al. 2022), and elsewhere (Rodell et al. 2018) have been documented. These effects are not modeled in ERA5-Land and could explain why certain regional TWS declines observed by GRACE/FO are substantially weaker or missing in ERA5-Land. Competing regions of significant TWS gain over the African Sahel and Rift Valley, eastern North America, Amazon basin, and many parts of Asia also exist. Notable discrepancies include the pronounced ERA5 drying in central Africa and the interior of the Amazon basin (Fig. 6b), which are opposite to those observed by GRACE/FO. These are areas that have few surface observations and are dominated by parameterized moist physics in the ERA5 atmospheric model. Reanalysis discrepancies that are likely related to advancements in satellite atmospheric temperature and moisture profiling in the assimilation data stream between 1998 and 2002 have been reported (Nogueira 2020; Hersbach et al. 2020). These improvements in observing

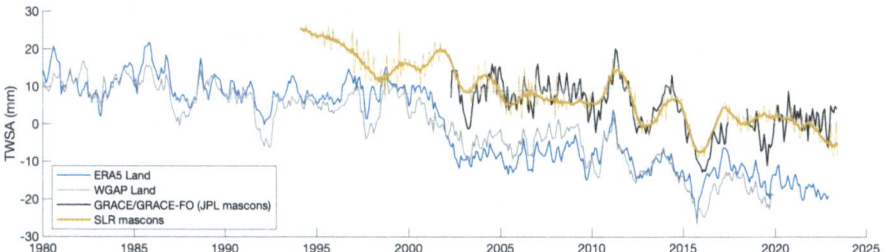

Fig. 5 Deseasonalized time series of TWS anomalies (TWSA) from ERA5, WaterGAP, SLR, and GRACE/FO, excluding Greenland, Antarctica, and the gulf coast of Alaska region. Both unfiltered monthly SLR (thin gold line) and filtered (Complete Ensemble Empirical Mode Decomposition with Adaptive Noise) SLR data (thick gold line) are plotted. The SLR data represent the sum of mascons 1, 2, and 5–19 as shown in Figure S1. The model time series are plotted as TWS anomalies relative to a zero mean. The vertical offset of the observational time series is artificial and was inserted for visual clarity, as only the temporal variations are meaningful (the absolute anomaly values are not)

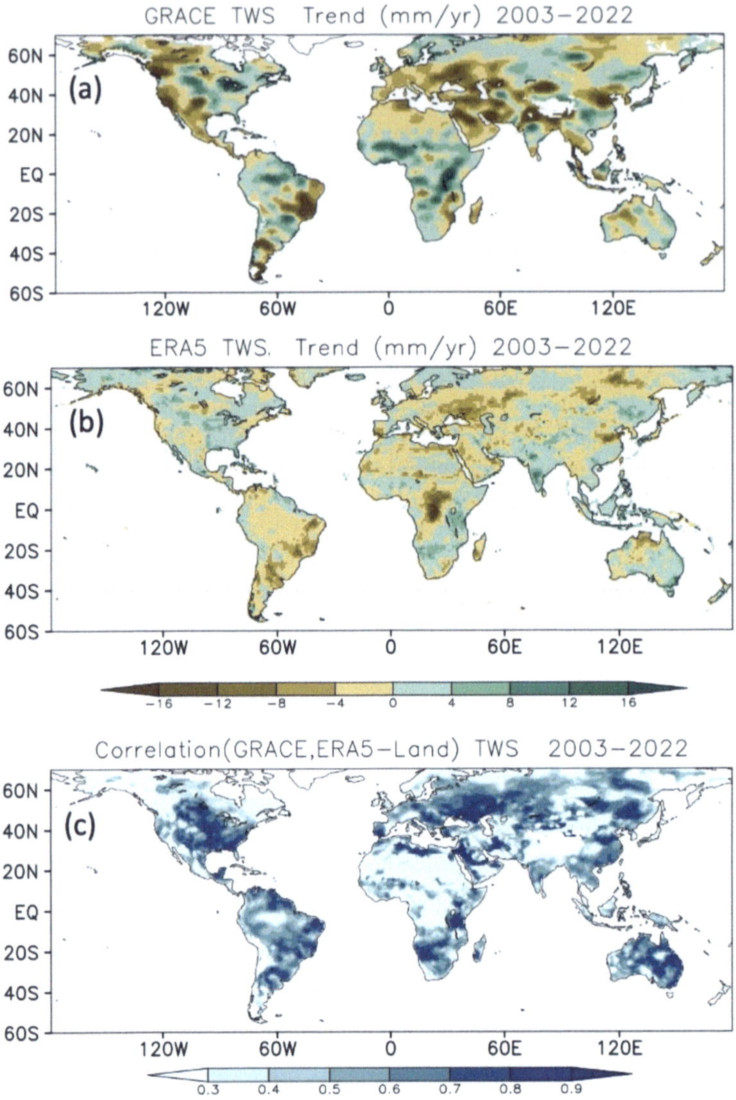

Fig. 6 TWS trends (mm/yr) over the period 2003–2022 in **a** GRACE/FO and **b** ERA5-Land; and **c** correlations of monthly TWS anomalies between GRACE/FO and ERA5-Land. Ice sheets and areas of significant glacial extent have been masked (see Figure S2)

capabilities continue during the GRACE/FO tenure (Hersbach et al. 2020). Ironically, improving the ability of atmospheric observations to offset assimilating model biases may have affected ERA5-Land TWS trends. There is also a junction in 2002 of the multiple streams in which the ERA5 and ERA5-Land reanalyses were produced.

Maps of coefficients of correlation between monthly GRACE/FO and ERA5-Land TWS anomalies (Fig. 6c) illustrate their degree of month-to-month agreement. Broad areas of high correlation include much of Eurasia and North America (r > 0.70), where observational networks are generally dense. In regions of sparse and variable rain gauge coverage, e.g., the Sahara, central Africa, and the interior of the Amazon, agreement is poor (Maidment et al. 2015; Nogueira 2020; Nicholson and Klotter 2021). Agreement in temporal variability (Fig. 6c.) is poor in regions of substantial seasonal snow cover or glacial extent (Alaska, western and eastern Canada, northern Siberia, and High Mountain Asia). In these regions, deficiencies in ERA5-Land physics (e.g., its lack of a glacier model (Muñoz-Sabater et al. 2021)), the sparseness of in situ and radiosonde observations for assimilation and calibration, and errors in the GRACE/FO TWS trends associated with GIA model uncertainty may contribute to lower time-series correlation (Hersbach et al. 2020; Mayer et al. 2021; Muñoz-Sabater et al. 2021).

To quantify the larger scale importance of these spatial patterns to the global mean trends and the extent to which GRACE/FO and ERA5-Land agree, we calculated area-weighted (continent area over global land area, excluding the white areas in Figure S2) time series for six near-continental regions (Fig. 7). Also shown for each plot is the near-global mean SST anomaly time series. While no simple relationship exists between SST anomalies and TWS changes averaged at continental scales, the SST record does exhibit prominent interannual signals that have continental scale influence. Asia, as a whole, was the dominant contributor to declines in global mean TWS, with major droughts beyond the 2016 El Niño being crucial to sustaining low global mean TWS despite the upward trend present in many parts of Africa (according to the GRACE/FO data). We suspect that disparities between the magnitudes of ERA5-Land and GRACE/FO TWS variations and trends, most prominently over Asia, stem from two related factors. First, ERA5-Land does not simulate groundwater storage, thus limiting the range of variability of TWS. Second, human water management and consumption, especially groundwater withdrawals, are not simulated by the ERA5-Land system. In Europe, it also exhibited a downward TWS trend which had a much smaller weighted rate owing to its proportionally smaller area. Australasia and South America were largely responsible for the 2011 global TWS peak (Boening et al. 2012), with North America, Europe, and Africa also playing a role. South America experienced two large declines and recoveries in the post-2015 period, and the lack of a large positive TWS anomaly after 2015 is remarkable. South America also exhibited the largest swings from positive to negative TWS anomalies over the 2011–2016 period,

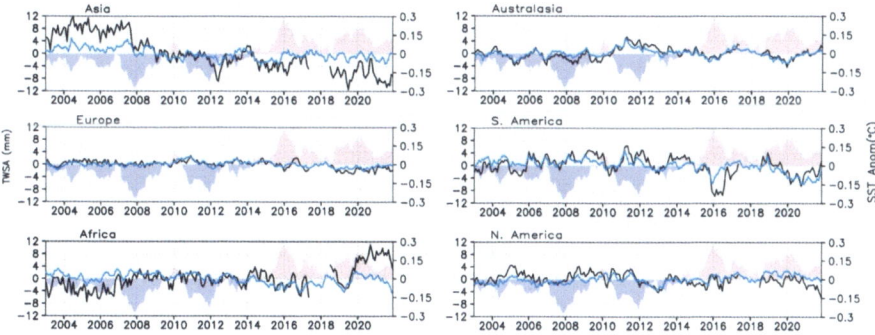

Fig. 7 Deseasonalized time series of area-weighted TWS (mm, left Y-axis) over six near-continental regions based on GRACE/FO (black line) and ERA5-Land (blue line). The area-weighted scaling ensures that the sum of these contributions is the global TWS time series as shown in Fig. 5. Shading indicates global SST anomalies (°C, right Y-axis), relative to a 2003–2022 base period. See Figure S2 for definition of continental averaging domains

but other continents, especially Asia, reflect this drawdown in TWS to varying degrees. South America is strongly influenced by ENSO and associated dislocations of moisture transport by the Walker circulation (Ropelewski and Halpert 1987; Castillo et al. 2014). After 2015, GRACE/FO and ERA5-Land agree that Asia and South America contributed most prominently to the diminished global TWS levels. The largest qualitative disparity between GRACE/FO and ERA5-Land TWS variations occurred in Africa, with GRACE/FO indicating continued accrual of water storage after a 2019 spike, while ERA5-Land TWS decreased substantially.

The emerging picture here is the dominance of Asia in sustaining a global, multi-decadal decline in TWS, largely driven by drought and exacerbated by human extraction of groundwater. A major additional contribution to the decline since 2015 comes from South America, whose hydrologic variations are large and strongly controlled by ENSO. In Asia, the drying trends are mostly found in the Middle East, Northern India, Northern China, and South-east Asia (Fig. 6). Partly counteracting those trends is a 20-year increase in TWS over Africa, which included the most intense extreme wet event (in central Africa) in the GRACE/FO data record (Rodell and Li 2023). Finally, as Fig. 6 shows, TWS variations within the continents are non-uniform, with consistency of trends being more the exception than the rule. Section 3.5 explores further the relationships between SST changes and TWS anomaly behavior at sub-continental scale.

3.4 Water Fluxes and Moisture Transport

In conjunction with a global SST rise of ~0.02 °C/yr during the GRACE/FO era, ERA5-Land TWS averages over 60S-70N (Fig. 8a) capture the increase and subsequent loss associated with the 2010/11 La Nina and 2015/2016 El Niño events. As noted above, ERA5-Land displays weaker TWS variability than GRACE/FO, especially over South America during the 2015/16 El Niño, and ERA5-Land TWS trends over the 2003–2022 period (−0.50 mm/yr) are also weaker relative to GRACE/FO (−0.74 mm/yr). Although ERA5-Land TWS is an outcome of its water budget (P-ET-RO), a semi-independent and strong constraint on this budget is provided by the vertically integrated moisture flux convergence over the global land which has previously been considered a more robust estimate of P-ET than the directly computed diagnostics (Landerer et al. 2010; Trenberth et al. 2011). ERA5 vertically integrated moisture flux convergence is correlated with, and systematically leads, ERA5 TWS by 1 month (r=0.41) due to the time scale of moisture import, precipitation, and hydrological response. ERA5 vertically integrated moisture flux convergence is anti-correlated with SST changes, owing to shifts in atmospheric circulation systems associated with internal climate oscillations such as ENSO and the Indian Ocean Dipole (IOD), which reduce precipitation over many tropical land areas during warm events (Ropelewski and Halpert 1987; Trenberth and Shea 2005; Trenberth et al. 2011; Bosilovich et al. 2020).

Figure 8b shows monthly anomalies of ERA5-Land P, ET, and RO, and Fig. 8c similarly compares observational estimates of the same from GPCP v3.2, GLEAM v3.6, and G-RUN RO. For both ERA5 and the observations, inputs of moisture from P are primarily balanced by losses through RO, although the response is relatively greater for ERA5 P and RO (r=0.79) than for GPCP P and G-RUN RO (r=0.64). For ERA5-Land, although ET is actually more strongly correlated with P (r=0.88) than is RO, the magnitude of variations in ET is smaller. The GLEAM ET correlation with GPCP P is much lower (r=0.22); yet, its comparison to ERA5 P (r=0.34) improves considerably, which is logical because ERA5 P is an input to the water balance module used by the GLEAM ET algorithm and is

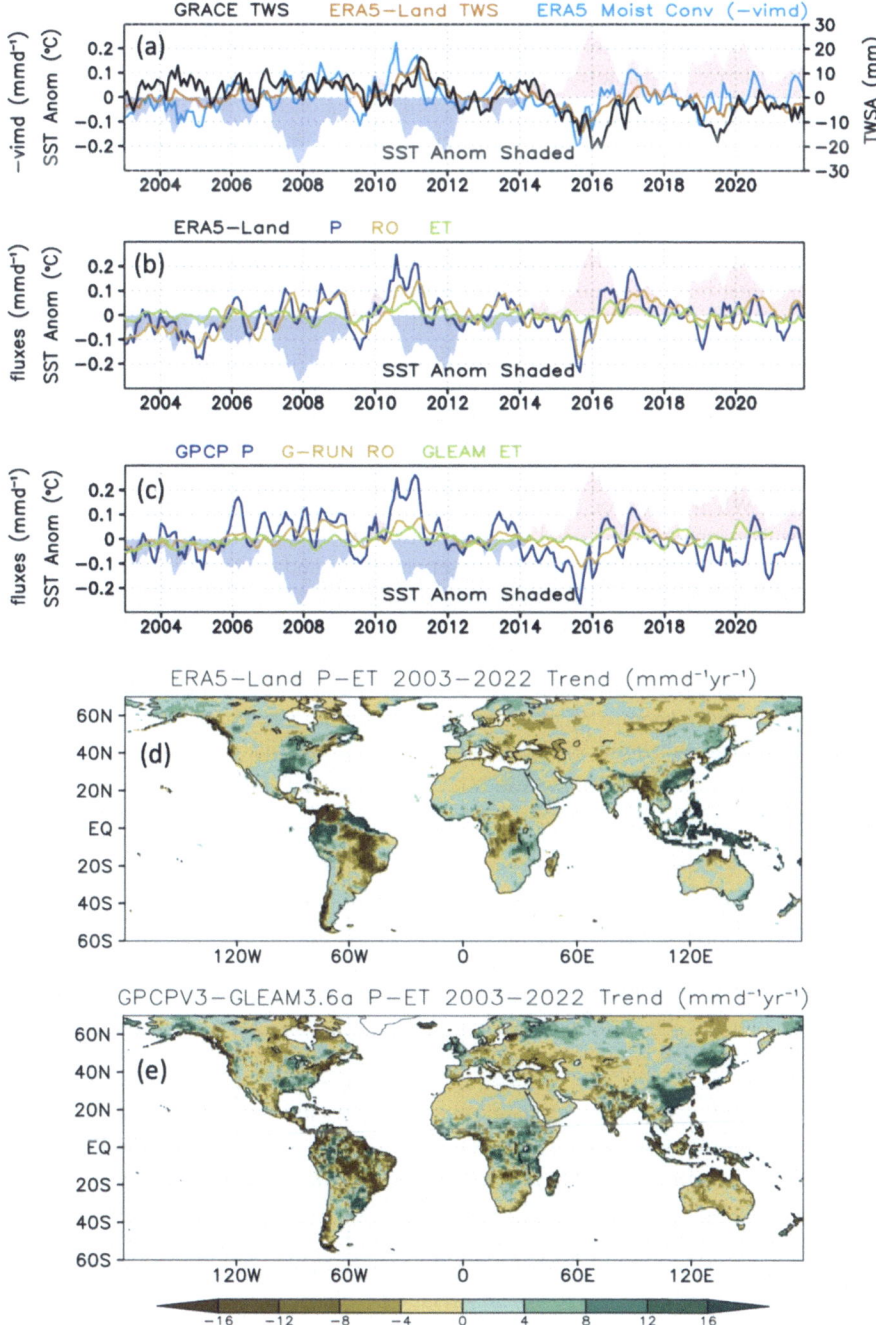

Fig. 8 Comparison of ERA5-Land, GRACE/FO, and complementary observations: **a** deseasonalized time series of global land average (60S-70N) TWS anomalies (mm) from GRACE/FO, ERA5-Land, and ERA5 vertically integrated moisture flux convergence, **b** ERA5-Land precipitation (*P*), evapotranspiration (ET), and runoff (RO), **c** same as **b** but for GPCP *P*, GLEAM ET and G-RUN RO. Units in **b** and **c** are mm d^{-1}, **d** 2003–2022 trend in ERA5-Land *P*-ET, **e** Same as **d** but for GPCP v3.2 *P*—GLEAM 3.6a ET. Units in **d** and **e** are mm d^{-1}/yr. All anomalies are relative to the 2003–2022 monthly resolved climatological mean

also the forcing for ERA5-Land. The internal consistency between water and energy fluxes enforced by ERA5-Land model physics is typically lacking in observational estimates of fluxes that are derived separately from each other. The longer term variations in P, ET, and RO in ERA5-Land since 1980 display similar inter-relationships with ENSO variability (Figure S5) and suggest that a long-term global decline in TWS began prior to the GRACE period (Fig. 5). However, the ever-changing mix of observations assimilated by atmospheric reanalyses limits confidence in their ability to simulate long-term trends (Allan et al. 2020; Hersbach et al. 2020).

On monthly time scales, P-ET serves as a proxy for moisture flux convergence in the atmospheric water budget. Trends in P-ET during 2003–2022 from ERA5 and from GPCP v3.2 minus GLEAM 3.6a (Figs. 8d, e) show strong consistency with TWS trends from GRACE/FO and ERA5-Land (Fig. 6a, b). Pattern agreement with GRACE/FO over western Eurasia is particularly striking, signifying the important role of decreasing moisture transport into this region. GPCP and GLEAM also capture much of the implied role for moisture convergence that support GRACE/FO TWS increases over the Sahel and Rift Valley portions of Africa. However, correspondence between ERA5-Land P-ET and GRACE/FO TWS trends over South America is weak. ERA5-Land P-ET over central Africa, though mapping to its own TWS decreases there, is inconsistent with GRACE/FO measurements. Again, changes in satellite observations around the turn of the 21st Century likely caused time-dependent biases to propagate through the ERA5-Land hydrology.

In summary, ERA5 and observational estimates of TWS changes and their driving fluxes confirm both the substantial decline in TWS associated with the strong 2015/16 El Niño event and the longer term trend in TWS, as observed by GRACE/FO. These changes in TWS are clearly related to prior variations in moisture convergence over global land and the resulting precipitation deficits in South America, Eurasia, and Australasia, though ERA5-Land simulates TWS changes that are smaller in magnitude than those of GRACE/FO and fails to capture TWS patterns and trends over sparsely observed regions such as Africa.

3.5 The Role of SST Changes

ERA5-Land and observational flux estimates demonstrate the fundamental relationship between variations in atmospheric moisture delivery to land and TWS changes as observed by GRACE/FO (Fig. 8). We now explore how SST variability partially mediates this relationship (Fig. 9). Significant SST variations characterizing the 20-year GRACE/FO record (Fig. 9) include (1) six warm and eight cold ENSO events (https://www.ncei.noaa.gov/access/monitoring/enso/sst) and (2) sustained global SST increases exhibiting a striking near-global pattern of warming except over the eastern Pacific and extreme Southern Ocean. An empirical orthogonal function (EOF) analysis (not shown) reveals that ENSO-related interannual variations and secular trends explain the leading 29% and 11% of the monthly variance, respectively. Variations in other SST indices such as the Pacific Decadal Oscillation (Newman et al. 2016), an Atlantic tripole pattern (Czaja and Frankignoul 2002), and the IOD (Saji et al. 1999) are also presented. Each of these is known to have some dependence on tropical Pacific variability (Zanchettin et al. 2008; Newman et al. 2016; Ham et al. 2017; Casselman et al. 2021). While debate continues as to how anthropogenic warming versus natural climate variations constrain this evolving structure (Seager et al. 2019; Watanabe et al. 2021; Heede and Fedorov 2023), the observed pattern of warming differs from the more uniform increases simulated by coupled climate models (Wills

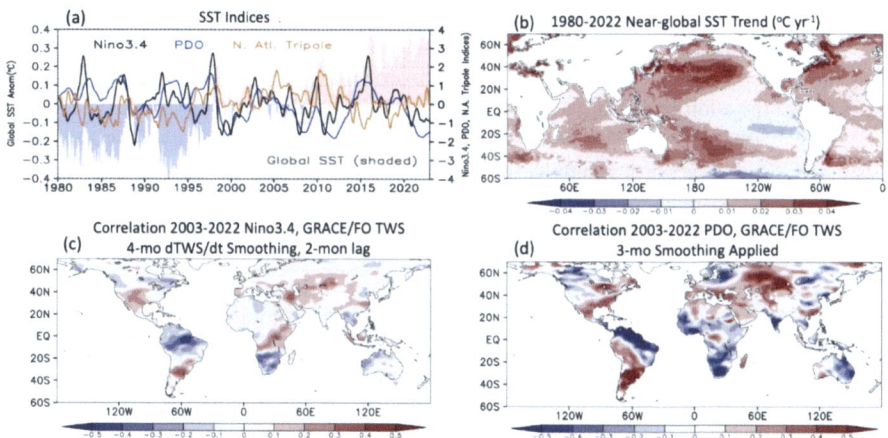

Fig. 9 **a** Time series of near-global (60° S–70° N) mean SST anomalies, Niño 3.4 (black), the Pacific Decadal Oscillation (PDO) index with 1-year smoothing (blue), and a North Atlantic tripole index (brown). Negative PDO corresponds to anomalous positive SSTs in the central N. Pacific. **b** Near-global ERA5 SST trend, 1980–2022, °C/yr. **c** Correlation between the Niño 3.4 index and GRACE/FO d(TWS)/dt, the latter lagged 2 months with 4-month smoothing applied. **d** Correlation between the PDO index and GRACE/FO TWS, 3-month smoothing applied

et al. 2022; Duffy and O'Gorman 2023). This "pattern effect" has been shown to reduce amplifying cloud feedbacks relative to the models through altering lapse rates over the global tropics (Ceppi and Gregory 2019; Andrews et al. 2022). It has long been known that human-induced global warming leads to a strong land–ocean warming contrast, with greater warming over land and associated reductions in relative humidity (Sutton et al. 2007; Byrne and O'Gorman 2015; Seneviratne et al. 2016; Wainwright et al. 2021; Intergovernmental Panel On Climate Change 2023).[1] But how this "pattern warming" is connected to regional trends in TWS (Fig. 6a, b) and the role of various processes needs further study.

Potential responses to SST pattern anomalies are not confined to long-term, secular trends but can also operate on interannual time scales (Izumo et al. 2019; Ceppi and Fueglistaler 2021). On the other hand, changes in land and ocean temperatures can be partially decoupled, such as during the so-called "hiatus" period of the early 2000s (Seneviratne et al. 2014), showing that SST changes do not necessarily affect land responses, in particular in the context of greenhouse gas forcing. In addition, any SST-induced adjustments over land are not uniformly distributed given regional hydrologic anomalies noted in previous studies (Hoerling and Kumar 2003; Lyon and DeWitt 2012; Schubert et al. 2016). Evidence for the influence of ENSO during the GRACE/FO era is seen particularly in correlations of d(TWS)/dt variations over South America, the southwestern U.S., southern and

---

[1] To maintain planetary radiative equilibrium in the face of increasing SSTs, atmospheric profiles stabilized by deep moist convection (lapse rate reductions) over ocean propagate to near-global extent, increasing energy transport to land. To equilibrate, the atmosphere/land system must lose energy radiatively which induces subsidence, inhibiting moisture import. The smaller land surface water content compared to the ocean surface means that evaporation becomes less effective as a cooling agent and land surface temperatures must elevate (contributing to energy loss to space; sensible heat loss becomes much more important than over ocean).

eastern Africa, and in the wave-like structure across northern Asia (Fig. 9c). Correlating SST anomalies with *P*-ET from GPCP and GLEAM ET (not shown) reveals similar patterns with even stronger correlations. The weaker GRACE/FO d(TWS)/dt correlations likely reflect the additional influence of RO on d(TWS)/dt behavior. Modes of tropical SST variability in other ocean basins (e.g., the Atlantic El Niño) also can assume importance in effecting TWS changes. For example, the IOD (Saji et al. 1999) can modulate rainfall over much of eastern Africa. A strong relationship (r = 0.66) between that mode and d(TWS)/dt over eastern Africa during 2003–2022 (Figure S6) exceeds that produced by ENSO (Fig. 9c).

Poleward propagating waves from ENSO events are also an important forcing for the Pacific Decadal Oscillation (Newman et al. 2016) as evidenced by their phasing in Fig. 9a (r = 0.53 with Niño 3.4 leading by 2 months). In Fig. 9d, the alternating pattern of correlation between the Pacific Decadal Oscillation index and GRACE/FO TWS extending across central Asia suggests modulation of storm tracks and their associated hydrologic processes. This pattern is replicated by replacing GPCP v3.2 *P* with TWS (not shown). The Pacific Decadal Oscillation, integrating the atmospheric forcing of SST, can be a marker of teleconnections extending around the higher latitudes. Likewise, TWS integrates *P*, ET, and RO anomalies in patterns reflecting the wave trains spanning the high latitudes.

Though direct greenhouse gas forcing and SST-induced anomalous moisture transport between land and ocean are essential in modulating TWS changes, it must be stressed that unforced climate system variability remains a major seasonal to interannual determinant of atmospheric circulation, weather, and, hence, hydrologic behavior. The influence of internal climate variability convolved with the preconditioned hydrologic state can often dominate (Miralles et al. 2019; Wehrli et al. 2019). A further cautionary note is that relationships between SST patterns and hydrologic response can be non-stationary (Torralba et al. 2015; Martija-Díez et al. 2023). More generally, interbasin SST connections via the atmosphere ensure that SST indices or modes are not independent (Cai et al. 2019), adding complexity to hydrologic responses and TWS evolution over land.

4 Discussion and Conclusions

The global mean variations in TWS can be examined at the continental level (Fig. 7) and also related to major droughts and pluvials. Referring to extreme TWS events in the 2002–2021 GRACE/FO data identified by Rodell and Li (2023), the 2014–2016 abrupt decline in global TWS was kicked-off by a drought in northern and central Brazil (intensity of − 10,513 GT months during August 2015 to January 2017) that was the most intense in the GRACE/FO record and dominated the TWS signal for South America in its entirety (Figs. 2, 7). In total, 13 of the 30 most intense droughts occurred during or after the 2014–2016 global TWS decline (their location, timing, and intensity are illustrated in Figure S7), helping to suppress global TWS since that time. Rodell and Li (2023) identified two sub-continental regions (sub-Saharan Africa and west central South America) with coherent tendencies of dry events being more common than wet events in the first half of the 2002–2021 study period and the opposite in the second half of the period, suggesting a tendency toward wetting in those regions since 2002. On the other hand, there were three regions of coherence (southwestern North America, south-eastern Brazil, and a large swath from southern Europe across the Middle East and Arabian Peninsula to south-western China and Bangladesh) where the frequency of wet events notably decreased while dry

events dominated during the second half of the study period. They noted both consistencies and inconsistencies between these tendencies and IPCC AR6 model predictions of precipitation change (Intergovernmental Panel On Climate Change 2023). Zonal mean TWS from GRACE/FO has been increasing between 5°S and 15°N and decreasing in the 10°–20° S and 25°–45° N bands (Dunn et al. 2024). There is much debate and little consensus about how patterns of wetting and drying will manifest in a warming world (Zaitchik et al. 2023); hence, it is difficult to evaluate whether the observed patterns are consistent with predictions and likely to persist.

Returning to Fig. 5 and the question of whether the 2014–2016 decline in global TWS was unusual, we evaluate three factors: magnitude of change (≤ -15 mm), slope of decline (≤ -1.0 mm/month), and whether there was a sustained recovery toward the pre-decline time series mean. As observed by GRACE, TWS decreased by 23 mm between May 2014 and March 2016 (22 months; slope $= -1.0$ mm/month). A decline of 22 mm between April 2011 and February 2013 (22 months; -1.0 mm/month) was nearly as large and steep. A third decline, 17 mm between May 2002 and March 2003 (10 months; -1.7 mm/month), was the only other in the GRACE/FO record to meet the magnitude and slope criteria. Following the 2011–2013 decline, TWS recovered to exceed the pre-decline mean 11 months later, during January–March 2014. Following the 2002–2003 decline, TWS exceeded the pre-decline mean 13 months later, in April 2004, and stayed above that level for a year. However, since the 2014–2016 decline ended, TWS has remained more than 1 mm below the pre-event mean in every month except September 2020 (54 months later), when it briefly spiked upward. Based on these three criteria as well as the statistical analysis presented in Sect. 3.1, we conclude that the 2014–2016 TWS decline was unique in the 2002–2023 GRACE/FO data record. As for the ERA5-Land and WaterGAP time series, there is agreement with GRACE regarding the depth and steepness of the 2011–2013 and 2014–2016 declines, as well as the recovery following the first and lack of recovery following the second. Prior to the GRACE/FO period, both models indicated steep declines that began in 1985, 1991, and 1997. However, none of these (in the case of either model) met both the magnitude and slope criteria, and all were followed within 1–2 years by sharp recoveries to above the pre-decline means. Of principal interest is a 22 mm decline in TWS indicated by ERA5-Land that began in November 2000 and reached a minimum in September 2002 (20 months; -1.1 mm/month), following which TWS never recovered to the pre-decline mean. The real world occurrence of such a decline in global TWS would invalidate our conjecture that the 2014–2016 TWS decline was unprecedented in the past 43 years. However, while both the WaterGAP and the SLR time series indicate that a decline in TWS occurred during that approximate timeframe (2000–2003) which met the magnitude and non-recovery criteria, both suggest it had a gentler slope, failing the ≤ -1.0 mm/month test. WaterGAP indicates a decline of 17 mm over July 2000 to April 2003 (-0.5 mm/month). SLR indicates a decline of 11 mm (less than the 15 mm threshold) over January 2002 to January 2003 (-0.9 mm/month). We are inclined to trust the SLR and WaterGAP data over ERA5-Land, and hence to conclude that the 2014–2016 decline in global TWS was, indeed, unprecedented during the past 4 decades. Our reasoning is twofold, First, if we break the period of Fig. 5 into three epochs, 1994–2001, 2003–2014, and 2015–2020, next average each of the four time series over those epochs, and finally compare the epoch-to-epoch changes, there is general consistency among all except ERA5 (Table 1), which has a steeper slope prior to 2003 and gentler slope after 2003 compared with the other time series. Second, as mentioned before, the multitude of consequential changes in meteorological observing systems between 1998 and 2002 that provided data to be assimilated by ERA5 during those years likely produced temporal discontinuities (Hersbach et al.

2020) which contributed to spurious precipitation trends (Nogueira 2020; Allan et al. 2020; Gleixner et al. 2020), thus casting doubt on any unusual events that occurred during that period, including ERA5-Land's November 2000 to September 2002 TWS decline. More generally, (Scanlon et al. 2018) called into question the ability of global models to represent large scale, multi-annual TWS variability.

While reanalyses and models are imperfect, they are helpful when attempting to infer physical mechanisms for observed phenomena. During the GRACE/FO era, global mean ocean temperature rose at a rate of 0.21 °C per decade, continuing, since 1980, a series of largest decadal scale trends seen since 1900 (see Fig. S3 in Xu et al. (2022); also https://www.ncei.noaa.gov/access/monitoring/climate-at-a-glance/global/time-series). At the largest scale, these warmer ocean temperatures cause more atmospheric upward motion and precipitation over the oceans, while in general, the opposite is true over land (Fasullo 2010; Lambert et al. 2011). On the other hand, land temperatures have also continued to rise rapidly and faster than those in the ocean, which is therefore unable to supply enough moisture to sustain relative humidity levels, leading to enhanced drying of the surface through evaporation (Sutton et al. 2007; Byrne and O'Gorman 2015; Intergovernmental Panel On Climate Change 2023). The resulting large-scale changes in vertically integrated moisture flux convergence and P-ET (shown in Fig. 8) in turn would be expected to drive declines in TWS, as observed by GRACE/FO (Figs. 6, 7). This leads to the supposition that a long-term downward trend in TWS is ultimately being driven by global warming. If true, the decline would not necessarily be gradual (as exemplified by recent global mean annual temperatures) and could instead manifest as an abrupt change like the one that occurred during 2014–2016. Variability internal to the climate system on interannual and longer scales (e.g., ENSO, PDO, IOD, and the Atlantic Meridional Oscillation) all exert significant influences on TWS changes. Underlying the global mean decline in TWS is a preponderance of extreme hydrological events since 2015 (Rodell and Li 2023), with the balance of those events being more dry than wet. At the continental scale, only Africa is currently resisting that trend (Fig. 7). The warmest 9 years in the modern global temperature record all have occurred since 2015, with 2024 certain to make it 10. Sudden TWS shifts can be expected to be convolved with a secular decline stemming from the erratic pace of warming.

Conclusively explaining the post-2015 divergence between GRACE/FO and altimetry minus thermosteric estimates of BSL will require further work. There are reasons to be uncertain about all three of the key observational inputs. Barnoud et al. (2023) discussed updates to the wet troposphere corrections for the Jason 2/3 altimeter time series that explain part of the gap. Argo profiling float measurements have been shown to contain spurious signals that can lead to biases in global mean thermosteric estimates when the salinity data after ~ 2015 are included (Hakuba et al. 2021). While we do not use the Argo salinity data for the global mean thermosteric sea level estimate

Table 1 Changes in global mean TWS (mm) between three epochs, 1994–2003, 2003–2014, and 2015–2020, based on GRACE/FO, SLR, ERA5, and WaterGAP

Dataset	dTWS (mm) [2003–2014] minus [1994–2003]	dTWS (mm) [2015–2020] minus [2003–2014]
GRACE/FO	N/A	− 9.1
SLR	− 11.0	− 7.4
ERA5	− 17.2	− 5.5
WaterGAP	− 11.5	− 11.9

here, undersampling the energetic ocean (i.e., eddies) can also introduce biases in the global mean thermosteric sea level variations (Lyman and Johnson 2023). GRACE/FO instrument measurements are in principle not prone to bias or drift, but the uncertainty levels of the monthly mass change observations vary over time, depending on ground-track coverage, thermal stability of the satellites, and external space-environmental factors (e.g., solar activity). Various studies have documented bias-free consistency of the GRACE and GRACE-FO data records, despite the 11-month gap between the missions (Landerer et al. 2020; Velicogna et al. 2020). To isolate surface mass (e.g., TWS or BSL) trends in the GRACE/FO observations, geophysical corrections are required to account for ongoing GIA; GRACE/FO uses as a standard the model of Peltier et al. (2015). Differences across an ensemble of GIA models equate to BSL trends of up to 0.5 mm/yr (Meyssignac et al. 2019). However, considering only state-of-the-art, global GIA models (such as Peltier et al. (2015), the actual resultant BSL trend uncertainty is estimated to be less than 0.2 mm/yr (Caron et al. 2018). Errors in accounting for GIA manifest as a constant bias in the GRACE/FO based TWS and BSL trend estimates, but they do not affect year-to-year variations. Ocean altimetry data are similarly corrected for GIA, with errors therein affecting sea-surface elevation change estimates in the same direction as they affect BSL change estimates from GRACE/FO. The two BSL time series (one from GRACE/FO, the other from altimetry-Argo; Fig. 4) are in good agreement prior to 2015 and diverge rapidly thereafter. This does not prove their accuracy prior to 2015, but considering that errors in the GRACE/FO based BSL time series would likely be steady (i.e., appear as a secular drift), the pre-2015 agreement and post-2015 divergence suggest an issue in the altimetry and/or Argo data as the root cause of that divergence.

As the planet continues to warm and ENSO cycles through its phases, it will be interesting to see if TWS will rebound to pre-2015 values, hold steady, or perhaps resume its decline. Future continuity (or enhancement) of the current suite of global sea level and terrestrial hydrological observations will be crucial for quantifying water cycle consequences of climate change, including long-term trends and shifts in seasonality and interannual variability, while constraining observing system biases and uncertainties.

Supplementary Information The online version contains supplementary material available at https://doi.org/10.1007/s10712-024-09860-w.

Acknowledgements This paper is an outcome of the Workshop "Challenges in Understanding the Global Water Energy Cycle and its Changes in Response to Greenhouse Gas Emissions" held at the International Space Science Institute (ISSI) in Bern, Switzerland (26–30 September 2022). The research was funded in part by NASA's GRACE-FO Science Team and NASA's Energy and Water Cycle Study (NEWS) program. GRACE and GRACE-FO were jointly developed and operated by NASA, DLR, and the GFZ German Research Centre for Geosciences. Portions of this research were conducted at the Jet Propulsion Laboratory, which is operated for NASA under contract with the California Institute of Technology.

Funding Open access funding provided by Swiss Federal Institute of Technology Zurich. National Aeronautics and Space Administration, Earth Sciences Division.

Data Availability GRACE/GRACE-FO data are available at http://grace.jpl.nasa.gov. The C3S altimetry data are available at https://doi.org/https://doi.org/10.24381/cds.4c328c78. The Argo data were collected and made freely available by the International Argo Program and the national programmes that contribute to it (https://argo.ucsd.edu, https://www.ocean-ops.org). The Argo Program is part of the Global Ocean Observing System.

Declarations

Conflict of interest The authors declare no competing interests.

Open Access This article is licensed under a Creative Commons Attribution 4.0 International License, which permits use, sharing, adaptation, distribution and reproduction in any medium or format, as long as you give appropriate credit to the original author(s) and the source, provide a link to the Creative Commons licence, and indicate if changes were made. The images or other third party material in this article are included in the article's Creative Commons licence, unless indicated otherwise in a credit line to the material. If material is not included in the article's Creative Commons licence and your intended use is not permitted by statutory regulation or exceeds the permitted use, you will need to obtain permission directly from the copyright holder. To view a copy of this licence, visit http://creativecommons.org/licenses/by/4.0/.

References

Allan RP, Barlow M, Byrne MP et al (2020) Advances in understanding large-scale responses of the water cycle to climate change. Ann N Y Acad Sci 1472:49–75. https://doi.org/10.1111/nyas.14337

Allan RP, Willett KM, John VO, Trent T (2022) Global changes in water vapor 1979–2020. JGR Atmos 127:e2022JD036728. https://doi.org/10.1029/2022JD036728

Andrews T, Bodas-Salcedo A, Gregory JM et al (2022) On the effect of historical SST patterns on radiative feedback. JGR Atmos 127:e2022JD036675. https://doi.org/10.1029/2022JD036675

Argo (2023) Argo float data and metadata from Global Data Assembly Centre (Argo GDAC)

Barnoud A, Pfeffer J, Guérou A et al (2021) Contributions of altimetry and argo to non-closure of the global mean sea level budget since 2016. Geophys Res Lett 48:e2021GL092824. https://doi.org/10.1029/2021GL092824

Barnoud A, Pfeffer J, Cazenave A et al (2023) Revisiting the global mean ocean mass budget over 2005–2020. Ocean Sci 19:321–334. https://doi.org/10.5194/os-19-321-2023

Bhanja SN, Mukherjee A, Rodell M (2020) Groundwater storage change detection from in situ and GRACE-based estimates in major river basins across India. Hydrol Sci J 65:650–659. https://doi.org/10.1080/02626667.2020.1716238

Boening C, Willis JK, Landerer FW et al (2012) The 2011 La Ni na: so strong, the oceans fell. Geophys Res Lett 39:19

Bosilovich MG, Robertson FR, Stackhouse PW (2020) El Niño-related tropical land surface water and energy response in MERRA-2. J Clim 33:1155–1176. https://doi.org/10.1175/JCLI-D-19-0231.1

Bounama C, Franck S, Von Bloh W (2001) The fate of Earth's ocean. Hydrol Earth Syst Sci 5:569–576. https://doi.org/10.5194/hess-5-569-2001

Byrne MP, O'Gorman PA (2015) The response of precipitation minus evapotranspiration to climate warming: why the "wet-get-wetter, dry-get-drier" scaling does not hold over land. J Clim 28:8078–8092. https://doi.org/10.1175/JCLI-D-15-0369.1

Cáceres D, Marzeion B, Malles JH et al (2020) Assessing global water mass transfers from continents to oceans over the period 1948–2016. Hydrol Earth Syst Sci 24:4831–4851. https://doi.org/10.5194/hess-24-4831-2020

Cai W, Wu L, Lengaigne M et al (2019) Pantropical climate interactions. Science 363:eaav4236. https://doi.org/10.1126/science.aav4236

Caron L, Ivins ER, Larour E et al (2018) GIA model statistics for GRACE hydrology, cryosphere, and ocean science. Geophys Res Lett 45:2203–2212. https://doi.org/10.1002/2017GL076644

Casselman JW, Taschetto AS, Domeisen DIV (2021) Non-linearity in the pathway of el niño-southern oscillation to the tropical north atlantic. J Cli 34(17):7277–7296. https://doi.org/10.1175/JCLI-D-20-0952.1

Castillo R, Nieto R, Drumond A, Gimeno L (2014) The role of the ENSO cycle in the modulation of moisture transport from major oceanic moisture sources. Water Resour Res 50:1046–1058. https://doi.org/10.1002/2013WR013900

Cazenave A, Llovel W (2010) Contemporary sea level rise. Annu Rev Mar Sci 2:145–173. https://doi.org/10.1146/annurev-marine-120308-081105

Ceppi P, Fueglistaler S (2021) The el niño-southern oscillation pattern effect. Geophys Res Lett 48:e2021GL095261. https://doi.org/10.1029/2021GL095261

Ceppi P, Gregory JM (2019) A refined model for the Earth's global energy balance. Clim Dyn 53:4781–4797. https://doi.org/10.1007/s00382-019-04825-x

Chambers DP, Wahr J, Nerem RS (2004) Preliminary observations of global ocean mass variations with GRACE. Geophys Res Lett 31:2004GL020461. https://doi.org/10.1029/2004GL020461

Chambers DP, Cazenave A, Champollion N et al (2017) Evaluation of the global mean sea level budget between 1993 and 2014. Surv Geophys 38:309–327. https://doi.org/10.1007/s10712-016-9381-3

Chandanpurkar HA, Reager JT, Famiglietti JS et al (2021) The seasonality of global land and ocean mass and the changing water cycle. Geophys Res Lett 48:e2020GL091248. https://doi.org/10.1029/2020GL091248

Chang L, Tang H, Wang Q, Sun W (2019) Global thermosteric sea level change contributed by the deep ocean below 2000 m estimated by Argo and CTD data. Earth Planet Sci Lett 524:115727. https://doi.org/10.1016/j.epsl.2019.115727

Chao N, Luo Z, Wang Z, Jin T (2018) Retrieving groundwater depletion and drought in the tigris-euphrates basin between 2003 and 2015. Groundwater 56:770–782. https://doi.org/10.1111/gwat.12611

Chen JL, Wilson CR, Tapley BD et al (2005) Seasonal global mean sea level change from satellite altimeter, GRACE, and geophysical models. J Geodesy 79:532–539. https://doi.org/10.1007/s00190-005-0005-9

Chen J, Tapley B, Seo K et al (2019) Improved quantification of global mean ocean mass change using grace satellite gravimetry measurements. Geophys Res Lett 46:13984–13991. https://doi.org/10.1029/2019GL085519

Chen J, Tapley B, Wilson C et al (2020) Global ocean mass change from GRACE and GRACE follow-on and altimeter and argo measurements. Geophys Res Lett 47:e2020GL090656. https://doi.org/10.1029/2020GL090656

Cheng L, Zhu J, Cowley R et al (2014) Time, probe type, and temperature variable bias corrections to historical expendable bathythermograph observations. J Atmos Oceanic Tech 31:1793–1825. https://doi.org/10.1175/JTECH-D-13-00197.1

Cheng L, Trenberth KE, Fasullo J et al (2017) Improved estimates of ocean heat content from 1960 to 2015. Sci Adv 3:e1601545. https://doi.org/10.1126/sciadv.1601545

Cheng L, Trenberth KE, Gruber N et al (2020) Improved estimates of changes in upper ocean salinity and the hydrological cycle. J Clim 33:10357–10381. https://doi.org/10.1175/JCLI-D-20-0366.1

Cowley R, Wijffels S, Cheng L et al (2013) biases in expendable bathythermograph data: a new view based on historical side-by-side comparisons. J Atmos Oceanic Tech 30:1195–1225. https://doi.org/10.1175/JTECH-D-12-00127.1

Cucchi M, Weedon GP, Amici A et al (2020) WFDE5: bias-adjusted ERA5 reanalysis data for impact studies. Earth Syst Sci Data 12:2097–2120. https://doi.org/10.5194/essd-12-2097-2020

Czaja A, Frankignoul C (2002) Observed impact of atlantic sst anomalies on the north atlantic oscillation. J Clim 15:606–623. https://doi.org/10.1175/1520-0442(2002)015%3c0606:OIOASA%3e2.0.CO;2

Decharme B, Delire C, Minvielle M et al (2019) Recent changes in the ISBA-CTRIP land surface system for use in the CNRM-CM6 climate model and in global off-line hydrological applications. J Adv Model Earth Syst 11:1207–1252. https://doi.org/10.1029/2018MS001545

Dobslaw H, Bergmann-Wolf I, Dill R et al (2017) A new high-resolution model of non-tidal atmosphere and ocean mass variability for de-aliasing of satellite gravity observations: AOD1B RL06. Geophys J Int 211:263–269. https://doi.org/10.1093/gji/ggx302

Douglas EM, Vogel RM, Kroll CN (2000) Trends in floods and low flows in the United States: impact of spatial correlation. J Hydrol 240:90–105. https://doi.org/10.1016/S0022-1694(00)00336-X

Duffy ML, O'Gorman PA (2023) Intermodel spread in walker circulation responses linked to spread in moist stability and radiation responses. JGR Atmos 128:e2022JD037382. https://doi.org/10.1029/2022JD037382

Dunn RJH, Blannin J, Gobron N et al (2024) Global climate. Bull Am Meteor Soc 105:S12–S155. https://doi.org/10.1175/BAMS-D-24-0116.1

Esser BK, Turekian KK (1988) Accretion rate of extraterrestrial particles determined from osmium isotope systematics of pacific pelagic clay and manganese nodules. Geochim Cosmochim Acta 52:1383–1388. https://doi.org/10.1016/0016-7037(88)90209-8

Fasullo JT (2010) Robust land-ocean contrasts in energy and water cycle feedbacks. J Clim 23:4677–4693. https://doi.org/10.1175/2010JCLI3451.1

Fasullo JT, Boening C, Landerer FW, Nerem RS (2013) Australia's unique influence on global sea level in 2010–2011. Geophys Res Lett 40:4368–4373

Flechtner F, Reigber C, Rummel R, Balmino G (2021) Satellite gravimetry: a review of its realization. Surv Geophys 42:1029–1074. https://doi.org/10.1007/s10712-021-09658-0

Frederikse T, Riva REM, King MA (2017) Ocean bottom deformation due to present-day mass redistribution and its impact on sea level observations. Geophys Res Lett 44:12–306. https://doi.org/10.1002/2017GL075419

Gaillard F, Reynaud T, Thierry V et al (2016) In situ-based reanalysis of the global ocean temperature and salinity with ISAS: variability of the heat content and steric height. J Clim 29:1305–1323. https://doi.org/10.1175/JCLI-D-15-0028.1

Garcia HE, Boyer TP, Baranova OK, et al (2019) World Ocean Atlas 2018: Product Documentation.

Ghiggi G, Humphrey V, Seneviratne SI, Gudmundsson L (2021) G-RUN ENSEMBLE: a multi-forcing observation-based global runoff reanalysis. Water Resour Res 57:e2020WR028787. https://doi.org/10.1029/2020WR028787

Gleixner S, Demissie T, Diro GT (2020) Did ERA5 improve temperature and precipitation reanalysis over east Africa? Atmosphere 11:996. https://doi.org/10.3390/atmos11090996

Good SA, Martin MJ, Rayner NA (2013) EN4: quality controlled ocean temperature and salinity profiles and monthly objective analyses with uncertainty estimates. JGR Oceans 118:6704–6716. https://doi.org/10.1002/2013JC009067

Gouretski V, Cheng L (2020) Correction for systematic errors in the global dataset of temperature profiles from mechanical bathythermographs. J Atmos Oceanic Tech 37:841–855. https://doi.org/10.1175/JTECH-D-19-0205.1

Gouretski V, Reseghetti F (2010) On depth and temperature biases in bathythermograph data: development of a new correction scheme based on analysis of a global ocean database. Deep Sea Res Part I 57:812–833. https://doi.org/10.1016/j.dsr.2010.03.011

Gregory JM, Lowe JA (2000) Predictions of global and regional sea-level rise using AOGCMs with and without flux adjustment. Geophys Res Lett 27:3069–3072. https://doi.org/10.1029/1999GL011228

Gregory JM, Griffies SM, Hughes CW et al (2019) Concepts and terminology for sea level: mean, variability and change, both local and global. Surv Geophys 40:1251–1289. https://doi.org/10.1007/s10712-019-09525-z

Guérou A, Meyssignac B, Prandi P et al (2023) Current observed global mean sea level rise and acceleration estimated from satellite altimetry and the associated measurement uncertainty. Ocean Sci 19:431–451. https://doi.org/10.5194/os-19-431-2023

Hakuba MZ, Frederikse T, Landerer FW (2021) Earth's energy imbalance from the ocean perspective (2005–2019). Geophys Res Lett 48:e2021GL093624. https://doi.org/10.1029/2021GL093624

Ham Y-G, Choi J-Y, Kug J-S (2017) The weakening of the ENSO–indian ocean dipole (IOD) coupling strength in recent decades. Clim Dyn 49:249–261. https://doi.org/10.1007/s00382-016-3339-5

Heede UK, Fedorov AV (2023) Colder eastern equatorial pacific and stronger walker circulation in the early 21st century: separating the forced response to global warming from natural variability. Geophys Res Lett 50:e2022GL101020. https://doi.org/10.1029/2022GL101020

Hersbach H, Bell B, Berrisford P et al (2020) The ERA5 global reanalysis. QJR Meteorol Soc 146:1999–2049. https://doi.org/10.1002/qj.3803

Hoerling M, Kumar A (2003) The perfect ocean for drought. Science 299:691–694. https://doi.org/10.1126/science.1079053

Hosoda S, Ohira T, Sato K, Suga T (2010) Improved description of global mixed-layer depth using Argo profiling floats. J Oceanogr 66:773–787. https://doi.org/10.1007/s10872-010-0063-3

Huffman GJ, Adler RF, Behrangi A, et al (2021) Algorithm Theoretical Basis Document (ATBD) for Global Precipitation Climatology Project Version 3.1 Precipitation Data.

Humphrey V, Zscheischler J, Ciais P et al (2018) Sensitivity of atmospheric CO_2 growth rate to observed changes in terrestrial water storage. Nature 560:628–631

Humphrey V, Rodell M, Eicker A (2023) Using satellite-based terrestrial water storage data: a review. Surv Geophys. https://doi.org/10.1007/s10712-022-09754-9

Intergovernmental Panel On Climate Change (2023) Climate Change 2021—The Physical Science Basis: Working Group I Contribution to the Sixth Assessment Report of the Intergovernmental Panel on Climate Change, 1st edn. Cambridge University Press

Ishii M, Fukuda Y, Hirahara S et al (2017) Accuracy of global upper ocean heat content estimation expected from present observational data sets. SOLA 13:163–167. https://doi.org/10.2151/sola.2017-030

Izumo T, Lengaigne M, Vialard J et al (2019) On the physical interpretation of the lead relation between warm water volume and the El Niño Southern Oscillation. Clim Dyn 52:2923–2942. https://doi.org/10.1007/s00382-018-4313-1

Jayne SR, Wahr JM, Bryan FO (2003) Observing ocean heat content using satellite gravity and altimetry. J Geophys Res 108:2002JC001619. https://doi.org/10.1029/2002JC001619

Joodaki G, Wahr J, Swenson S (2014) Estimating the human contribution to groundwater depletion in the Middle East, from GRACE data, land surface models, and well observations. Water Resour Res 50:2679–2692. https://doi.org/10.1002/2013WR014633

L'Ecuyer TS, Beaudoing HK, Rodell M et al (2015) The observed state of the energy budget in the early twenty-first century. J Clim 28:8319–8346. https://doi.org/10.1175/JCLI-D-14-00556.1

Lambert FH, Webb MJ, Joshi MM (2011) The relationship between land-ocean surface temperature contrast and radiative forcing. J Clim 24:3239–3256. https://doi.org/10.1175/2011JCLI3893.1

Landerer FW, Dickey JO, Güntner A (2010) Terrestrial water budget of the Eurasian pan-Arctic from GRACE satellite measurements during 2003–2009. J Geophys Res Atmos. https://doi.org/10.1029/2010JD014584

Landerer FW, Flechtner FM, Save H et al (2020) Extending the global mass change data record: GRACE follow-on instrument and science data performance. Geophys Res Lett 47:e2020GL088306. https://doi.org/10.1029/2020GL088306

Lange S (2019) WFDE5 over land merged with ERA5 over the ocean (W5E5). GFZ Data Service. https://doi.org/10.5880/pik.2019.023

Legeais J-F, Meyssignac B, Faugère Y et al (2021) Copernicus sea level space observations: a basis for assessing mitigation and developing adaptation strategies to sea level rise. Front Mar Sci 8:704721. https://doi.org/10.3389/fmars.2021.704721

Levitus S, Antonov JI, Boyer TP et al (2009) Global ocean heat content 1955–2008 in light of recently revealed instrumentation problems. Geophys Res Lett 36:2008GL037155. https://doi.org/10.1029/2008GL037155

Levitus S, Antonov JI, Boyer TP et al (2012) World ocean heat content and thermosteric sea level change (0–2000 m), 1955–2010. Geophys Res Lett 39:2012GL051106. https://doi.org/10.1029/2012GL051106

Lickley MJ, Hay CC, Tamisiea ME, Mitrovica JX (2018) Bias in estimates of global mean sea level change inferred from satellite altimetry. J Clim 31:5263–5271. https://doi.org/10.1175/JCLI-D-18-0024.1

Llovel W, Purkey S, Meyssignac B et al (2019) Global ocean freshening, ocean mass increase and global mean sea level rise over 2005–2015. Sci Rep 9:17717. https://doi.org/10.1038/s41598-019-54239-2

Llovel W, Balem K, Tajouri S, Hochet A (2023) Cause of substantial global mean sea level rise over 2014–2016. Geophys Res Lett 50:e2023GL104709. https://doi.org/10.1029/2023GL104709

Loomis BD, Luthcke SB (2014) Optimized signal denoising and adaptive estimation of seasonal timing and mass balance from simulated GRACE-like regional mass variations. Adv Adapt Data Anal 06:1450003. https://doi.org/10.1142/S1793536914500034

Loomis BD, Luthcke SB, Sabaka TJ (2019) Regularization and error characterization of GRACE mascons. J Geodesy 93:1381–1398

Love SG, Brownlee DE (1993) A direct measurement of the terrestrial mass accretion rate of cosmic dust. Science 262:550–553. https://doi.org/10.1126/science.262.5133.550

Lyman JM, Johnson GC (2023) Global high-resolution random forest regression maps of ocean heat content anomalies using in situ and satellite data. J Atmos Oceanic Tech 40:575–586. https://doi.org/10.1175/JTECH-D-22-0058.1

Lyon B, DeWitt DG (2012) A recent and abrupt decline in the East African long rains. Geophys Res Lett 39:2011GL050337. https://doi.org/10.1029/2011GL050337

Maidment RI, Allan RP, Black E (2015) Recent observed and simulated changes in precipitation over Africa. Geophys Res Lett 42:8155–8164. https://doi.org/10.1002/2015GL065765

Martens B, Miralles DG, Lievens H et al (2017) GLEAM v3: satellite-based land evaporation and root-zone soil moisture. Geosci Model Dev 10:1903–1925. https://doi.org/10.5194/gmd-10-1903-2017

Martija-Díez M, López-Parages J, Rodríguez-Fonseca B, Losada T (2023) The stationarity of the ENSO teleconnection in European summer rainfall. Clim Dyn 61:489–506. https://doi.org/10.1007/s00382-022-06596-4

Mayer J, Mayer M, Haimberger L (2021) Consistency and homogeneity of atmospheric energy, moisture, and mass budgets in ERA5. J Clim 34:3955–3974. https://doi.org/10.1175/JCLI-D-20-0676.1

Meyssignac B, Boyer T, Zhao Z et al (2019) Measuring global ocean heat content to estimate the earth energy imbalance. Front Mar Sci 6:432. https://doi.org/10.3389/fmars.2019.00432

Miralles DG, Gentine P, Seneviratne SI, Teuling AJ (2019) Land–atmospheric feedbacks during droughts and heatwaves: state of the science and current challenges. Ann N Y Acad Sci 1436:19–35. https://doi.org/10.1111/nyas.13912

Müller Schmied H, Cáceres D, Eisner S et al (2021) The global water resources and use model WaterGAP v2.2d: model description and evaluation. Geosci Model Dev 14:1037–1079. https://doi.org/10.5194/gmd-14-1037-2021

Muñoz-Sabater J, Dutra E, Agustí-Panareda A et al (2021) ERA5-Land: a state-of-the-art global reanalysis dataset for land applications. Earth Syst Sci Data 13:4349–4383. https://doi.org/10.5194/essd-13-4349-2021

Nerem RS, Beckley BD, Fasullo JT et al (2018) Climate-change–driven accelerated sea-level rise detected in the altimeter era. Proc Natl Acad Sci USA 115:2022–2025. https://doi.org/10.1073/pnas.1717312115

Newman M, Alexander MA, Ault TR et al (2016) The pacific decadal oscillation, revisited. J Clim 29:4399–4427. https://doi.org/10.1175/JCLI-D-15-0508.1

Nicholson SE, Klotter DA (2021) Assessing the reliability of satellite and reanalysis estimates of rainfall in equatorial Africa. Remote Sens 13:3609. https://doi.org/10.3390/rs13183609

Nikraftar Z, Parizi E, Saber M et al (2024) Groundwater sustainability assessment in the Middle East using GRACE/GRACE-FO data. Hydrogeol J 32:321–337. https://doi.org/10.1007/s10040-023-02717-3

Nogueira M (2020) Inter-comparison of ERA-5, ERA-interim and GPCP rainfall over the last 40 years: process-based analysis of systematic and random differences. J Hydrol 583:124632. https://doi.org/10.1016/j.jhydrol.2020.124632

Peltier WR (2004) Global glacial isostasy and the surface of the ice-age Earth: the ICE-5G (VM2) model and GRACE. Annu Rev Earth Planet 32:111–149

Peltier WR, Argus DF, Drummond R (2015) Space geodesy constrains ice age terminal deglaciation: the global ICE-6G_C (VM5a) model. JGR Solid Earth 120:450–487. https://doi.org/10.1002/2014JB011176

Peltier RW, Argus DF, Drummond R (2018) Comment on "an assessment of the ICE-6G_C (VM5a) glacial isostatic adjustment model" by Purcell et al. JGR Solid Earth 123:2019–2028. https://doi.org/10.1002/2016JB013844

Ramillien G, Bouhours S, Lombard A et al (2008) Land water storage contribution to sea level from GRACE geoid data over 2003–2006. Global Planet Change 60:381–392

Ray RD, Luthcke SB (2006) Tide model errors and GRACE gravimetry: towards a more realistic assessment. Geophys J Int 167:1055–1059. https://doi.org/10.1111/j.1365-246X.2006.03229.x

Reager JT, Gardner AS, Famiglietti JS et al (2016) A decade of sea level rise slowed by climate-driven hydrology. Science 351:699–703

Rietbroek R, Brunnabend S-E, Kusche J et al (2016) Revisiting the contemporary sea-level budget on global and regional scales. Proc Natl Acad Sci 113:1504–1509. https://doi.org/10.1073/pnas.1519132113

Riva REM, Bamber JL, Lavallée DA, Wouters B (2010) Sea-level fingerprint of continental water and ice mass change from GRACE. Geophys Res Lett 37:2010GL044770. https://doi.org/10.1029/2010GL044770

Rodell M, Li B (2023) Changing intensity of hydroclimatic extreme events revealed by GRACE and GRACE-FO. Nat Water 1:241–248. https://doi.org/10.1038/s44221-023-00040-5

Rodell M, Reager JT (2023) Water cycle science enabled by the GRACE and GRACE-FO satellite missions. Nat Water 1:47–59. https://doi.org/10.1038/s44221-022-00005-0

Rodell M, Velicogna I, Famiglietti JS (2009) Satellite-based estimates of groundwater depletion in India. Nature 460:999–1002. https://doi.org/10.1038/nature08238

Rodell M, Beaudoing HKK, L'Ecuyer TSS et al (2015) The observed state of the water cycle in the early twenty-first century. J Clim 28:8289–8318. https://doi.org/10.1175/JCLI-D-14-00555.1

Rodell M, Famiglietti JS, Wiese DN et al (2018) Emerging trends in global freshwater availability. Nature 557:651–659. https://doi.org/10.1038/s41586-018-0123-1

Roemmich D, Gilson J (2009) The 2004–2008 mean and annual cycle of temperature, salinity, and steric height in the global ocean from the Argo Program. Prog Oceanogr 82:81–100. https://doi.org/10.1016/j.pocean.2009.03.004

Ropelewski CF, Halpert MS (1987) Global and regional scale precipitation patterns associated with the el niño/southern oscillation. Mon Wea Rev 115:1606–1626. https://doi.org/10.1175/1520-0493(1987)115%3c1606:GARSPP%3e2.0.CO;2

Rowlands DD, Luthcke SB, Klosko SM et al (2005) Resolving mass flux at high spatial and temporal resolution using GRACE intersatellite measurements. Geophys Res Lett 32:1–4. https://doi.org/10.1029/2004GL021908

Saji NH, Goswami BN, Vinayachandran PN, Yamagata T (1999) A dipole mode in the tropical Indian Ocean. Nature 401:360–363. https://doi.org/10.1038/43854

Save H, Bettadpur S, Tapley BD (2016) High-resolution CSR GRACE RL05 mascons. J Geophys Res Solid Earth 121:7547–7569

Save H (2020) CSR GRACE and GRACE-FO RL06 Mascon Solutions v02

Scanlon BR, Zhang Z, Save H et al (2018) Global models underestimate large decadal declining and rising water storage trends relative to GRACE satellite data. Proc Natl Acad Sci USA 115:E1080–E1089. https://doi.org/10.1088/1748-9326/ac16ff

Schubert SD, Stewart RE, Wang H et al (2016) Global meteorological drought: a synthesis of current understanding with a focus on SST drivers of precipitation deficits. J Clim 29:3989–4019. https://doi.org/10.1175/JCLI-D-15-0452.1

Seager R, Cane M, Henderson N et al (2019) Strengthening tropical Pacific zonal sea surface temperature gradient consistent with rising greenhouse gases. Nat Clim Chang 9:517–522. https://doi.org/10.1038/s41558-019-0505-x

Seneviratne SI, Donat MG, Mueller B, Alexander LV (2014) No pause in the increase of hot temperature extremes. Nat Clim Change 4:161–163. https://doi.org/10.1038/nclimate2145

Seneviratne SI, Donat MG, Pitman AJ et al (2016) Allowable CO_2 emissions based on regional and impact-related climate targets. Nature 529:477–483. https://doi.org/10.1038/nature16542

Sutton RT, Dong B, Gregory JM (2007) Land/sea warming ratio in response to climate change: IPCC AR4 model results and comparison with observations. Geophys Res Lett 34:2006GL028164. https://doi.org/10.1029/2006GL028164

Swain S, Taloor AK, Dhal L et al (2022) Impact of climate change on groundwater hydrology: a comprehensive review and current status of the Indian hydrogeology. Appl Water Sci 12:120. https://doi.org/10.1007/s13201-022-01652-0

Tapley BD, Bettadpur S, Watkins M, Reigber C (2004) The gravity recovery and climate experiment: mission overview and early results. Geophys Res Lett. https://doi.org/10.1029/2004GL019920

Tapley BD, Watkins MM, Flechtner F et al (2019) Contributions of GRACE to understanding climate change. Nat Clim Chang 9:358–369. https://doi.org/10.1038/s41558-019-0456-2

Thomas AC, Reager JT, Famiglietti JS, Rodell M (2014) A GRACE-based water storage deficit approach for hydrological drought characterization. Geophys Res Lett 41:1537–1545

Torralba V, Rodríguez-Fonseca B, Mohino E, Losada T (2015) The non-stationary influence of the Atlantic and Pacific Niños on North Eastern South American rainfall. Front Earth Sci 3:55. https://doi.org/10.3389/feart.2015.00055

Torres ME, Colominas MA, Schlotthauer G, Flandrin P (2011) A complete ensemble empirical mode decomposition with adaptive noise. In: 2011 IEEE International Conference on Acoustics, Speech and Signal Processing (ICASSP). IEEE, Prague, Czech Republic, pp 4144–4147.

Trenberth KE, Shea DJ (2005) Relationships between precipitation and surface temperature. Geophys Res Lett 32:2005GL022760. https://doi.org/10.1029/2005GL022760

Trenberth KE, Fasullo J, Smith L (2005) Trends and variability in column-integrated atmospheric water vapor. Clim Dyn 24:741–758. https://doi.org/10.1007/s00382-005-0017-4

Trenberth KE, Fasullo JT, Mackaro J (2011) Atmospheric moisture transports from ocean to land and global energy flows in reanalyses. J Clim 24:4907–4924. https://doi.org/10.1175/2011JCLI4171.1

Trent T, Schroeder M, Ho S-P, et al (2023) Evaluation of Total Column Water Vapour Products from Satellite Observations and Reanalyses within the GEWEX Water Vapor Assessment.

Velicogna I, Mohajerani YAG et al (2020) Continuity of ice sheet mass loss in greenland and antarctica from the GRACE and GRACE follow-on missions. Geophys Res Lett 47:e2020GL087291. https://doi.org/10.1029/2020GL087291

Wada Y, Lo M-H, Yeh PJ-F et al (2016) Fate of water pumped from underground and contributions to sea-level rise. Nat Clim Change 6:777–780. https://doi.org/10.1038/nclimate3001

Wahr J, Molenaar M, Bryan F (1998) Time variability of the Earth's gravity field: hydrological and oceanic effects and their possible detection using GRACE. J Geophys Res Solid Earth 103:30205–30229

Wainwright CM, Finney DL, Kilavi M et al (2021) Extreme rainfall in East Africa, October 2019–January 2020 and context under future climate change. Weather 76:26–31. https://doi.org/10.1002/wea.3824

Watanabe M, Dufresne J-L, Kosaka Y et al (2021) Enhanced warming constrained by past trends in equatorial Pacific sea surface temperature gradient. Nat Clim Chang 11:33–37. https://doi.org/10.1038/s41558-020-00933-3

Watkins MM, Wiese DN, Yuan DN et al (2015) Improved methods for observing Earth's time variable mass distribution with GRACE using spherical cap mascons. J Geophys Res Solid Earth 120:2648–2671

Wehrli K, Guillod BP, Hauser M et al (2019) Identifying key driving processes of major recent heat waves. JGR Atmos 124:11746–11765. https://doi.org/10.1029/2019JD030635

Wiese DN, Landerer FW, Watkins MM (2016) Quantifying and reducing leakage errors in the JPL RL05M GRACE mascon solution. Water Resour Res 52:7490–7502

Wills RCJ, Dong Y, Proistosecu C et al (2022) Systematic climate model biases in the large-scale patterns of recent sea-surface temperature and sea-level pressure change. Geophys Res Lett 49:e2022GL100011. https://doi.org/10.1029/2022GL100011

Wong APS, Wijffels SE, Riser SC et al (2020) Argo data 1999–2019: two million temperature-salinity profiles and subsurface velocity observations from a global array of profiling floats. Front Mar Sci 7:700. https://doi.org/10.3389/fmars.2020.00700

World Climate Research Programme Global Sea Level Budget Group (2018) Global sea-level budget 1993–present. Earth Syst Sci Data, vol 10, pp1551–1590. https://doi.org/10.5194/essd-10-1551-2018

Wouters B, Gardner AS, Moholdt G (2019) Global glacier mass loss during the GRACE satellite mission (2002–2016). Front Earth Sci 7:96

Xu Z, Huang G, Ji F et al (2022) Robustness of the long-term nonlinear evolution of global sea surface temperature trend. Geosci Lett 9:25. https://doi.org/10.1186/s40562-022-00234-x

Zaitchik BF, Rodell M, Biasutti M, Seneviratne SI (2023) Wetting and drying trends under climate change. Nat Water. https://doi.org/10.1038/s44221-023-00073-w

Zanchettin D, Franks SW, Traverso P, Tomasino M (2008) On ENSO impacts on European wintertime rainfalls and their modulation by the NAO and the Pacific multi-decadal variability described through the PDO index. Intl J Clim 28:995–1006. https://doi.org/10.1002/joc.1601

Zhao K, Wulder MA, Hu T et al (2019) Detecting change-point, trend, and seasonality in satellite time series data to track abrupt changes and nonlinear dynamics: a Bayesian ensemble algorithm. Remote Sens Environ 232:111181. https://doi.org/10.1016/j.rse.2019.04.034

Publisher"s Note Springer Nature remains neutral with regard to jurisdictional claims in published maps and institutional affiliations.

Authors and Affiliations

Matthew Rodell[1] · Anne Barnoud[2] · Franklin R. Robertson[3] · Richard P. Allan[4] · Ashley Bellas-Manley[5] · Michael G. Bosilovich[1] · Don Chambers[6] · Felix Landerer[7] · Bryant Loomis[1] · R. Steven Nerem[5] · Mary Michael O'Neill[1,8] · David Wiese[7] · Sonia I. Seneviratne[9]

✉ Sonia I. Seneviratne
  Sonia.Seneviratne@ethz.ch

[1] NASA Goddard Space Flight Center, Greenbelt, MD 20771, USA

[2] Magellium, 31520 Ramonville Saint-Agne, France

[3] NASA Marshall Space Flight Center, Huntsville, AL 35808, USA

[4] Department of Meteorology and National Centre for Earth Observation, University of Reading, Reading RG6 6UR, UK

[5] University of Colorado, Boulder, CO 80309, USA

[6] University of South Florida, Tampa, FL 33620, USA

[7] Jet Propulsion Laboratory, California Institute of Technology, Pasadena, CA 91011, USA

[8] University of Maryland, College Park, MD 20742, USA

[9] ETH Zurich, 8092 Zurich, Switzerland

Tropical Deep Convection, Cloud Feedbacks and Climate Sensitivity

Graeme L. Stephens[1,2] · Kathleen A. Shiro[3] · Maria Z. Hakuba[1] · Hanii Takahashi[1] · Juliet A. Pilewskie[4] · Timothy Andrews[5] · Claudia J. Stubenrauch[6,7] · Longtao Wu[1]

Received: 3 November 2023 / Accepted: 19 February 2024 / Published online: 31 May 2024
© California Institute of Technology 2024

Abstract

This paper is concerned with how the diabatically-forced overturning circulations of the atmosphere, established by the deep convection within the tropical trough zone (TTZ), first introduced by Riehl and (Malkus) Simpson, in Contr Atmos Phys 52:287–305 (1979), fundamentally shape the distributions of tropical and subtropical cloudiness and the changes to cloudiness as Earth warms. The study first draws on an analysis of a range of observations to understand the connections between the energetics of the TTZ, convection and clouds. These observations reveal a tight coupling of the two main components of the diabatic heating, the cloud component of radiative heating, shaped mostly by high clouds formed by deep convection, and the latent heating associated with the precipitation. Interannual variability of the TTZ reveals a marked variation that connects the depth of the tropical troposphere, the depth of convection, the thickness of high clouds and the TOA radiative imbalance. The study examines connections between this convective zone and cloud changes further afield in the context of CMIP6 model experiments of climate warming. The warming realized in the CMIP6 SSP5-8.5 scenario multi-model experiments, for example, produces an enhanced Hadley circulation with increased heating in the zone of tropical deep convection and increased radiative cooling and subsidence in the subtropical regions. This impacts low cloud changes and in turn the model warming response through low cloud feedbacks. The pattern of warming produced by models, also influenced by convection in the tropical region, has a profound influence on the projected global warming.

Keywords Tropical convection · Cloud feedbacks · Climate change

Article Highlights

- The heating of the tropical trough zone drives circulations that affect remote cloud feedbacks and in turn global climate sensitivity
- The heating of the tropic zone is established by feedbacks between high cloud radiative heating and latent heating that are each by-products of convection
- Interannual variability of the high clouds in the tropics determines the interannual variability of the top-of-atmosphere (TOA) radiative imbalance of the tropics

Extended author information available on the last page of the article

1 Introduction

As the heated air of low latitudes rises and moves poleward, it cools and sinks in the subtropics creating a massive overturning of the atmosphere that stretches from the equator to the poles. This overturning circulation establishes both the main climate zones of Earth and the cloud patterns of these zones. That this rising, heated air at low latitudes is accomplished by the transports of many narrow plumes in deep convection, referred to as hot towers, was one of the most remarkable inferences about the atmosphere made in the mid-20th century by Riehl and Malkus (1958; hereafter R–M). Riehl and Simpson (1979; hereafter R–S) later revisited that study and estimated that about 1600–2400 hot towers must exist at any one time in the deep tropics under an assumption of a fixed size of 5 km.

Figure 1, adapted from Boucher et al. (2013), is a schematic depiction of the major cloud patterns of Earth and, more importantly, a schematic of their connection to the large-scale overturning circulations of the atmosphere. These connections begin with the area of deep convection at lower latitudes in the region referred to as the tropical trough zone (TTZ) by R–M. It is a zone where moisture is transported from higher latitudes at low levels and where heat is moved poleward in the upper levels. The TTZ is fundamental to the way that Earth's heat engine moves heat from low latitudes to higher latitudes and establishes the latitudes where heat transport is toward the winter pole. Deep convection forms in this zone, and its seasonal changes are intimate parts of this transport which relates to the other major cloud types in ways hinted in Fig. 1. This includes high clouds that are directly produced by the convection and the low cloud regimes of the subtropics governed

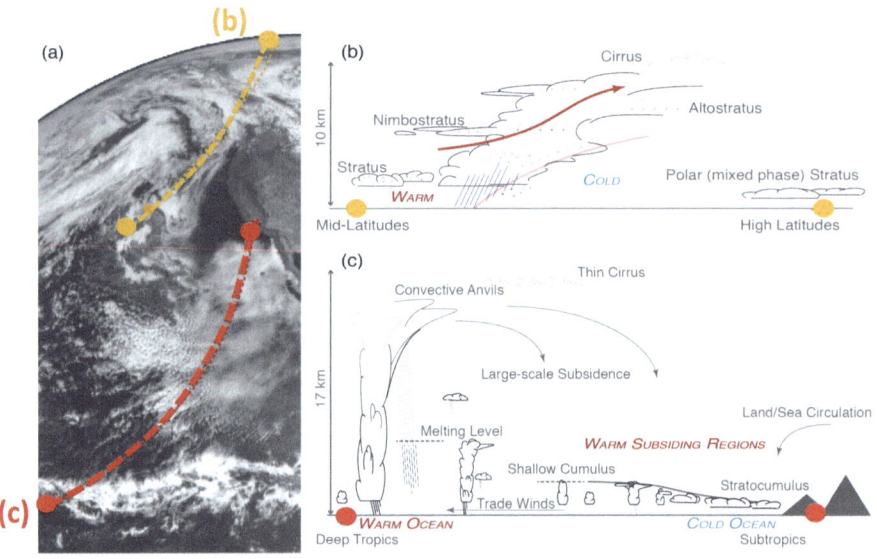

Fig. 1 **a** Visible imagery from geostationary satellite radiometer that shows clouds associated with extratropical cyclones, subtropical coastal stratocumulus near Baja California breaking up into shallow cumulus clouds in the central Pacific and mesoscale convective systems outlining the Pacific intertropical convergence zone (ITCZ) that resides in the tropical trough zone described later in the text. **b** A schematic section along the orange dashed lines highlighted in **a**, through a typical warm front of an extratropical cyclone. **c** A schematic section along the red dashed line in **a**, along the low-level trade wind flow from a subtropical west coast of a continent to the ITCZ [from Boucher et al. (2013)]

by the overturning circulation convection helps create. The influences of this circulation on clouds and, in turn, the influence of clouds on this circulation are a basic feedback cycle that frames all essential cloud feedbacks that regulate Earth's climate and its response to the forcing of climate change.

Some of the most important cloud feedbacks that are thought to operate on Earth can be traced to the circulations portrayed in Fig. 1 in a way that underscores how these feedbacks are indeed connected. This paper deals with only limited aspects of such connections, that of the influence of the deep convection that forms in the convergence zones of the tropics on cloud feedbacks that then occur both in these zones and further afield, all ultimately influencing Earth's global climate sensitivity, as expressed as the equilibrium climate sensitivity (ECS, Knutti et al. 2017). The paper first outlines data sources used to examine these influences. The initial focus of the paper begins in Sect. 3 with a review of the energy balance of the tropical convergence zone and radiative properties of clouds in that zone. Observations are used to frame our understanding about processes that are the bedrock of the cloud feedbacks widely discussed in the literature. Addressing these processes within the context of the current observed state and its variability, such as described in Sect. 4, and assessing model representations of these processes is an activity, although not specifically pursued in this paper in any detail, that is necessary to establish credibility in any model projection of change. Dessler (2010) is an example of how observations might offer some constraint or test of model feedback processes. Understanding gleaned from observed variabilities of Earth alone, while necessary, is an insufficient test of understanding as interactions of processes responsible for shorter-term observed variabilities do not always project onto longer term changes and do not always define how cloud feedbacks necessarily operate on this longer term. This is a point discussed more fully in later sections of the paper. Sections 5 and 6 expand the focus outward from the TTZ with an examination of the sensitivity of subtropical low cloud feedbacks on tropical convection and then on the sensitivity to deep convection of broader patterns of global warming and model estimates of ECS.

2 Sources of Data

A variety of different data types is examined throughout this study. These include:

(i) The CERES Energy Balance and Filled (EBAF) Ed. 4.1 top-of-atmosphere (TOA) flux product (Loeb et al. 2018) with matching surface radiative flux data (Kato et al. 2018). These matched fluxes are made from a combination of observations and radiative transfer calculations archived in the form of monthly means for March 2000–March 2020 set on a 1° × 1° latitude–longitude grid.

(ii) Cloud properties consistent with the CERES EBAF flux product (Loeb et al. 2018) and matched in time are derived using the CERES-MODIS cloud retrieval algorithm (https://ceres-tool.larc.nasa.gov/ord-tool/jsp/EBAF4Selection.jsp). The cloud area fraction (CF) is the fraction of cloud identified in MODIS pixels within a CERES footprint based on the CERES-MODIS cloud mask algorithm and weighted by the total number of pixels within the footprint. The cloud effective pressure (we refer to it as cloud top pressure (CTP) although it does not necessarily denote the geometric top of the cloud) is derived from the cloud effective temperature (corresponding to the effective emission from clouds), which is based on the cloud's emissivity derived from the MODIS window channel temperature. The cloud visible optical depth (COD) is defined by the 0.65-μm channel radiances from cloudy pixels during daytime; during nighttime, COD is determined with the MODIS infrared channels, but this is limited to a narrow range of COD. We complement this cloud information with direct atmosphere opacity observations from CALIPSO (Guzman et al. 2017) to deduce temporal variability in tropical mean opaque cloud cover which are notionally clouds that attenuate the CALIPSO lidar approximately corresponding to COD > 3.

(iii) The IMERG V06B monthly precipitation data are also used in the analysis, and these data are matched to the CERES gridded data at the same resolution of a 1 × 1 degree latitude-longitude (Huffman et al. 2020).

(iv) CloudSat radar data in the form of the 2B-Geoprof-lidar product (e.g., Mace et al. 2009). These data are used to characterize convective core intensity in this study based on past studies of Luo et al (2008) and Takahashi and Luo (2012). The deep convective core (DCC) is identified in the CloudSat profile data as profiles having (1) continuous radar echo from cloud top to within 2 km of the surface (i.e., the target cloud is rooted in the planetary boundary layer) and (2) echo top height (ETH) of the 10 dBZ reflectivity above 10 km. This core information is examined in more detail in related papers (Pilewskie et al. 2024; Derras-Chouk and Luo 2024).

(v) The multi-mission climate data record (CDR) of zonally averaged, monthly tropopause heights used here originates from the EUMETSAT Satellite Application Facility on Radio Occultation (RO) Meteorology (https://doi.org/10.15770/EUM_SAF_GRM_0001) and is derived from sounder measurements on Metop, COSMIC, GRACE and CHAMP. The dry temperature lapse-rate tropopause height is based on the standard WMO lapse-rate criterion which defines the tropopause at the lowest level at which (1) the lapse rate is less than 2 K km^{-1} and (2) the average lapse rate between this lower level and all levels within 2 km above it that does not exceed

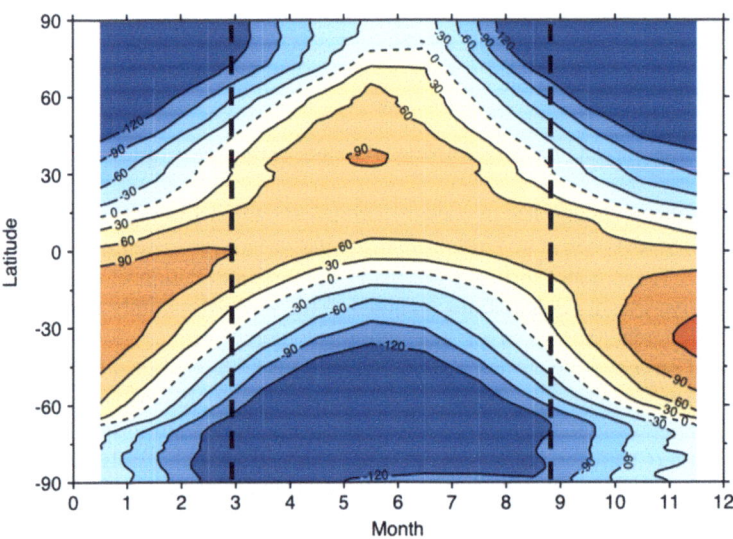

Fig. 2 The annual progression of the net radiation balance of the Earth-atmosphere system deduced from CERES radiative flux data (refer to discussion of data sources) where the fine dashed lines delineate the zero next flux. The vertical dashed lines identify times in the annual cycle used to define the extreme positions of the tropical trough zone. The contours of next flux are in Wm^{-2}

2 K km^{-1}. Further information on the data product can be found here: https://rom-saf.eumetsat.int/product_documents/romsaf_atbd_tph.pdf

(vi) Radiative heating profile data of Stubenrauch et al. (2021) for the period 2004 to 2018. These radiative heating rate profiles have been extended in space and time from CloudSat FLXHR-lidar data (Henderson et al. 2013) by artificial neural network (ANN) regression models applied on cloud properties retrieved from AIRS (Stubenrauch et al. 2021) and atmospheric and surface properties from ECMWF re-analyses (Dee et al. 2011).

In addition to these observational data records, we also report on analysis of a number of CMIP6 experiments in Sects. 5 and 6. These analyses make use of data taken from CMIP6 model simulations (Eyring et al. 2016) from both the historical simulation experiment, from the amip-piForcing simulations, the abrupt-4xCO$_2$ simulations and the Shared Socio-Economic Pathway 5 (SSP5-8.5) simulations.

3 The Tropical Trough Zone (TTZ)

The seminal study of R–M provided the first estimates of the energy balance of the equatorial low-pressure trough zone in coordinates following the trough around the globe. Out of that study evolved the concept of convective hot towers and their fundamental role in the heat balance of this zone. Twenty years later, Riehl and Simpson (1979) revisited that study using updated global data available at that time, and noting some significant differences in the radiation balance between the two studies. Dramatic improvements in our ability to observe the global atmosphere and create globally consistent data records of the atmosphere have occurred since then with new satellite observations and re-analysis appearing at the apex of this improvement. Here, we provide a partial update only to these earlier studies applying the same analysis of R–S to define the seasonal location of this heated zone of the tropics with a brief description of the convection that forms within it. This latter aspect of the tropical trough zone is specifically discussed in more detail in a companion paper (Pilewskie et al. 2024).

As in R–S, the place to begin is the net radiation balance at the top of the Earth's atmosphere as shown in Fig. 2. This balance represents the entire column of the atmosphere, the Earth's surface and sub surface combined. If latitudinal energy transports are to be derived using these data on any basis other than annual, then the local heat storages of the atmosphere and oceans are needed. The atmospheric heat storage is small. Estimating the oceanic heat storage has been historically challenging, although ocean analyses approaches continue to improve (e.g., Forget et al. 2015) and could be considered as a source of heat storage information. Here, we adopt the simple approach introduced by R–M and compute the transports at those times when the ocean temperatures are at their maximum or minimum as these are times when local changes are near zero implying near zero ocean heat storage at these times depicted in Fig. 2 by the dashed vertical lines.

3.1 Meridional Energy Transports and the Positions of the TTZ

The meridional transports of energy corresponding to the times identified in Fig. 2 are presented in Fig. 3a. R–S argue that the transports derived at these times can be taken to be representative of longer periods and representative of transports to the respective summer and winter poles. When averaged together, as done in Fig. 3b, these transports should equal

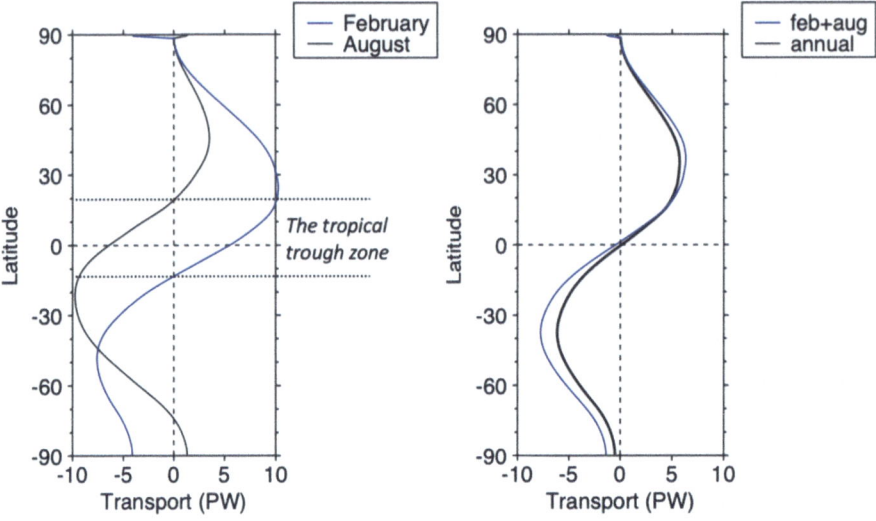

Fig. 3 **a** The atmospheric heat transports for February and August and **b** their combined transports compared to the annual total mean. The latter basically verifies the R–M approach/hypothesis that the February and August transports are largely atmospheric with their combination equivalent to the annual mean transport

that of the annual mean for the arguments of R–S to be valid. The small residual of this combined transport at the pole implies that a small amount of internal storage exists in the February analysis. Nevertheless, the close agreement between the summed transport and the annual mean transport revealed in Fig. 3b broadly supports the original arguments of R–S.

The dotted lines of Fig. 3a identify the latitudes that divide energy transports to summer and winter poles for both seasons. In combination, these define the position of the tropical trough for each season being the latitude of zero transport that is the latitude at which heat is moved in opposite directions to both poles. From analysis of these transports derived from the CERES radiative flux data, the tropical trough is determined to move annually between latitudes of 19 N and 13 S, which is a similar range to that originally identified in the R–S study although R–S subsequently argued for a more equatorial limit of the southern extent of this zone putting it at 5 S based on wind field analysis. We hereafter refer to this 19 N to 13 S region as the tropical trough zone (TTZ) and note a few important observations about it:

(i) It is a fundamental zone containing deep convection, including the inter tropical convergence zone (ITCZ), and the net energy of it is always exported to the winter pole.

(ii) The zone is asymmetrically positioned around the equator, being more weighted to the northern hemisphere (NH). This observation is consistent with our observation of the climatological annual mean position of the inter tropical convergence zones correspondingly lie in the NH, about which much has been written and is further discussed below (e.g., Philander et al 1996).

(iii) This hemispheric asymmetry can be traced to the asymmetric heat losses of the respective NH and SH winter hemispheres. R–S note that it is the heating require-

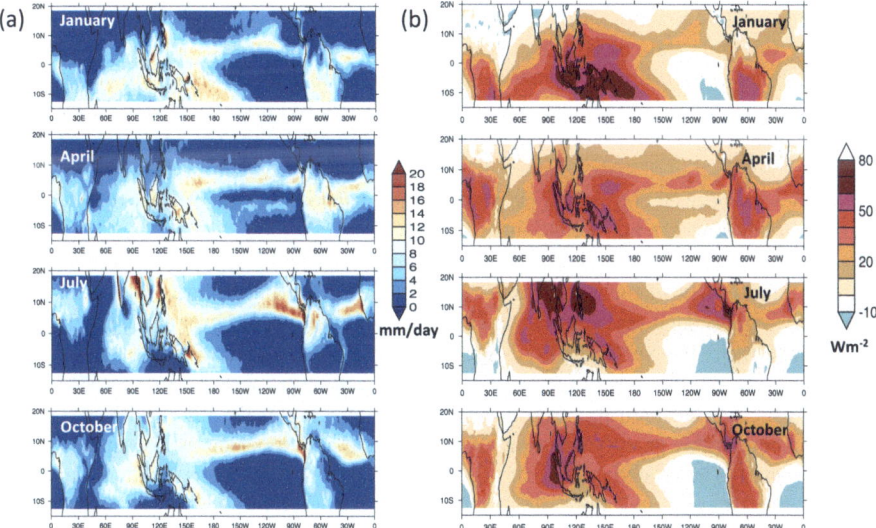

Fig. 4 **a** Left panels are January, April, July and October 22-year monthly mean distributions of IMERG precipitation within the TTZ and **b** right panels are the matching distributions of ACRE shown for comparison and discussed later in relation to Fig. 8

ment of the high latitude winter hemisphere that furnishes a constraint on the position of the trough and its range of displacement. A significant source of the greater heat loss from the SH winter hemisphere can be simply traced to differences in the amount of solar energy received by the respective winter hemispheres determined by differences in winter season Sun-Earth distance. This hemispheric asymmetry of the TTZ is consistent with our broad understanding of the displacement of the ITCZ favoring the warmer hemisphere (e.g., Broccoli et al. 2006). It is also consistent with studies that show a southward shift in convection as the energy transport northward increases to balance the increased loss of energy from the NH when increased cooling occurs at the higher NH latitudes (e.g., Kang et al. 2008, 2009; Frierson and Hwang 2012).

3.2 Convection in the TTZ

The variation of convection within the TTZ is often described as following the Sun. The pattern of change over the annual cycle is, however, more complex than this, being clearly shaped by different mechanisms of forcing that come into play. This point is well illustrated with reference to Fig. 4a which presents the 22-year mean IMERG precipitation climatology for 4 months. These months illustrate the shifting patterns of precipitation, and by inference convection, over the mean annual cycle. It is noteworthy that the largest total precipitation is not always aligned with the most intense convection, a point Zipser et al. (2006) noted and further elaborated on in the study of Pilweskie et al. (2023). The precipitation, and the convection that produces it, can be thought of in terms of three regimes governed by different forcing mechanisms. One is the regime of convection over the large southern land masses of S. America and the African continent. A second regime is that of convection confined to narrow convergence zones more typical of the classical view of the ITCZ over the mid- and eastern Pacific and tropical Atlantic oceans, and a third regime encapsulates the convection of the maritime continent. The physical characteristics of convection within each of these regimes are further described in Pilweskie et al. (2023), while Derras-Chouk and Luo (2024) provide independent estimates of the mass transported vertically by the convection of these regimes. In a related paper, Roca et al. (2024) describe how the morphology of deep convection over Africa has changed over the 30 years of Meteosat observations shifting to smaller but longer-lived systems.

Convection over the tropical land masses indeed moves annually north and south following the movement of the Sun. This is evident in the shifting patterns of precipitation observed over these land masses and shown in Fig. 4a, being forced by surface heating that more-or-less tracks the Sun's position. According to Zipser et al. (2006), convection over these land masses is the most intense of all tropical convection according to TRMM radar data-based metrics of intensity, metrics that do not geographically align with maximum rainfall accumulations or other measures of convective activity, such as average IR cloud top temperature. The related studies of Pilweskie et al. (2013) and Derras-Chouk and Luo (2024), using different and more direct measures of convective intensity, including estimation of vertical motions, find convection over Africa to be the most intense of all deep convection on the planet.

The Maritime Continent, on the other hand, is an archipelago that includes Malaysia, Indonesia, the Philippines and Papua New Guinea and hundreds of islands of varying shapes and sizes. This complex geography, together with the fact it lies in some of the warmest waters on Earth, creates a complex set of interactions and circulations generated between ocean and land producing complicated precipitation patterns. The complexity of these patterns is exemplified in the pattern of the diurnal cycle of precipitation such as described by Wei et al. (2020). The maritime continent, also broadly referred to as the tropical western pacific (TWP), is a major center for tropical atmospheric convection, and the latent heat released by the condensation of water vapor into cloud water and rain in this convection drives large-scale tropical circulations. Takahashi et al. (2023) offer a mechanistic explanation for why land-based deep convection is more intense than oceanic deep convection. The importance of the TWP to the global circulation is also reinforced in Derras-Chouk and Luo (2024) who find the TWP is a region of greatest mass fluxes of anywhere on the planet. As a consequence of this heating, and this mass flux produced by it, this region is a principal source of rising air in the tropics being the ascending branch of the Hadley and Walker circulations. As with convection over land, convection in the region of the maritime continent also undergoes a distinct seasonal cycle in a manner that also appears to follow the movement of the sun across the TTZ and is a key element of Earth's major monsoon systems.

The third regime is the convection that forms over the warmest waters of the Middle and Eastern Pacific and Atlantic oceans. Convection in this regime, unlike in the other regimes, does not follow the insolation maximum but sits north of the equator both in the Atlantic and Pacific being quasi-stationary throughout the year. Over the Pacific, convection remains largely anchored to the SST maxima that reside north of the equator that are more or less geographically fixed, according to Philander et al. (1996), by the asymmetry in positions of continents relative to the equator. The sea surface temperatures (SSTs) of the tropical Atlantic Ocean experience a small annual cycle with migration from the equator during the Boreal winter slightly northward of the equator during the Boreal summer (Crespo et al. 2019) with convection and precipitation following this migration.

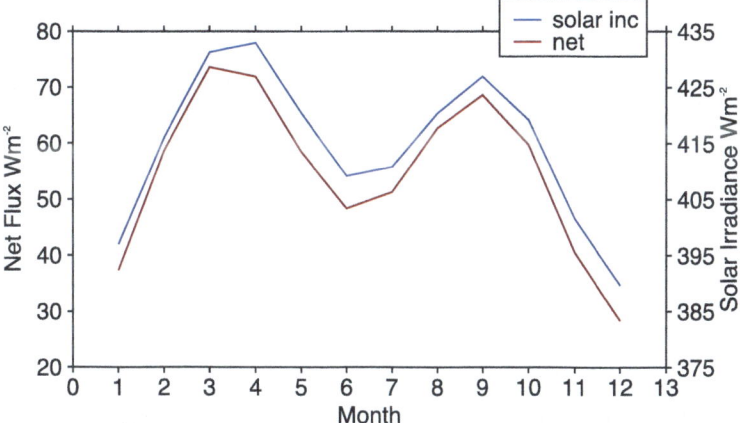

Fig. 5 The 22-year averaged annual cycles of TOA net radiation and daily mean solar irradiance of the TTZ

3.3 Radiative Properties of the TTZ

The radiative properties of the TTZ are now examined in the context of the convection that occurs in that zone. The connections between these properties and convection are elements of important cloud-convection climate feedbacks described later. All radiative properties reviewed here are derived from CERES EBAF data. In addition to the TOA net flux (Fig. 2), two sets of radiative flux properties are the focus of this section. First are the cloud radiative effects (CREs) defined at the TOA, and second are the atmospheric CREs (hereafter ACREs) representing the net absorption and emission by and from the atmosphere being the difference between the TOA and surface CREs. The monthly mean distributions of the ACRE are shown in Fig. 4b, and its comparison to the precipitation distributions of Fig. 4a and the implication of these comparisons are discussed below.

3.3.1 The Annual Cycle

The 20-year mean annual cycle of net radiative flux averaged over the TTZ is presented in Fig. 5 along with the annual cycle of the daily mean flux of incoming solar radiation. The annual cycle of net radiation, both in its phasing and amplitude, merely follows the incoming solar annual cycle that has a bimodal behavior established by the bimodal character of the daily mean TTZ averaged cosine of solar elevation. The tight correlation between the incoming solar and the net TOA flux ($R^2 = 0.98$) is a remarkable result in that all the factors that can be considered to affect the annual cycles of outgoing long and shortwave radiation, most notably cloud radiative effects discussed next, cancel although not entirely as discussed below. The extent of this cancelation is reflected both in the correlation between these two cycles and the rms flux difference between the two cycles which might be taken as an indicator of the degree to which internal processes influence the outgoing TOA fluxes on this time scale. This rms flux difference is 1.7 Wm^{-2} and, as we will see later in Sect. 4, is also of this same order as the interannual variations of net TOA flux forced by interannual changes in convection in the TTZ.

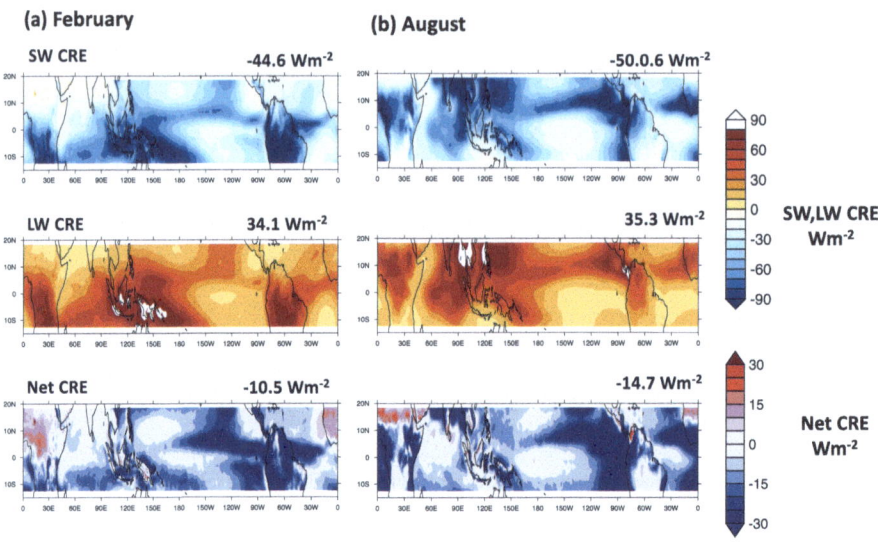

Fig. 6 a, b A 20-year climatology of the TTZ TOA cloud radiative effects for the 2 transition months (February and August) that define the latitudinal extent of the TTZ

3.3.2 Cloud Radiative Effects (CREs)

The discussion above notes a fact that has long been known—that the TOA longwave (LW) and shortwave (SW) components of outgoing radiation defined by clouds, and expressed as the net TOA CRE, cancel each other to a remarkable degree in lower latitudes. Figures 6 further illustrates this same point. Shown in Fig. 6, are the distributions of the SW and LW TOA CREs for the 2 months used to deduce the limits of the TTZ. Whether the regime where the CREs negatively mirror each other and cancel over the bulk of the TTZ that is purely coincidental or is part of an overall but as yet unknown constraint on the system still remains a mystery despite studies that have looked into the topic (e.g., Kiehl 1994). The degree to which this cancelation can be expected as Earth warms is also particularly pertinent (e.g., Sokul and Hartmann 2020) because small changes to one component without compensating changes to the other would constitute potential significant feedbacks (e.g., Stephens 2005). Existing tropical high cloud feedback concepts described in the next section are so posed, mostly hypothesized around changes to the LW component of the CREs or around the effects of high clouds on the ACRE.

While the majority of the climate change literature frames climate feedbacks almost exclusively in terms of changes to the TOA CREs, a number of studies over the years call out the importance of the in-atmospheric radiative heating by clouds not only for cloud-climate radiative feedbacks (e.g., Stephens et al. 2018; Needham and Randall 2021b) but also for feedbacks on global precipitation change (e.g., Stephens and Hu 2010). A number of studies have noted the influence of this heating on tropical convection and the ITCZ by, for example, its effect on the strength of the Hadley and Walker circulations (e.g., Slingo and Slingo 1988; Sherwood et al. 1994; Li et al. 2015; Harrop and Hartmann 2016; Popp and Silvers 2017). More recently Needham and Randall (2021b) argue that the ACRE of clouds associated with tropical convection is an important source of tropical heating that balances the horizontal export of energy out of the humid tropics by the large-scale

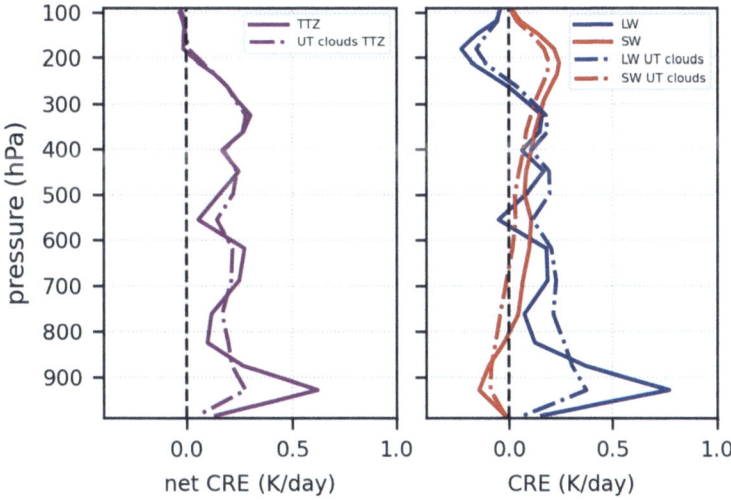

Fig. 7 The 2004–2018 averaged profiles of cloud radiative heating in the TTZ. The effect of all clouds (solid lines) is given by the radiative heating rate difference between all scenes and clear sky, while the effect of upper tropospheric clouds at a scale of 0.5° (broken lines) is given by the radiative heating rate difference between these clouds and those of clear sky, weighted by these clouds

overturning circulations. They recognized that this heating can be significant in inducing thermally direct circulations connecting humid and dry regions and further noted how the heating of the ACRE correlates to precipitation in the humid regions. The suggestion of a circulation driven in part by horizontal gradients of ACRE is not new and is a notion first suggested by Gray and Jacobsen (1977) and examined specifically by Slingo and Slingo (1988). Webster (1994) call it out as a mechanism central to the convection in the Western Pacific Warm Pool, and Stephens et al. (2008a) note its fundamental role in convective aggregation that occurs in numerical radiative convection equilibrium experiments. Before the CloudSat mission (e.g., Stephens et al. 2008b), however, the only way to estimate the ACRE was via model simulations of TOA and surface radiative fluxes. With the advent of cloud profile data, CloudSat observations provided the means to produce a more observationally-robust estimate of the global ACRE. These observations, along with others, were subsequently used to infer magnitudes of this differential heating associated with the warm phase of ENSO (Stephens et al. 2018).

The A-train data have also provided a mechanism for unambiguously defining the cloud structures responsible for the observed behavior of the ACRE as shown in Fig. 4b and explored further in discussion of Fig. 8. The effects of clouds and in particular of upper tropospheric clouds above 440 hPa on the radiative heating rates in the TTZ are presented in Fig. 7. The averages are given separately for the 24 h net CRE, LW CRE and 24 h SW CRE. The important point demonstrated by this figure is that it supports the assertions made above that it is the high clouds that most define the observed patterns of ACRE in the TTZ.

CloudSat measurements have also revealed how tropical cyclones that intensify versus weaker storms that do not contain greater amounts of ice in the high clouds in the vicinity of the storm center. This, in turn, is associated with larger ACRE and thus stronger radiative heating there. Similar to other studies mentioned above, Wu et al. (2023) show that an enhanced thermally direct circulation driven by a differential

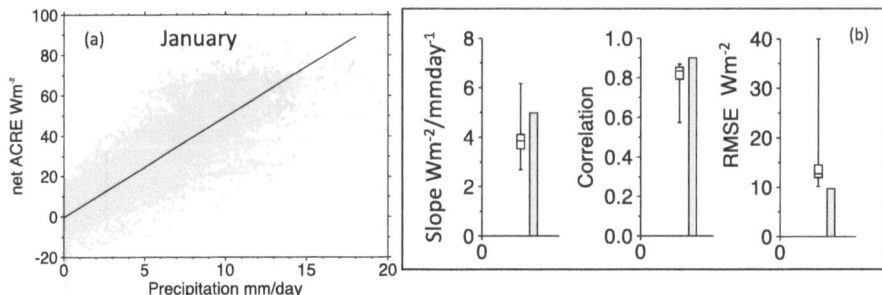

Fig. 8 **a** The correlation of monthly mean TTZ precipitation and the net ACRE. **b** The comparisons between the linear fit parameters of the data shown in **a** and the analogous fit to 26 the CMIP6 model produced data taken from the historical experiments. Model data are represented by the Box–Whisker diagrams

radiative heating formed between this enhanced heating in the vicinity of the center of the storm in contrast to the outer environment that further sustains the storm and is an important mechanism in cyclone intensification. They subsequently demonstrated that forecasts of intensity of tropical cyclones could be improved significantly when the forecast model better reproduces the observed spatial structure of this radiative heating.

An important point implicit to most of these past studies but often overlooked is how the radiative heating associated with high clouds in regions of convection reinforces the latent heating associated with the precipitation produced by the same convection thereby strengthening this thermally direct circulation. This is qualitatively evident in comparison of Fig. 4a, b noted previously and further reinforced with the scatter diagrams of Fig. 8. These scatter diagrams illustrate how the two main modes of diabatic heating of the TTZ atmosphere, in the form of the monthly mean diabatic ACRE and the monthly mean IMERG precipitation, are tightly coupled. The results shown in Fig. 8a apply to January only but are typical of the correlations observed for other months not shown. A linear-least-squares fit of the data is also provided for reference with a slope of 4.96 $Wm^{-2}/mmday^{-1}$ and the prediction of net ACRE using precipitation as the predictor yields an ACRE rmse of 9.7 Wm^{-2}. The same analysis was applied to the CMIP6 historical experiments for 26 models. These simulations span the period of extensive instrumental temperature measurements from 1850 to the present, and for the analysis performed only the model, January data for latter years of the simulations between 1995 and 2014 were used for comparison being nearer the period of the observational record. The comparisons are summarized in Fig. 8b in the form of box-quantile plots that define the maximum and minimum values of the 26 models, the 25th and 75th quantiles and the median compared to the observations of Fig. 8a that are summarized as a histogram bar. Except for one model outlier, all models produce similar high correlations between precipitation and the ACRE although the correlations of models are lower than that observed. On average, the model slope is also about 20% lower than the observed slope suggesting a weaker coupling between ACRE and precipitation in models than observed which hints at weaker couplings between convection (precipitation) and high clouds (ACRE).

4 The Interannual Variability of the Tropical Trough Zone (TTZ)

The feedback system involving the differential heating described above is also hypothesized to operate on interannual time scales (e.g., Stephens et al. 2018). Here the interannual variability of the energetics of the TTZ are further examined, with a focus of the total energy balance of the TTZ. Such variability hints at potential mechanisms of feedback associated with convection and the high clouds of the TTZ. As previously noted, almost all discussions of tropical high cloud feedbacks are framed around changes to the longwave component, that is the LW CRE, but the discussion above and the analysis now described also suggests that influences of high cloud changes on both the internal radiative heating of the atmosphere and on the TOA SW fluxes reflected to space cannot simply be overlooked. The importance of the latter is revealed in the data plotted in Fig. 9. The figure is a time series of a number of variables averaged over the TTZ all expressed as de-seasonalized anomalies. The time series of Fig. 9a illustrate the interannual variabilities of the TTZ-mean tropopause height (TROH) derived from COSMIC data contrasted against the anomalies of the CERES observed TOA energy imbalance (TEI). The time series of Fig. 9a reveals a distinct correlation ($R = -0.6$) between the interannual anomalies of the TTZ TROH and the corresponding changes to the TEI during the first half of the time series but with a lower correlation overall ($R = -0.36$) which is a reminder that different processes operate over the time period of the observations producing different responses in TEI. Figure 9b presents a measure of changes in convective intensity, similar in philosophy to that introduced by Zipser et al (2006), being a measure of the heights of the deep convective cores as defined by CloudSat observations (Takahashi and Luo 2012, 2014). This time series is limited by the availability of CloudSat observations. Figure 9c presents the matching time series of thick high clouds of the TTZ as defined by the opaque cloud product of CALIPSO (Guzman et al. 2017) again limited by data availability contrasted against the CERES net CRE.

The variabilities observed before 2017 presented in Fig. 9 in particular offer a consistent picture of the interannual variability of the TTZ. The depth of the TTZ troposphere undergoes a multi-year quasi-cyclic variation with periods of deepening and shallowing which have been connected both to ENSO and to the quasi—biennial oscillation and interannual variation of the Brewer Dobson Circulation (e.g., Davis et al. 2013; Li and Thompson 2013, Fueglistaler et al. 2009). The periods of deeper TTZ correspond to periods of deeper and more intense convection as suggested by a greater frequency of deep core convection observed by the CloudSat radar. Associated with this period of enhanced deeper convection is an increase of thick high cloud frequency which correlates to increased reflected sunlight from the TTZ during these periods. Further analysis of cloud property changes over the time frame of Fig. 9 is presented in Fig. 10 and supports the conclusion that the changes to the observed TEI that correlates to the TROH changes before 2017 are driven by changes to reflected sunlight from increasingly think anvil clouds that form during times of deeper convection.

The spatiotemporal correlation analysis applied to monthly data for the period 2002–2022 identifies the cloud properties most responsible for the observed interannual variabilities and confirms the conclusion stated above. The properties examined include cloud fraction (CF), cloud top height/pressure (CTP) and cloud optical depth (COD). The anomalies of these three properties were normalized for comparison and correlated to the LW CRE, SW CRE and the net CRE anomalies. The results of this analysis as summarized in Fig. 10 lead to the following conclusion:

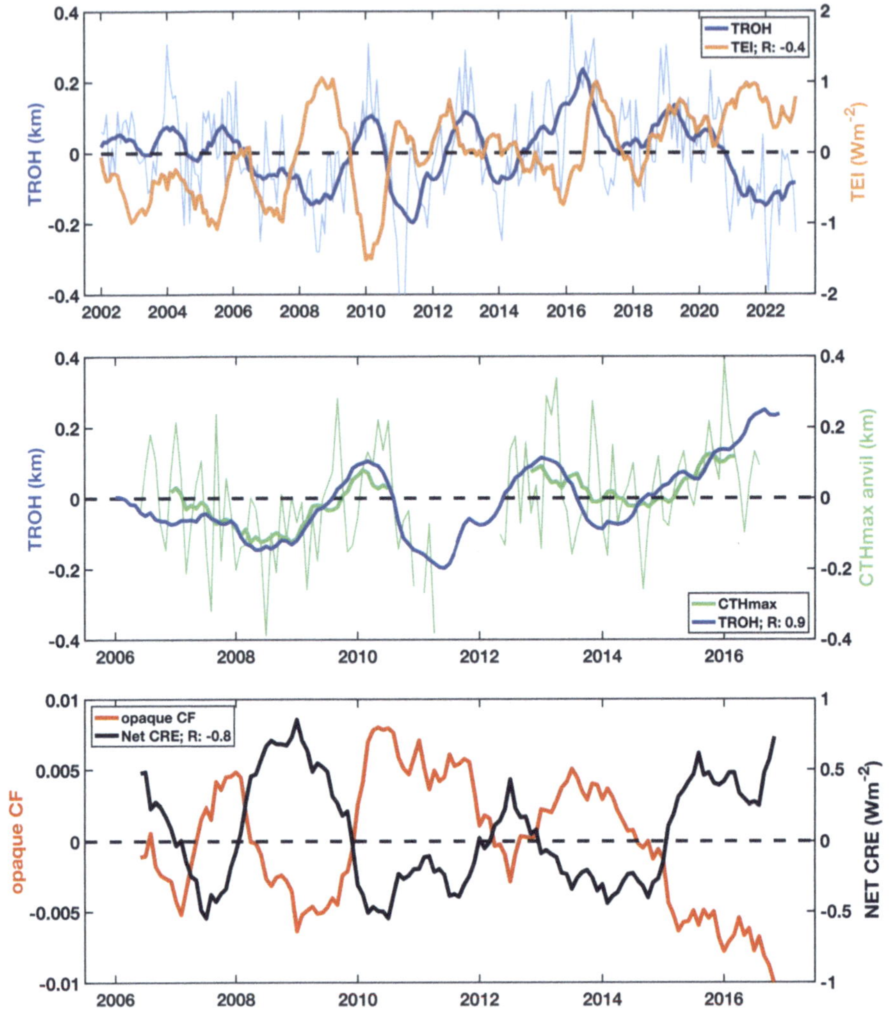

Fig. 9 a Time series of TROH (blue) and TEI (red) de-seasonalized monthly anomalies that are smoothed using a 5-month running mean filter, **b** time series of TROH (blue) and the convective cove maximum height (green) de-seasonalized monthly anomalies for a shorter time series that **a**. **c** Standardized monthly anomalies (unitless) smoothed with a 5-month running mean filter of tropical mean net CRE (black) and opaque cloud fraction (orange) with a correlation of − 0.8. The net CRE change is largely defined by the SW CRE (correlation of 0.7, not shown), and the correlation SW CRE with COD (not shown) is − 0.83 (Fig. 9f)

(i) CF (Fig. 10 a, d, g) positively correlates ($R = 0.75$) to the LW CRE and negatively correlates to the SW CRE ($R = 0.7$) leaving a weak correlation overall ($R = -0.27$). This hints at the role of CF changes as a major factor in the LW-SW cancelation described above.

(ii) An expected negative correlation exists between the LW CRE and CTP ($R = -0.75$; Fig. 10b) and a correlation of opposite sign exists between the SW CRE and CTP ($R = 0.52$; Fig. 10e). A significant relationship between CF and CTP ($R = -0.7$)

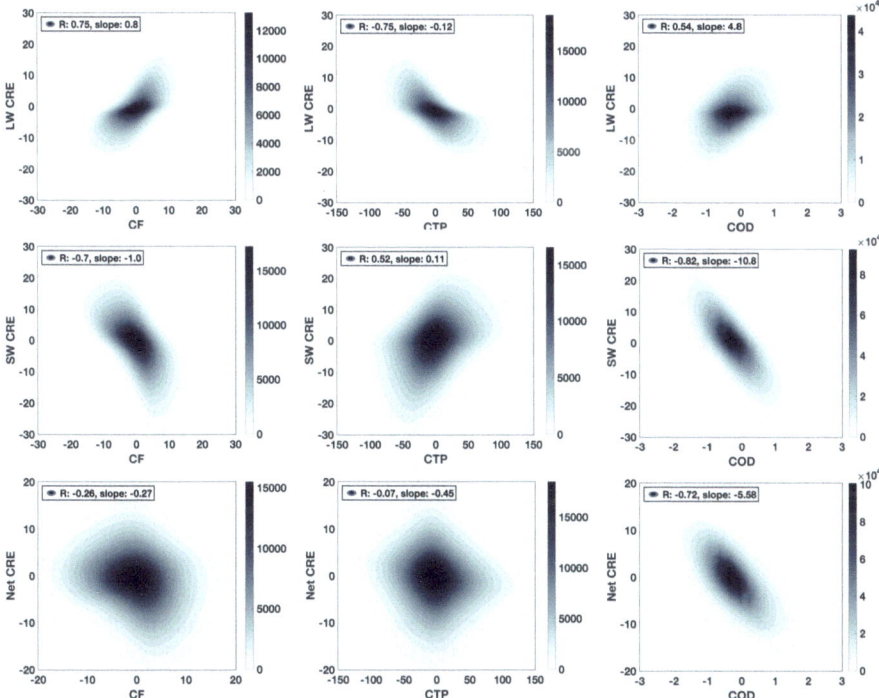

Fig. 10 Two-dimensional density scatter plots considering all space–time points in the TTZ over mid 2006 to end 2014 of cloud properties against radiative fluxes under cloudy conditions. **a** LWCRE versus cloud fraction (CF), **b** LWCRE versus cloud top pressure (CTP), **c** LWCRE versus cloud optical depth (COD), **d** SWCRE versus CF, **e** SWCRE versus CTP, **f** SWCRE versus COD, **g** Net CRE versus CF, **h** Net CRE versus CTP, and **j** Net CRE versus COD. Figure legends indicate the R correlations and regression slopes of the respective analyses. Shading indicates number density of data points

also exists (not shown) such that higher clouds are associated with enhanced CFs. The effect of CTP on the net CRE is nearly zero (Fig. 10h) with the spurious SW and LW responses to CTP again compensating one another. This strong correlation between CF and CTP implies that more intense convection (indicative of lower CTP) develop into larger convective systems (higher CF) as called out in number of previous publications have shown this relationship (e.g., Machado and Rossow (1993), Takahashi et al. (2017), among others).

(iii) The most dominant cloud parameter of influence on the observed variability of the TEI for the period up to 2017 is the cloud optical depth (COD). The covariabilities of COD and CF ($R=0.4$) and COD and CTP ($R=-0.2$), respectively, are smaller than is the covariability of CF and CTP. While the LW CRE increases with COD, a result that suggests even the thinner, non-radiative black-body high clouds also thicken during periods of deeper convection, this response is much weaker than the overwhelming negative relationship between COD and the SW CRE ($R=-0.82$) that determines the net CRE response ($R=-0.72$). The strong correlation of the CALIPSO measure of thick high clouds with the timeseries of SW CRE ($R=-0.83$) and the net CRE ($R=-0.76$) further supports the conclusions that it is a thickening of anvil clouds and increased SW reflection that determines the interannual variation of the spatially distributed TEI.

5 Cloud-Radiative-Convective Feedbacks in a Changing Climate

We consider now the influence of tropical convection within the context of climate change and do so with a focus on cloud feedbacks mostly relying of a number of climate model climate change experiments archived under CMIP6.

5.1 Tropical High Cloud Feedbacks

There is a growing literature on the topic of high cloud feedbacks that involve the interactions of convection, clouds and radiation. Some of that literature was described above in relation to the role of these clouds in heating the atmosphere. Other concepts exist that focus either on feedbacks at the surface or feedbacks involving the entire column as viewed through from the perspective of TOA CREs. Ramanathan and Collins (1991), for example, argued that the shielding of solar radiation by thick anvil clouds overwhelm the enhanced greenhouse effect that they induce in a moist, convectively active atmosphere over warm oceans, and suggested that these clouds act as a natural thermostat regulating the sea surface temperature (SST). Stephens et al. (2004) interpreted TOGA COARE observations as showing that the longwave radiative heating of the upper troposphere, induced by enhanced amounts of high clouds during active phases of the Madden-Julian oscillation (MJO), coupled with the solar radiation reduction at the ocean surface act together as a self-regulating radiative-convective system. A number of other ideas center on the longwave effect of anvil clouds. Lindzen et al. (2001) proposed an "iris" mechanism, built upon the assumption that precipitation formed in convection over warmer oceans will rain more efficiently with proportionally less water being transported aloft and detrained into anvil clouds. Anvil clouds produced by convection are then hypothesized to shrink in area as sea surface temperatures (SSTs) warm. The result is an increased emission to space with warming implying a negative cloud anvil feedback on increasing SSTs. The thermostat and iris hypotheses, although based on different mechanisms, both hypothesize that anvil clouds produced by detrainment from deep convection are essentially components of negative feedbacks that buffer the warming of the oceans.

A separate set of hypotheses has been framed around the temperatures of anvil clouds as these clouds rise with the deepening of convection as SSTs warm. In the hypotheses known as fixed anvil temperature (FAT) and proportionally higher anvil temperature (PHAT), anvil cloud temperature remains nearly constant regardless of a surface warming thus resulting in no real change in the OLR as the clouds rise. In this case, the outgoing longwave radiation (OLR) cannot efficiently remove the extra heat associated with the warming (Hartmann and Larson 2002; Zelinka and Hartmann 2010). The clear-sky radiative cooling of the convective environment, however, is fundamental to this idea as the curvature of this cooling in the upper troposphere, in turn defined by the profile of upper tropospheric humidity, defines the heights of convergence required to balance the subsidence driven by the profile of cooling (e.g., Mapes 2001). The feedback involves more than lifting of anvil clouds with deeper convection over warmer SSTs but includes the influence of convection on the environmental upper tropospheric water vapor and the latter, in turn, on radiative cooling which then determines the detrainment level of deep convection. Some aspects of the FAT feedback are contained in the "stability iris" hypothesis of Bony et al. (2016) who argue, like Lindzen et al. (2001), for a reduction in the detrained anvil cloud cover from convection as SST's warm effectively producing an IRIS effect but shaped by changes to upper tropospheric stability with warming as opposed to changes in precipitation efficiency of convective storms as proposed by Lindzen et al. (2001).

It is a challenge to use observations to assess the reality of such feedback concepts in any definitive way given the difficulty of assessing the long-term regulation of SSTs with typically short-term observations. Nevertheless, observations must be scrutinized to claim credibility for these feedback ideas. One approach is to test basic understanding of the central processes underlying the hypothesized feedbacks (e.g., Stephens 2005). The results of Figs. 9 and 10 above, for example, offer a test of our understanding of cloud anvil property changes as convection deepens albeit in the context of interannual variability. The observations point to the thickening of anvils associated with deepening convection as most influential on the CREs on those time scales. The study of Ito and Masunaga (2022) is another example of how satellite observations matched to re-analysis data can be used to interrogate hypotheses inherent to feedback concepts. They found that the inter-relationships between anvil cloud fraction, upper tropospheric stability and upper tropospheric convergence are consistent with expectations of the stability iris theory of Bony et al. (2016). By contrast, they further showed that an enhancement in precipitation efficiency is unlikely to have appreciable consequences on the horizontal extent of anvil clouds although admittedly defining the actual precipitation efficiency with observations is challenging and analysis of CMIP6 model data suggest an increased precipitation efficiency with warming with associated reductions in high cloud (e.g., Li et al. 2022).

5.2 Deep Convection, High and Low Clouds and Circulation Driven Feedbacks

Critical to many of the cloud feedback ideas proposed, it is the underlying relation between convection and high clouds detrained from it and how these clouds then modulate the local heating and the circulation that, in turn, affects convection and clouds further afield. Aspects of the topics described above now come together with a discussion framed around Fig. 11a–c. Figure 11a is a schematic summary of the changes in convection, high clouds, OLR, precipitation and far field influences on low clouds via changes to the Hadley circulation that are consequence of the warming expressed as the ECS of climate models. This figure is a conceptual synthesis of 23 model simulations forced by the CMIP6 SSP5-8.5 scenario contrasted against the historical simulations for the period 2000–2014 (e.g., Schiro et al. 2022). The figure is a graphical summary of the discussions of Su et al. (2017) and Schiro et al (2022) and is a synthesis of model responses for the convective zone between latitudes of 10 N and 10 S (Fig. 11b) and responses in the subtropical regions in latitude ranges 10–2 0 N/S (Fig. 11c). A prominent feature of the projected tropical circulation change in a warmer climate is the intensification of zonal-mean equatorial ascent flanked by the weakening of upward motion to its north and south resulting in a narrowing of the deep convection zone (e.g., Lau and Kim 2015). This feature is simulated by most climate models and is referred to by Su et al. (2017) as the tightening of Hadley ascent in contrast to the well-known widening of the Hadley cell (e.g., Seidel et al. 2008; Tao et al. 2016). There is an expansion of the surrounding dry regions which are regions that acts as analogous to radiator fins (Pierrehumbert 1995) or are regions described as an expanding iris as proposed by Lindzen et al. (2001) and discussed by Mauritsen and Stevens (2015). This is also a region in which the upper troposphere both dries, the stability increases and

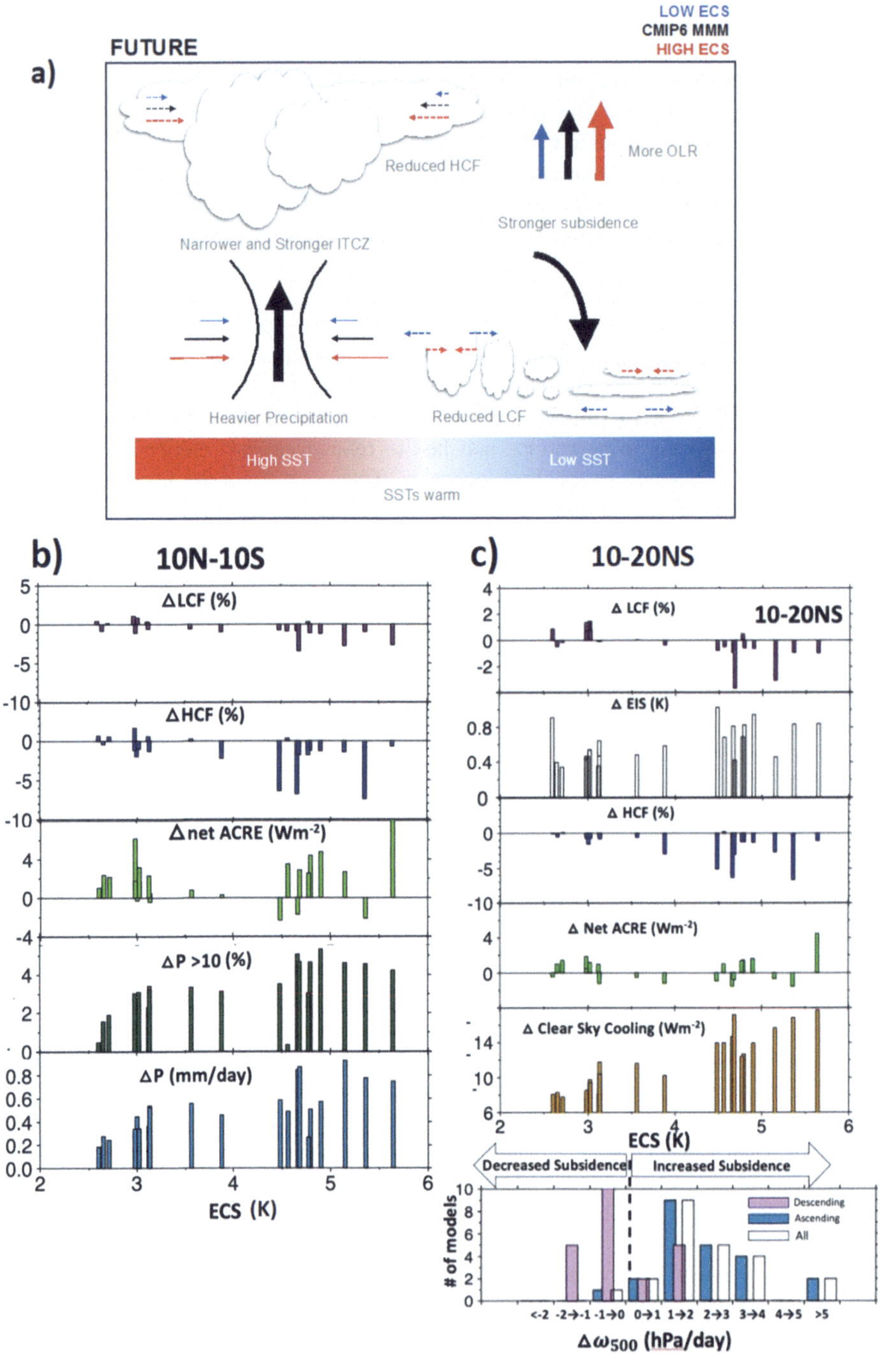

Fig. 11 a A schematic of model simulated changes in the Hadley circulation and tropical clouds along with the OLR and precipitation changes experienced in a warmer climate. The different color arrows signify the degree of warming expressed by the model ECS. The length of the arrows symbolically displays the strength of the process change. **b, c** Specific changes to a number of process defining quantities that reflect on the changes called out in **a** (see text)

high clouds decrease which is a combination consistent with the stability iris hypothesis of Bony et al. (2016). Changes to low clouds that occur in the subtropical regions are linked to the changes that occur in the convective zone.

Study of the model responses conveyed in Fig. 11b and c reveals the following:

(i) In the 10 N/S latitude convective zone, the precipitation increases both in total amount and in intensity as indicated by the increases in the fraction of precipitation that exceeds 10 mm/day. The ACRE also shows a greater column warming in a manner that might have been anticipated from the analysis of the observational record (e.g., Figure 8). The combination of decreased high cloud cover in this zone of convection and enhanced precipitation, including the more intense precipitation presumably from the deeper convection, is at least broadly consistent with that posited by Lindzen et al. (2001).

(ii) Most models produce an overall strengthened subsidence in the subtropical regions indicated by positive changes in the 500 hPa vertical velocity (ω_{500}). This enhanced subsidence is consistent with the enhanced clear-sky radiative cooling that results under warming that highly correlates with the ECS ($r = 0.93$) and with the increased OLR from the region. The increased radiative cooling is expected given this column cooling proportionally depends on column water vapor (e.g., Stephens et al. 1994). The net changes to vertical motion of the region, however, are realized in a more complex manner than is suggested by bulk changes of the region. The vertical motions in the ascending portions of the region weaken, presumably associated with reduced deep convection in the fringes of the deep convective zone as described above, whereas the vertical motions in the descending regions also weaken as reported by Schiro et al (2022); thus, creating a more complicated response than is conveyed simply in the area-mean vertical motion change (bottom panel of Fig. 11c). Lapse-rate changes that occur result in an increase in the estimated inversion strength (EIS, e.g., Wood and Bretherton 2006) in all models. Although this increase in EIS favors a response of increased low cloud in the subtropical regions (e.g., Klein et al. 2017), the opposite response occurs with most models producing reductions in low clouds.

6 Deep Convection and the Pattern of Warming

The previous sections were concerned with the connections between the large-scale heat transport by the overturning circulations to tropical convection within the tropical region defined as the TTZ. These sections further focus on mechanisms that connect convection in the TTZ to cloudiness and to the radiative responses associated with this cloudiness. This section now expands this view to even larger scales as represented by patterns of change associated with global warming.

It is natural to expect that changes to the overturning circulations induced by a global warming and shaped by the deep convection responses in the tropics would impact and respond to the regional patterns of the warming because of the associated changes in cloud patterns that result. This is a topic now broadly referred to as the "pattern effect" (e.g., Stevens et al. 2016) which encapsulates the finding that the regional details of the patterns of the warming influence the global mean ECS. This influence is expected since climate feedbacks that define the ECS, such as those associated with cloud and lapse-rate changes

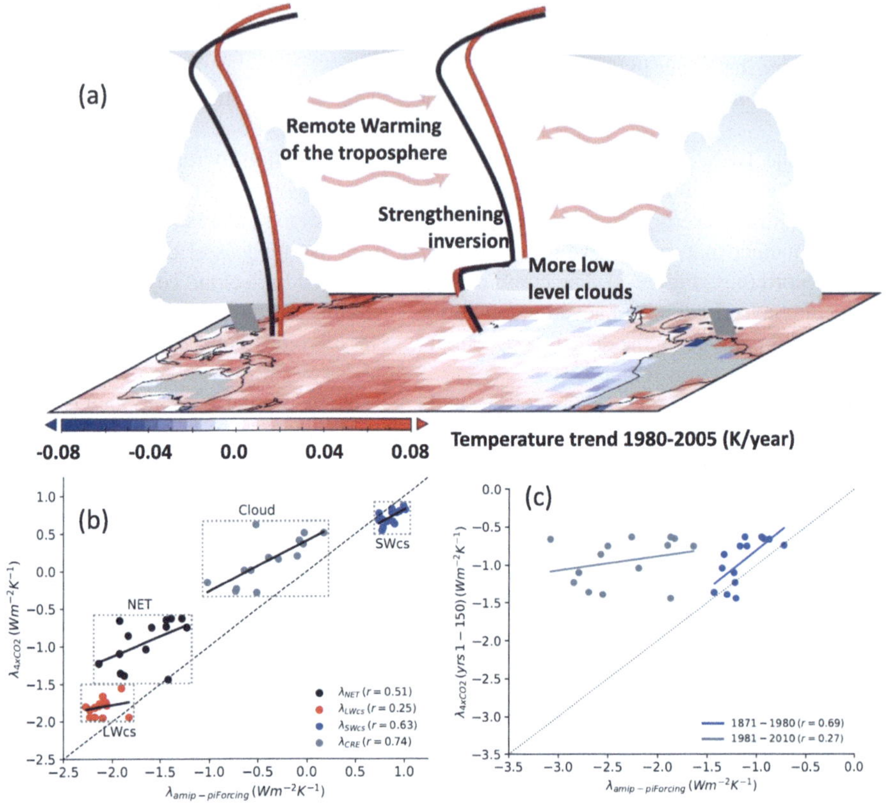

Fig. 12 **a** A schematic of the way warming spreads across the tropical region, **b** relationship across models (dots) between the feedback parameter in amip-piForcing (calculated over years 1871–2010) and abrupt-4xCO$_2$ simulation (calculated over years 1–150). The net feedback parameter is decomposed into its longwave clear-sky, shortwave clear-sky and cloud radiative effect components and **c** relationships between model simulated feedbacks in amip-piForcing over years 1871–1980 (blue) or 1981–2010 (gray)—adapted from Mauritsen and Stevens (2015) and Andrews et al. (2022)

(Zhou et al. 2016; Andrews and Webb 2018), vary with the pattern of surface warming. This pattern effect is exemplified in the pattern of warming that occurs across the tropical Pacific Ocean as schematically portrayed in Fig. 12. Warming in the ascent regions of tropical deep convection, particularly in the TWP region, influences the warming throughout the tropical atmosphere. This remote warming gives rise to a potential strong negative lapse rate feedback that results under a scenario of an upper-level warming that extends widely across the tropics that exceeds the surface warming below it. When this occurs, the low cloud controlling factors of EIS, for example, are inductive to increased low cloud amount. This creates a west to east warming-cooling dipole pattern in the tropics that results in a smaller ECS and thus smaller global warming (e.g., Ceppi and Gregory 2017). Under the scenario of the reversed circumstance of stronger surface warming, reduced EIS and reduced low cloud cover, the positive feedback associated with these low cloud reductions drives higher values of ECS and greater warming. We now see how these two contrasting scenarios of patterns effects are found in the historical record of change.

Andrews et al. (2018) quantify the pattern effect using the global "climate feedback parameter" λ as diagnosed from the linearization of the global mean energy budget with global mean surface temperature (e.g., Gregory et al. 2004). They introduce the difference, $\Delta\lambda$, between the feedback deduced from the atmospheric GCM (AGCM) simulation forced by the observed historical SST and sea-ice variations (i.e., the amip-piForcing simulation; λ_{hist}) and the λ inferred from 150 years of a coupled AOGCM abrupt-4xCO$_2$ simulation with the same AGCM (λ_{4xCO2}) and use $\Delta\lambda = \lambda_{\text{4xCO2}} - \lambda_{\text{hist}}$ as a measure of this pattern effect. The contributions by clouds on $\Delta\lambda$ can be gleaned in Fig. 12b (after Andrews et al. 2022). The figure is in the form of λ_{4xCO2} versus λ_{hist} such that deviations from the one-to-one line express the pattern effect. The decomposition of λ into long, shortwave clear-sky contributions and the net radiative effect of clouds clearly illustrates how the bulk of the pattern effect arises from the changing pattern of cloud feedbacks. This analysis reveals net cloud feedbacks that are more negative and less positive in the amip-piForcing experiments than in the abrupt-4xCO$_2$ experiments.

Figure 12c is presented in the same format as Fig. 12b but now separates the historical responses into two epochs: before and after 1980. The rationale for defining these two epochs, described in more detail in Andrews et al. (2022), is largely revealed in Fig. 12c. The spread in feedback across models over the earlier (1871–1980) time period in amip-piForcing experiments is well correlated with the spread in feedbacks across models in abrupt-4xCO$_2$ ($r = 0.69$). In contrast, feedbacks over the most recent decades (1981–2010) weakly correlate with λ_{4xCO2} ($r = 0.27$) which implies the feedbacks operating in 1871–1980 historical period differ from the feedbacks that govern the responses in 1981–2010. It also implies that inferences about feedbacks in this period do not apply to the feedbacks that establish the longer historical record of change. This is somewhat ironical as it is this latter period that is more intensely observed. It also explains why observational based estimates of ECS from the more recent historical climate change are lower that the ECS inferred from the abrupt experiments: Feedbacks in this more recent period are thought to be more negative in response to observed SST patterns that show a lack of Southern Ocean and Eastern Pacific warming, in contrast to longer term warming seen in paleo and AOGCM data. This further underscores a point called out above—feedbacks that play out on one time scale, such as that suggested from analysis of the current observational period, might not necessarily define the most influential feedbacks associated with longer term, historical climate change.

7 Summary Discussion

This paper revisits the energetics of the tropical trough zone as defined by Riehl and Simpson (1979). A theme of the paper is concerned with how the diabatically-forced overturning circulations of the atmosphere, forced by the deep convection within this tropical zone, fundamentally shape the distributions of tropical and subtropical cloudiness and the changes to cloudiness as Earth warms. The study at first draws on a range of observations to understand the connections between the energetics of the TTZ, convection and clouds. This focus on the TTZ is then expanded with a more global view based on model responses to a forced warming. The main findings of the research described in this paper include:

(1) The characteristics of convection across the TTZ and the patterns of change over the annual cycle are shown to be more complex than the simple, more classical view of a narrow, quasi-stationary convergence zone typically thought of as the ITCZ such as conveyed in simple depictions like that of Fig. 11. Seasonal changes to precipitation within the TTZ (Fig. 4a) reveal shifting patterns of precipitation, and by inference convection, over the mean annual cycle that is argued to exist as three different regimes within the TTZ. One is the regime of convection over the large southern land masses of S. America and the African continent that follow the movement of the Sun through its annual cycle. A second regime is that of convection confined to narrow convergence zones more typical of the classical view of the ITCZ over the Mid- and Eastern Pacific and tropical Atlantic oceans that are quasi-stationary, locked to the maximum SSTs of these regions. A third regime encapsulates the convection of the maritime continent that by virtue of the complex geography of the region, together with the fact it lies in some of the warmest waters on Earth, creates a complex set of interactions and circulations generated between ocean and land producing complicated precipitation patterns and a broad seasonal cycle tied to the monsoon systems of the region.

(2) Observations of the two main components of the diabatic heating, the radiative heating as characterized by the ACRE and the latent heating taken to be proportional to the surface precipitation, while additive, reveal a very tight coupling between the two (e.g., Fig. 8) in the TTZ. The reason for such a tight coupling between them and for the magnitude of the slopes of the relationship found in Fig. 8 requires further investigation. This coupling points to a feedback system where one form of heating reinforces the other. Associated far field changes, typically in the form of increased radiative cooling of the surrounding environment, drive thermally direct circulations that also directly influence convection. This has been called out in a number of past studies that argue, for example, that the ACRE can both feedback on convection directly (e.g., Stephens et al. 2004), create a thermally direct circulation that reinforces convection and thus the latent heating (e.g., Wu et al. 2023; Stephens et al. 2008a, 2018), and, in turn, strengthen the larger scale circulation (e.g., Slingo and Slingo 1988; Dixit et al. 2018, among many others). This circulation is also an important mechanism of convective aggregation with further consequences to the TOA energy budget (e.g., Bony et al. 2020).

(3) Observations of convection, clouds and the energy balance of the TTZ were examined through the perspective of their observed interannual variations over a 20-year period approximately between 2002 and 2022. It was found that:

(i) A clear anti-correlation exists between the interannual anomalies of mean tropopause height and the net energy imbalance of the TTZ for a considerable portion of this observed period. The higher the tropopause, the deeper is the convection (Fig. 9). Associated with the deepening of convection is an increase in thickening of anvil clouds with an associated increase of optical depth.

(ii) Deeper convection and optically thicker anvils produce an increased reflection of solar radiation that dominates the interannual changes of the net radiation input into the zone (Fig. 10) giving rise to the observed anti-correlation between tropopause height anomalies and TOA radiation budget anomalies.

(4) One important influence of global warming on the circulation occurs via changes to convection and increased convective heating that strengthen the circulation resulting in a positive low cloud feedback in some models and an enhanced global ECS of these models. Figure 11 provides a summary of the multi-model changes observed under the

scenario of warming realized in the CMIP6 SSP5-8.5 scenario multi-model experiments with the finding that:

(i) A pronounced narrowing of the ITCZ occurs with an associated increase in precipitation including increased heavy precipitation. Such a projected narrowing of the ITCZ is also supported in precipitation observations from 1979 to 2014 (Wodzicki and Rapp 2019). Whether contraction of convection such as observed in the Eastern Pacific applies to convection formed over the maritime continent or over tropical land masses is yet to be studied although the related study of Roca et al. (2024) finds that convection over Africa has changed in its morphology over four decades of Meteosat observations with a significant shift of the occurrence of deep convective systems to less frequent large and short-lived systems to more frequent smaller and longer-lived systems. Twentieth century observations also show that, during the last 50 years, the sea surface temperature (SST) of the tropical oceans has increased by 0.5C and the areas of SST > 26.5 and 28C by 15 and 50%, respectively, in association with an increase in greenhouse gas concentrations (Hoyas and Webster 2012). We do not know how convection responds to this expanded area of warmed SSTs and whether there will be any contraction of convection in this zone.

(ii) Although precipitation increases in the tropical convective zone in models, the high clouds do not (Fig. 11b) but the column radiative heating does via reduction on low cloud that act as net coolers of the column. In the subtropical regions, the clear-sky radiative cooling increases as expected from increased water vapor, thus setting up a differential heating patterns consistent with a strengthening of the meridional circulation as speculated in a number of studies (e.g., Needham and Randall 2021a, 2021b; Stephens et al 2018; Dixit et al. 2018) and consistent with the projected increased subsidence (Fig. 11c).

(iii) The multi-model response of a combined reduction in high clouds and increased precipitation in the convection zone is also consistent with the Lindzen et al. (2001) iris mechanism that the precipitation efficiency of convection should increase with warming with proportionally more water falling out as precipitation than is vertically advected and detrained aloft. This still remains an unproven concept (e.g., Ito and Masunaga 2022) although other model studies support it (e.g., Li et al. 2022).

(iv) In the subtropical regions (Fig. 11c), most models produce a strengthened subsidence, constrained overall by clear-sky radiative cooling increases but how this increased subsidence is realized is complicated with reduced vertical motion in both ascending and descending regions of the subtropics. Lapse rate changes also result in an increase in the EIS. While this increase in EIS favors increased low cloud responses in the subtropical regions, most models, and notably those with highest ECS, experience reduced low clouds in this region which appears to be an important feedback mechanism that sustains the higher ECS (e.g., Schiro et al. 2022). These model low cloud changes involve more than changes to simple cloud controlling factors like EIS, at least in the CMIP6 SSP5-8.5 scenario experiments (e.g., Koshiro et al. 2022).

(5) Specific details of the pattern of warming influence the global mean response of models: This influence, referred to as the pattern effect, is exemplified in the pattern of warming across the tropical Pacific Ocean which is influenced by the low cloud responses on the Eastern Pacific SSTs. Andrews et al. (2022) quantify this effect as a difference between the feedback metrics derived as the difference in the feedback measures of the CMIP6 amip-piForcing experiments from the CMIP6 abrupt-4xCO$_2$, the rationale being that feedbacks calculated from 150 years of the abrupt experiments that strongly correlate to longer term projections of climate change will have a weaker pattern effect

than when the correlation is much weaker. They found that over the period 1871–1980, the Earth warmed with feedbacks largely consistent and strongly correlated with long-term climate sensitivity feedbacks—that is the historical pattern of warming mirrors that change expected of the future (as diagnosed from corresponding atmosphere–ocean GCM abrupt-4xCO_2 simulations). Post 1980, however, the Earth warmed with patterns of warming that differ from past historical warming being weakly correlated to feedbacks of the abrupt experiments (Fig. 12). The basic conclusion is the feedbacks operating in 1871–1980 historical period differs from the feedbacks that govern the responses in 1981–2010. This is somewhat ironical as it is this latter period that is more intensely observed and processes that undergird feedbacks are better understood. Andrews et al. (2022) further reason that it also explains why observational based estimates of ECS from the more recent historical climate change are likely biased low.

The science of climate change and the study of feedbacks that define the overall climate sensitivity appear to face a fundamental dilemma. There are a number of examples presented throughout this paper where mechanisms observed over shorter time spans may not be the dominant mechanisms of feedbacks that play out on longer time times scales. For example, the thickening and brightening of anvil clouds described above and observed on interannual time scales are not reflected in the high cloud responses of climate models as described in Sect. 5. A second example was illustrated in the study of Andrews et al. (2022), where the more recent, shorter period observational record points to mechanisms and feedbacks involving cloud and convection that apparently do not define the feedbacks that establish longer term projected changes diagnosed from models. One could argue that the processes examined during the more recent intensely observed period are intrinsically those that must operate over the longer period of change albeit with different degrees of interaction. How can the credibility of projections be claimed while dismissing the observed natural state and its behaviors of the shorter past? What then is the role of observing the natural world and studying how processes interact from that vantage point if these are not deemed relevant to a longer term climate change? What then are the relevant observations that are unequivocal tests of feedbacks most pertinent to climate change for without them model projections will remain purely untested hypotheses? Regardless of these challenges, the need to quantify the role of deep convection in the Earth system remains acute.

The coming decade is also a decade that brings a focus on atmospheric convection. The desire to resolve convection is now a major focus of next generation Earth models (e.g., Slingo et al. 2022). From an observational perspective, major inroads to observing convection on the global scale are also about to happen. These observational programs are meant to address a high priority needs as called out in the 2017 Decadal Survey (NAS 2018). Two important new measurement initiatives of the current decade that address convection specifically are: i) the INvestigation of Convective UpdraftS (INCUS) mission of NASA that is to provide basic global information about convective motions and the transports produced by them, and ii) NASA's Earth System Observatory (the Atmospheric Observing System (AOS) https://aos.gsfc.nasa.gov),[1] aimed at addressing the Decadal priority by providing information about convective updraft and downdraft motions with space-borne Doppler radars. AOS will also include the tandem microwave radiometer measurements aimed at further quantifying vertical transports in deep convection (e.g., Brogneiz et al. 2022).

---

[1] The decadal strategy identified observations of aerosol, cloud, convection and precipitation as a priority for the decade. This priority has led to the definition of NASA's atmospheric observing system (AOS).

Acknowledgements This paper is an outcome of the Workshop "Challenges in Understanding the Global Water Energy Cycle and its Changes in Response to Greenhouse Gas Emissions" held at the International Space Science Institute (ISSI) in Bern, Switzerland (26–30 September 2022). © 2024. All rights reserved. Government sponsorship acknowledged.

Funding The lead author was supported in part by INCUS, Libera and by CloudSat, each being NASA Earth Venture Missions, funded by NASA's Science Mission Directorate and managed through the Earth System Science Pathfinder Program. Portions of the research described in this paper were performed at the Jet Propulsion Laboratory, California Institute of Technology under contract with the National Aeronautics and Space Administration. T.A. was supported by the UK Meteorological Office Hadley Centre Climate Program funded by the Department of Business, Energy and Industrial Strategy and received funding from the European Union's Horizon 2020 research and innovation program under Grant Agreement 820829. K.A.S. is supported by the National Science Foundation Grant # 2225954.

Data availability The radiative heating rate data will be available via https://gewex-utcc-proes.aeris-data.fr/.

Declarations

Conflict of Interest The authors declare no conflicts of interest.

Open Access This article is licensed under a Creative Commons Attribution 4.0 International License, which permits use, sharing, adaptation, distribution and reproduction in any medium or format, as long as you give appropriate credit to the original author(s) and the source, provide a link to the Creative Commons licence, and indicate if changes were made. The images or other third party material in this article are included in the article's Creative Commons licence, unless indicated otherwise in a credit line to the material. If material is not included in the article's Creative Commons licence and your intended use is not permitted by statutory regulation or exceeds the permitted use, you will need to obtain permission directly from the copyright holder. To view a copy of this licence, visit http://creativecommons.org/licenses/by/4.0/.

References

Andrews T, Webb MJ (2018) The dependence of global cloud and lapse rate feedbacks on the spatial structure of tropical Pacific warming. J Climate 31(2):641–654. https://doi.org/10.1175/JCLI-D-17-0087.1

Andrews T, Gregory JM, Paynter D, Silvers LG, Zhou C, Mauritsen T et al (2018) Accounting for changing temperature patterns increases historical estimates of climate sensitivity. Geophys Res Lett 45(16):8490–8499. https://doi.org/10.1029/2018GL078887

Andrews T, Bodas-Salcedo A, Gregory JM, Dong Y, Armour KC, Paynter D et al (2022) On the effect of historical SST patterns on radiative feedback. J Geophys Res: Atmos 127:e2022JD036675. https://doi.org/10.1029/2022JD036675

Bony S et al (2016) Thermodynamic control of anvil cloud amount. Proc Natl Acad Sci 113:8927–8932

Bony S, Semie A, Kramer RJ, Soden B, Tompkins AM, Emanuel KA (2020) Observed modulation of the tropical radiation budget by deep convective organization and lower-tropospheric stability. Aguadvances 1:e2019AV00155. https://doi.org/10.1029/2019AV000155

Boucher O, Randall D, Artaxo P, Bretherton C, Feingold G, Forster P, Kerminen V-M, Kondo Y, Liao H, Lohmann U, Rasch P, Satheesh SK, Sherwood S, Stevens B, Zhang XY (2013) Clouds and aerosols. In: Stocker TF, Qin D, Plattner G-K, Tignor M, Allen SK, Boschung J, Nauels A, Xia Y, Bex V, Midgley PM (eds) Climate change 2013: the physical science basis. contribution of working group I to the 5th assessment report of the intergovernmental panel on climate change. Cambridge University Press, Cambridge, New York

Broccoli AJ, Dahl KA, Stouffer RJ (2006) Response of the ITCZ to Northern Hemisphere cooling. Geophys Res Lett 33:01702. https://doi.org/10.1029/2005GL024546

Brogniez H, Roca R, Auguste F, Chaboureau J-P, Haddad Z, Munchak SJ, Li X, Bouniol D, Dépée A, Fiolleau T, Kollias P (2022) Time-delayed tandem microwave observations of tropical deep convection: overview of the C2OMODO mission. Front Remote Sens 3:854735. https://doi.org/10.3389/frsen.2022.854735

Ceppi P, Gregory JM (2017) Relationship of tropospheric stability to climate sensitivity and Earth's observed radiation budget. Proc Natl Acad Sci USA 114(50):13126–13131. https://doi.org/10.1073/pnas.1714308114

Crespo LR, Keenlyside N, Koseki S (2019) The role of sea surface temperature in the atmospheric seasonal cycle of the equatorial Atlantic. Clim Dyn 52:5927–5946. https://doi.org/10.1007/s00382-018-4489-4

Davis SM, Liang CK, Rosenlof KH (2013) Interannual variability of tropical tropopause layer clouds. Geophys Res Lett 40(11):2862–2866

Dee DP et al (2011) The ERA-Interim reanalysis: configuration and performance of the data assimilation system. Q J R Meteorol Soc 137:553–597. https://doi.org/10.1002/qj.828

Derras-Chouk A, Luo ZJ (2024) Revisiting Riehl and Malkus (1958) and Riehl and Simpson (1979): characterizing tropical hot towers and estimating convective mass fluxes using geostationary satellite data, Survs Geophys (ISSI special issue) (to submitted)

Dessler AE (2010) A determination of the cloud feedback from climate variations over the past decade. Science 330:1523–1527. https://doi.org/10.1126/science.1192546

Dixit V, Geoffroy O, Sherwood SC (2018) Control of ITCZ width by low-level radiative heating from upper-level clouds in aqua-planet simulations. Geophys Res Lett 45:5788–5797. https://doi.org/10.1029/2018GL078292

Eyring V, Bony S, Meehl GA, Senior CA, Stevens B, Stouffer RJ, Taylor KE (2016) Overview of the coupled model intercomparison project phase 6 (CMIP6) experimental design and organization. Geosci Model Dev 9:1937–1958. https://doi.org/10.5194/gmd-9-1937-2016

Forget G, Campin J-M, Heimbach P, Hill CN, Ponte RM, Wunsch C (2015) ECCO version 4: an integrated framework for non-linear inverse modeling and global ocean state estimation. Geosci Model Dev. https://www.geosci-model-dev.net/8/3071/2015/

Frierson DM, Hwang YT (2012) Extratropical influence on ITCZ shifts in slab ocean simulations of global warming. J Clim 25(2):720–733

Fueglistaler S, Dessler AE, Dunkerton TJ, Folkins I, Fu Q, Mote PW (2009) Tropical tropopause layer. Rev Geophys 47:1004. https://doi.org/10.1029/2008RG000267

Gray WM, Jacobsen Jr RW (1977) Diurnal variation of deep cumulus convection. Mon Wea Rev 105:1171–1188, https://doi.org/10.1175/1520-0493(1977)105<1171:DVODCC>2.0.CO;2

Gregory JM, Ingram WJ, Palmer MA, Jones GS, Stott PA, Thorpe RB et al (2004) A new method for diagnosing radiative forcing and climate sensitivity. Geophys Res Lett 31(3):L03205. https://doi.org/10.1029/2003GL018747

Guzman R, Chepfer H, Noel V, Vaillant de Guélis T, Kay JE, Raberanto P, Cesana P, Vaughan MA, Winker DM (2017) Direct atmosphere opacity observations from CALIPSO provide new constraints on cloud-radiation interactions. J Geophys Res Atmos 122:1066–1085. https://doi.org/10.1002/2016JD025946

Harrop BE, Hartmann DL (2016) The role of cloud radiative heating in determining the location of the ITCZ in aqua-planet simulations. J Climate 29(8):2741–2763

Hartmann DL, Larson K (2002) An important constraint on tropical cloud-climate feedback. Geophys Res Lett 29(20):1951–1954

Hoyos CD, Webster PJ (2012) Evolution and modulation of tropical heating from the last glacial maximum through the twenty-first century. Clim Dyn 38:1501–1519. https://doi.org/10.1007/s00382-011-1181-3

Huffman GJ, Bolvin DT, Braithwaite D, Hsu KL, Joyce RJ, Kidd C, Nelkin EJ, Sorooshian S, Stocker EF, Tan J, Wolff DB (2020) Integrated multi-satellite retrievals for the global precipitation measurement (GPM) mission (IMERG). In: Levizzani V, Kidd C, Kirschbaum D, Kummerow C, Nakamura K, Turk F (eds) Satellite precipitation measurement. Advances in global change research, vol 67, pp 343–353. https://doi.org/10.1007/978-3-030-24568-9_19

Ito M, Masunaga H (2022) Process level assessment of the iris effect over tropical oceans. Geophys Res Lett 49:e2022GL097997. https://doi.org/10.1029/2022GL097997

Kang SM, Held IM, Frierson DM, Zhao M (2008) The response of the ITCZ to extratropical thermal forcing: idealized slab-ocean experiments with a GCM. J Clim 21(14):3521–3532

Kang SM, Frierson DM, Held IM (2009) The tropical response to extratropical thermal forcing in an idealized GCM: The importance of radiative feedbacks and convective parameterization. J Atmos Sci 66(9):2812–2827

Kato S et al (2018) Surface irradiances of edition 4.0 clouds and the earth's radiant energy system (CERES) energy balanced and filled (EBAF) data product. J Climate 31:4501–4527

Kiehl JT (1994) On the observed near cancellation between longwave and shortwave cloud forcing in tropical regions. J Climate 7:559–565

Klein SA, Hall A, Norris JR, Pincus R (2017) Low-cloud feedbacks from cloud-controlling factors: a review. Surv Geophys 38:1307–1329. https://doi.org/10.1007/s10712-017-9433-3

Knutti R, Rugenstein M, Hegerl G (2017) Beyond equilibrium climate sensitivity. Nature Geosci 10:727–736. https://doi.org/10.1038/ngeo3017

Koshiro T, Kawai H, Noda AT (2022) Estimated cloud-top entrainment index explains positive low-cloud-cover feedback. PNAS 119:e2200635119. https://doi.org/10.1073/pnas.2200635119

Lau WK, Kim KM (2015) Robust Hadley circulation changes and increasing global dryness due to CO2 warming from CMIP5 model projections. Proc Natl Acad Sci 112:3630–3635

Li Y, Thompson DW (2013) The signature of the stratospheric Brewer–Dobson circulation in tropospheric clouds. J Geophys Res Atmos 118(9):3486–3494

Li Y, Thompson DW, Bony S (2015) The influence of atmospheric cloud radiative effects on the large-scale atmospheric circulation. J Clim 28(18):7263–7278

Li RL, Studholme JH, Fedorov AV, Storelvmo T (2022) Precipitation efficiency constraint on climate change. Nat Clim Chang 12(7):642–648

Lindzen RS, Chou M-D, Hou AY (2001) Does the Earth have an adaptive infrared iris? Bull Am Meteorol Soc 82(3):417–432. https://doi.org/10.1175/1520-0477

Loeb NG et al (2018) Clouds and the Earth's radiant energy system (CERES) energy balanced and filled (EBAF) top-of-atmosphere (TOA) edition 4.0 data product. J Clim. https://doi.org/10.1175/jcli-d-17-0208.1

Luo Z, Liu GY, Stephens GL (2008) CloudSat adding new insight into tropical penetrating convection. Geophys Res Lett 35:L19819. https://doi.org/10.1029/2008GL035330

Mace GG, Zhang Q, Vaughan M, Marchand R, Stephens G, Trepte C, Winker D (2009) A description of hydrometeor layer occurrence statistics derived from the first year of merged cloudsat and CALIPSO data. J Geophys Res. https://doi.org/10.1029/2007JD009755

Machado LAT, Rossow WB (1993) Structural characteristics and radiative properties of tropical cloud clusters. Mon Wea Rev 121:3234–3260

Mapes BE (2001) Water Vapor two scale heights: the moist adiabat and the radiative troposphere. Q.J.R. Meteorol Soc 127:2353–2366

Mauritsen T, Stevens B (2015) Missing iris effect as a possible cause of muted hydrological change and high climate sensitivity in models. Nat Geosci 8(5):346–351. https://doi.org/10.1038/ngeo2414

National Academies of Sciences, Engineering, and Medicine (2018) Thriving on our changing planet: a decadal strategy for earth observations from space. The National Academies Press, Washing. https://doi.org/10.17226/24938

Needham MR, Randall DA (2021a) Linking atmospheric cloud radiative effects and tropical precipitation. Geophys Res Lett 48:e2021GL094004. https://doi.org/10.1029/2021GL094004

Needham MR, Randall DA (2021b) Riehl and Malkus revisited: the role of cloud-radiative effects. https://doi.org/10.1002/essoar.10506726.1

Philander SGH, Gu D, Lambert G, Li T, Halpern D, Lau NC, Pacanowski RC (1996) Why the ITCZ is mostly north of the equator. J Clim 9(12):2958–2972

Pierrehumbert RT (1995) Thermostats, radiator fins, and the local runaway greenhouse. J Atmos Sci 52:1784–1806

Pilewskie J, Stephens G, L'Ecuyer T, Takahashi H (2024) A multi-satellite perspective on "hot tower" characteristics in the equatorial trough zone, Surv Geophys

Popp M, Silvers LG (2017) Double and single ITCZs with and without clouds. J Clim 30(22):9147–9166

Ramanathan V, Collins W (1991) Thermodynamic regulation of ocean warming by cirrus clouds deduced from observations of the 1987 El Niño. Nature 351:27–32

Riehl H, (Malkus) Simpson J (1979) The heat balance of the equatorial trough zone, revisited. Contr Atmos Phys 52:287–305

Riehl H, Malkus JS (1958) On the heat balance in the equatorial trough zone. Geophysica 6:503–537

Roca R, Fiolleau T, John VO, Schulz J (2024) METEOSAT long term observations reveal changes in convective organization over tropical Africa and Atlantic Ocean. Surv Geophys

Schiro KA, Su H, Ahmed F et al (2022) Model spread in tropical low cloud feedback tied to overturning circulation response to warming. Nat Commun 13:7119. https://doi.org/10.1038/s41467-022-34787-4

Seidel DJ, Fu Q, Randel WJ, Reichler TJ (2008) Widening of the tropical belt in a changing climate. Nat Geosci 1:21–24

Sherwood SC, Ramanathan V, Barnett TP, Tyree MK, Roeckner E (1994) Response of an atmospheric general circulation model to radiative forcing of tropical clouds. J Geophys Res 99(D10):20829–20845

Slingo JM, Bates P, Belcher S, Palmer T, Stephens G, Stevens B (2022) Ambitious partnership needed for reliable climate prediction. Nat Clim Chang. https://doi.org/10.1038/s41558-022-01384-8

Slingo A, Slingo JM (1988) The response of a general circulation model to cloud longwave radiative forcing. I: introduction and initial experiments. Q J R Meteorol Soc 114(482):1027–1062. https://doi.org/10.1002/qj.49711448209

Sokol AB, Hartmann DL (2020) Tropical anvil clouds: Radiative driving toward a preferred state. J Geophys Res Atmos 125:e2020JD033107. https://doi.org/10.1029/2020JD033107

Stephens GL (2005) Cloud feedbacks in the climate system: a critical review. J Clim 18:237–273

Stephens GL, Webster PJ, Johnson RH, Engelen R, L'Ecuyer TS (2004) Observational evidence for the mutual regulation of the tropical hydrological cycle and tropical sea surface temperatures. J Climate 17:2213–2224

Stephens GL, van den Heever S, Pakula LA (2008) Radiative convective feedback in idealized states of radiative-convective equilibrium. J Atmos Sci 65:3899–3916

Stephens GL, Hakuba MZ, Webb MJ, Lebsock M, Yue Q, Kahn BH et al (2018) Regional intensification of the tropical hydrological cycle during ENSO. Geophys Res Lett 45:4361–4370. https://doi.org/10.1029/2018GL077598

Stephens GL, Hu Y (2010) Are climate-related changes to the character of global precipitation predictable? Environ Res Lett. https://doi.org/10.1088/1748-9326/5/2/025209

Stephens GL, Slingo A, Webb MJ, Minnett PJ, Daum PH, Kleinman L, Wittmeyer I, Randall DA (1994) Observations of the earth's radiation budget in relation to atmospheric hydrology 4. Atmospheric column radiative cooling over the world's oceans. J Geophys Res 99(D9):18585–18604

Stephens GL et al (2008) CloudSat mission: performance and early science after the first year of operation. J Geophys Res 113:D00A18. https://doi.org/10.1029/2008JD009982

Stevens B, Sherwood SC, Bony S, Webb MJ (2016) Prospects for narrowing bounds on Earth's equilibrium climate sensitivity. Earth's Fut 4(11):512–522. https://doi.org/10.1002/2016EF000376

Stubenrauch CJ, Caria G, Protopapadaki SE, Hemmer F (2021) 3D radiative heating of tropical upper tropospheric cloud systems derived from synergistic A-Train observations and machine learning. Atmos Chem Phys 21:1015–1034. https://doi.org/10.5194/acp-21-1015-2021

Su H et al (2017) Tightening of tropical ascent and high clouds key to precipitation change in a warmer climate. Nat Commun 8:1–9

Takahashi H, Luo ZJ (2012) Where is the level of neutral buoyancy for deep convection? Geophys Res Lett 39:L15809. https://doi.org/10.1029/2012GL052638

Takahashi H, Luo ZJ (2014) Characterizing tropical overshooting deep convection from joint analysis of CloudSat and geostationary satellite observations. J Geophys Res Atmos 119:112–121. https://doi.org/10.1002/2013JD020972

Takahashi H, Luo ZJ, Stephens GL (2017) Level of neutral buoyancy, deep convective outflow, and convective core: new perspectives based on 5 years of CloudSat data. J Geophys Res Atmos 122:2958–2969. https://doi.org/10.1002/2016JD025969

Takahashi H, Luo ZJ, Stephens G, Mulholland JP (2023) Revisiting the land-ocean contrasts in deep convective cloud intensity using global satellite observations. Geophys Res Lett 50:e2022GL102089

Tao L, Hu Y, Liu J (2016) Anthropogenic forcing on the Hadley circulation in CMIP5 simulations. Clim Dyn 46:3337–3350

Webster P (1994) The role of hydrological processes in ocean-atmosphere interaction. Rev of Geophys 32:427–476

Wei Y, Pu Z, Zhang C (2020) Diurnal cycle of precipitation over the maritime continent under modulation of MJO: perspectives from cloud-permitting scale simulations. J Geophys Res Atmos 125:e2020JD032529. https://doi.org/10.1029/2020JD032529

Wodzicki KR, Rapp AD (2019) Long-term characterization of the Pacific ITCZ using TRMM, GPCP, and ERA-Interim. J Geophys Res Atmos 121:3153–3170

Wood R, Bretherton CS (2006) On the relationship between stratiform low cloud cover and lower-tropospheric stability. J Clim 19(24):6425–6432

Wu S-N, Soden BJ, Alaka, Jr GJ (2023) The influence of radiation on the prediction of tropical cyclone intensification in a forecast model. Geophys Res Lett 50:e2022GL099442. https://doi.org/10.1029/2022GL099442

Zelinka MD, Hartmann DL (2010) Why is longwave cloud feedback positive? J Geophys Res Atmos 115:D16–D117

Zhou C, Zelinka MD, Klein SA (2016) Impact of decadal cloud variations on the Earth's energy budget. Nat Geosci 9(12):871–875. https://doi.org/10.1038/ngeo2828

Zipser EJ, Cecil DJ, Liu C, Nesbitt SW, Yorty DP (2006) Where are the most intense thunderstorms on Earth? Bull Am Meteorol Soc. https://doi.org/10.1175/BAMS-87-8-I057

Publisher's Note Springer Nature remains neutral with regard to jurisdictional claims in published maps and institutional affiliations.

Authors and Affiliations

Graeme L. Stephens[1,2] · Kathleen A. Shiro[3] · Maria Z. Hakuba[1] · Hanii Takahashi[1] · Juliet A. Pilewskie[4] · Timothy Andrews[5] · Claudia J. Stubenrauch[6,7] · Longtao Wu[1]

✉ Graeme L. Stephens
graeme.stephens@jpl.nasa.gov

[1] Jet Propulsion Laboratory, California Institute of Technology, Pasadena, CA 91011, USA

[2] Department of Physics, University of Oxford, Oxford, OX1 3PJ, UK

[3] Department of Environmental Sciences, University of Virginia, Charlottesville, VA 22904-4123, USA

[4] Department of Atmospheric and Oceanic Sciences, University of Wisconsin-Madison, Madison, WI 53706, USA

[5] Met Office Hadley Centre, Exeter EX1 3PB, UK

[6] Laboratoire de Météorologie Dynamique/Institute Pierre-Simon Laplace, (LMD/IPSL), CNRS, Sorbonne Universities, University Pierre and Marie Curie (UPMC) Paris, University of Paris, Paris Cedex 5, Paris 75252, France

[7] Laboratoire de Météorologie Dynamique/Institute Pierre-Simon Laplace, (LMD/IPSL), CNRS, Ecole Polytechnique, Université Paris-Saclay, 91128 Palaiseau, France

A Multi-satellite Perspective on "Hot Tower" Characteristics in the Equatorial Trough Zone

Juliet Pilewskie[1,2] · Graeme Stephens[3,4] · Hanii Takahashi[3] · Tristan L'Ecuyer[5,6]

Received: 18 December 2023 / Accepted: 9 October 2024 / Published online: 6 November 2024
© The Author(s) 2024

Abstract

In 1979, Herbert Riehl and Joanne Simpson (Malkus) analytically estimated that 1600–2400 undilute convective cores vertically transport energy to the tropopause at any given time within a region where upper-tropospheric energy is only exported from the tropics. The focus of this paper is to update this estimate using modern satellite observations, compare hot tower frequency and intensity characteristics to all deep convective cores that reach the upper troposphere, and document hot tower spatiotemporal variability in relation to precipitation and high cloud properties within the tropical trough zone (between 13 °S and 19 °N). Cloud vertical profiles from CloudSat and CALIPSO measurements supply convective core diameters and proxies for intensity and convective activity, and these proxies are augmented with brightness temperature data from geostationary satellite observations, precipitation information from IMERG, and cloud radiative properties from CERES. Less than 35% of all deep cores are classified as hot towers, and we estimate that 800–1700 hot towers occur at any given time over the course of a day, with the mean maximum core and hot tower frequency occurring at the time of year when peak convective intensity and precipitation occur. Convective objects that contain hot towers frequently contain multiple cores, and the largest systems with five or more distinct cores most frequently occur in regions where organized mesoscale convective systems and the highest climatological mean rain rates are known to occur. Analysis of co-located radar and infrared brightness temperatures reveals that passive observations alone are not sufficient to unambiguously distinguish hot towers using simple brightness temperature thresholds.

Keywords Tropical deep convection · Satellite observations · Global energy budget · Vertical mass transport · Large-scale circulation

---

✉ Juliet Pilewskie
jap2335@columbia.edu

[1] Center for Climate Systems Research, Columbia University, New York, NY 10025, USA

[2] NASA Goddard Institute for Space Studies, New York, NY 10025, USA

[3] Jet Propulsion Laboratory, California Institute of Technology, Pasadena, CA 91011, USA

[4] Department of Physics, University of Oxford, Oxford OX1 3PJ, UK

[5] Department of Atmospheric and Oceanic Sciences, University of Wisconsin-Madison, Madison, WI 53706, USA

[6] Cooperative Institute for Meteorological Satellite Studies, Madison, WI 53706, USA

Article Highlights

- Convective cores that reach the tropopause make up 10% of all deep convective cores between 13 °S and 19 °N
- 800–1700 cores that reach the tropopause simultaneously occur using CloudSat/CALIPSO and geostationary satellite observations
- Convective core frequency, intensity, and precipitation peak during September, October, and November

1 Introduction

The hot tower hypothesis posited by Herbert Riehl and Joanne Simpson (Malkus) in 1958 and revisited in 1979 (hereafter, RM58 and RS79, respectively) has largely shaped our understanding of how tropical deep convective systems impact large-scale circulations. They reasoned that individual updrafts, or hot towers, transport boundary layer air to the upper troposphere that not only balances local radiative cooling, but also supplies a surplus of energy to be transported to higher latitudes. RM58 and RS79 estimated that 1500–5000 hot towers (modified to 1600–2400 hot towers in RS79), each having a diameter of 3–5 km, populate 30 synoptic disturbances at any given time within a region known as the equatorial trough zone (5 °S–15 °N) in which energy is exported to higher latitudes within the upper troposphere. However, their estimates were based on analytical calculations with limited access to observations. The focus of this study is to provide an updated hot tower estimate using a novel approach that combines multiple satellite observing perspectives of hot towers.

The hot tower hypothesis propelled both field campaigns, such as Tropical Ocean Global Atmospheres Coupled Ocean Atmosphere Response Experiment (TOGA COARE), and satellite missions such as the Tropical Rainfall Measuring Mission (TRMM), which was proposed and led by Joanne Simpson, to characterize atmospheric deep convective systems, their energetics, and their coupled interactions with the local and large-scale environment (e.g., Chen and Houze 1997; Johnson et al. 1999; Nesbitt et al. 2000; Reed and Recker 1971; Yanai et al 1973; Zipser et al. 2006; among others). These observations have aided in modifying the definition of a hot tower. The initial hot tower estimate was made by assuming that updrafts ascend adiabatically and do not entrain any surrounding air, which imposed restrictions on updraft sizes and vertical velocities (RM58). Early measurements from Global Atmospheric Research Program's (GARP) Atlantic Tropical Experiment (GATE) found that updraft core sizes and vertical velocities are much smaller than what would be required with adiabatic ascent (LeMone and Zipser 1980; Zipser and LeMone 1980). It has since been suggested that while updrafts entrain air, they can be reinvigorated through latent heat released from condensational freezing that enables them to supply the necessary moist static energy to the upper troposphere (Zipser et al. 2003). In the final publication that Joanne Simpson co-authored, idealized model simulations showed that updrafts that reach the upper troposphere and entrain air could supply sufficient energy and mass to the upper troposphere (Fierro et al. 2009). A hot tower was redefined as "any deep convective cloud with a base in the planetary boundary layer (PBL) and reaching near the upper-tropospheric outflow layer". In light of these updates, it is important to quantify the frequency and characteristics of hot towers that fit both the initial and updated definitions.

From an observational perspective, convective core sizes and vertical velocities provide insight into the energetic contributions of updrafts. Historically, convective mass flux has been defined using field campaign measurements to understand how much mass is transported vertically within updraft regions (e.g., Byer and Braham 1948; Giangrande et al. 2016; LeMone and Zipser 1980; Lucas et al. 1994). However, information from field campaigns is limited to specific regions and times, so they do not capture the statistical properties on a tropics-wide scale. Satellite observations have been implemented to study convective mass flux on a tropics-wide scale (Masunaga and Luo 2016), but it has proven challenging to quantify convective mass flux because of the inability to measure vertical velocities. As a workaround, several active satellite-based studies have defined convective vertical intensity with precipitation- and cloud-based metrics that indirectly relate to updraft vertical velocities. Convective core vertical structures from A-Train measurements have been used to estimate the height at which ascending motion ceases and detrainment begins, as well as the height at which precipitation-sized particles extend in the atmosphere as proxies related to intensity and convective mass flux (Luo et al. 2008, 2010, 2014; Masunaga and Luo 2016; Takahashi and Luo 2014). More recently, Pilewskie and L'Ecuyer 2022 (hereafter, PL22) calculated the convective core center of gravity (CoG) from 94-GHz Cloud Profiling Radar (CPR) reflectivity measurements on board CloudSat to serve as another index of vertical intensity. As noted in PL22, the CoG 'is not the true center of mass of the convective core but rather a robust proxy of intensity derived directly from unadjusted CloudSat observations'. Thus, it captures the observed vertical hydrometeor distribution above the point at which the radar is attenuated (PL22; Storer et al. 2014).

Previous literature has noted that, in addition to condensation within convective updrafts, both the thermodynamic impact of phase changes within stratiform areas as well as the radiative effects by cloud shields heat and influence large-scale circulations (e.g., Moncrieff and Miller 1976; Houze 1981, 2004, 2018; Schumacher et al. 2004; Stephens et al. 2018). These theories highlight the importance of mesoscale convective system (MCS) circulations in transporting energy, and latent heat release in the stratiform portion of MCSs largely determines the environment's net vertical profile of heating (Houze 2004; Schumacher et al. 2004). It has been found in single-core systems that wider convective cores that extend deeper into the atmosphere tend to produce larger anvil extents compared to weaker and smaller cores (Takahashi et al. 2017; 2021). Furthermore, Hamada et al. (2015) indicates that the heaviest precipitation is more often attributed to convective rain in organized systems that contain both convective and stratiform rain compared to isolated, intense convection. These studies motivate revisiting precipitation and cloud productivity in relation to convective core features.

Complexities in relating such features are often attributed to environmental influences that vary both spatially and temporally. For example, surface temperatures and atmospheric conditions over the Congo Basin and Amazon are ideal for generating the deepest, or most intense, convective systems across the tropics (Heymsfield et al. 2010; LeMone and Zipser 1980; Lucas et al. 1994; Takahashi and Luo 2014; Takahashi et al. 2023; Zipser et al. 2006). While the Intertropical Convergence Zone (ITCZ), which defines the region where precipitation prevails in the tropics, shifts latitudinally over the course of the year over land, there is little latitudinal shift in precipitation over some oceanic regions such as the East Pacific and Atlantic Oceans (Stephens et al. 2024). There also exists a strong diurnal cycle in convective intensity over land but not over ocean. However, there is a noticeable diurnal cycle in convective frequency, anvil horizontal extent, and precipitation that peaks in the early morning over the ocean (Chen and Houze 1997; Nesbitt and Zipser 2003; Liu and Zipser 2008; PL22). Over the Maritime Continent islands, land–ocean interactions that induce land-sea-breeze circulations contribute to a strong diurnal cycle of precipitation (e.g., van Bemmelen 1922; Miller et al. 2003; Yang and Slingo 2001). These examples motivate the need for a detailed analysis studying the spatiotemporal variability in convective cores, cloud extent, and precipitation on a tropics-wide scale.

The focus of this analysis is to update the frequency and spatial occurrence of all deep convective cores and more specifically hot towers, and to relate their characteristics to cold cloud and precipitation features using a new multi-year dataset that combines CloudSat and CALIPSO observations with passive satellite estimates of precipitation, outgoing longwave radiation (OLR), and brightness temperatures (TBs). In the next section, we define assumed deep convective updraft regions in two ways: 1) hot towers that have cloud tops reaching the tropopause as posited initially by RM58 and 2) deep convective cores that only require cloud top heights to extend beyond the minimum detrainment height (Takahashi and Luo 2012). Convective features are defined using combined CloudSat and CALIPSO measurements. The following section compares hot tower frequency, intensity, and bulk activity to that of all deep convective cores within the tropical trough zone (TTZ; Stephens et al. 2024), which is an updated boundary to the equatorial trough zone. The temporal and regional variabilities of all deep convective core, and exclusively hot tower, properties are compared to climatological precipitation and high cloud measurements to relate the strength and location of convective activity to TTZ-wide cloud and precipitation productivity. Finally, a new estimate of how many hot towers occur at any given time using a combination of Low Earth Orbit and Geostationary Equatorial Orbit (i.e., leo-geo) observations is provided.

2 Data and Methods

2.1 Defining the Tropical Trough Zone (TTZ)

The hot tower hypothesis was proposed based on the understanding that there is a surplus of energy that exists within the tropics that must be exported to higher latitudes where there is an energy deficit. The ocean supplies most of the heat transported out of the tropics, which makes it challenging to partition between the atmospheric and oceanic contributions of meridional energy transport. As a workaround, RM58 and RS79 reasoned that the oceanic component of meridional heat transport is negligible at the times of year when the sea surface temperature (SST) is at a maxima or minima (local rate of change of SST is zero) as no energy can be gained nor exported. They found that these local maxima and minima occur at the end of February at 5 °S and August at 15 °N. These latitudes became known as the bounds of the equatorial trough zone, which is a 10-degree latitudinal band that shifts 10 degrees over the course of a year. Thus, the equatorial trough zone is defined by the latitudes in which the meridional heat transport is from the tropics to the respective winter poles.

Stephens et al. (2024) used CERES radiative flux measurements to update this analysis and found that 13 °S and 19 °N are the latitudes at the end of February and August, respectively, that define the dividing lines between energy transfer to both summer and winter poles. This updated region is renamed as the Tropical Trough Zone (TTZ) to match the fact that the latitude bounds extend to higher latitudes within the tropics (i.e., not confined to the equator) compared to the bounds of the equatorial trough zone. The present study focuses on convective characteristics within the 32-degree TTZ at all times of year.

2.2 Deep Convective Cores and Hot Towers from an A-Train Perspective

This study uses a combination of measurements from satellite members of the A-Train constellation (L'Ecuyer and Jiang 2010). Launched in 2006, CloudSat is a nadir-pointing satellite with a 94 GHz Cloud Profiling Radar (CPR) that made twice-daily measurements at 0130 and 1330 local solar time (LST) between August 2006 to March 2011 and transitioned to operating only during the daytime beginning in 2012. The CPR is sensitive to cloud droplets and light precipitation as it began with a minimum sensitivity of −30.5 dBZ that approached −26 dBZ by the end of its lifetime (Stephens et al. 2002). Because of the short wavelength, the radar attenuates when encountering precipitation-sized hydrometeors. It has a 1.4 km cross-track resolution and 1.8 km along-track resolution, with a vertical resolution of 480 m that is oversampled to 240 m that enables it to capture vertical characteristics of convective systems (Tanelli et al. 2008). Stephens et al. (2008) provides an overview of the CloudSat mission, and detailed information on 2B-GEOPROF (Marchand et al. 2008), 2B-CLDCLASS-lidar (Sassen et al. 2008), 2C-PRECIP-COLUMN (Haynes et al. 2009), and ECMWF-AUX (Partain 2022) data that are used in this analysis can be found from the CloudSat Data Processing Center at https://www.cloudsat.cira.colostate.edu.

Convective characteristics are taken from a convective object (CO) database, which consists of identified convective systems using a two-dimensional image detection approach (PL22). Deep convective cores are identified along CloudSat curtains using the "Conv_strat_flag" variable from the 2C-PRECIP-COLUMN data product. Profiles are flagged as "convective" if there is an inflection in the reflectivity profile at least 500 m above the freezing level. It is inferred that convective updrafts vertically transport hydrometeors to levels above the freezing level, and their ability to attenuate suggests that these hydrometeors are liquid precipitation droplets. Contiguous convective flags along a profile are grouped to define a deep convective core in which the bulk cloud microphysical response to convective updrafts are captured within identified core regions (Takahashi et al. 2014; PL22).

In this study, deep convective cores are required to have a cloud top height of at least 10 km following buoyancy reasonings outlined in Takahashi et al. (2012, 2014, 2017). This also follows the updated hot tower definition in Fierro et al. (2009). To be consistent with RM58 and RS79, we define a hot tower as a deep convective core that overshoots the environmental tropopause defined below. In this study, we make no assumptions based on whether cores that reach the environmental tropopause entrain surrounding air. The following criteria are employed to define hot towers along a CloudSat overpass:

1. A convective core is identified following the methods outlined above (see also PL22).
2. Use the 'cloudtype' and 'cloudtopheight' variables from 2b-CLDCLASS-lidar to select the height of the cloud layer defined as deep convection from each profile within the core to avoid any multi-layer clouds.
3. Calculate the lapse rate tropopause height (height in the atmosphere when the lapse rate is less than 2 km; LRT) and cold point tropopause height (height of the minimum temperature; CPT) at each convective core location using state variables that are interpolated to each CPR bin from the ECMWF-AUX data product.
4. If the core cloud top height (CTH) has a height greater than 16 km, which is considered the lower limit of the tropopause height in the tropics, and is greater than both the LRT and CPT heights, then it is considered a hot tower.

Pixels adjacent to the deep convective cores and/or hot towers are considered part of the convective object if they have reflectivity values greater than -28 dBZ, which is the mean between the CPR's initial (-30.5 dBZ) and final (-26 dBZ) minimum detectable signal or are identified by Cloud-Aerosol Lidar with Orthogonal Polarization (CALIOP) on board CALIPSO. Figure 1 shows a combined CPR-CALIOP cross-section of a synoptic event located over the South Pacific Ocean on January 1, 2009. Several deep convective cores are identified within this system, but only two of them are flagged as hot towers as denoted by the shaded blue columns. Within the largest hot tower, there is noticeable attenuation that begins above the freezing level and continues downwards through the column, signifying that this is an active updraft region. In this case, the updraft is tilted and CALIOP detects an overshooting top just south of the hot tower. Since the algorithm is not designed to account for very strongly tilted updrafts, it could potentially underestimate their cloud top heights and frequency.

2.3 A CloudSat Metric of Convective Intensity

To capture the vertical intensity of convective cores and hot towers, we calculate the convective core center of gravity (CoG), which, as noted previously, is the center of mass of the convective core weighted by unadjusted CloudSat reflectivities (i.e., $CoG = \sum_i Z_i H_i / \sum_i Z_i$; PL22; Storer et al. 2014). Reflectivity values decrease below where precipitation-sized particles occur due to attenuation as shown in Fig. 1. The equation shows that an attenuated signal alters both the numerator and denominator as both $Z_i H_i$ (i.e., the numerator) at lower levels and the summed reflectivity (i.e., the denominator) over the full column are reduced compared to if the signal were not attenuated. Therefore, from a conceptual standpoint, attenuation does not drastically impact the qualitative metric of intensity. It is worthwhile to compare attenuated CoG values against simulated profiles that are not attenuated to fully address this issue, but this is the focus of a future study.

To establish the utility of CoG as a proxy for convective intensity, it can be compared to other indices of convective intensity. Precipitation-based features, such as the radar echo-top height (ETH) and rain rates within identified convective regions, have often been used as indices of intensity particularly in TRMM studies (e.g., Hamada et al. 2015; Liu et al.

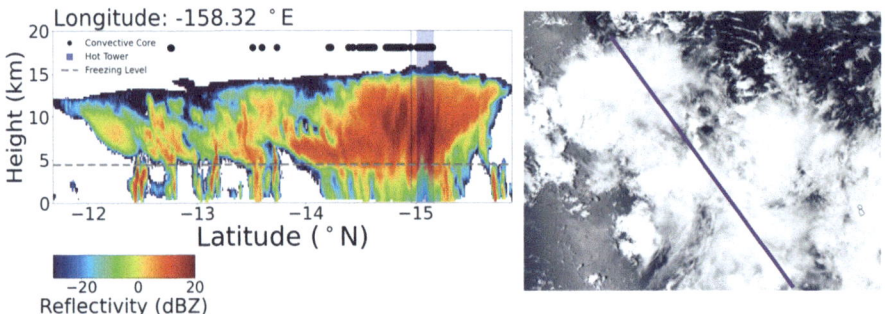

Fig. 1 (Left) CloudSat CPR radar reflectivity profile of a convective system found to have several convective cores (black dots), two of which are flagged as hot towers defined by the shaded blue columns. The area between CALIOP-detected cloud base and heights are filled in with "fake" -30 dBZ values for anything not detected by CPR. (Right) Aqua MODIS true color corrected reflectances of storm from EOSDIS Worldview. The purple line indicates the location of CloudSat overpass

2007; Liu and Zipser 2008; 2013; 2015; Nesbitt et al. 2000; Zipser et al. 2006; among others). Following a similar framework, Luo et al. (2008) and Takahashi and Luo (2014) defined an ETH greater than 10 dBZ from the CloudSat CPR as a metric of intensity. They note that radar reflectivity attenuation does not interfere with identifying ETH.

Figure 2a demonstrates that CoG and ETH > 10 dBZ are well correlated. On the x-axis in Fig. 2b is the relative CoG, or the mean freezing level of the convective object subtracted from the core CoG (hereafter considered relative CoG, or rCoG) to fully capture the lofted effect of the largest hydrometeors (PL22). Like the CoG-ETH relationship, ETH and rCoG are approximately linearly related. Figure 2c and d shows how CoG and rCoG relate to AMSR-E/AMSR2 convective core rain rates. As CoG increases in altitude, the full distribution of convective rain rates increases exponentially. Likewise, convective rain rates increase as rCoG increases, suggesting that there exists a relationship between the altitude extent of the largest hydrometeors and rain rates within convective cores. Based on these comparisons, it appears that attenuation does not significantly influence our ability to infer convective intensity using CoG or rCoG. Furthermore, CloudSat's relatively high spatial resolution and ability to directly sense hydrometeors within the convective system mitigate some of the challenges associated with using alternate proxies.

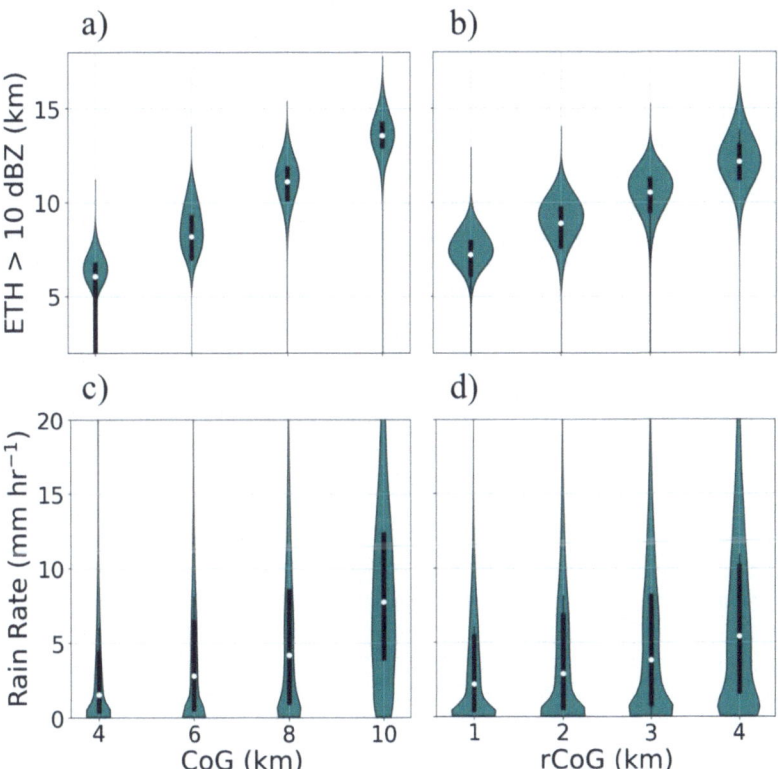

Fig. 2 **a** and **b** PDFs of CPR echo-top heights (ETH) greater than 10 dBZ of each convective core sorted by the convective core **a** CoG and **b** rCoG. **c** and **d** PDFs of rain rates averaged over each convective core from AMSR-E (2006–2010) and AMSR2 (2012–2016) sorted by the convective core **c** CoG and **d** rCoG. The black lines designate the interquartile range (25th to 75th percentiles) and the white dots are the means of the distributions. Data are over the tropical (30 °S/ °N) ocean between 2006 and 2016

A unique index used to capture the energetic impact of tropical storms is the Accumulated Cyclone Energy (ACE), which is defined as the summed squared estimated 6-hourly maximum sustained wind speed (Bell et al. 2000). Although ACE is not a direct measure of energy, it is credited for capturing the overall impact of cyclone activity on climate by weighing storm intensity by its lifetime (Bell et al. 2000; Camargo and Sobel 2005). Following similar reasoning, we define *accumulated rCoG* as the product between the core or hot tower rCoG, core length along the CloudSat overpass (hereafter, diameter; D), and number (N) of convective cores or hot towers ($N \times D \times rCoG$) within a given domain. Like ACE, these are measures that can be related to other climate variables (Camargo and Sobel 2005), such as precipitation and high cloud amount. Thus, accumulated rCoG can serve as an index of cumulative convective activity in which systems or regions with a large accumulated rCoG describe the occurrence of highly active convection.

2.4 Contextual Precipitation, Cloud, and OLR Datasets

Although CloudSat and CALIPSO measurements provide detailed information on the instantaneous vertical structures of convective clouds, they are limited in capturing the horizontal high cloud extent both spatially and temporally. Furthermore, these measurements are not able to capture heavy precipitation that often accompanies convective systems. Therefore, additional satellite observations are used to characterize the tropics-wide distribution of high cloud and the heaviest precipitation to relate to the CloudSat/CALIPSO-based metrics of convective activity.

The Global Precipitation Measurement (GPM) mission is a network of satellites designed jointly between the Japan Aerospace Exploration Agency (JAXA) and the United States National Aeronautics and Space Administration (NASA) to be a follow-on to the Tropical Rainfall Measuring Mission (TRMM) (Lu et al. 2018). The GPM "core" satellite carries a Dual-frequency Precipitation Radar (DPR) with both Ku (13 GHz) and Ka (35 GHz) bands in addition to a multichannel (10–183 GHz) GPM Microwave Imager (GMI) (Hou et al. 2014). A fleet of microwave imagers completes the constellation. Integrated Multi-satellitE Retrievals (IMERG) is a level-3 precipitation algorithm that combines all microwave measurements from the GPM constellation with infrared estimates and precipitation gauge analyses to create a global uniformly-gridded climatology of surface precipitation through present-day. The GPM IMERG V06B monthly data product provides monthly mean rain rates at a $1° \times 1°$-degree resolution between January 2006 and September 2021 (Huffman et al. 2020).

The Clouds and the Earth's Radiant Energy System (CERES) project combines radiometric measurements from six instruments hosted on EOS Terra, Aqua, Suomi National Polar-orbiting Partnership (NPP), and National Oceanic and Atmospheric Administration's (NOAA) Joint Polar Satellite System 1 (JPSS-1) satellites to provide a record of top-of-atmosphere and surface radiative fluxes between March 2000 through present-day. CERES Energy Balanced and Filled (EBAF) Edition 4.2 data product is a level-4 product at $1° \times 1°$-degree resolution that provides monthly and climatological mean clear-sky and all-sky TOA fluxes (Loeb et al. 2018). For this study, we select monthly mean rain rates and all-sky TOA longwave fluxes to capture the precipitation and high cloud distribution over the TTZ between 2007 and 2016.

Our aim for updating the hot tower estimate is to consider both the spatial and temporal variability of convective core activity. Because of the variability in convective core activity over the diurnal cycle, we supplement CloudSat-identified convective cores with passive infrared satellite observations that capture inferred convective cloud activity at cloud top. First, Moderate Resolution Imaging Spectroradiometer (MODIS) 11-micron TBs from the MOD06-5KM-AUX P1 R05 data product are matched to hot towers identified by CloudSat to provide a distribution of infrared TBs. The MOD06-5KM-AUX P1 R05 data product contains a subset of Collection 6 Aqua MODIS cloud properties (Toller et al. 2013) that are oversampled to match CloudSat CPR footprints at a 5 km resolution using a nearest neighbors approach (Cronk and Partain 2018). The convenience of using these observations is that MODIS flies on both Terra and Aqua within the A-Train constellation.

In order to infer convective activity at cloud top on a half-hourly basis, 11-micron TBs from the NCEP/CPC L3 Half Hourly 4 km Global (60S–60N) Merged IR V1 (GPM_MERGIR) data product are employed (Janowiak et al. 2017). GPM_MERGIR is a product created by the Goddard Earth Sciences Data and Information Services Center and provides half-hourly near-global 4-km gridded IR TB data from European (METEOSAT-5/7/8/9/10), Japanese (GMS-5/MTSat-1R/2/Himawari-8), and US (GOES-8/9/10/11/12/13/14/15/16) geostationary satellites. Data have been corrected for zenith angle dependence to reduce discontinuities at the boundaries of each satellite observing domain (Janowiak et al. 2017). At every hour and in each grid box within the TTZ, GPM-MERGIR TBs are averaged over the course of a year as well as for each season of 2013. The portion of the TTZ with GPM-MERGIR TBs that fall within the distribution of the MODIS-derived TBs of hot towers supply the basis for estimating the number of hot towers that occur at each hour of the day. Section 3.3 provides a detailed explanation of the calculation.

3 Results

3.1 Convective Core and Hot Tower Characteristics in Relation to High Clouds and Precipitation Within the TTZ

Convective updrafts often produce extensive cloud shields that can spatially organize on scales up to thousands of kilometers (Zipser 1969; Nakazawa 1988), denoted by a significant reduction in outgoing longwave radiation (OLR) as shown in Fig. 3a. These cloud shields are associated with heavy precipitation (Fig. 3b), particularly in the Intertropical Convergence Zone (ITCZ) and over the Maritime Continent.

Figure 4a depicts the distribution of deep convective core counts at 1:30 am/pm LST within the TTZ. As a reminder, the deep convective cores follow the relaxed hot tower definition of Fierro et al. (2009). Even though the diurnal variability of convective core frequency is not fully represented in Fig. 4a, a relationship between core frequency and surface precipitation accumulation is exhibited over the ocean as the highest climatological mean rain rates (Fig. 3b) occur where deep convective cores are most frequent. However, over land this relationship is not as apparent seeing that, for example, deep convective cores most frequently occur over the Amazon Basin but do not produce the heaviest rain rates. Likewise, the Congo Basin contains a concentration of convective cores similar to that found over the West Pacific Warm Pool and South Pacific Convergence Zone (SPCZ) at the two times of day but contributes nearly half the climatological mean rain rates.

Cores over land, particularly over Africa, have the largest rCoG values (Fig. 4b) which suggests that they are the most vertically intense compared to over other regions, as was noted by Zipser et al. (2006) and more recently Takahashi et al. (2023). The sizes of cores are fairly homogeneous across the TTZ—ranging between 6 and 8 km and are only slightly

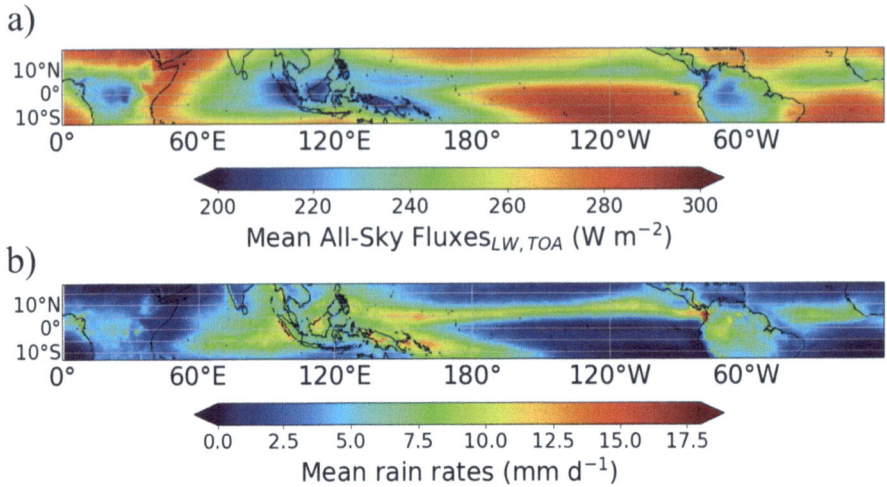

Fig. 3 **a** CERES EBAF (Energy Balanced and Filled) monthly mean top-of-atmosphere all-sky longwave radiative fluxes. **b** Global Precipitation Measurement (GPM) Integrated Multi-satellitE Retrievals (IMERG) monthly mean rain rates. Both are averaged between January 2007-December 2016

larger (~9 km) over Africa (Fig. 4c). The distribution of accumulated rCoG is found in Fig. 4e and is primarily weighted by core counts; however, despite there not being as many cores over the Congo Basin as over South America, convective activity appears to be equivalent in these regions due to cores in the Congo Basin being more intense and larger. Furthermore, continental convection has a higher accumulated rCoG per area than over the ocean and Maritime Continent.

Figure 5 presents convective properties of hot towers (i.e., deep convective cores that reach the environmental tropopause). The highest fraction of hot towers relative to all cores preferentially occurs over regions where convection is thermodynamically driven by warm surface temperatures, such as over India and the Bay of Bengal above 10 °N, West Africa, the Congo Basin, and off the coasts of the Maritime Continent Islands (Fig. 5a). Figure 5b-c shows that the spatial distribution of rCoG and core sizes for hot towers follows the distribution for all cores. However, hot towers have, on average, a 128% higher rCoG and 146% larger diameter compared to all deep convective cores (Fig. 4). Although the hot tower count never exceeds 35% of the total convective core count (Fig. 5a), in regions where the fraction of hot towers to all cores is largest, hot towers contribute over 40% of the total accumulated rCoG (Fig. 5e) owing to their enhanced size and intensity relative to all cores.

Vertical pressure velocity profiles are climatologically top-heavy (i.e., elevated upper-tropospheric upward motion and condensate), over the West Pacific Ocean compared to over the East Pacific Ocean, which is associated with bottom heavy profiles (Back and Bretherton 2006; Bretherton and Hartmann 2009). Bretherton et al. (2005) note that for a given virtual temperature profile, top-heavy profiles are less susceptible to drying from entrainment, which allows for clouds to extend deeper into the atmosphere. These characteristics are apparent in Figs. 4 and 5, which show that both the frequency of deep convective cores (Fig. 4a) and the fraction of cores that are hot towers (Fig. 5a) over the East Pacific Ocean is less than that over the West Pacific Ocean. Figures 4b and 5b show that cores have a higher accumulated rCoG per area over the West Pacific compared to over the

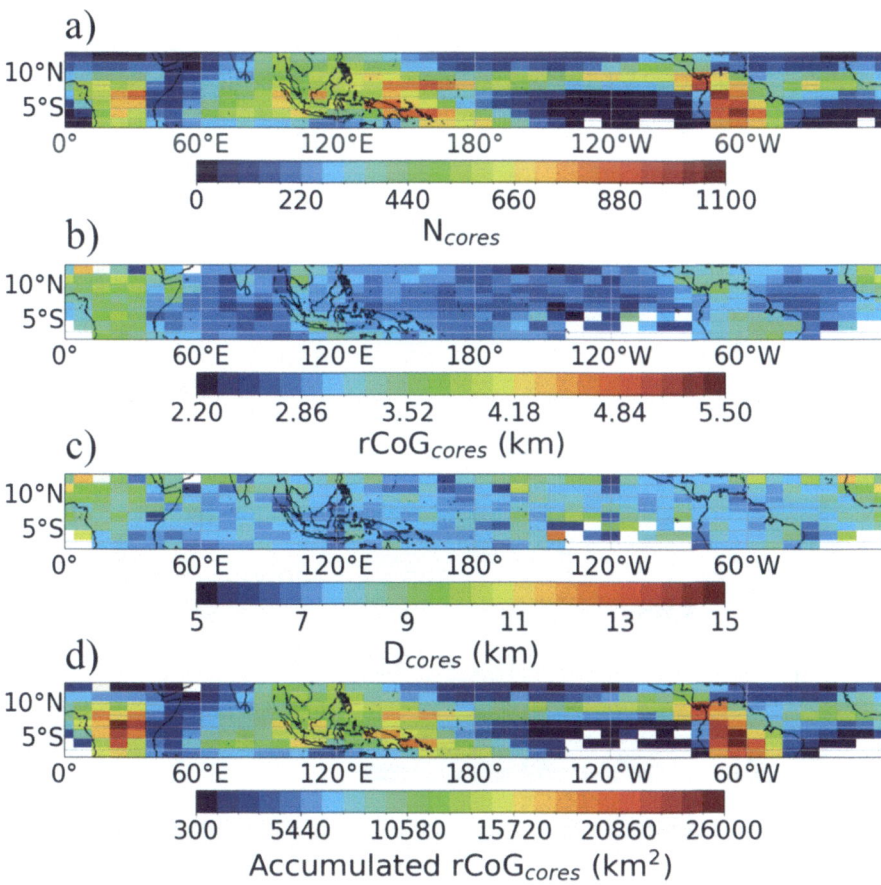

Fig. 4 **a** Number of convective cores, **b** mean core relative Center of Gravity (rCoG), **c** mean core diameter (D), and **d** accumulated rCoG defined as the product between the rCoG, core diameter, and the number of cores in each grid box. At least 20 cores are required in each 8×4 degree grid box for a mean to be calculated. The cores are identified at 1:30 am/pm LST between 2007 and 2016

East Pacific, and the accumulated rCoG contribution by hot towers is at least three times greater over the West Pacific Ocean than over the East Pacific Ocean. The West Pacific Ocean is also accompanied by heavier rain rates compared to the East Pacific Ocean, which is supported by the fact that deeper clouds in moist environments tend to also produce deeper stratiform anvils that can further promote top-heavy profiles (Bretherton et al. 2005; Houze 1982; Houze 2018; Schumacher et al. 2004).

RM58 reasoned that a synoptic disturbance with a wavelength of ~1350 km is the main mode of convection within the tropics—10% of which is occupied by active rain and 10% of the active rain is occupied by undilute towers. Figure 6 shows the distribution and length scales of the convective objects in which convective cores and hot towers reside. To clarify, non-precipitating detraining anvil clouds are not separated from deep stratiform clouds within the COs. COs containing hot towers are on the order of 1000 km in length and are most prevalent over the Maritime Continent. Previous literature has noted that this region is dominated by mesoscale convective systems (MCSs) that have merged (non-precipitating

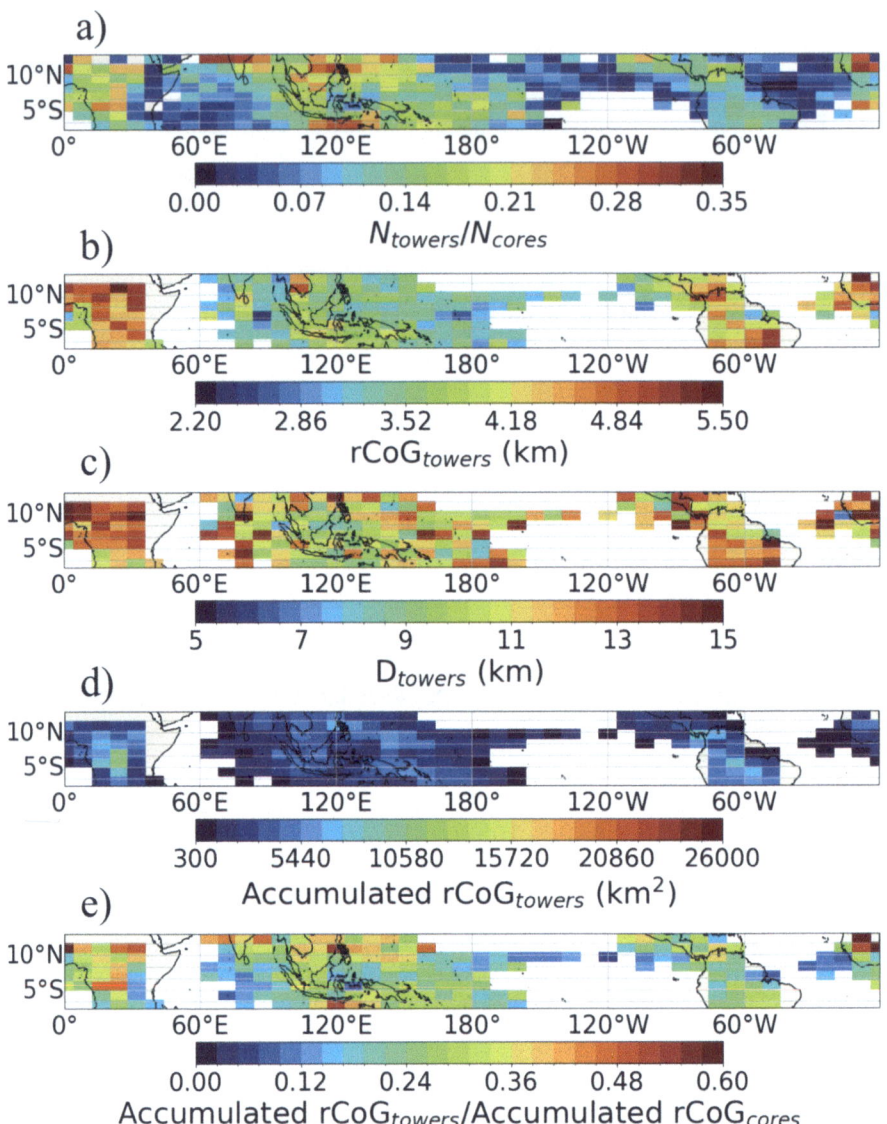

Fig. 5 **a** Hot tower counts divided by counts of all cores, **b-d** Like Fig. 4b-d but for hot towers, and **e** accumulated rCoG of hot towers divided by accumulated rCoG of all deep convective cores. At least 20 hot towers are required in each 8×4 degree grid boxes for the mean or fraction to be calculated

and precipitating) anvil clouds, otherwise known as mesoscale convective complexes or superclusters (Maddox 1980; Mapes and Houze 1993; Yuan and Houze 2010).

Figure 7a indicates that over 50% of these COs contain multiple cores or hot towers, which reinforces the concept that energy and mass are not being distributed primarily through isolated convective systems. Figure 7b-c shows the distribution of convective objects sorted by how many convective cores the objects contain. COs with greater than five distinct cores most frequently occur over the West Pacific Warm Pool and off

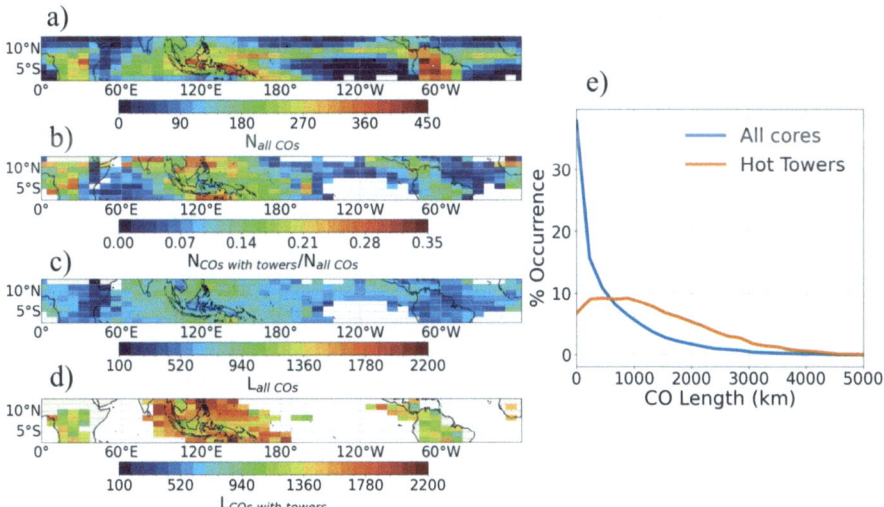

Fig. 6 **a** Counts of all convective objects (COs), **b** Ratio of counts of COs that contain hot towers to counts of all COs, **c** mean CO length and **d** mean length of COs that contain hot towers. At least 20 cores or hot towers are required in each grid box for a mean to be calculated. **e** PDF showing the frequency of convective object length sorted by systems that contain (blue) either hot towers or deep convective cores or (orange) must contain at least one hot tower

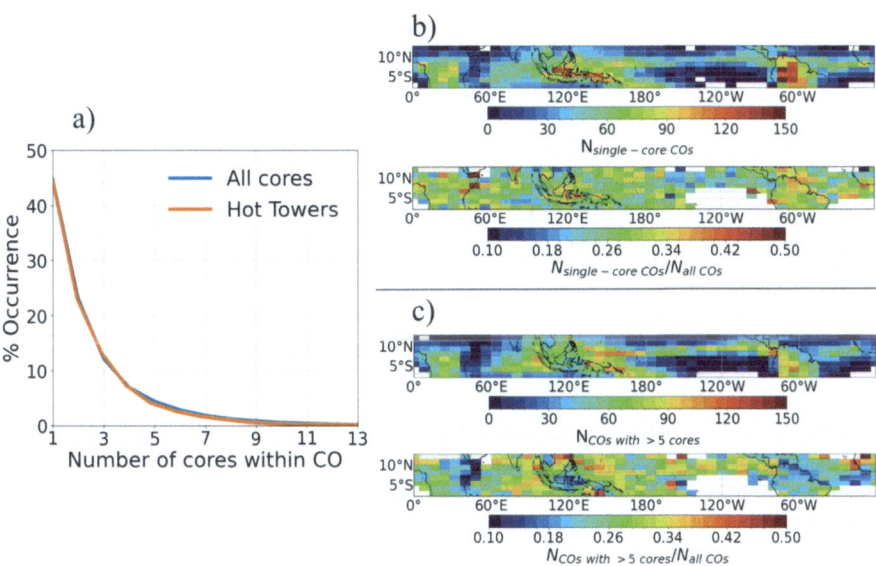

Fig. 7 **a** Percent occurrence of the number of cores within convective objects sorted by systems that contain (blue) any type of deep convective core or (orange) must contain at least one hot tower. **b** Distributions of (top) the number of convective objects with one core and (bottom) single-core convective objects divided by the number of all convective objects (Fig. 6a). **c** Distributions of (top) the number of convective objects with more than five cores and (bottom) convective objects with more than five cores divided by Fig. 6a

the Pacific coast of South America and Indonesian coasts, where both organized systems (Maddox 1980; Mapes and Houze 1993; Yuan and Houze 2010) and the highest climatological mean rain rates occur (Fig. 3b). The combined frequency and size of such organized convective events, which is correlated with duration, leads to large rainfall accumulations as noted by Hamada et al. (2015).

3.2 Spatiotemporal Variability of Convective Activity Within the TTZ

RM58 does not discuss how spatiotemporal variability of deep convection could influence the hot tower estimate. As mentioned in previous literature and from the characteristics outlined above, convective behavior varies regionally. We define three dominant regimes of convective activity that are loosely based on regions defined in Williams and Stanfill (2002) and elaborated upon in Stephens et al. (2024):

(1) *Continental land* in which convection is driven by several factors such as orography, land heterogeneity (e.g., rainforest, desert, etc.), and the diurnal and seasonal variability in surface heating from solar insolation;
(2) *ITCZ ocean* consisting of the East Pacific, South Pacific, Central Pacific, and Atlantic Oceans that are driven by sea surface temperature gradients and moisture convergence in the low levels; and
(3) *Maritime Continent* that includes the Indian and West Pacific Oceans in addition to land. Convection is often concentrated over the West Pacific Ocean due to anomalously warm sea surface temperatures compared to other parts of the ocean. The islands have their own complex terrain that, in addition to the land-sea-breeze circulations generated, influence convection unlike anywhere else within the TTZ.

The full distribution of convective core and hot tower activity within these three regimes are shown in Fig. S1 in the Supplementary Information. Hot towers contribute 5%, 7%, and 15% to the full convective activity over continental land, ITCZ ocean, and the Maritime Continent, respectively.

Figure 8 presents the distribution in monthly (a) total core and (b) hot tower counts, mean intensity, and accumulated rCoG between 2007 and 2016 within the different regions. The seasonal cycle in convective core frequency is bimodal across all regions, with peaks in convective core frequency doubling the minima. The peaks occur in May and October for the ocean and Maritime Continent. Over land, the peak in core frequency occurs during October with a second maximum in April. Monthly mean rCoG decreases and plateaus for the remainder of the year over the ocean and Maritime Continent but is bimodal over land and maximizes in September. Because both core frequency and intensity exhibit seasonal variability, while convective core size does not (not shown), there is a bimodal distribution in the accumulated rCoG. Thus, it is expected that convective activity from all convective cores peaks in May and October over all ocean and the Maritime Continent, and primarily in October with a secondary peak in April over land.

Hot tower counts do not have an obvious bimodal seasonal cycle over the ITCZ ocean and Maritime Continent. Over the Maritime Continent, hot towers are consistently the most prevalent between June to December compared to other times of year. The seasonality of hot tower occurrence over the full ocean primarily tracks behavior over the Indian and West Pacific Oceans due to the tendency for hot towers to occur in these regions as opposed to over the ITCZ ocean (see Fig. 5). The annual cycle in mean rCoG for hot towers is

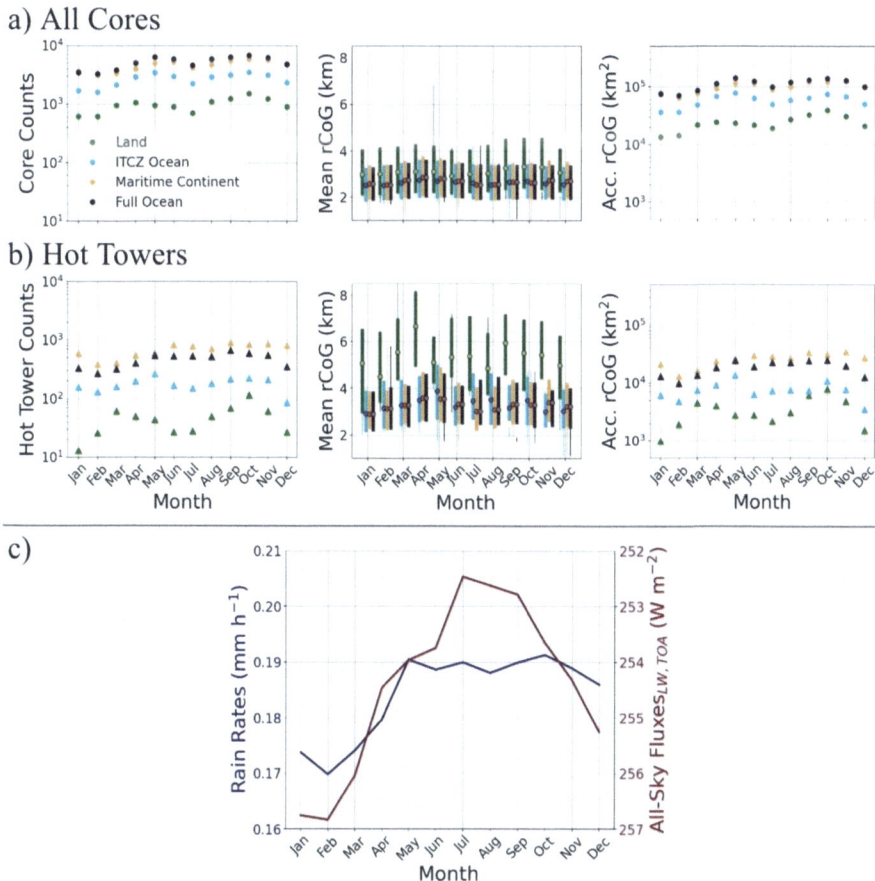

Fig. 8 Monthly CloudSat-detected **a** core and **b** hot tower counts (left column), rCoG (middle column), and accumulated rCoG (core counts × mean rCoG × mean size; right column) sorted by land (green), ITCZ ocean (light blue), the Maritime Continent (orange), and the full ocean (black) between January 2007–December 2010 and January 2012–December 2016. Full Ocean is the ITCZ ocean plus the Indian and West Pacific Oceans. The rCoG figures show the distribution of the 25% to 75% rCoG values with the dots representing the monthly mean rCoG. **c** Monthly mean CERES EBAF all-sky TOA LW fluxes (dark red) and GPM IMERG rain rates (dark blue) averaged between January 2007–December 2010 and January 2012–December 2016 over the full TTZ

consistent with that of all cores between January to May over the ocean and Maritime Continent. Interestingly, hot tower accumulated rCoG over the full ocean does not vary significantly between June to November due to the combined increase and decrease in convective activity over the Maritime Continent Oceans and ITCZ ocean, respectively. Meanwhile, hot towers over the land maintain the seasonal cycle in activity that is reflected in all deep convective cores over land.

Figure 8c shows the monthly mean rain rates and OLR averaged over the full TTZ. The most rain is produced at the same time of year as when the most numerous and intense convective cores occur. Despite rain rates remaining nearly constant (and reaching a local minimum) between May to October, mean OLR reaches a minimum during that time suggesting that clouds deepen during July through September. This could be capturing tropospheric expansion during the boreal summer months. However, this large-scale deepening of clouds is not observed in the annual cycle of individual core rCoG likely because rCoG considers the hydrometeor extent relative to the freezing level. Therefore, the annual cycle of CoG (i.e., including freezing level height) might capture how the height of the freezing level changes over the course of the year. For a more detailed look at the spatiotemporal variability between convective core, high cloud, and precipitation activity, refer to the 10-year time series in Fig. S2 and seasonal maps in Fig. S3.

3.3 How Many Hot Towers Occur at Any Given Time?

With these geographic and temporal variations in convective activity in mind, we can update the RS79 estimate of how many hot towers occur at any given time in the TTZ. In particular, we want to capture how the estimate might change over the course of a day. Because it is not possible to estimate how many hot towers occur at any given time solely using A-Train observations, we leverage the spatiotemporal resolution of geostationary satellite observations from GPM IMERG.

First, the area of deep convective cores within the TTZ from a GPM IMERG perspective is defined by:

$$A_{TTZ,allcores} = F_{TTZ,TB<X} A_{TTZ} \tag{1}$$

where $F_{TTZ,TB<X}$ is the fraction of the TTZ containing observed TBs less than a desired threshold (X) and A_{TTZ} is the area of the TTZ. An appropriate TB threshold needs to be applied for $A_{TTZ,allcores}$ to represent the total area of the TTZ covered by deep convective cores. A TB threshold of 208 K has commonly been used to identify regions of deep convective precipitation from geostationary satellite data (e.g., Mapes and Houze 1993; Rickenbach 1999; Williams and Houze 1987). However, the present analysis considers exclusively cores that reach the tropopause, and it has been found across multiple reanalysis datasets that the tropical tropopause temperature is less than 195 K (Tegtmeier et al. 2020). In fact, cores that overshoot the tropopause can have TBs nearing 190 K (e.g., Fiolleau and Roca 2013).

To establish a TB threshold consistent with convective cores and hot towers observed by CloudSat, Aqua MODIS 11-micron TBs are co-located to deep convective core and hot tower coordinates between 2007 and 2016 (Fig. 9). Figure 9a shows that the peak TB for hot towers occurs at 195 K, and TBs are slightly lower (i.e., colder) for higher cloud top heights. A 45-K range encompasses 90% of observed hot towers, with 70% of hot towers having TBs less than 208 K. However, Fig. 9b indicates that the distribution of TBs of all deep convective cores (i.e., including cores that are not hot towers) peak at 210 K, which makes it challenging to determine a single TB threshold that captures a majority of the deepest core (hot tower) activity without including potential information from cores that are not hot towers.

Dividing the total number of hot towers by the total number of all cores below any given TB threshold (purple curve in Fig. 9b) yields a method to scale Eq. 1 by the anticipated fraction of cores below any assumed TB threshold that are hot towers. This method assumes that the TBs from Aqua Modis oversampled at a 5-km resolution reasonably match the 4-km resolution GPM IMERG data. The equation to estimate the area of towers within the TTZ can then be defined as:

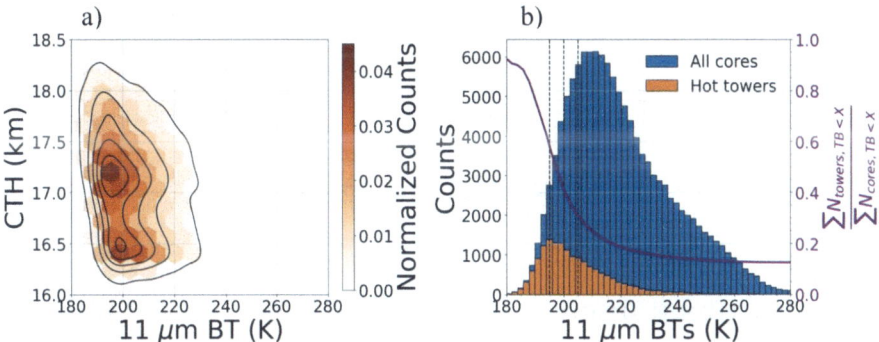

Fig. 9 a Distribution showing the relationship between the hot tower cloud top height (y-axis) from CloudSat-CALIPSO and MODIS 11-micron TBs (x-axis). The innermost line contour corresponds to 10% of the data and each contour is an additional 20% of the data. The shaded hexbins present another distribution perspective. Each hexbin contains a hot tower count that is divided by the total hot tower count. **b** Counts of all cores (blue) and hot towers (orange) binned by their mean 11-micron brightness temperatures (TBs) from Aqua MODIS. The purple curve represents the summed number of hot towers divided by the summed number of all cores below each TB bin, as shown by the y-axis on the right. Black dashed lines are TB intervals 195 K, 200 K, and 205 K

$$A_{TTZ,towers} = bA_{TTZ,allcores} \qquad (2a)$$

where

$$b = \left[\frac{\sum N_{towers,TB<X}}{\sum N_{allcores,TB<X}} \right] \frac{\overline{A}_{tower}}{\overline{A}_{core}} \qquad (2b)$$

The term in brackets is given by the purple curve in Fig. 9b. $\overline{A}_{tower}$ and $\overline{A}_{core}$ are the mean areas of hot towers and all cores, and diameters used to calculate the areas are presented in Table 1. Therefore, b is a scaling factor considering the hot tower-to-core count ratio (term in brackets) normalized by the hot tower-to-core mean area ratio. Finally, the number of hot towers is estimated by dividing $A_{TTZ,towers}$ by the mean area of a hot tower ($\overline{A}_{tower}$).

Figure 10 presents results using thresholds of 195 K, 200 K, and 205 K. To characterize the diurnal cycle of convective activity, TBs are averaged at each hour of the day for 2013 within the TTZ. Given the prior discussion, we reason that TBs above 205 K are no longer representative of updrafts that reach the tropopause (e.g., Fiolleau

Table 1 Mean hot tower diameter within each region averaged between 2007 and 2016, excluding 2011. The uncertainties listed are the standard errors of the mean diameters

	Land	ITCZ Ocean	Maritime Continent	Full Ocean	Full TTZ
Mean convective core diameter (km)	7.24 ± 0.06	7.71 ± 0.05	7.53 ± 0.03	7.60 ± 0.03	**7.72 ± 0.02**
Mean hot tower diameter (km)	13.3 ± 0.4	11.9 ± 0.3	10.6 ± 0.1	11.4 ± 0.2	**11.4 ± 0.1**

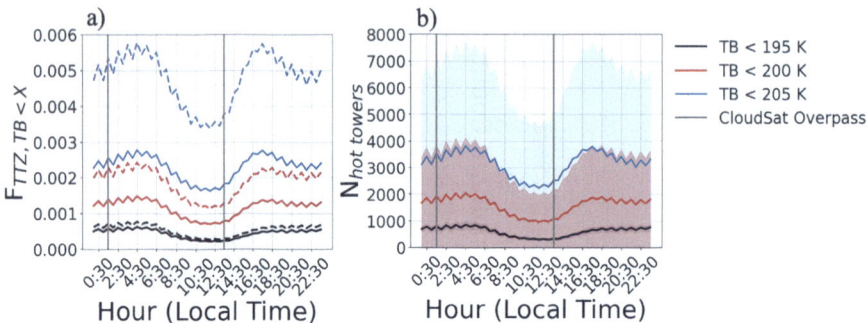

Fig. 10 a Dashed curves are the fraction of the TTZ containing GPM IMERG TBs < 195 K (black), TB < 200 K (red), and TB < 205 K (blue) over the TTZ averaged over 2013 at each hour. Solid curves are the fractions defined by the solid lines multiplied by the scaling factor, b, to represent the fraction of hot towers that cover the TTZ. **b** The estimate of hot towers at each hour of the day for each TB threshold (solid curves). Shaded regions represent calculated uncertainties from random variability in hot tower and convective core lengths. Vertical gray lines represent the times of CloudSat flyovers

and Roca 2013; Mapes and Houze 1993; Rickenbach 1999; Tegtmeier et al. 2020; Williams and Houze 1987). The solid curves in Fig. 10a represent the fractions of the TTZ below each TB threshold (i.e., $F_{TTZ,TB<X}$, as in Eq. 1 and the dashed curves in Fig. 10a) multiplied by b (Eq. 2b) to represent the fraction of the TTZ containing hot towers. As indicated by the solid curves, TBs less than 195 K cover between 0.02% to 0.06% of the TTZ over the course of a day, with each percentage doubling with an increase in 5 K. Thus, hot tower TBs < 205 K represent nearly 0.2% of the TTZ (blue solid curve in Fig. 10a). These percentages are comparable to the RM58 estimate that hot towers cover nearly 0.1% at any given time in the equatorial trough zone. Figure 10b provides estimates for how many hot towers occur over the full diurnal cycle for each TB threshold. The estimates range between 300–850 (TB < 195 K), 950–2050 (TB < 200 K), and 2250–3850 (TB < 205 K) over the full diurnal cycle, with the minima occurring at noon LST and the maxima at 0430 LST.

Following the above methodology, we provide estimates for hot towers within each region (i.e., land, ITCZ ocean, Maritime Continent, and full ocean). Equations 1 and 2 are modified to provide an estimate of hot towers for a given TB threshold within each region as follows:

$$N_{towers,region,TB<X} = \frac{b_{region} F_{region,TB<X} A_{region}}{\overline{A}_{tower,region}} \quad (3)$$

where each value in the equation is calculated for each region instead of the full TTZ. The hot tower-to-core ratio below each TB bin is calculated for each region and shown in Fig. 11a. The ratio of hot towers to cores with TBs less than 195 K is nearly the same across all regions, but the spread among regions increases as TB thresholds increase. For TBs greater than 195 K, it is less likely that cores are hot towers over land and the ITCZ Ocean compared to over the Maritime Continent.

The hot tower estimate for TB < 195 K, TB < 200 K, and TB < 205 K are then calculated for each region using Eq. 3. To obtain a single estimated range for each region, we weight the estimates for each TB threshold by their hot tower-to-core ratio:

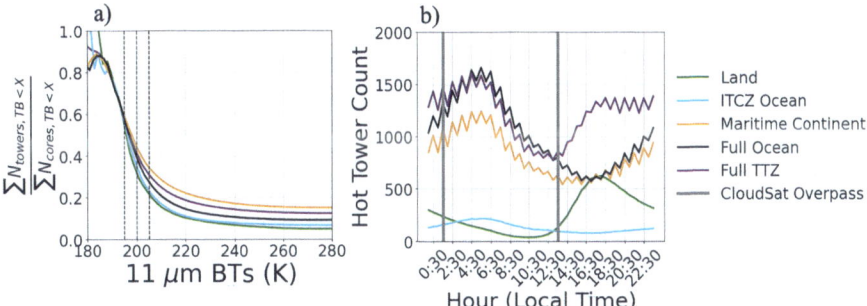

Fig. 11 **a** Summed number of hot towers divided by the summed number of all cores below each TB value (as in Fig. 9b) separated by region: full TTZ (purple), land (green), ITCZ ocean (blue), the Maritime Continent (orange), and the full ocean (black). Black dashed lines are TB intervals 195 K, 200 K, and 205 K. **b** Estimated hot tower count over the diurnal cycle separated by region. The vertical gray lines represent the times of CloudSat flyovers

$$\overline{N}_{\text{towers}} = \sum_{i=1}^{3} w_i N_{\text{towers},i} / \sum_{i=1}^{3} w_i \quad (4)$$

where w_i are the weights given by the curves in Fig. 11a at each of the three TB thresholds and $N_{towers,i}$ are the hot tower estimates below each TB threshold. Figure 11b presents the weighted average hot tower estimate for each region. The Full TTZ curve is now defined as the combined estimates over the Maritime Continent, ITCZ Ocean, and Land, with an updated and final estimate of ***800–1700*** hot towers. Most of the hot towers contributing to the Full TTZ curve occur over the Indian Ocean and West Pacific Warm Pool (represented in both the Maritime Continent and Full Ocean curves) since these regions are the most convectively active. Continental hot tower counts peak in the afternoon, which supplies the secondary peak in the TTZ-wide count.

Fig. S4 in the Supplementary Information shows the estimated number of hot towers sorted by both region and season. The TTZ-wide early morning peak remains relatively consistent across the year, aside from the weakened peak over the Maritime Continent in March, April, and May (MAM). Conversely, the afternoon peak is highest over MAM because continental land is the most convectively active during this season. Fig. S2, which presents the monthly core and hot tower estimate from solely CloudSat observations, indicates that 2013 might have been an anomalous year of convective activity, as the highest estimated hot tower counts across the 10-year time period occurred around September 2013. However, these anomalies are not represented in either the IMERG precipitation or CERES OLR data.

4 Discussion of Errors and Uncertainties

The focus of this section is to highlight known systematic and probabilistic errors and to document uncertainties in the hot tower estimate. It is partitioned into two parts: 1) systematic errors associated with the A-Train's spatiotemporal resolution and from combining measurements from multiple satellite platforms, and 2) random error from the variability of convective core sizes that propagates into the hot tower estimate.

4.1 Systematic Errors

A potential source of uncertainty arises from the CPR sampling geometry. Since CloudSat does not always profile through the center of cores, the diameters listed in this analysis could be systematically underestimated by ~21% (please refer to Takahashi et al. 2021 for a thorough derivation of this error). Systematically increasing the hot tower size might reduce the hot tower estimate by up to 4%. However, it is unclear whether this modification should be applied, as the underestimation of core size is only a concern for core widths larger than the CPR 1.4 × 1.8 km footprint resolution. Field campaign results over the Amazon and near the U.S. Virgin Islands have revealed that individual updrafts are often under a kilometer in diameter (Giangrande et al. 2013; Wang et al. 2020). It is uncertain whether CloudSat can detect such updrafts that are smaller than the CPR footprint. If they are being detected, the width being stored is the along-track resolution of the CPR footprint. It could also be that the CPR does not identify sufficiently small updrafts, which means that cores are under-detected. Both results actually lead to a potential systematic overestimation in core sizes. However, CloudSat is the highest resolution radar flown to date, so no other currently available dataset could improve upon the estimates presented here.

Another possible caveat to the hot tower estimate is not considering the potential diurnal cycle of convective core sizes. Our methodology averages core sizes across the twice-daily (0130 and 1330 LST) measurements. Yet, Figs. 10, 11, and S4 indicate that 0130 LST is at an onset of peak convective activity, while convective activity is at a near minimum at 1330 LST. Given the correlation between intensity and core widths (e.g., Takahashi et al. 2017; 2021), convective core sizes may vary at other times of day. It would be beneficial to study differences in convective core sizes at both times of day and supplement with other active remote sensing observations that have the temporal resolution to capture the diurnal cycle of convective core activity.

We also must address limitations when taking a multi-satellite approach. Although the MODIS data is oversampled to a 5-km resolution that is similar to the GPM IMERG data product, MODIS is likely more sensitive at detecting small regions of cold TBs compared to GPM IMERG. Furthermore, there are possible caveats when using a product containing merged observations from multiple geostationary satellites. As noted in Fiolleau et al. (2020), potential inconsistencies in spatiotemporal resolution and calibration methodologies between satellite measurements could lead to challenges in adequately capturing cold clouds in a blended product.

4.2 Random Error

The main source of random error in our hot tower estimate is due to the spatial variability of convective core sizes. Table 1 indicates that the standard error of the mean diameters are two orders of magnitude smaller than the means. These uncertainties are used to calculate the relative uncertainty of the hot tower estimate. Since the relative uncertainty scales with the area of the TTZ containing TBs less than a specified threshold, a larger TB threshold leads to a larger relative uncertainty (as shown by the shaded regions in Fig. 10b). Conversely, there is little random error for a hot tower estimate using TBs < 195 K.

5 Conclusions

We provide an observational update on convective core and system characteristics using satellite observations within the tropical trough zone (13 °S–19 °N), which is known to be the region where energy is primarily exported to higher latitudes. RM58 and RS79 postulated that convective updrafts reaching the tropopause, or hot towers, transport boundary layer air to the upper troposphere that aid in both balancing local radiative cooling and supplying a surplus of energy to be transported laterally. With combined CloudSat and CALIPSO observations, we identify deep convective cores that surpass both the cold point tropopause and lapse rate tropopause heights, or hot towers, and document their geographic and temporal frequency. The main findings of the research described in the paper are as follows:

(1) Hot towers make up, on average, 10% of all convective cores, but are between 125 and 150% larger in diameter and more intense than the full distribution of deep convective cores.
(2) RM58 noted that such hot towers are embedded within synoptic systems. Based on our results, less than 50% of the updrafts are isolated cases, while the remaining are embedded within systems that are likely MCSs or MCCs and occur in regions that have the highest climatological mean rain rates.
(3) The convective core reflectivity-weighted relative Center of Gravity multiplied by the core diameter provides a sufficient proxy for convective activity of a convective core. Multiplying this value by the total number of cores in any given area can provide an understanding of the cumulative convective activity of a region, or accumulated rCoG. Along the ITCZ band (stretching from the West Pacific Warm Pool to the Atlantic Ocean) as well as over the SPCZ, precipitation and high cloud cover are the most productive. Over these regions there is also a large accumulated rCoG by all convective cores. However, the hot tower accumulated rCoG is smaller per grid box over these regions compared to over land.
(4) The seasonal cycle in convective core and hot tower frequency and intensity are surveyed across three distinct regimes within the TTZ: ITCZ ocean, continental land, and the Maritime Continent (including the West Pacific and Indian Oceans). Convective core frequency, and to a lesser extent, intensity, exhibit a bimodal spread that peaks first in April (land) or May (ITCZ ocean and Maritime Continent) and October (all regions). Over the ocean and Maritime Continent where rain rates are the highest, core frequency follows the seasonal cycle of precipitation suggesting a strong relationship between rain rates and the frequency of convective cores.
(5) We estimate *800–1700* hot towers occur at any given time within the TTZ. The estimate is calculated using the probability that a given brightness temperature below 205 K is a hot tower from geostationary satellite observations averaged during 2013.
(6) We must also discuss the probabilistic nature of the hot tower estimate. The co-located MODIS TBs to hot tower coordinates reveals a 45-K spread in hot tower TBs. Furthermore, the 205 K threshold captures nearly 70% of the hot towers, meaning that this threshold does not represent the full distribution of hot tower TBs. However, approximately 25% of cores not considered hot towers have TBs less than 205 K. For these reasons, it is apparent that there is no single TB threshold sufficient for representing hot towers, let alone the full range of deep convective cores.

We acknowledge that our study differs from RM58 and RS79 in that we estimate hot tower activity over the full extent of the TTZ, which is 32-degrees (Stephens et al. 2024), as opposed to RM58 and RS79 that considers a 10-degree band that shifts latitudinally. The reason for this difference is to determine if convective activity is primarily confined to a 10-degree band. Our results indicate that the area of hot tower activity contributes 0.2% of the TTZ compared to the 0.1% approximation from RM58 and RS79. Interestingly, our estimate is at the lowest end of the 1600–2400 estimate made by RS79 despite the fact that our reported area of hot towers is double that of RM58 and RS79.

Derras-Chouk and Luo (2024) provide a hot tower estimate using instead a convective mass flux approach with geostationary satellite observations. They report 600–800 hot towers at any given time within a 10-degree band, also smaller than what RM58 and RS79 report. The differences between the original and updated estimates could be due to differences in the mean hot tower diameter used for the calculation. The observed mean hot tower diameter reported in our study is 11.4 km, which is two to three times larger than the RM58 estimate of 3–5 km. Derras-Chouk and Luo (2024) and other recent studies using CloudSat data have also reported convective core diameters ranging between 10 and 14 km (e.g., Takahashi and Luo 2012, 2023).

Finally, it is worth mentioning whether the assumption of undilute adiabatic ascent might play a role in the hot tower estimate. Because only dilute convective cores have so far been observed and convective updrafts are not the sole contributors to energy and mass transport, the estimate from RS79 could be an overestimate. Nevertheless, the precision of these estimates is quite remarkable given the limited data available at the time.

Supplementary Information The online version contains supplementary material available at https://doi.org/10.1007/s10712-024-09868-2.

Acknowledgements This research has been supported by the NASA Science Mission Directorate through the Future Investigators in NASA Earth and Space Science and Technology program (Award no. 80NSSC20K1651) and the NASA Jet Propulsion Laboratory (CloudSat Grant no. G-39690-1). This paper is an outcome of the Workshop "Challenges in Understanding the Global Water Energy Cycle and its Changes in Response to Greenhouse Gas Emissions" held at the International Space Science Institute (ISSI) in Bern, Switzerland (26–30 September 2022). The authors would like to thank Zhengzhao Johnny Luo and Angela Rowe for providing feedback on prior manuscript drafts.

Funding This study was funded by the NASA Science Mission Directorate through the Future Investigators in NASA Earth and Space Science and Technology program (Award no. 80NSSC20K1651) and the NASA Jet Propulsion Laboratory (CloudSat Grant no. G-39690-1).

Data availability The combined CloudSat and CALIPSO data were obtained from the NASA CloudSat data processing center (DPC) at https://www.cloudsat.cira.colostate.edu/. The specific data products used were the R05 versions of 2B-GEOPROF (Marchand et al. 2008), 2B-CLDCLASS-LIDAR Sassen et al. 2008), 2C-PRECIP-COLUMN (Haynes et al. 2009), ECMWF-AUX (Partain 2022), and MOD06-5KM-AUX P1 R05 (Cronk and Partain 2018). These data products were used to create a convective object database, which will be made publicly available at the CloudSat DPC. GPM IMERG V06B monthly data product (Huffman et al. 2020) was found at https://disc.gsfc.nasa.gov/. CERES EBAF Level 4 Edition 4.2 data (Loeb et al. 2018) can be found at https://asdc.larc.nasa.gov/project/CERES/CERES_EBAF_Edition4.2. Finally, the NCEP/CPC L3 Half Hourly 4 km Global (60S–60N) Merged IR V1 (GPM_MERGIR) data product (Janowiak et al. 2017) were found at https://catalog.data.gov/dataset/ncep-cpc-l3-half-hourly-4km-global-60s-60n-merged-ir-v1-gpm-mergir-at-ges-disc. We also acknowledge the use of imagery from the NASA Worldview application (https://worldview.earthdata.nasa.gov/), which is part of the NASA Earth Observing System Data and Information System (EOSDIS).

Declarations

Conflict of interest The authors Juliet Pilewskie, Graeme Stephens, Hanii Takahashi, and Tristan L'Ecuyer declare that they have no conflict of interest.

Ethical approval This article does not contain any studies with human participants or animals performed by any of the authors.

Open Access This article is licensed under a Creative Commons Attribution-NonCommercial-NoDerivatives 4.0 International License, which permits any non-commercial use, sharing, distribution and reproduction in any medium or format, as long as you give appropriate credit to the original author(s) and the source, provide a link to the Creative Commons licence, and indicate if you modified the licensed material. You do not have permission under this licence to share adapted material derived from this article or parts of it. The images or other third party material in this article are included in the article's Creative Commons licence, unless indicated otherwise in a credit line to the material. If material is not included in the article's Creative Commons licence and your intended use is not permitted by statutory regulation or exceeds the permitted use, you will need to obtain permission directly from the copyright holder. To view a copy of this licence, visit http://creativecommons.org/licenses/by-nc-nd/4.0/.

References

Back LE, Bretherton CS (2006) Geographic variability in the export of moist static energy and vertical motion profiles in the tropical pacific. Geophys Res Lett 33(17):L17810. https://doi.org/10.1029/2006GL026672

Bell GD, Halpert MS, Schnell RC, Higgins RW, Lawrimore J, Kousky VE, Tinker R, Thiaw W, Chelliah M, Artusa A (2000) Climate assessment for 1999. Bull Am Meteor Soc 81:S1–S50. https://doi.org/10.1175/1520-0477(2000)81[s1:CAF]2.0.CO;2

Bretherton CS, Hartmann DL (2009) Large-scale controls on cloudiness. Clouds Perturbed Climate Syst: Their Relationship Energy Balance, Atmospheric Dyn Precipitation 217:234. https://doi.org/10.7551/mitpress/9780262012874.003.0010

Bretherton CS, Blossey PN, Khairoutdinov M (2005) An energy-balance analysis of deep convective self-aggregation above uniform SST. J Atmos Sci 62:4273–4292. https://doi.org/10.1175/JAS3614.1

Byers HR, Braham RR (1948) Thunderstorm structure and circulation. J Atmos Sci 5:71–86. https://doi.org/10.1175/1520-0469(1948)005%3c0071:TSAC%3e2.0.CO;2

Camargo SJ, Sobel AH (2005) Western north pacific tropical cyclone intensity and enso. J Climate 18(15):2996–3006. https://doi.org/10.1175/JCLI3457.1

Chen SS, Houze RA (1997) Diurnal variation and life-cycle of deep convective systems over the tropical pacific warm pool. Quart J Royal Meteorol Soc 123(538):357–388. https://doi.org/10.1002/qj.49712353806

Cronk H, Partain P (2018) Cloudsat mod06-aux auxiliary data set. Technical report, CloudSat Data Processing Center Cooperative Institute for Research in the Atmosphere, Fort Collins, Colorado USA.

Derras-Chouk A, Luo ZJ (2024) Revisiting Riehl and Malkus (1958) and Riehl and Simpson (1979): Characterizing tropical hot towers and estimating convective mass fluxes using geostationary satellite data. Surv Geophys., in press. https://doi.org/10.1007/s10712-024-09856-6

Fierro AO, Simpson J, LeMone MA, Straka JM, Smull BF (2009) On how hot towers fuel the hadley cell: an observational and modeling study of line-organized convection in the equatorial trough from TOGA COARE. J Atmos Sci 66:2730–2746. https://doi.org/10.1175/2009JAS3017.1

Fiolleau T, Roca R (2013) An algorithm for the detection and tracking of tropical mesoscale convective systems using infrared images from geostationary satellite. IEEE Trans Geosci Remote Sens 51:4302–4315. https://doi.org/10.1109/TGRS.2012.2227762

Fiolleau T, Roca R, Cloché S, Bouniol D, Raberanto P (2020) Homogenization of geostationary infrared imager channels for cold cloud studies using megha-tropiques/ScaRaB. IEEE Trans Geosci Remote Sens 58(9):6609–6622. https://doi.org/10.1109/TGRS.2020.2978171

Giangrande SE, Collis S, Straka J, Protat A, Williams C, Krueger S (2013) A summary of convective-core vertical velocity properties using ARM UHF wind profilers in Oklahoma. J Appl Meteorol Climatol 52:2278–2295. https://doi.org/10.1175/JAMC-D-12-0185.1

Giangrande SE, Toto T, Jensen MP, Bartholomew MJ, Feng Z, Protat A, Williams CR, Schumacher C, Machado L (2016) Convective cloud vertical velocity and mass-flux characteristics from radar wind profiler observations during goamazon2014/5. J Geophys Res: Atmospheres 121:12891–12913. https://doi.org/10.1002/2016JD025303

Hamada A, Takayabu YN, Liu C, Zipser EJ (2015) Weak linkage between the heaviest rainfall and tallest storms. Nat Commun 6:6213. https://doi.org/10.1038/ncomms7213

Haynes JM, L'Ecuyer TS, Stephens GL, Miller SD, Mitrescu C, Wood NB, Tanelli S (2009) Rainfall retrieval over the ocean with spaceborne W-band radar. J Geophys Res: Atmospheres 114(D8):D00A22-1. https://doi.org/10.1029/2008JD009973

Heymsfield GM, Tian L, Heymsfield AJ, Li L, Guimond S (2010) Characteristics of deep tropical and subtropical convection from nadir-viewing high-altitude airborne doppler radar. J Atmos Sci 67:285–308. https://doi.org/10.1175/2009JAS3132.1

Hou AY, Kakar RK, Neeck S, Azarbarzin AA, Kummerow CD, Kojima M, Oki R, Nakamura K, Iguchi T (2014) The global precipitation measurement mission. Bull Am Meteorol Soc 95:701–722. https://doi.org/10.1175/BAMS-D-13-00164.1

Houze RA Jr (1977) Structure and dynamics of a tropical squall-line system. Mon Weather Rev 105:1540–1567

Houze RA Jr (1981) Structures of atmospheric precipitation systems: a global survey. Radio Sci 16(5):671–689. https://doi.org/10.1029/RS016i005p00671

Houze RA (2018) 100 years of research on mesoscale convective systems. Meteorol Monogr 59:17.1-17.54. https://doi.org/10.1175/AMSMONOGRAPHS-D-18-0001.1

Huffman GJ, Bolvin DT, Braithwaite D, Hsu K-L, Joyce RJ, Kidd C, Nelkin EJ, Sorooshian S, Stocker EF, Tan J, Wolff DB, Xie P (2020) Integrated Multi-satellite Retrievals for the Global Precipitation Measurement (GPM) Mission (IMERG). In: Levizzani V, Kidd C, Kirschbaum DB, Kummerow CD, Kenji Nakamura F, Turk J (eds) Satellite Precipitation Measurement: Volume 1. Springer International Publishing, Cham, pp 343–353. https://doi.org/10.1007/978-3-030-24568-9_19

Janowiak J, Joyce B, Xie P (2017) Ncep/cpc l3 half hourly 4km global (60s - 60n) merged ir v1. Technical report, Goddard Earth Sciences Data and Information Services Center (GES DISC), Greenbelt, MD.

Johnson RH, Rickenbach TM, Rutledge SA, Ciesielski PE, Schubert WH (1999) Trimodal characteristics of tropical convection. J Clim 12:2397–2418. https://doi.org/10.1175/1520-0442(1999)012%3c2397:tcotc%3e2.0.co;2

L'Ecuyer TS, Jiang J (2010) Touring the atmosphere aboard the A-train. Phys Today 63(7):36–41

LeMone MA, Zipser EJ (1980) Cumulonimbus vertical velocity events in GATE. Part I: diameter, intensity and mass flux. J Atmos Sci 37:2444–2457. https://doi.org/10.1175/1520-0469(1980)037%3c2444:CVVEIG%3e2.0.CO;2

Liu C, Zipser EJ (2008) Diurnal cycles of precipitation, clouds, and lightning in the tropics from 9 years of TRMM observations. Geophys Res Lett. https://doi.org/10.1029/2007GL032437

Liu C, Zipser E (2013) Regional variation of morphology of organized convection in the tropics and subtropics. J Geophys Res Atmos 118:453–466. https://doi.org/10.1029/2012JD018409

Liu C, Zipser EJ (2015) The global distribution of largest, deepest, and most intense precipitation systems. Geophys Res Lett 42:3591–3595. https://doi.org/10.1002/2015GL063776

Liu C, Zipser EJ, Nesbitt SW (2007) Global distribution of tropical deep convection: different perspectives from TRMM infrared and radar data. J Climate 20:489–503. https://doi.org/10.1175/JCLI4023.1

Loeb NG, Doelling DR, Wang H, Su W, Nguyen C, Corbett JG, Liang L, Mitrescu C, Rose FG, Kato S (2018) Clouds and the earth's radiant energy system (ceres) energy balanced and filled (ebaf) top-of-atmosphere (toa) edition-4.0 data product. J Clim 31:895–918. https://doi.org/10.1175/JCLI-D-17-0208.1

Lu D, Yong B (2018) Evaluation and hydrological utility of the latest gpm imerg v5 and gsmap v7 precipitation products over the tibetan plateau. Remote Sensing 10(12):2022. https://doi.org/10.3390/rs10122022

Lucas C, Zipser EJ, Lemone MA (1994) Vertical velocity in oceanic convection off tropical Australia. J Atmos Sci 51:3183–3193. https://doi.org/10.1175/1520-0469(1994)051%3c3183:VVIOCO%3e2.0.CO;2

Luo Z, Liu GY, Stephens GL (2008) CloudSat adding new insight into tropical penetrating convection. Geophys Res Lett 35:L19819. https://doi.org/10.1029/2008GL035330

Luo ZJ, Liu GY, Stephens GL (2010) Use of a-train data to estimate convective buoyancy and entrainment rate. Geophys Res Lett 37:L09804. https://doi.org/10.1029/2010GL042904

Luo ZJ, Jeyaratnam J, Iwasaki S, Takahashi H, Anderson R (2014) Convective vertical velocity and cloud internal vertical structure: an a-train perspective. Geophys Res Lett 41:723–729. https://doi.org/10.1002/2013GL058922

Maddox R (1980) Mesoscale convective complexes. Bull Am Meteorol Soc 61:1374–1387

Mapes BE, Houze RA (1993) Cloud clusters and superclusters over the oceanic warm pool. Mon Weather Rev 121:1398. https://doi.org/10.1175/1520-0493(1993)121%3c1398:CCASOT%3e2.0.CO;2

Marchand R, Mace GG, Ackerman T, Stephens G (2008) Hydrometeor detection using cloudsat–an earth-orbiting 94-ghz cloud radar. J Atmos Oceanic Tech 25(4):519–533. https://doi.org/10.1175/98720 07JTECHA1006.1

Masunaga H, L'Ecuyer TS (2014) A mechanism of tropical convection inferred from observed variability in the moist static energy budget. J Atmos Sci 71:3747–3766. https://doi.org/10.1175/JAS-D-14-0015.1

Masunaga H, Luo ZJ (2016) Convective and large-scale mass flux profiles over tropical oceans determined from synergistic analysis of a suite of satellite observations. J Geophys Res: Atmospheres 121(13):7958–7974. https://doi.org/10.1002/2016JD024753

Miller STK, Keim BD, Talbot RW, Mao H (2003) Sea breeze: structure, forecasting, and impacts. Rev Geophys 41(3):1011. https://doi.org/10.1029/2003RG000124

Moncrieff MW, Miller MJ (1976) The dynamics and simulation of tropical cumulonimbus and squall lines. Q.J.R Meteorol Soc 102:373–394. https://doi.org/10.1002/qj.49710243208

Nakazawa T (1988) Tropical super clusters within intraseasonal variations over the western pacific. J Meteorol Soc Japan Ser II 66:823–839. https://doi.org/10.2151/jmsj1965.66.6_823

Needham MR, Randall DA (2021) Riehl and Malkus revisited: the role of cloud radiative effects. J Geophys Res: Atmospheres 126(16):e2021JD035019. https://doi.org/10.1029/2021JD035019

Nesbitt SW, Zipser EJ (2003) The diurnal cycle of rainfall and convective intensity according to three years of trmm measurements. J Clim 16:1456–1475. https://doi.org/10.1175/1520-0442(2003)016%3c1456:tdcora%3e2.0.co;2

Nesbitt SW, Zipser EJ, Cecil DJ (2000) A census of precipitation features in the tropics using TRMM: radar, ice scattering, and lightning observations. J Clim 13:4087–4106. https://doi.org/10.1175/1520-0442(2000)013%3c4087:ACOPFI%3e2.0.CO;2

Partain P (2022). Cloudsat ECMWF-AUX auxiliary data product process description and interface control document. Technical report, CloudSat Data Processing Center Cooperative Institute for Research in the Atmosphere, Fort Collins, Colorado USA.

Pilewskie JA, L'Ecuyer TS (2022) The global nature of early-afternoon and late-night convection through the eyes of the A-train. J Geophys Res: Atmospheres 127(13):e2022JD036438. https://doi.org/10.1029/2022JD036438

Reed RJ, Recker EE (1971) Structure and properties of synoptic-scale wave disturbances in the equatorial western pacific. J Atmospheric Sci 28:1117–1133. https://doi.org/10.1175/1520-0469(1971)028%3c1117:SAPOSS%3e2.0.CO;2

Rickenbach TM (1999) Cloud top evolution of tropical oceanic squall lines from radar reflectivity and infrared satellite data. Mon Wea Rev 127:2951–2976

Riehl H, Malkus JS (1958) On the heat balance in the equatorial trough zone. Geophysica 6:503–538

Riehl H, Simpson JS (1979) The heat balance of the equatorial trough zone, revisited. Beitr Phys Atmos 52:287–305

Robert A, Houze Jr (1982) Cloud clusters and large-scale vertical motions in the tropics. J Meteorol Soc Japan. Ser. II 60(1):396–410. https://doi.org/10.2151/jmsj1965.60.1_396

Sassen K, Wang Z, Liu D (2008) Global distribution of cirrus clouds from CloudSat/Cloud-Aerosol lidar and infrared pathfinder satellite observations (CALIPSO) measurements. J Geophys Res: Atmospheres 113(D8):D00A12. https://doi.org/10.1029/2008JD009972

Schumacher C, Houze RA Jr (2003) Stratiform rain in the tropics as seen by the TRMM precipitation radar. J Climate 16(11):1739–1756

Schumacher C, Houze RA, Kraucunas I (2004) The tropical dynamical response to latent heating estimates derived from the TRMM precipitation radar. J Atmos Sci 61:1341–1358. https://doi.org/10.1175/1520-0469(2004)061%3c1341:TTDRTL%3e2.0.CO;2

Stephens GL, Vane DG, Boain RJ, Mace GG, Sassen K, Wang Z, Illingworth AJ, O'connor EJ, Rossow WB, Durden SL, Miller SD, Austin RT, Benedetti A, Mitrescu C (2002) The cloudsat mission and the a-train: a new dimension of space-based observations of clouds and precipitation. Bull Am Meteor Soc 83:1771–1790. https://doi.org/10.1175/BAMS-83-12-1771

Stephens GL, Vane DG, Tanelli S, Im E, Durden S, Rokey M, Reinke D, Partain P, Mace GG, Austin R, L'Ecuyer T, Haynes J, Lebsock M, Suzuki K, Waliser D, Wu D, Kay J, Gettelman A, Wang Z, Marchand R (2008) Cloudsat mission: performance and early science after the first year of operation. J Geophys Res: Atmospheres 113:D00A18. https://doi.org/10.1029/2008JD009982

Stephens GL, Hakuba MZ, Webb MJ, Lebsock M, Yue Q, Kahn BH, Hristova-Veleva S, Rapp AD, Stubenrauch CJ, Elsaesser GS, Slingo J (2018) Regional intensification of the tropical hydrological cycle during ENSO. Geophys Res Lett 45:4361–4370. https://doi.org/10.1029/2018GL077598

Stephens GL, Hakuba M, Takahashi H, Pilewskie J, Andrews T, Shiro K, Stubenrauch C, Wu L (2024) Tropical deep convection, cloud feedbacks and climate sensitivity. Surv Geophys. https://doi.org/10.1007/s10712-024-09831-1

Storer RL, van den Heever SC, L'Ecuyer TS (2014) Observations of aerosol-induced convective invigoration in the tropical east Atlantic. J Geophys Res Atmos 119:3963–3975. https://doi.org/10.1002/2013JD020272

Takahashi H, Luo ZJ (2012) Where is the level of neutral buoyancy for deep convection? Geophys Res Lett 39:L15809. https://doi.org/10.1029/2012GL052638

Takahashi H, Luo ZJ (2014) Characterizing tropical overshooting deep convection from joint analysis of CloudSat and geostationary satellite observations. J Geophys Res: Atmospheres 119:112–121. https://doi.org/10.1002/2013JD020972

Takahashi H, Luo ZJ, Stephens GL (2017) Level of neutral buoyancy, deep convective outflow, and convective core: New perspectives based on 5 years of CloudSat data. J Geophys Res: Atmospheres 122(5):2958–2969

Takahashi H, Luo ZJ, Stephens G (2021) Revisiting the entrainment relationship of convective plumes: a perspective from global observations. Geophys Res Lett 48(6):e2020GL092349. https://doi.org/10.1029/2020GL092349

Takahashi H, Luo ZJ, Stephens G, Mulholland JP (2023) Revisiting the land-ocean contrasts in deep convective cloud intensity using global satellite observations. Geophys Res Lett 50(5):e2022GL102089. https://doi.org/10.1029/2022GL102089

Tanelli S, Durden SL, Im E, Pak KS, Reinke DG, Partain P, Haynes JM, Marchand RT (2008) Cloudsat's cloud profiling radar after two years in orbit: performance, calibration, and processing. IEEE Trans Geosci Remote Sens 46:3560–3573. https://doi.org/10.1109/TGRS.2008.2002030

Tegtmeier S, Anstey J, Davis S, Dragani R, Harada Y, Ivanciu I, Pilch Kedzierski R, Krüger K, Legras B, Long C, Wang JS, Wargan K, Wright JS (2020) Temperature and tropopause characteristics from reanalyses data in the tropical tropopause layer. Atmospheric Chem Phys 20:753–770. https://doi.org/10.5194/acp-20-753-202010.5194/acp-20-753-2020

Toller G, Xiong XJ, Sun J, Wenny BN, Geng X, Kuyper J, Angal A, Chen H, Madhavan S, Wu A (2013) Terra and aqua moderate-resolution imaging spectroradiometer collection 6 level 1B algorithm. J Appl Remote Sens 7:073557. https://doi.org/10.1117/1.JRS.7.073557

van Bemmelen W (1922) Land-und seebrise in batavia. Beitr Phys Frei Atmos 10:169–177

Wang D, Giangrande SE, Feng Z, Hardin JC, Prein AF (2020) Updraft and downdraft core size and intensity as revealed by radar wind profilers: MCS observations and idealized model comparisons. J Geophys Res: Atmospheres 125(11):e2019JD031774. https://doi.org/10.1029/2019JD031774

Williams M, Houze RA (1987) Satellite-observed characteristics of winter monsoon cloud clusters. Mon Wea Rev 115:505–519

Williams E, Stanfill S (2002) The physical origin of the land–ocean contrast in lightning activity. C R Phys 3(10):1277–1292. https://doi.org/10.1016/s1631-0705(02)01407-x

Yanai M, Esbensen SK, Chu J-H (1973) Determination of bulk properties of tropical cloud clusters from large-scale heat and moisture budgets. J Atmos Sci 30:611–627

Yang G-Y, Slingo J (2001) The diurnal cycle in the tropics. Mon Weather Rev 129:784–801. https://doi.org/10.1175/1520-0493(2001)129%3c0784:TDCITT%3e2.0.CO;2

Yuan J, Houze RA (2010) Global variability of mesoscale convective system anvil structure from A-train satellite data. J Climate 23:5864–5888. https://doi.org/10.1175/2010JCLI3671.1

Zipser EJ (1969) The role of organized unsaturated convective downdrafts in the structure and rapid decay of an equatorial disturbance. J Appl Meteorol Climatol 8:799–814. https://doi.org/10.1175/1520-0450(1969)008%3c0799:TROOUC%3e2.0.CO;2

Zipser EJ (2003) Some Views On "Hot Towers" after 50 Years of Tropical Field Programs and Two Years of TRMM Data. In: Tao W-K, Adler R (eds) Cloud Systems, Hurricanes, and the Tropical Rainfall Measuring Mission (TRMM). American Meteorological Society, Boston, MA, pp 49–58. https://doi.org/10.1007/978-1-878220-63-9_5

Zipser EJ, LeMone MA (1980) Cumulonimbus vertical velocity events in gate. part II: synthesis and model core structure. J Atmospheric Sci 37:2458–2469. https://doi.org/10.1175/1520-0469(1980)037%3c2458:CVVEIG%3e2.0.CO;2

Zipser EJ, Cecil DJ, Liu C, Nesbitt SW, Yorty DP (2006) Where are the most: intense thunderstorms on Earth? Bull Am Meteor Soc 87:1057–1071. https://doi.org/10.1175/BAMS-87-8-1057

Publisher's Note Springer Nature remains neutral with regard to jurisdictional claims in published maps and institutional affiliations.

A Geostationary Satellite-Based Approach to Estimate Convective Mass Flux and Revisit the Hot Tower Hypothesis

Amel Derras-Chouk[1] · Zhengzhao Johnny Luo[1]

Received: 27 December 2023 / Accepted: 31 July 2024 / Published online: 10 October 2024
© The Author(s) 2024

Abstract

This study aims to revisit the classic "hot tower" hypothesis proposed by Riehl and Simpson (Malkus) in 1958 and revisited in 1979. Our investigation centers on the convective mass flux of hot towers within the tropical trough zone, using geostationary (GEO) satellite data and an innovative analysis technique, known as ML16, which integrates various data sources, including hot tower heights, ambient profiles, and a plume model, to determine convective mass flux. The GEO-based ML16 approach is evaluated against collocated ground-based radar wind profiler observations, showing broad agreement. Our GEO-based estimate of hot tower convective mass flux, 2.8×10^{11}–3.4×10^{11} kg s^{-1}, is similar to the revisited estimate in Riehl and Simpson (1979), 2.6–3.0×10^{11} kg s^{-1}. Additionally, our analysis gives a median count of around 550 hot towers with a median size of about 11 km, in contrast to the previous estimates of 1600–2400 hot towers, each characterized by a fixed size of 5 km. We discuss the causes of these discrepancies, emphasizing the fundamental differences between the two approaches in characterizing tropical hot towers. While both approaches have various uncertainties, the evidence suggests that greater credibility should be placed on results derived from direct satellite observations. Finally, we identify future opportunities in Earth Observations that will provide more accurate measurements, enabling further evaluation of the role played by tropical hot towers in mass transport.

Keywords Tropical convection · "Hot tower" hypothesis · Convective mass flux

Article Highlights

- Using geostationary satellite observations and an observation-constrained plume model, tropical hot towers are identified, and their convective mass fluxes are estimated
- Results are evaluated with collocated A-Train and ground-based estimates, showing broad agreement
- Estimates of convective mass flux, mean size, and total number of hot towers are compared with those made in Riehl and Malkus (1958) and Riehl and Simpson (1979)

✉ Zhengzhao Johnny Luo
zluo@ccny.cuny.edu

[1] Department of Earth and Atmospheric Sciences, The City College of the New York of City University of New York (CUNY), New York, NY 10031, USA

1 Introduction

The "hot tower" hypothesis was originally introduced by Riehl and Malkus (1958) to clarify the process of heat transport from the surface to the upper troposphere within the tropical trough zone, providing an explanation for the vertical movement of air against energy gradient to reach high altitudes. At that time, the term "hot tower" had not yet been coined; the authors used the terms "undilute cloud towers" and "cumulonimbus chimneys."

Subsequent studies revised the undilute requirement for cloud towers defined as deep convective clouds originating in the planetary boundary layer and extending near the upper tropospheric outflow layer, allowing for some entrainment of ambient air (Zipser 2003; Fierro et al. 2009), although the requirement remains in Riehl (1979) revision.

Nevertheless, the core idea of the hypothesis remains: Vertical heat transport, required for energy balance of the tropical trough zone, is achieved through hot towers and not through gradual, large-scale ascent of air masses, which fails to overcome the mid-tropospheric energy minimum. Furthermore, through energy balance and continuity considerations, Riehl and Malkus (1958) estimated the mass flux in undilute hot towers, and, with assumptions about the sizes and vertical velocities within cloud towers, they postulated that a total of 1500–5000 active hot towers embedded within the tropical trough zone maintain the tropical energy balance. A later study by Riehl (1979) revised the hot tower count to 1600–2400.

The "hot tower" concept and subsequent inferences about the properties of tropical deep convective cloud were ground-breaking achievements at a time when global observations were scarce. Nevertheless, any quantitative conclusion about mass and energy transport by hot towers in those early studies should be considered tentative. Indeed, Riehl and Malkus (1958) acknowledged that "lack of [global] data prevents a definitive treatment" and their attempt should be treated as "an initial study proceeding a much more extensive analysis." Presently, with the wide availability of satellite observations of clouds and precipitation, alongside the development of new analysis techniques in recent years (Masunaga 2022; Luo et al. 2022), it is of interest to revisit the subject. The primary objective of this current paper is to reassess the "hot tower" hypothesis using modern satellite data, particularly from geostationary satellites, and hone in on a key component of the hypothesis—convective mass flux in hot towers.

Convective mass flux is a fundamental parameter for characterizing convective clouds. It controls the amount of heating and drying imposed by convection on the environment (Yanai et al. 1973; Arakawa and Schubert 1974). A large number of global climate model (GCM) cumulus parameterization schemes are based on the concept of convective mass flux (Arakawa 2004). Despite its foundational significance, measuring convective mass flux remains a serious challenge, primarily due to the difficulty of obtaining vertical air motion within intense convective clouds. Historically, storm-penetrating aircraft have been used to directly measure the vertical air velocity inside convective cores (Byers and Brahma 1948; LeMone & Zipser 1980a, 1980b; Lucas et al. 1994). To increase spatial coverage and to address safety concerns, high-altitude aircraft have been deployed to fly above convective storms to remotely retrieve convective air motions using Doppler radars (e.g., Heymsfield et al. 2010). Similarly, ground-based Doppler radars and wind profilers have also been utilized to estimate convective vertical velocities (Giangrande et al. 2013, 2016; Kumar et al. 2015; May and Rajopadhyaya 1999; North et al. 2017; Wang et al. 2019, 2020; Williams 2012).

While aircraft and ground-based observations offer valuable insights into convective mass flux, their scope is limited in both spatial extent and temporal duration. Satellite observations capture information on much larger spatial and temporal scales, but no current satellite is designed to measure the vertical air motions inside convective clouds. To fill the vacuum, Masunaga and Luo (2016; hereinafter ML16) introduced a novel, satellite-based approach, which blends information across scales to globally characterize convective mass flux. Initially, this technique was applied to A-Train satellite data (ML16; Jeyaratnam et al. 2020). Comparisons of ML16's mass flux computations with collocated ground-based radar wind profiler observations reveal broad agreement (Jeyaratnam et al. 2020). A-Train observations restrict the sampled cases in the tropics to approximately 1:30 AM and 1:30 PM local time. This temporal sampling does not capture the full diurnal variations in convection, which is particularly pronounced over tropical and subtropical land areas (e.g., Bowman et al. 2005; Liu and Zipser 2008). Moreover, the limited temporal sampling makes it difficult to obtain a comprehensive depiction of cloud life cycle. A solution to these challenges lies in geostationary satellites (GEOs), which offer continuous tracking of convective clouds, especially those in low- and mid-latitudes, from inception to dissipation. In addition, GEO platforms capture diurnal variations by providing high-temporal-resolution global observations.

In this study, we apply the ML16 approach to GEO observations, transitioning it away from A-Train-based retrieval. Section 2 describes the data and methodology, followed by an assessment against earlier estimates utilizing A-Train data and ground-based radar wind profiler observations (Sect. 3). Section 4 revisits the "hot tower" hypothesis discussed in Riehl and Malkus (1958; hereafter RM58) and Riehl (1979; hereafter RS79), focusing on convective mass flux in hot towers, as well as the size and counts of the hot towers in tropical trough zones. Some initial results about diurnal variations are also presented (Sect. 5). Finally, we discuss limitations, future improvements, and potential applications.

2 Data and Methodologies

The primary data used in this work are infrared (IR) brightness temperatures (BTs), precipitation rates, and atmospheric profiles of temperature and moisture. The IR BT and precipitation rates were used to identify hot towers or deep convective cores, while the collocated atmospheric profiles of temperature and moisture were used to drive a convective plume model and compute vertical velocities of hot towers, as part of the ML16 approach. (Refer to Sect. 2.4 for details.) IR BT and vertical temperature profiles are also used to help constrain the plume model outputs toward an optimal estimate of convective mass flux.

2.1 Brightness Temperatures and Precipitation Rates

IR BTs are obtained from the NCEP/CPC's Global Merged IR V1 data (MERGIR) (Janowiak et al. 2017; https://disc.gsfc.nasa.gov/datasets/GPM_MERGIR_1/summary). The MERGIR data combine IR brightness temperatures from all available operational GEO satellites, providing continuous global coverage from 60° S to 60° N with a horizontal resolution of about 4 km and a temporal resolution of 30 min. Precipitation rates are taken from the IMERG dataset, with half-hourly temporal resolution and 0.1°×0.1° spatial resolution (Huffman et al. 2019).

2.2 Ambient Profiles of Temperature and Moisture

Two datasets provided temperature and moisture information in this study. The primary profiles come from ECMWF Reanalysis v5 (ERA5) data (Hersbach et al. 2023) with 37 pressure levels between 1000 and 1 hPa, temporal resolution of one hour, and spatial resolution of 0.25°×0.25°. To test the sensitivity of our results to the ambient profile data, we also use the Modern-Era Retrospective analysis for Research and Applications data (MERRA-2; Global Modeling and Assimilation Office (GMAO), Goddard Earth Sciences Data and Information Services Center (GES DISC), n.d.) in certain sections of the paper and compare results with those obtained using ERA5. MERRA-2 data are provided at 41 pressure levels between 1000 and 0.3 hPa. The temporal and spatial resolutions are 3 h and 0.625°×0.5°, respectively.

2.3 Identifying Hot Towers

Hot towers, or deep convective cores, usually ascend above nearby cirrus anvils, forming a dome-like shape which appear as local minima in IR BT images (Machado and Rossow 1993; Takahashi and Luo 2012, 2014). They produce substantial surface precipitation (Takahashi et al. 2021; Pfister et al. 2022). Consequently, our definition of hot towers involves a combined analysis of IR BTs and surface precipitation rates.

To identify hot towers in GEO IR BT data, we first apply temperature thresholds of 208 K and 215 K to filter for clouds at or above the height of tropical anvils. We chose the 208K threshold partly because it has been used in several studies to isolate cold tropical clouds (Mapes and Houze 1992; Chen et al. 1996). We select the slightly higher temperature of 215 K to test the sensitivity of our vertical mass flux estimates to the choice of brightness temperature threshold, while still restricting to the tall hot towers that RM58/RS79 consider. While these temperatures might seem somewhat subjective, they correspond to the altitude of 13–14 km, typical heights of cirrus anvils in tropical regions (Takahashi and Luo 2012). Hence, towers extending above these heights are likely dome-shaped deep convective cores or hot towers.

Given the cloudy region at or below tropical anvil cloud temperatures, we identify local minima in the filtered scene using the determinant of the Hessian, a metric that quantifies the curvature of two-dimensional images (Lindeberg 1998). This metric provides the concavity of each BT pixel, making it straightforward to isolate local minima. We refine the resulting blobs by approximating each of them as an ellipse (Machado et al. 1998). Figure 5 in the Appendix provides some examples of the hot towers identified.

We then filter the BT local minima to keep only the ones with surface precipitation exceeding criteria for deep convection, following Pfister et al. (2022). Their study analyzed collocated CloudSat/CALIPSO radar/lidar profiles, IR BTs, and surface precipitation rates (from combined radar/microwave/IR TRMM 3B42) and established region-by-region precipitation rate thresholds for tropical deep convection. (Refer to Table 1 in Pfister et al. 2022 for details.)

Table 1 Convective mass flux at 500 hPa averaged within a latitudinal belt extending from the trough to a distance of 10° on the winter side. DJF is December, January, February, and JJA is June, July, August. Units: kg s^{-1}

	DJF	JJA	February	August
ML16 applied to GEO	3.1×10^{11}	2.8×10^{11}	3.4×10^{11}	2.8×10^{11}
RM58	1.8×10^{11}			
RS79	$2.6\text{–}3.0 \times 10^{11}$			

Note that only mass fluxes associated with updrafts are reported for RM58/RS79

2.4 ML16 Applied to Geostationary Satellite Observations

ML16 is a hybrid approach which blends information across scales, leveraging plume model computations, to estimate convective mass flux. Here, we provide a concise overview of the method.

To apply the ML16 method to GEO data, we start by identifying deep convective cores (O(1 km)) in GEO data, as described in Sect. 2.3, and calculating the height of those cores. The cloud-top height is found by matching the cloud-top temperature with the collocated temperature profile from ERA5 or MERRA-2. While the use of IR BT underestimates cloud-top height due to the non-blackbody effect (Minnis et al. 2008; Sherwood et al. 2004), the bias is generally negligible for deep convective cores. Unlike thin cirrus and cirrus anvils whose cloud tops are "fuzzy," deep convective cores contain "packed" cloud tops that emit radiation in the IR, almost like blackbodies. Using collocated CloudSat radar, CALIPSO lidar, and MODIS IR TB observations, Wang et al. (2014) demonstrated that the difference between the physical cloud-top (determined from CloudSat and CALIPSO vertical profiles) and the IR emission level is, on average, approximately 200 m.

For each hot tower or deep convective core identified, the ML16 method is used to compute a vertical velocity (w_c) profile. Only upward motion is quantified in this method, so w_c does not include any contribution from downdrafts. ML16 involves using ambient temperature and moisture profiles representative of the hot tower's environment (O(10 km)–O(100 km)) to drive a convective plume model. (For details on the plume model equations, see Appendix 1.) The model generates a range of in-cloud vertical velocity (w_c) profiles, each corresponding to a prescribed turbulent entrainment rate—the most important, yet unknown, parameter in the model. The plume model outputs are then constrained using a Bayesian weighting procedure, with cloud-top height serving as the primary constraint for plume model simulations, leading to an optimal estimate of the w_c profile. (For details on the plume model equations and implementation flowchart, see Appendix 1.)

Finally, convective mass flux M_c is calculated in a region on the order of O(100 km) (similar to the current GCM grids) as $M_c = \sigma \rho w_c$, where σ represents the fractional coverage of the convective core, defined as the convective core area divided by the grid area, and ρ is the air density. When multiple cores exist within a single grid box, the total convective mass flux is calculated as the sum of all the individual mass fluxes.

3 Comparisons with Previous Studies and Sensitivity Tests

In previous work, estimates of convective mass flux based on A-Train observations were evaluated using collocated ground-based radar wind profiler observations made by the DOE ARM program at Manacapuru, Brazil, during the "Observations and Modeling of the Green Ocean Amazon 2014–2015" (GoAmazon 2014/15; Martin et al. 2016) field campaign. Comparisons reveal broad agreements between ML16's computations of convective vertical velocity and mass flux and the radar wind profiler's estimates (Jeyaratnam et al. 2020). It should be noted that, unlike validating satellite retrieval of precipitation, which can be achieved by matching satellite observations with certain "ground truth" (usually radar or rain gauge observations) pixel-by-pixel, there is no such procedure for evaluating convective mass flux due to lack of "ground truth" observations. A meaningful comparison is only possible in a statistical sense, as is done in Jeyaratnam et al. (2020). Consequently, our estimate of convective mass flux through the ML16 approach should represent the mass flux from an ensemble of convective clouds within a mesoscale region, rather than depicting mass transport by an individual cumulonimbus. This concept is similar to GCM's cumulus parameterization. Following the approach in Jeyaratnam et al. (2020), we compile statistics of convective mass flux based on GEO data at the site of the GoAmazon 2014/15 field campaign.

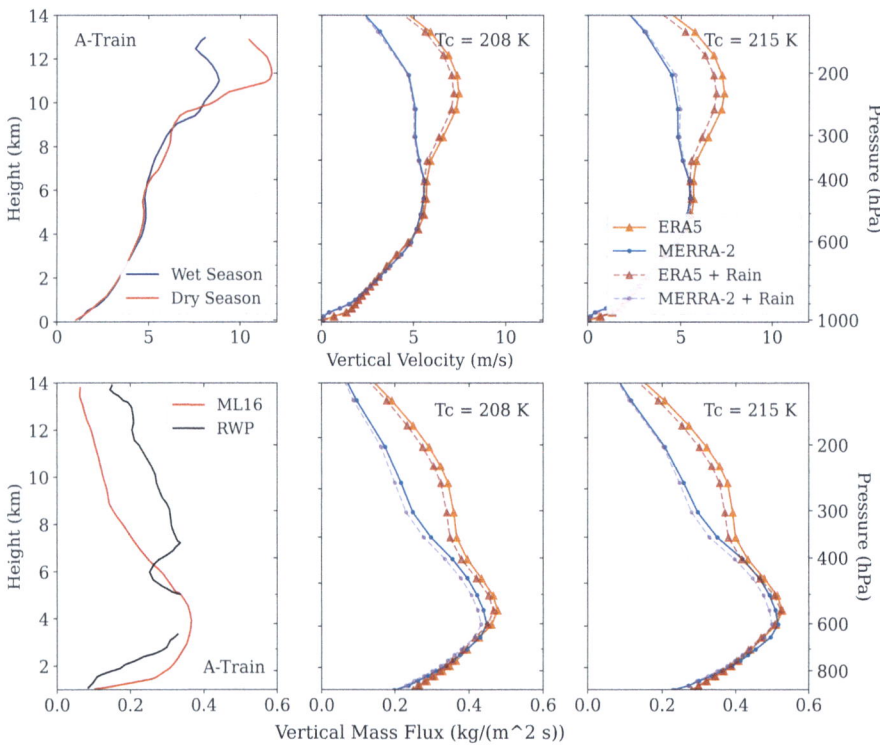

Fig. 1 Mean vertical profiles of convective vertical velocity (w_c; upper panels) and convective mass flux (M_c; lower panels) computed within a 0.5°×0.5° domain surrounding the GoAmazon site from 2014–2015. The left two panels are adopted from Jeyaratnam et al. (2020), showing the estimates from ground-based radar wind profiler (RWP; black) and satellite estimates based on A-Train data (ML16; red). The two right columns show results from the current study. The middle column is for IR BT threshold of 208 K, and the right column is for IR BT threshold of 215 K. Triangles and dots represent results based on ERA5 and MERRA-2, respectively. Solid and dashed lines represent results based on hot tower selections with and without surface rainfall filtering, respectively

Figure 1 shows the mean vertical profiles of w_c and M_c computed within a 0.5°×0.5° domain surrounding the GoAmazon site from 2014 to 2015. These means are derived from scenes when deep convection occurs. The domain analyzed and methodology used were chosen to match the approach of Jeyaratnam et al. (2020). Comparing our current results with Fig. 3 in Jeyaratnam et al. (2020), which is reproduced in the left panels of Fig. 1, we first note that w_c and M_c are broadly comparable in magnitude between the GEO-based estimates and those based on A-Train and radar wind profiler observations.

We tested the sensitivity of the mean vertical velocity and mass flux to choices made in the IR BT threshold, rain filter, and temperature and moisture profile data. The choice of different IR BT thresholds and the inclusion or exclusion of a rain filter influences the mean convective area, resulting in differences of less than 7%. Different ambient profile data affect w_c primarily through the plume model simulations. The main differences occur above 400 hPa, where MERRA-2 produces a w_c that is weaker by approximately 30–40% compared to ERA5. These sensitivity tests provide an estimate of the uncertainties for our GEO-based determination of convective mass flux in tropical hot towers. There is slightly better agreement between ground-based wind profiler observations and results obtained using the IR BT threshold of 208 K. Hence, we use this threshold value in the remainder of the paper.

4 Revisiting the "Hot Tower" Hypothesis

Figure 2 shows the global distribution of convective mass flux in hot towers at 500 hPa for two solstice seasons (DJF and JJA). It should be noted that the mass fluxes depicted here are specifically linked to hot towers as defined in this study, namely, deep convective cores that reach altitudes exceeding 13–14 km. Shallower convection is not included. To create this figure, mass fluxes within grid boxes of size 1°×1° are computed at hourly intervals for each MERGIR image in 2018–2019. The years 2018–2019 were chosen to reduce missing data biases that are present over the West Pacific in 2015–2016. The plots show the mean mass flux across all scenes within a season, including instances where deep convection does not occur. We focus on 500 hPa to facilitate a comparison with RM58 and RS79, as their hot tower statistics were derived from mass flux at this level. Across the globe, the

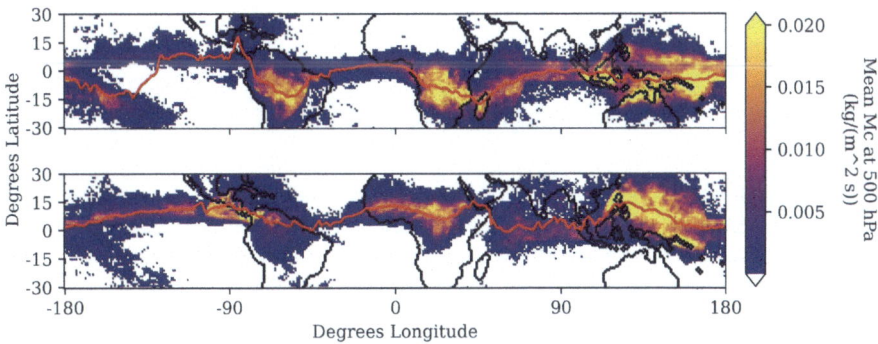

Fig. 2 Global distribution of convective mass flux in hot towers at 500 hPa for two solstice seasons based on ML16-GEO. The upper panel shows DJF, and the lower panel shows JJA. The red lines show the latitude corresponding to the "center of mass" of convective mass flux at each longitude

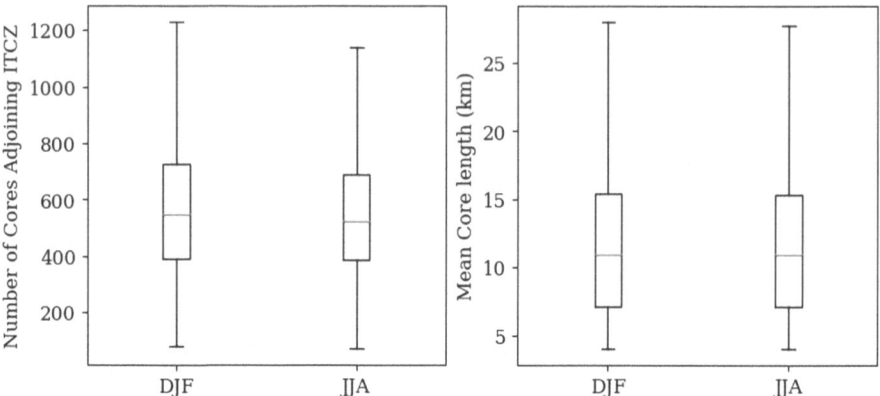

Fig. 3 Box-and-whisker plots showing the distribution of the number of hot towers cores within the tropical trough zone (left) and the core size (right)

geographical distributions of convective mass flux in hot towers closely match the locations where deep convective clouds are prevalent in each season. Prior studies have extensively documented the climatology and variability of these tropical deep convective regions using various satellite-based methodologies, such as the analysis of infrared images (Gettelman et al. 2002; Rossow and Pearl 2007), precipitation radar (Liu et al. 2007), and cloud radar observations (Takahashi et al. 2017), and the subject is being expanded upon in a related paper by Pilewskie et al. (2024). In this study, we introduce convective mass flux within hot towers, contributing to an enhanced understanding of the roles played by tropical deep convection, particularly its impact on heat and moisture budget. As demonstrated by Yanai et al. (1973), the apparent convective heating and drying of hot towers are proportional to convective mass flux.

To compare our estimates with those in RM58 and RS79, we employ a coordinate system that follows the ITCZ. In their studies, RM58 and RS79 used data available at the time to define a constant trough line at about 10° N in northern summer; convective mass flux and hot tower statistics were estimated for a latitudinal belt extending from the trough to a distance of 10° on the winter side (i.e., toward north during the northern hemispheric winter and toward south during the northern hemispheric summer). Following their approach, we define the tropical trough position as the "center of mass" of the convective mass flux for each longitude using the formula:

$$\text{Trough(lon)} = \frac{\sum_{\text{lat}} M_{\text{lat,lon}} \times \text{lat}}{\sum_{\text{lat}} M_{\text{lat,lon}}}$$

Here, Trough (lon) is the latitude (in degrees) of the equatorial trough as a function of longitude, and $M_{\text{lat,lon}}$ is the mean mass flux as a function of latitude and longitude. The red lines in Fig. 2 show the calculated trough lines or Trough (lon). Notably, the trough lines closely follow the meander of the ITCZ. (Note that the term "ITCZ" was not referenced in RM58; this concept might not have been fully established back then due to a lack of a comprehensive global view of the Earth.)

Table 1 shows our GEO-based estimates of the mean convective mass transport within hot towers at 500 hPa for the 10°-belt on the winter side of the tropical trough for the years

2018–2019. We present the total mass transport (units of kg s^{-1}) within the latitudinal belt and not the mass flux (units of kg m^{-2} s^{-1}) because this is the quantity that RM58/RS79 estimated. The selection of hot towers was based on an IR BT threshold of 208 K. Results are presented for two solstice seasons and for the months of February and August. These months (February and August) are chosen to follow the procedure employed in RM58/RS79, which focused on the months during which the position of the equatorial trough reaches its extremes. By doing so, the authors could assume that the ocean heat storage term in their energy budget analysis was negligible, thereby allowing them to neglect this uncertain value. RM58 and RS79 further combined results from February and August to improve the statistics of the sparsely available data at the time. This is acceptable because the winter side of the 10°-trough belt is mostly symmetric for February and August. Moreover, RM58 and RS79 estimated both total convective mass flux and convective mass flux attributed to updrafts, with the latter being more relevant to the ML16 estimates.

According to Table 1, the GEO-based estimate of convective mass transport at 500 hPa (2.8–3.4×10^{11} kg s^{-1}) is similar to the mass transport predicted in RS79 (2.6–3.0×10^{11} kg s^{-1}), although it lies on the upper end of the predicted range of values. Because RS79 presents amended values based on satellite data available at the time, we consider this value to likely be closer to the true mean mass transport in the ITCZ than RM58 (which gives 1.8×10^{11} kg s^{-1}). Considering the ML16 and RM58/RS79 methods represent fundamentally different approaches to estimating convective mass flux, it is remarkable that both methods lead to similar values of mass transport. The RM58/RS79 method implements a "top-down" approach, where constraints on energy balance define vertical motion. The estimation of total vertical mass transport in the tropical trough zone is derived indirectly. In contrast, the ML16 method, applied to GEO data, functions as a "bottom-up" approach. It utilizes satellite observations of clouds and precipitation to define the locations and sizes of hot towers; plume model simulations driven by ambient profiles and constrained by satellite observations provide their vertical velocities, which are directly used to estimate convective mass flux. Both methods possess their intrinsic uncertainties. The main source of uncertainty in the RM58/RS79 approach lies in the accuracy of radiative fluxes within and at the top of the atmosphere, as well as surface latent and sensible heat fluxes. Knowledge of these energy fluxes at a global scale was understandably limited during the 1950s and even the 1970s. For example, the top-of-the-atmosphere radiative flux used in RM58 was only about 50% of its modern value (Hartmann 2016). Conversely, the ML16 approach depends on the proper identification of convective cores and accurate estimates of their vertical velocities.

In Sect. 3, we assessed GEO-based estimates by comparing them with those presented in Jeyaratnam et al. (2020) at the GoAmazon site. Our findings indicate that at 500 hPa, our estimates of convective mass flux are around 10–20% larger. If we interpret this as the overall bias assessment of the GEO-based estimate of convective mass flux, it suggests that convective mass flux in hot towers is slightly overestimated in our study. Nevertheless, the absence of global direct measurements of vertical air motions inside convective clouds hinders our ability to draw a definitive conclusion. Future Earth Observation missions with a focus on cloud dynamics are expected to provide more accurate global observations to revisit and refine our understanding of this subject.

A difference between ML16 method and the hot tower hypothesis comes from the different ways that entrainment is treated. In RM58/RS79, the authors emphasize that cloud towers are undilute. Conversely, the ML16 method solves an entraining and detraining plume model, which means that all results are presented for dilute cloud towers. Entrainment of ambient air impacts cumulus growth and vertical motion, so there will inevitably be some differences between the mass transport estimated by the ML16 method and RM58/RS79.

Another potential source of uncertainty arises from our reliance on GEO-based local minima in IR BTs to detect the presence of hot towers above the anvils. It is possible that the convective core within the cloud, where maximum mass flux occurs, might have a different size than the section rising above the anvils. Updrafts may be smaller than the overshooting cloud area, which could explain why our estimate of vertical mass transport is skewed to the upper end of RS79's estimate. Depending on observations of clouds to make inferences about updrafts may lead to a systematic bias. Vertical profiles of the convective cores obtained through radar measurements would provide insight into this potential uncertainty. In this regard, several studies based on CloudSat radar data (Takahashi et al. 2017, 2023 and Pilewskie et al. 2024) consistently offer estimates of hot tower size that are similar to the GEO-based results presented here. This consistency suggests that utilizing above-anvil IR BT features provides a reasonable estimate of the size of hot towers. In our ongoing research, we use a large-eddy simulation in an Observing System Simulation Experiment (OSSE) framework to further investigate this potential uncertainty.

Figure 3 presents the statistics for the number of hot towers within the tropical trough zone and their sizes. Following RM58 and RS79, the hot tower size is defined as the square root of the hot tower area. Figure 3 shows a median count of approximately 550 hot towers and a median size of around 11 km. In comparison, RM58 and RS79 provided an estimate of 1500–5000 and 1600–2400 hot towers, respectively, each with a fixed length of 5 km. As mentioned earlier, RM58 and RS79 employed a "top-down" approach, deducing the number of hot towers entirely from their mass flux estimate and a provisional value for hot tower size (5 km), without using any global cloud observations, due to their unavailability at that time. Both the provisional hot tower size and the assumption that the size stays fixed greatly influence the number of towers that Riehl and Simpson estimate. Our GEO-based results (not shown) reveal that the distribution of convective cores sizes follows a power-law, in line with data from aircraft measurements (LeMone and Zipser 1980a, b). The parameters in the power-law remain unchanged with temperature threshold and season. By assuming that the total area of convective cores within the ITCZ amounts to 0.1% of the area of the ITCZ, the integral of the convective core size distribution can be constrained. Using the constrained size distribution to compute the number of convective cores, we find a value of roughly 600, matching our direct measurements. If, however, we assume that all convective cores have a fixed area of 25 km^2, we find around 1800 cores, similar to what Riehl and Simpson found. Therefore, the factor of 3 difference between the number of cores that we report and the one found in RM58/RS79 could be explained by the assumption about the size distribution of convective cores. Because the ML16 approach leverages satellite observations, more credibility should be attributed to the results derived from this method. It directly observes and derives statistics from cloud observations, providing a more robust foundation for the analysis.

It should be noted that the GEO-based analysis of hot tower counts and sizes in this study comes with inherent biases and uncertainties. One factor contributing to these biases and uncertainties is the resolution of the GEO data, which is 4 km. This resolution may potentially result in an overestimation of hot tower sizes, as any core smaller than 4 km would be assumed to be 4 km. There is some evidence from aircraft-based measurements suggesting that the cores of many tropical deep convective clouds are smaller than our estimated value of 11 km (LeMone and Zipser 1980a, b; Heymsfield et al. 2010). However, aircraft-based measurements likely introduce a bias toward weaker convection due to safety concerns, and weaker convection generally features smaller cores, as indicated by multiple years of CloudSat radar observations (Takahashi et al. 2017). Surveys conducted by space-borne cloud radar suggest that the median diameter of deep convection cores is approximately 10–15 km (Takahashi et al. 2017, 2023). A related paper on hot towers by Pilewskie et al. (2024), which primarily used CloudSat radar observations, also arrived at an estimate of hot tower size of 11 km.

5 Diurnal Variations

An important advantage of GEO satellites over polar-orbiting counterparts is their capability to observe the complete diurnal cycle. We present a brief exploration of diurnal variations in convective mass flux. In Fig. 4, the two-dimensional histogram illustrates diurnal changes in convective mass flux at 500 hPa over land and ocean, with overlaid curves indicating mean values. These plots show the diurnal cycle of convective mass flux when convection occurs, omitting all grid boxes that do not contain convective cores. Across the ocean, there is only a marginal diurnal change in convective mass flux, with a minor peak in the early morning. Over land, the diurnal cycle is prominent, showing well-defined minima in the early morning and maxima in the late afternoon. The mean convective mass flux over land is approximately twice as much during the late afternoon compared to the early morning.

These diurnal variations in convective mass flux generally follow the diurnal pattern of convective cloud activities, which have been well documented in prior studies analyzing IR data (Hendon and Woodberry 1993; Machado and Rossow 1993) and TRMM precipitation radar observations (Bowman et al. 2005; Liu and Zipser 2008). However, to our knowledge, this is the first explicit estimation of convective mass flux throughout the entire diurnal cycle. Considering that convective mass flux is a critical parameter controlling the impact of convection on heat and moisture budgets, detailing its diurnal variations is an important initial step toward examining how convection affects the surrounding environment on a diurnal time scale. It is important to note that Fig. 4 represents the average across the whole tropics. The diurnal variation in tropical convection exhibits substantial regional differences, which we intend to investigate in more detail in a future study.

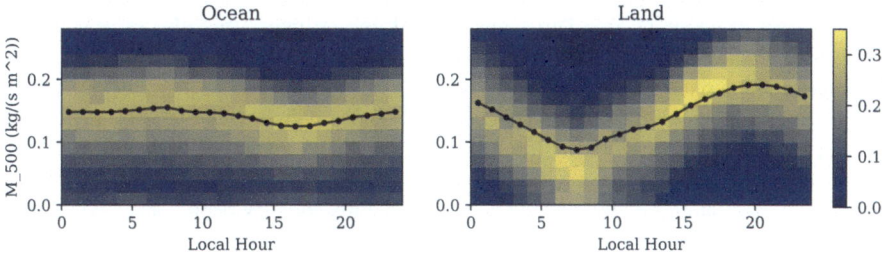

Fig. 4 Distribution of convective mass flux at 500 hPa as a function of local time. Color represents relative occurrence frequency (normalized to 1 at each local hour), and the black lines show the mean. Statistics are collected separately for the ocean (left) and the land (right)

Figure A.1 Flowchart of the ML16 method applied to GEO data. (**a**) Convective cores are identified as local minima in brightness temperature in geostationary satellite images. (**b**) Atmospheric sounding information, which is collocated with the convective cores, drives a plume model. (**c**) The plume model outputs a range of possible buoyancy profiles for a range of entrainment rates. (**d**) Using a Bayesian weighting procedure, we select the plume that best models the observed cloud-top height and buoyancy. (**e**) Identify the best-fit vertical velocity profile by the same Bayesian weight as selected from step (**d**)

6 Summary and Discussions

The "hot tower" hypothesis is a crucial concept in tropical meteorology that was originally introduced by RM58 and RS79. It explains how heat is transported from Earth's surface to the upper troposphere in the tropical trough zone, bypassing the mid-tropospheric minimum of total energy. RM58 and RS79 deduced mass fluxes associated with the hot towers and estimated their total numbers based on constraints of energy balance, a remarkable achievement at a time when no global observations of clouds were available. Nevertheless, these estimates should be considered tentative, as cautioned by the authors. This study aims to reevaluate the "hot tower" hypothesis, focusing on convective mass flux, using geostationary (GEO) satellite data and an innovative analysis technique, known as ML16. This satellite-based method integrates various data sources, including hot tower heights, ambient profiles, and a plume model, to determine convective mass flux globally.

The ML16 approach applied to A-Train data was previously evaluated against collocated ground-based radar wind profiler observations during the GoAmazon field campaign, showing overall favorable comparisons. This study expands on the evaluation by comparing the GEO-based implementation of the ML16 method with earlier estimates at the same GoAmazon site. Both vertical velocities and convective mass fluxes are broadly comparable in magnitude among these different estimates. The use of GEO-based observations provides an opportunity to investigate the entire diurnal cycle of hot towers, briefly explored in this paper, with a more comprehensive analysis reserved for a future study. Sensitivity tests are conducted to quantify the uncertainties arising from choices regarding IR BT thresholds, rain filter, and ambient profile data. The largest uncertainty occurs above 400 hPa, where MERRA-2 produces vertical velocities approximately 30–40% weaker compared to those from ERA5.

In accordance with RM58 and RS79, we define the tropical trough zone as a latitudinal belt extending from the trough line to a distance of 10° on the winter side. Comparisons show that our GEO-based estimate of convective mass flux for hot towers within this tropical trough zone, which ranges between 2.8×10^{11} kg s^{-1} and 3.4×10^{11} kg s^{-1}, is similar to the estimate of RS79, which was around 2.6–3.0×10^{11} kg s^{-1}. Meanwhile, our analysis gives a median count of approximately 550 hot towers within the tropical trough zone with a median size of about 11 km. This contrasts with the estimates provided by RM58 and RS79, which indicated a range of 1500–5000 and 1600–2400 hot towers, respectively, each characterized by a fixed size of 5 km.

It is crucial to highlight the fundamental differences between the ML16 and RM58/RS79 approaches in characterizing tropical hot towers. The RM58/RS79 determines convective mass flux of the hot towers based on energy balance constraints, whereas the ML16 method employs satellite cloud observations to explicitly identify deep convective towers and then utilizes plume model computations, constrained by observations, to derive their vertical velocities and mass fluxes. The RM58/RS79 method takes a "top-down," indirect approach in characterizing hot towers, while the ML16 method adopts a "bottom-up," direct approach. Concerning the counts and sizes of hot towers, more credibility should be

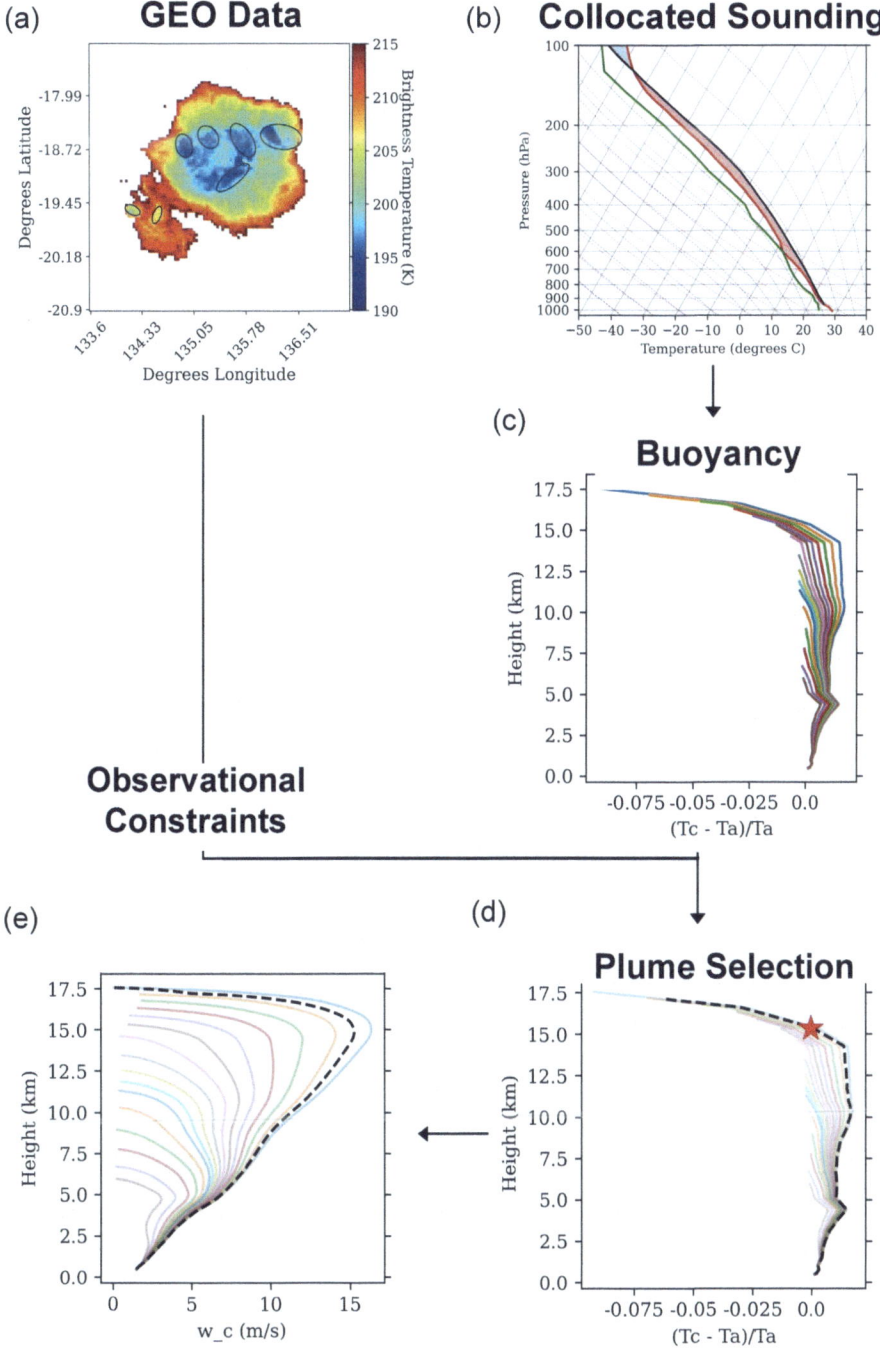

Table 2 Variable definitions for the plume model equations

Symbol	Variable name
B	Buoyancy, $g\left(\frac{(T_{vc}-T_{va})}{T_{va}} - q_w\right)$
dT_i	7 °C
e_s	Saturated vapor pressure
g	Gravitational acceleration, 9.81 m s^{-2}
H	Heaviside theta function
h_a	Ambient moist static energy
h_c	In-cloud moist static energy
L_i	Latent heat of freezing
M	Vertical mass flux, $\sigma \rho w_c$
p	Pressure
q_i	Ice mixing ratio
q_{vc}	In-cloud saturated water vapor mixing ratio
q_{va}	Ambient saturated water vapor mixing ratio
q_w	Ambient water vapor mixing ratio
$q_{w,crit}$	10^{-3} kg/kg
$T_{0,i}$	-20 °C
T_c	In-cloud temperature
T_{vc}	In-cloud virtual temperature
T_{va}	Ambient virtual temperature
w_c	In-cloud vertical velocity
ϵ	Entrainment rate, the sum of turbulent and dynamic components, $\epsilon_T + \epsilon_D$
ε	$\frac{R_d}{R_v} \approx 0.622$
ρ	Density
τ_{auto}	10^3 s

placed on results derived from direct satellite observations. While the use of GEO IR data may introduce some uncertainties, comparisons with prior surveys of tropical deep convective cores conducted by space-borne cloud radar show broad consistency, suggesting that the GEO-based analysis is reliable.

An aspect of RM58/RS79 that is missing from our analysis is downdrafts. The plume model we solve does not include downdrafts and instead represents convection as a cylinder of upward moving air, a model similar to what RM58/RS79 proposed in their hot tower hypothesis. In reality, convection is made up of updrafts and downdrafts at scales similar to or smaller than the convective cores we observe. It is likely that the hot towers that we observe in GEO contain some fraction of downdrafts, leading us to overestimate convective area coverage. While the original ML16 paper provides a way to estimate downdrafts as the residual between total mass flux and updraft mass flux, we focus only on updrafts in this study because the hot tower hypothesis is mainly concerned with upward moving air.

While both the ML16 approach using GEO data and RS79 give similar estimates of convective mass flux associated with updrafts, the lack of direct measurements of vertical air motions inside convective clouds leaves room for uncertainties and hampers our ability to draw a definitive conclusion. Nevertheless, our current study represents a significant first step toward utilizing modern-day satellite observations, along with a new analysis technique, to quantify the role played by tropical hot towers in mass transport. This is a crucial issue, as convective mass flux controls the effects of convective clouds on energy and water budgets. Looking ahead, it is anticipated that future Earth Observation mission, particularly those focused on cloud dynamics and vertical mass fluxes (e.g., EarthCARE, Illingworth et al. 2015; INCUS, van den Heever et al. 2023; C2OMODO, Brogniez et al. 2022; and future AOS, aos.gsfc.nasa.gov), will be instrumental in providing more accurate data to address this question.

Appendix 1: Numerical Method and Implementation Flowchart

The ML16 method combines remote sensing observations with plume model computations to estimate the vertical velocity and convective mass flux in a procedure shown on the flowchart in Fig. A1. Broadly, the method involves identifying convective cores in geostationary satellite data as local minima in brightness temperature, solving a one-dimensional plume model for a range of entrainment rates, and finally constraining the plume model outputs using the observed cloud-top height and cloud-top buoyancy.

The system of differential equations that make up the plume model used in the ML16 method is summarized in Eqs. (1)–(7). Table2 defines all of the variables used in these equations. For clarity, z-dependence of variables has been explicitly written except where it is obvious. These equations provide the vertical rate of change of the entrainment, vertical velocity, temperature, and water vapor and water/ice mixing ratios within a cloud. It assumes all cloud properties are horizontally homogeneous.

We solve this system of equations by initializing each function at the lifting condensation level (LCL) and stepping forward in height using a standard finite difference method. Ambient temperature, moisture, and geopotential profiles are used to compute the ambient moist static energy. At the LCL, the in-cloud temperature, density, and relative humidity are assumed to be equal to the ambient ones at the same height. The initial vertical velocity is 1.5 m/s. The model assumes that the pressure difference between the environment and the cloud is small, so the ambient pressure is used wherever pressure is referenced. All of the observables on the right side of the equations are known (from either the initial conditions or the previous integration step) except for $\frac{dT_c}{dz}$, the in-cloud temperature lapse rate. We deal with this numerically, guessing an initial value of $\frac{dT_c}{dz} = -0.1$ K km^{-1} and checking for consistency with the rest of the equations, that is, we use this preliminary lapse rate to compute the in-cloud MSE and T_c one step above the current height, checking whether our initial guess of $\frac{dT_c}{dz}$ agrees with the finite difference derivative of T_c. If the difference between the finite difference derivative and the initial guess is smaller than some tolerance, then we accept this value and move on. If not, we update the guess with the finite difference derivative and try again until the value converges. When convergence is reached, $\frac{dT_c}{dz}$ follows moist adiabat, as expected for a saturated environment inside clouds.

The system of equations is solved for turbulent entrainment rates ranging from 0 to 0.4 km^{-1} to obtain a range of possible plumes, following ML16. The optimal plume is the one with cloud-top height and cloud-top buoyancy closest to one obtained from GEO observations. We estimate the cloud-top buoyancy using a mean climatological buoyancy profile, which defines the cloud-top buoyancy as a function of cloud-top height.

The equations presented here are written in a form somewhat different from the ones originally presented in ML16. We have combined and reformulated expressions to rewrite the total entrainment rate for the $\frac{dM}{dz} > 0$ case in terms of the buoyancy, density, and vertical velocity. This step offers an advantage because if we assume that $\frac{dM}{dz} > 0$ at the LCL, a reasonable assumption for the deep convective clouds that we model, we can compute the total entrainment rate directly, simplifying the numerical procedure (Table 2 and Fig. A1).

$$\frac{1}{M(z)}\frac{dM}{dz} = f(z) - \delta(z) \tag{1}$$

$$\frac{dw_c}{dz} = \frac{1}{w_c(z)}\left(a_B B(z) - f(z)\left(w_c(z)\right)^2\right) \tag{2}$$

$$\varepsilon(Z) = \begin{cases} \frac{1}{2}\left(\epsilon_T + \frac{a_B B(z)}{w_c(z)^2} + \frac{1}{\rho(z)}\frac{d\rho}{dz}\right) \text{for } \frac{dM}{dz} \geq 0 \\ \epsilon_T \text{for } \frac{dM}{dz} < 0 \end{cases} \tag{3}$$

$$\frac{dq_{vc}}{dz} = \varepsilon\left(\frac{1}{p(z)}\frac{de_s}{dT_c}\frac{dT_c}{dz} - e_s(T_c)\frac{1}{p(z)^2}\frac{dp}{dz}\right) \tag{4}$$

$$\frac{dq_i}{dz} = \frac{1}{2}\left(\frac{dq_w}{dz}\left(1 - \tanh\left(\frac{T_c(z) - T_{0i}}{dT_i}\right)\right) + \frac{w_c(z)}{dT_i}\text{sech}^2\left(\frac{T_c(z) - T_{0i}}{dT_i}\right)\frac{dT_c}{dz}\right) \tag{5}$$

$$\frac{dq_w}{dz} = -\varepsilon(z)q_w(z) - \left(\frac{dq_{vc}}{dz} + \varepsilon(z)\left(q_{vc}(z) - q_{va}(z)\right)\right) \\ - \frac{1}{\tau_{auto}w_c(z)}\left(q_w(z) - q_{w,crit}\right)H\left(q_w(z) - q_{w,crit}\right) \tag{6}$$

$$\frac{dh_c}{dz} = L_i\frac{dq_i}{dz} - \varepsilon\left(h_c(z) - L_i q_i(z) - h_a(z)\right) \tag{7}$$

Acknowledgments The authors would like to acknowledge funding support from NASA Grants 80NSSC18K1600 and 80NSSC23K0116 awarded to CUNY, as well as funding support from INCUS project under Grant 80LARC22DA011. The authors are grateful to Drs. William B. Rossow and Graeme L. Stephens for constructive comments and suggestions. This paper is an outcome of the Workshop "Challenges in Understanding the Global Water Energy Cycle and its Changes in Response to Greenhouse Gas Emissions" held at the International Space Science Institute (ISSI) in Bern, Switzerland (September 26–30, 2022).

Declarations

Conflict of interest The authors have no conflicts of interest to declare that are relevant to the content of this article.

Open Access This article is licensed under a Creative Commons Attribution-NonCommercial-NoDerivatives 4.0 International License, which permits any non-commercial use, sharing, distribution and reproduction in any medium or format, as long as you give appropriate credit to the original author(s) and the source, provide a link to the Creative Commons licence, and indicate if you modified the licensed material. You do not have permission under this licence to share adapted material derived from this article or parts of it. The images or other third party material in this article are included in the article's Creative Commons licence, unless indicated otherwise in a credit line to the material. If material is not included in the article's Creative Commons licence and your intended use is not permitted by statutory regulation or exceeds the permitted use, you will need to obtain permission directly from the copyright holder. To view a copy of this licence, visit http://creativecommons.org/licenses/by-nc-nd/4.0/.

References

Arakawa A, Schubert WH (1974) Interaction of a cumulus cloud ensemble with the large-scale environment, part I. J Atmos Sci 31:674–701
Bowman KP, Collier JC, North GR, Wu Q, Ha E, Hardin J (2005) Diurnal cycle of tropical precipitation in tropical rainfall measuring mission (TRMM) satellite and ocean buoy rain gauge data. J Geophys Res 110:D21104. https://doi.org/10.1029/2005JD005763
Brogniez H, Roca R, Auguste F, Chaboureau J-P, Haddad Z, Munchak SJ, Li X, Bouniol D, Dépée A, Fiolleau T, Kollias P (2022) Time-delayed tandem microwave observations of tropical deep convection: overview of the C2OMODO mission. Front Remote Sens 3:854735. https://doi.org/10.3389/frsen.2022.854735
Byers HR, Brahma RR Jr (1948) Thunderstorm structure and circulation. J Meteorol 5:71–86
Chen S, Houze R, Mapes B (1996) Multiscale variability of deep convection in relation to large-scale circulation in TOGA COARE. J Atmos Sci 53:1380–1409
Fierro AO, Simpson JM, Lemone MA, Straka JM, Smull BF (2009) On how hot towers fuel the Hadley cell: an observational and modelling study of line-organized convection in the equatorial trough from TOGA COARE. J Atmos Sci 66:2730–2746. https://doi.org/10.1175/2009JAS3017.1
Gettelman A, Salby ML, Sassi F (2002) Distribution and influence of convection in the tropical tropopause region. J Geophys Res 107(D10):4080. https://doi.org/10.1029/2001JD001048
Giangrande SE, Collis S, Straka J, Protat A, Williams C, Krueger S (2013) A summary of convective-core vertical velocity properties using ARM UHF wind profilers in Oklahoma. J Appl Meteorol Climatol 52(10):2278–2295
Giangrande SE, Toto T, Jensen MP, Bartholomew MJ, Feng Z, Protat A, Williams CR, Schumacher C, Machado L (2016) Convective cloud vertical velocity and mass-flux characteristics from radar wind profiler observations during GoAmazon2014/5. J Geophys Res: Atmos 121(21):12–891. https://doi.org/10.1002/2016JD025303
Instantaneous 3-Hourly. https Global modeling and assimilation office (GMAO), Goddard Earth Sciences Data and Information Services Center (GES DISC) n.d. MERRA-2 inst3_3d_asm_Cp, MERRA-2 3d IAU State, Meteorology: https://doi.org/10.5067/9SC1VNTWGWV3
Hartmann DL (2016) Global physical climatology, Elsevier Inc. pp 485
Hendon HH, Woodberry K (1993) The diurnal cycle of tropical convection. J Geophys Res: Atmos 98(D9):16623–16637
Hersbach H, Bell B, Berrisford P, Biavati G, Horányi A, Muñoz Sabater J, Nicolas J, Peubey C, Radu R, Rozum I, Schepers D, Simmons A, Soci C, Dee D, Thépaut JN (2023) ERA5 hourly data on pressure levels from 1940 to present. Copernicus Climate Change Service (C3S) Climate Data Store (CDS). https://doi.org/10.24381/cds.bd0915c6 (Accessed on 01–08–2023)
Heymsfield GM, Tian L, Heymsfield AJ, Li L, Guimond S (2010) Characteristics of deep tropical and subtropical convection from nadir-viewing high-altitude airborne Doppler radar. J Atmos Sci 67:285–308
Huffman GJ, Stocker EF, Bolvin DT, Nelkin EJ, Jackson Tan (2019) GPM IMERG final precipitation L3 half hourly 0.1 degree x 0.1 degree V06, Greenbelt, MD, Goddard Earth Sciences Data and Information Services Center (GES DISC). Accessed: 2023. 10.5067/GPM/IMERG/3B-HH/06
Illingworth AJ, Barker HW, Beljaars A, Ceccaldi M, Chepfer H, Clerbaux N, Cole J, Delanoë J, Domenech C, Donovan DP, Fukuda S (2015) The EarthCARE Satellite: the next step forward in global measurements of clouds, aerosols, precipitation, and radiation. Bull Am Meteorol Soc 96(8):1311–1332

Janowiak J, Joyce B, Xie P (2017) NCEP/CPC L3 half hourly 4km global (60S-60N) merged IR V1, Edited by Andrey Savtchenko, Greenbelt, MD, Goddard Earth Sciences Data and Information Services Center (GES DISC). Accessed: 2023, 10.5067/P4HZB9N27EKU

Jeyaratnam J, Luo ZJ, Giangrande SE, Wang D, Masuanga H (2020) A satellite-based estimate of convective vertical velocity and convective mass flux: global survey and comparison with radar wind profiler observations. Geophys Res Lett. https://doi.org/10.1029/2020GL090675

Kumar VV, Jakob C, Protat A, Williams CR, May PT (2015) Mass-flux characteristics of tropical cumulus clouds from wind profiler observations at Darwin, Australia. J Atmos Sci 72:1837–1855. https://doi.org/10.1175/JAS-D-14-0259.1

LeMone MA, Zipser EJ (1980a) Cumulonimbus vertical velocity events in GATE. Part I: diameter, intensity, and mass flux. J Atmos Sci 37:2444–2457

LeMone MA, Zipser EJ (1980b) Cumulonimbus vertical velocity events in GATE. Part II: synthesis and model core structure. J Atmos Sci 37:2458–2469

Lindeberg T (1998) Feature detection with automatic scale selection. Int J Comput Vision 30(2):79–116. https://doi.org/10.1023/A:1008045108935

Liu C, Zipser EJ (2008) Diurnal cycles of precipitation, clouds, and lightning in the tropics from 9 years of TRMM observations. Geophys Res Lett 35:L04819. https://doi.org/10.1029/2007GL032437

Liu C, Zipser EJ, Nesbitt SW (2007) Global distribution of tropical deep convection: different perspectives from TRMM infrared and radar data. J Clim 20:489–503

Lucas C, Zipser EJ, LeMone MA (1994) Vertical velocity in oceanic convection off tropical Australia. J Atmos Sci 51:3183–3193

Luo ZJ, Tselioudis G, and Rossow WB (Eds) (2022) Studies of cloud, convection and precipitation processes using satellite observations. Lectures in Climate Change: Volume 3. World Scientific. https://doi.org/10.1142/12862

Machado LAT, Rossow WB (1993) Structural characteristics and radiative properties of tropical cloud clusters. Mon Wea Rev 121:3234–3260

Machado LAT, Rossow WB, Guedes RL, Walker AW (1998) Life cycle variations of mesoscale convective systems over the Americas. Mon Weather Rev 126:1630–1654. https://doi.org/10.1175/1520-0493(1998)126%3c1630:LCVOMC%3e2.0.CO;2

Mapes B, Houze RA Jr (1992) An integrated view of the 1987 Australian monsoon and its mesoscale convective systems. I: horizontal structure. Quart J Royal Meteorol Soc 118(507):927–963

Martin ST, Artaxo P, Machado LAT, Manzi AO, Souza RAF, Schumacher C, Wang J, Andreae MO, Barbosa HMJ, Fan J, Fisch G, Goldstein AH, Guenther A, Jimenez JL, Pöschl U, Silva Dias MA, Smith JN, Wendisch, M (2016) Introduction: Observations and Modeling of the Green Ocean Amazon (GoAmazon2014/5). Atmospheric Chemistry and Physics, 16(8):4785–4797. https://doi.org/10.5194/acp-16-4785-2016

Masunaga H, Luo ZJ (2016) Convective and large-scale mass flux profiles over tropical oceans determined from synergistic analysis of a suite of satellite observations. J Geophys Res: Atmos 121(13):7958–7974. https://doi.org/10.1002/2016JD024753

Masunaga H (2022) Satellite measurements of clouds and precipitation, theoretical basis, springer remote sensing/photogrammetry. Springer, Singapore, pp 297

May PT, Rajopadhyaya DK (1999) Wind profiler observations of vertical motion and precipitation microphysics of a tropical squall line. Mon Wea Rev 124(4):621–633

Minnis P, Yost CR, Sun-Mack S, Chen Y (2008) Estimating the top altitude of optically thick ice clouds from thermal infrared satellite observations using CALIPSO data. Geophys Res Lett 35:L12801. https://doi.org/10.1029/2008GL033947

North KW, Oue M, Kollias P, Giangrande SE, Collis SM, Potvin CK (2017) Vertical air motion retrievals in deep convective clouds using the ARM scanning radar network in Oklahoma during MC3E. Atmos Meas Tech 10:2785–2806. https://doi.org/10.5194/amt-10-2785-2017

Pfister L, Ueyama R, Jensen EJ, Schoeberl MR (2022) Deep convective cloud top altitudes at high temporal and spatial resolution. Earth Space Sci 9(11):e2022EA002475. https://doi.org/10.1029/2022EA002475

Pilewskie J, Stephens G, L'Ecuyer T and Takahashi H (2024) A multi-satellite perspective on 869 "hot tower" characteristics in the equatorial trough zone, submitted to Surv of Geophys

Riehl H (1979) The heat balance of the equatorial trough zone, revisited. Beitr Phys Atmos 52:287–305

Riehl H, Malkus JS (1958) On the heat balance in the equatorial trough zone. Geophysica 6:503–537

Rossow WB, Pearl C (2007) 22-Year survey of tropical convection penetrating into the lower stratosphere. Geophys Res Lett 34:L04803. https://doi.org/10.1029/2006GL028635

Sherwood SC, Chae J-H, Minnis P, McGill M (2004) Underestimation of deep convective cloud tops by thermal imagery. Geophys Res Lett 31:L11102. https://doi.org/10.1029/2004GL019699

Takahashi H, Luo Z (2012) Where is the level of neutral buoyancy for deep convection? Geophys Res Lett 39:L15809. https://doi.org/10.1029/2012GL052638

Takahashi H, Luo ZJ (2014) Characterizing tropical overshooting deep convection from joint analysis of CloudSat and geostationary satellite observations. J Geophys Res Atmos 119:112–121. https://doi.org/10.1002/2013JD020972

Takahashi H, Luo ZJ, Stephens GL (2017) Level of neutral buoyancy, deep convective outflow, and convective core: new perspectives based on 5 years of CloudSat data. J Geophys Res: Atmos 122(5):2958–2969. https://doi.org/10.1002/2016JD025969

Takahashi H, Lebsock M, Luo ZJ, Masunaga H, Wang C (2021) Detection and tracking of tropical convective storms based on globally gridded precipitation measurements: algorithm and survey over the tropics. J Appl Meteorol Climatol 60(3):403–421. https://doi.org/10.1175/JAMC-D-20-0171.1

Takahashi H, Luo ZJ, Stephens G, Mulholland JP (2023) Revisiting the land-ocean contrasts in deep convective cloud intensity using global satellite observations. Geophy Res Lett 50(5):e2022GL102089. https://doi.org/10.1029/2022GL102089

van den Heever S, Haddad Z, Dolan B, Freeman S, Grant L, Kollias P, Leung G, Luo J, Marinescu P, Posselt D, Rasmussen K, Sai P, Schulte R, Stephens G, Storer R, Takahashi H (2023) Tropical Convection through the Lens of the INCUS Mission (EGU23–11285). EGU23. Copernicus Meetings.https://doi.org/10.5194/egusphere-egu23-11285

Wang C, Luo ZJ, Chen X, Zeng X, Tao W-K, Huang X (2014) A physically based algorithm for non-blackbody correction of cloud-top temperature and application to convection study. J Appl Meteor Climatol 53:1844–1857. https://doi.org/10.1175/JAMC-D-13-0331.1

Wang D, Giangrande SE, Schiro K, Jensen MP, Houze RA (2019) The characteristics of tropical and midlatitude mesoscale convective systems as revealed by radar wind profilers. J Geophys Res: Atmos 124:4601–4619. https://doi.org/10.1029/2018JD030087

Wang D, Giangrande SE, Feng Z, Hardin JC, Prein AF (2020) Updraft and downdraft core size and intensity as revealed by radar wind profilers: MCS observations and idealized model comparisons. J Geophys Res: Atmos 125(11):e2019JD031774. https://doi.org/10.1029/2019JD031774

Williams CR (2012) Vertical air motion retrieved from dual-frequency profiler observations. J Atmos Oceanic Technol 29:1471–1480. https://doi.org/10.1175/JTECH-D-11-00176.1

Yanai M, Esbensen S, Chu J-H (1973) Determination of bulk properties of tropical cloud clusters from large-scale heat and moisture budgets. J Atmos Sci 30:611–627

Zipser EJ (2003) Some views on "hot towers" after 50 years of tropical field programs and two years of TRMM data. Meteorol Monogr 29:49

Publisher's Note Springer Nature remains neutral with regard to jurisdictional claims in published maps and institutional affiliations.

Surveys in Geophysics (2024) 45:1979–1998
https://doi.org/10.1007/s10712-024-09862-8

METEOSAT Long-Term Observations Reveal Changes in Convective Organization Over Tropical Africa and Atlantic Ocean

Rémy Roca[1] · Thomas Fiolleau[1] · Viju O. John[2] · Jörg Schulz[2]

Received: 24 October 2023 / Accepted: 5 September 2024 / Published online: 4 October 2024
© The Author(s) 2024

Abstract

In the tropics, deep convection, which is often organized into convective systems, plays a crucial role in the water and energy cycles by significantly contributing to surface precipitation and forming upper-level ice clouds. The arrangement of these deep convective systems, as well as their individual properties, has recently been recognized as a key feature of the tropical climate. Using data from Africa and the tropical Atlantic Ocean as a case study, recent shifts in convective organization have been analyzed through a well-curated, unique record of METEOSAT observations spanning four decades. The findings indicate a significant shift in the occurrence of deep convective systems, characterized by a decrease in large, short-lived systems and an increase in smaller, longer-lived ones. This shift, combined with a nearly constant deep cloud fraction over the same period, highlights a notable change in convective organization. These new observational insights are valuable for refining emerging kilometer-scale climate models that accurately represent individual convective systems but struggle to realistically simulate their overall arrangement.

Keywords Deep convective systems · Satellite observations · Climate change · Africa

Article Highlights

- Long-term observations of deep convective cloud systems over continental Africa and the Atlantic Ocean reveal a robust shift in the frequency of these systems' occurrences
- Relatively small and long-lived systems tend to increase over the last three decades
- Relatively large and short-lived systems tend to decrease over the last three decades

✉ Rémy Roca
remy.roca@cnrs.fr

[1] Observatoire Midi-Pyrénées, Laboratoire d'Études en Géophysique et Océanographie Spatiales, Université de Toulouse III, CNRS, CNES, IRD, 14 Avenue Edouard Belin, 31400 Toulouse, France

[2] EUMETSAT, Eumetsat Allee 1, 64295 Darmstadt, Germany

1 Introduction

In the tropics, deep convection—a buoyancy-driven dynamical process—transports moist air from the lower levels of the atmosphere upwards into the free troposphere, producing ice-topped upper-level clouds. These clouds are at the heart of climatic feedbacks (Stephens et al. 2024). Deep cloud systems (DCS), ranging from individual cumulonimbus to extensive cloud decks, cover scales from a few to 300 km and last from several hours to a few days (Houze 1993). These features of the tropical atmosphere, scattered throughout the intertropical belt, are pivotal for understanding climate dynamics (Fig. 1).

Due to their significant societal impacts and their anticipated evolution under climate change, Mesoscale convective systems (MCS)—a major category of these DCS—have received renewed attention. This focus is supported by advances in convection-permitting simulations of climate change (Prein et al. 2022) and due to the maturity and length of the observational record. Indeed recent changes in the occurrence and various intensity metrics of MCS have been documented, using satellite observations mainly, so far for the whole tropics (Tan et al. 2015) as well as for numerous large regions: United States (Feng et al. 2016; Prein et al. 2017; Hu et al. 2020), East Asia (Guo et al. 2023), Africa (Hart et al. 2019), the Atlantic Ocean (Bennartz and Schroeder 2012) or with specific regional emphasis such as the Sahel during monsoon months (Bennartz and Schroeder 2012; Taylor et al. 2017), the Congo basin (Alber et al. 2021; Raghavendra et al. 2018), eastern South Africa (Morake et al. 2021). All of these studies point of more intense MCS metrics that have occurred in recent decades, even if the attribution of these changes and the relative role of thermodynamics vs. dynamical drivers of such trends are still disputed (Maybee et al. 2024; Zhao et al. 2024).

In contrast, few analyses, if any, document the recent observed evolution of the full DCS spectrum as well as the arrangement between the DCS (Tan et al. 2015). As illustrated in Fig. 1, the DCS are spatially arranged in diverse geometries, often loosely referred to as *convective organization*. Climate and its future evolution is influenced by the two facets of *convective organization*: the collective effect of the deep cloud systems population and the resulting upper-level cloudiness, as well as the properties of the individual systems (Stephens et al. 2023). *Convective organization* is crucial for extreme precipitation (Pendergrass 2020) and the tropical radiation budget (Bony et al. 2020), among others. More generally, for a given water cycle intensity, *convective organization* can modify the associated energy cycle, like radiation and surface fluxes (Tobin et al. 2012) offering a key element of the ongoing efforts discussed in this Special Issue. Furthermore, modeling efforts have recently highlighted that the spread in climate sensitivity estimates among a large number of models can be related to each model's handling of *convective organization* (Becker and Wing 2020; Wing et al. 2020). Despite its well-recognized importance, the theoretical foundations for *convective organization* are still being intensely debated.

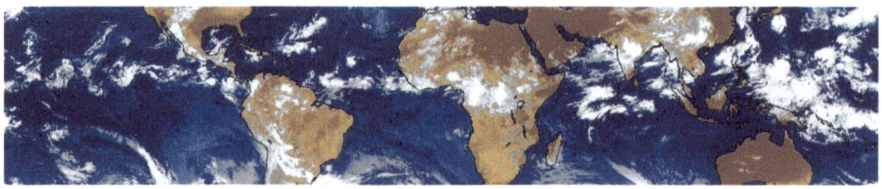

Fig. 1 A multi satellite mosaic depicting the ubiquitous deep clouds, shown as white clusters, on August 8, 2019 at 06 UT. *Source* https://satmos.aeris-data.fr/products/xsat/

Current frameworks addressing the distribution of the individual convective systems range from wave dynamics (Mapes et al. 2006) to radiatively driven self-aggregation (Muller et al. 2022) or statistical mechanics paradigm (Peters and Neelin 2006), but no definitive proposition has yet emerged. Frameworks addressing the evolution of upper level cloud cover include the fixed anvil temperature (Hartmann and Larson 2002) or the "iris effect" (Lindzen and Choi 2021; Mauritsen and Stevens 2015). One reason for this diversity of perspectives may lie in the lack of observational data to confront these theoretical developments (Stephens et al. 2024).

The main objective of this study is to propose an observational constraint on the evolution of both cold cloud cover and the distribution of individual systems. Utilizing geostationary infrared imagery and a cloud tracking algorithm, we document changes in *convective organization* over tropical Africa and the Atlantic Ocean by analyzing the evolution of both upper-level cloud cover and the arrangement of the full spectrum of DCS over the last four decades. This region has been extensively surveyed by the METEOSAT system since the early 1980s, providing a unique long-term record of high-resolution (<5 km) and high-frequency (≤ 30 min) thermal infrared imagery. From these radiation measurements, deep convective clouds can be easily identified (Duvel 1989). High temporal sampling further allows us to delineate the life cycle of deep convective cloud systems (Fiolleau and Roca 2013a). The geostationary field of view spans a wide area, enabling not only the estimation of DCS morphology but also the analysis of their convective organization. Tropical Africa encompasses well-known regions of deep convective activity (Hart et al. 2019), as revealed by the 1983–2020 climatology of the cold cloud cover associated with deep convection (Fig. 2). The Congo basin hosts deep convective systems (Mohr and Zipser 1996; Nguyen and Duvel 2008); orography-linked deep convection spans the Guinea coast,

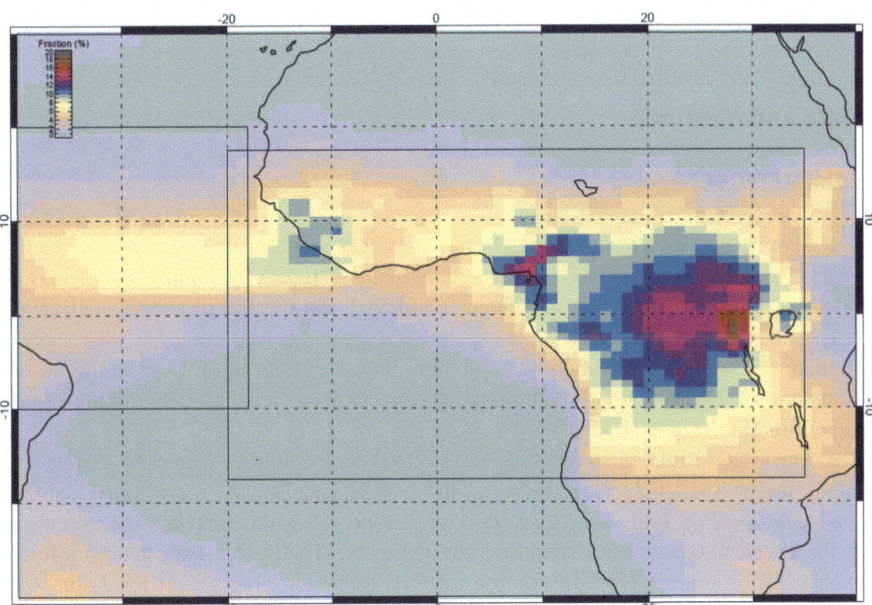

Fig. 2 The map of annual cold cloud fraction (in %) over 1983–2023 shows the active convective areas. The black rectangles delineate the two regions used in the study. Note that, for the African box, oceanic regions are excluded

Mount Cameroun (Vondou 2012) and the edge of the Ethiopian plateau (Jackson et al. 2009). The Sahel experiences monsoon deep convection (Lafore et al. 2017). The adjacent tropical Atlantic Ocean, where numerous DCS are found in the Intertropical Convergence Zone (Houze and Betts 1981) can further be used to assess the sensitivity of our results to the type of surface-atmosphere interactions. In short, this region is populated by a wide spectrum of deep convective systems and a very diverse geography, making it an appropriate area for investigating *convective organization*.

Various metrics have been proposed to quantify aspects of convective organization, such as the occurrence of convective structures, their morphological characteristics, and the arrangements among them (Weger et al. 1992). The simplest of these metrics scales with the occurrence of the convective clusters in the region of interest (Tobin et al. 2012), while others test the departure from a random arrangement to quantify clustering effects (Tompkins and Semie 2017). Some metrics emphasize the geometry of the scene, including or excluding the morphology of the cluster (White et al. 2018). Despite these various attempts, interpreting convective organization remains challenging, and no definitive definition or metric has emerged (Cheng et al. 2018; Retsch et al. 2020). Consequently, in this study, we opt for a straightforward approach by relating the occurrence of deep convective systems to the mean cold cloud fraction. Unlike previous efforts that only considered snapshots of deep cloud clusters, we include the time-related aspects of convective systems in our analysis. Our metric for convective organization incorporates the life cycle information of DCS.

The next section presents the dataset used in this study, along with the climatological features of cold cloud fraction and DCS distribution. The rationale for our characterization of convective organization is introduced in Sect. 3, where we also present the results of the linear trend analysis. The paper concludes with a discussion and suggestions for future research.

2 Data and Method

2.1 The METEOSAT Record

Table 1 summarizes the characteristics of the METEOSAT observing system over the last 42 years. The first-generation and second-generation METEOSAT satellites (Schmetz et al. 2002) share a broad thermal infrared channel, although the spectral response function has slightly changed over the decades. The spatial and temporal resolution has also evolved throughout the record. Consequently, combining measurements taken from multiple sensors on various platforms for climate monitoring necessitates recalibration and homogenization efforts (Fiolleau et al. 2020).

The Meteosat visible infra-red imager (MVIRI) and Spinning enhanced visible and infrared imager (SEVIRI) measurements have been consistently recalibrated using a method described by John et al. (2019). The method relies on reference measurements from the Infrared Atmospheric sounding interferometer (IASI), the Atmospheric infrared sounder (AIRS), and the High-Resolution Infrared Radiation Sounder (HIRS/2) on polar orbiting satellites. The recalibration method can be applied directly to the lowest level of geostationary measurements available, that is, digital counts, to obtain recalibrated radiances. A significant reduction in biases (~5%) have been observed for the IR channel, considered in this article, compared to the operational radiances. IASI hyperspectral

Table 1 The characteristics of the METEOSAT observing system and of the thermal infrared channel over the last four decades

period	Platform	Nadir location	Instrument	Central wavelength (µm)	Spectral interval (µm)	Spatial resolution at nadir (km)	Temporal resolution (min)
1981/08–1988/08	METEOSAT-2	0°	MVIRI	11.5	10.5–12.5	5	30
1988/08–1990/11	METEOSAT-3	0°	MVIRI	11.5	10.5–12.5	5	30
1989/06–1994/02	METEOSAT-4	0°	MVIRI	11.5	10.5 12.5	5	30
1991/11–1997/02	METEOSAT-5	0°	MVIRI	11.5	10.5–12.5	5	30
1997/02–1998/06	METEOSAT-6	0°	MVIRI	11.5	10.5–12.5	5	30
1998/06–2004/10	METEOSAT-7	0°	MVIRI	11.5	10.5–12.5	5	30
2004/10–2007/05	METEOSAT-8	0°	SEVIRI	10.8	9.8–11.8	3	15
2007/05–2013/02	METEOSAT-9	0°	SEVIRI	10.8	9.8–11.8	3	15
2013/02–2017/12	METEOSAT-10	0°	SEVIRI	10.8	9.8–11.8	3	15
2018/01–2023/03	METEOSAT-11	0°	SEVIRI	10.8	9.8–11.8	3	15
2023/03–2023/12	METEOSAT-10	0°	SEVIRI	10.8	9.8–11.8	3	15

measurements are used to determine spectral band adjustment factors (SBAF) that account for the spectral differences among the different instruments on geostationary satellites. Radiances for all instruments have been adjusted as if they had been measured by METEOSAT-5.

As a result, the record is expected to be homogeneous throughout, with a small residual ranging from 0.2 to 0.5 K, increasing with the age of the satellite. The stability of the resulting equivalent record of METEOSAT-5 is exemplified in Fig. 3, which shows the difference between the BT derived from METEOSAT-7 (the last MFG satellite) and METEOSAT-8 (the first MSG series). The analysis was conducted over 20 days between July 1 and 20, 2004, during which METEOSAT-7 and MSG-1 simultaneously monitored the African region and the Atlantic Ocean. The IR data from the two platforms have been first remapped from their native formats to a regular lon/lat 0.5° grid and the biases were computed on the region spanning 40° W and 40° E and between 0° and 30° S to minimize the impact of the time delay between the scanning of METEOSAT-7 and MSG-1 (Fiolleau et al. 2020). The results show an overall residual bias of -0.18 K for $BT_{MSG-1} < 235$ K and confirm the high stability of the record over this major change in the time series.

An alternative geostationary fundamental data record is the GRIDSAT-1B product (Knapp et al. 2011), derived from the ISCCP data and the operational HIRS data. However, the GRIDSAT-1B product does not benefit from reprocessed HIRS observations or the hyperspectral sounder calibrations that have reduced calibration errors in the HIRS/2 measurements. This limitation makes the GRIDSAT record less appealing for long-term analysis. Further discussion and comparison of this new Fundamental Climate Data Record (FCDR) with the GRIDSAT-1B record are provided in Supplementary Note 1.

Quality control is performed using a number of tests (Liefhebber et al. 2020) aimed at flagging corrupted images. Once flawed images are identified, sequence of missing

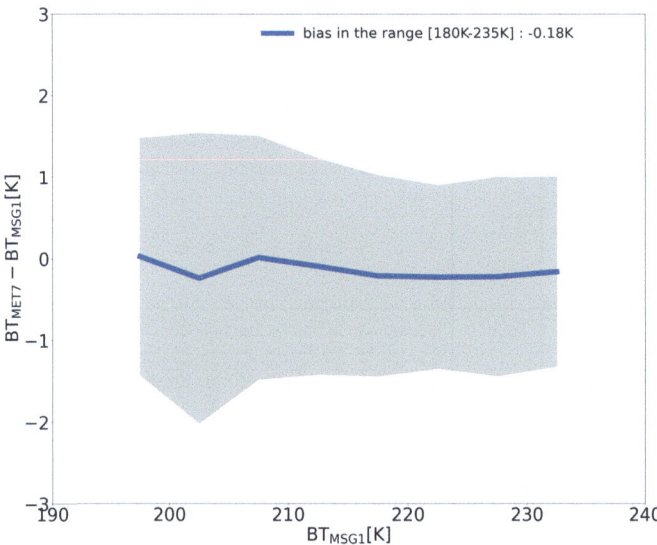

Fig. 3 Distribution of the differences of instantaneous brightness temperature between MSG1 and MET7 as a function of MSG1 brightness temperature for a common 20 days long period in July 2004 over Africa and the Atlantic Ocean. The blue line is the mean and the gray shading corresponds to plus minus one standard deviation

images which can be detrimental to the cloud tracking algorithm (see below) are sought. A very conservative approach is employed here: any corrupted image is considered missing, regardless of where the missing data are within the image. Any day with a missing sequence of images is simply removed from the analysis. Figure S1 illustrates how the quality of the monitoring improves over time, from the early preoperational phase of the METEOSAT up to the highly functional METEOSAT Second Generation era. To account for inhomogeneities in data availability, all the statistics are first computed based on the available imagery and scaled up with a simple coefficient. This coefficient corresponds to the ratio of the number of days in the year divided by the number of actually available days. The impact of using this scaling technique on our analysis is assessed in the Discussion section of the paper. The imagery is re-gridded to a 0.04° regular grid to ensure continuity from the MFG to MSG era. Finally, brightness temperatures are limb corrected following the method of Fiolleau et al. (2020).

In summary, the long-term METEOSAT observations have been carefully homogenized over the length of the record, making them compliant with climate monitoring standards. Possible remaining sources of inhomogeneities have been characterized and should be kept in mind when interpreting the trends analysis.

2.2 Deep Cloud Fraction

The deep cloud fraction is calculated by estimating the fraction of observations colder than 235 K, a well-suited threshold for delineating deep convective cloudiness in the tropics (Duvel 1989; Fiolleau and Roca 2013a; Bouniol et al. 2016). As discussed in the introduction, we use two contrasting regions in this study: tropical land Africa (referred to as AFR) and a large region of the Atlantic ITCZ representing open Atlantic Ocean conditions (ATL), as depicted in Fig. 2. Annual means are constructed from the full-resolution record over the 1983–2020 period.

2.3 Deep Cloud Systems (DCS)

Deep Cloud Systems are identified using the Tracking of organized convection algorithm through a 3D segmentatioN (TOOCAN) algorithm (Fiolleau and Roca 2013). TOOCAN is designed to overcome one major limitation of the traditional overlap-based tracking technique by analyzing the 3D (2D+time) imagery. It is based on a conceptual model of a convective system, where the cloud deck is composed of a convective and stratiform part (Houze 2004) particularly suited for tropical Africa and the Atlantic Ocean (Futyan and Del Genio 2007; Schumacher and Houze 2006; Lafore et al. 2017). The TOOCAN algorithm first performs a multithreshold, multistep screening on the space and time infrared image volume to detect convective seeds. Then, it grows these seeds towards the edges of their stratiform extension, defined the upper level cloudiness threshold of 235 K. The algorithm outputs various morphological parameters of the cold cloud shield (geolocation, size, brightness temperature distribution) as a function of the cloud system life cycle, with a temporal resolution of 30 min. From these raw outputs, numerous integrated DCS parameters are built: duration, distance of propagation, average speed of propagation. Earlier versions of this MCS database have been derived from various geostationary archives, and used in a number of studies to investigate the morphology of the cold cloud shield (Roca et al. 2017; Elsaesser et al. 2022), the distribution of DCS and tropical rainfall (Roca et al. 2014) or rainfall extreme (Roca and Fiolleau 2020); the radiative properties of the MCS

(Bouniol et al. 2016). In this study, we focus on the duration parameter and the maximum extent of the cloud shield to characterize the spectrum of DCS.

Due to the large volume of data under consideration, a data reduction step is first performed by projecting the raw systems data onto a regular 1°×1° grid, where all information about the systems is retained (see the Methods section of Roca and Fiolleau 2020 for more details). Statistics are then built from this reduced dataset, also known as CACA-TOES (http://toocan.ipsl.fr).

The annual climatology of the occurrence of systems, based on the 1983–2023 period, is shown in a phase diagram built using the systems' duration and maximum extent for the African and Atlantic ITCZ regions (Fig. 4). To first order, the distributions of each region are similar, with systems ranging from a few hours up to more than two days and from approximately 1000 km2 up to several million km^2. The mean system properties (~6.2 h; ~6300 km^2 for AFR and ~6.3 h; ~5000 km^2 for ATL) correspond to the most frequently observed systems. Comparatively, the larger and longer-lived systems are much less frequent, as observed for the entire tropical belt (Roca and Fiolleau 2020). Despite a strong relationship between maximum area and duration, the statistics highlight that for a given duration, the maximum area of the DCS cloud shield can vary by multiple orders of magnitude. Even systems lasting 24 h or more can span over two orders of magnitude in size. This result for African and Atlantic systems is consistent with previously reported tropical-wide analyses (Roca et al. 2017). This feature of the spectrum of DCS is often underappreciated, mainly due to the emphasis on MCS in the literature, but it remains an important aspect of convective organization in the tropics.

The distribution of the DCS cloud top characteristics can be summarized using the 90th percentile of the coldest pixels forming the cluster at the time of the maximum extent of the cold cloud shield (Roca and Ramanathan 2000). Overall, the larger the system, the deeper it is, with this trend being more pronounced over land than over ocean. This is particularly evident for systems exceeding 100,000 km^2 over land and those lasting longer than 15 h over the ocean (Fig. 5). A 'V' pattern characterizes both land and ocean distributions, with two branches associated with the less deep (warmer) systems: shorter-lived and slightly larger than the mean, and longer-lived and smaller than the mean. The largest and

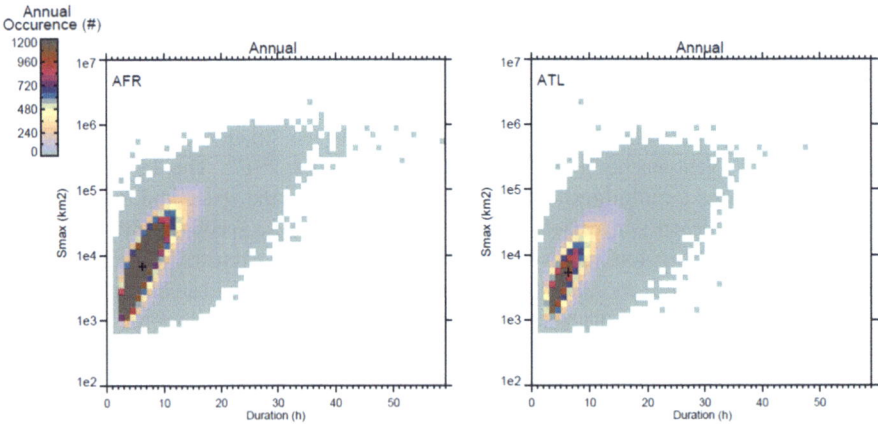

Fig. 4 The 1983–2023 climatology of annual occurrence of DCS. For Africa (left) and the Atlantic region (right). S_{max} is the maximum extension of the DCS along its life cycle. The cross corresponds to the average of the 2D distribution

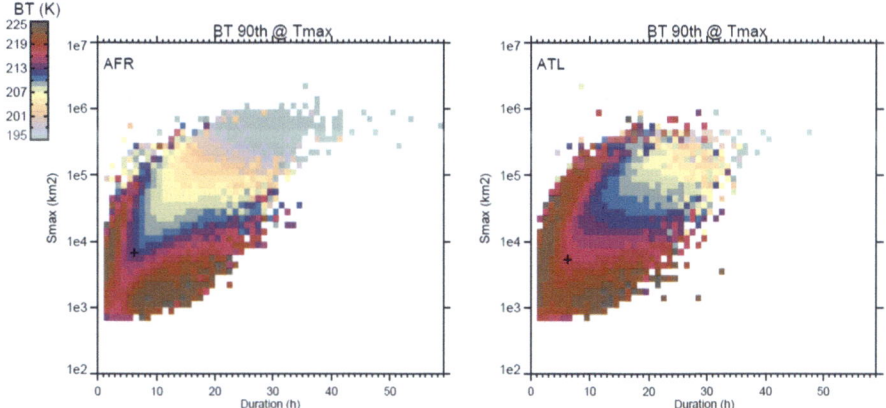

Fig. 5 The 1983–2023 climatology of the 90th percentile of the cluster brightness temperature at the time of maximum extent. For Africa (left) and the Atlantic region (right). The cross corresponds to the average of the 2D distribution of Fig. 2

longest-lived systems are deeper over land than over the ocean, primarily a feature associated with the Congo region (not shown), which is consistent with previous analyses (Liu et al. 2007).

An aspect of the distribution of the DCS concerns the shape of the cloud shield, which is summarized by computing the eccentricity of the cloud shield at the time of its maximum extent in the life cycle. The eccentricity is defined by fitting an equivalent ellipse to the cloud shield and is expressed as the ratio of the small to the large axis of the ellipse. The eccentricity distribution mirrors the cloud top characteristics (Fig. 6). Over both land and ocean, the two 'warm' top populations of DCS tend to be more linear systems (eccentricity < 0.5), while the 'coldest' top DCS are more circular, a feature often associated with Mesoscale Convective Complexes (Jirak et al. 2003).

Finally, DCS are characterized by their movement speed, computed as the propagated distance from beginning to end divided by the system duration. Fast systems, with speeds

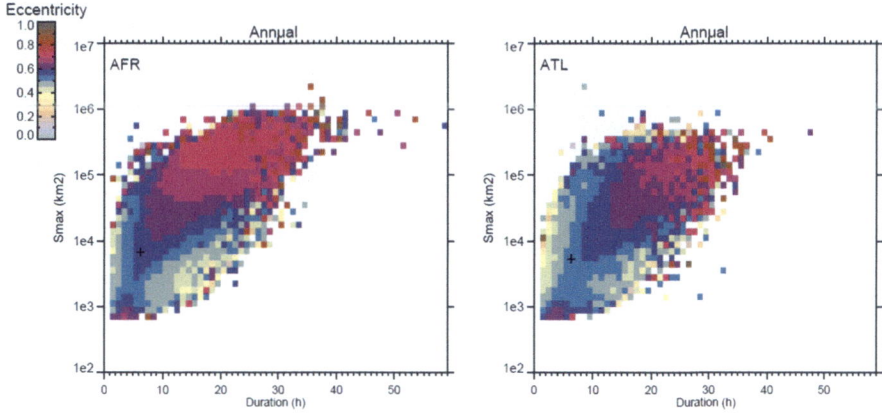

Fig. 6 The 1983–2023 climatology of eccentricity of the cluster at the time of maximum extent. For Africa (left) and the Atlantic region (right). The cross corresponds to the average of the 2D distribution of Fig. 2

up to 15 m/s, are seen on the outer upper edge of the phase space diagram (Fig. 7), while the inner lower edge corresponds to slower systems. These features are more pronounced over the ocean compared to land, where a secondary population of fast systems is observed at the upper tail of the distribution in both size and duration. Over land, the most significant, longest-lived, and fastest propagating systems are likely squall lines (Lafore et al. 2017).

In summary, the climatological features of the DCS distribution over both Africa and the Atlantic Ocean encompass a wide range of system duration, size, depth, shape, and speed. This diversity confirms the suitability of this region for analyzing *convective organization*.

3 Results

3.1 Rationale for Convective Organization Monitoring

For a given region, assuming a constant cloud fraction over a given time period, any changes in the size and duration distributions of DCS during that period indicate a change in convective organization. This is due not only to changes in the occurrence of systems with different morphologies and properties but also to the arrangement of DCS relative to each other. Therefore, we first identify a time period in the long-term record during which the deep cloud fraction remains steady. For this period, we explore the changes in the DCS morphology distribution using phase-space diagrams.

3.2 Identification of the Stable Period in the Deep Cloud Fraction

The time series of deep cloudiness for our two regions of interest exhibits up to 1% interannual variability, corresponding to ~10–20% variation in relative units (Fig. 8). The linear trend in deep cloudiness over time is calculated using the Theil-Sen slope estimator for various periods. The periods begin from a starting year and end in 2023 included. The result is shown in Fig. 8. The Theil-Sen slope estimator, based on the median of possible slopes

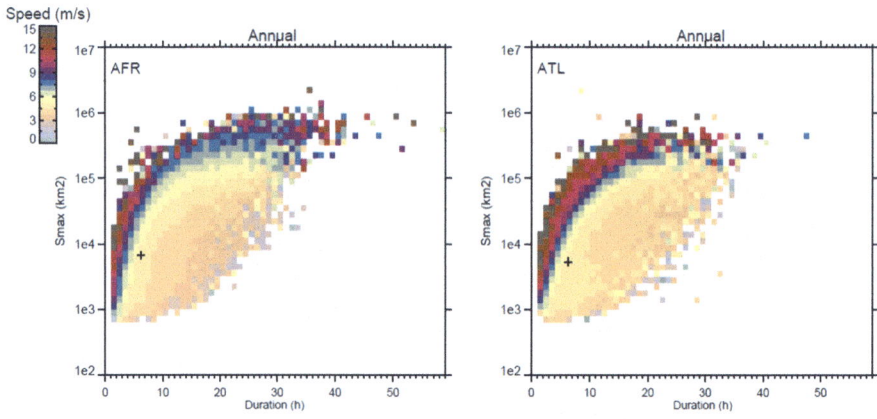

Fig. 7 The 1983–2023 climatology of movement speed of the system. For Africa (left) and the Atlantic region (right). The cross corresponds to the average of the 2D distribution of Fig. 2

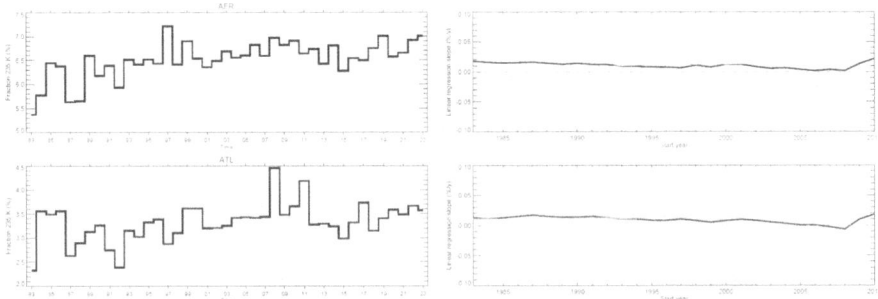

Fig. 8 Time series (left) of the upper level cloudiness and linear trends estimates over the start year to 2023. The thin gray lines on the right panels correspond to the confidence interval at 95%. For Africa (top) and Atlantic Ocean (bottom)

among pairs of points, is preferred over the classic ordinary least squares linear approach due to its robustness, although the results are not significantly sensitive to the choice of method (not shown). The significance of the trend is assessed by computing the 95% confidence interval of the Theil-Sen slope estimation (e.g., Helsel et al. 2020). The difference between the upper and lower confidence intervals reflects the slight skewness of the slope distribution. The results indicate that there are no significant trends for most of the record for both regions. Only for Africa does the analysis show a slightly significant positive trend if the record starts prior to 1990.

During the 1991–2023 period, deep cloudiness remains steady over our regions of interest. This stability is observed not only in the annual mean but also across each individual season (not shown). This finding aligns with previous regional studies focused on precipitation, which indicate a leveling off of annual Sahelian rainfall following a recovery period from the long drought of the 1980s (Biasutti 2019; Panthou et al. 2014). his trend is possibly linked to the evolution of sea surface temperature (SST) patterns (Chen et al. 2020). The 1991–2023 period will be examined in further detail in the next section to advance our analysis of convective organization.

3.3 Changes in the DCS Morphology Distributions

The linear trends in the DCS morphology in the phase space are also computed using the Theil-Sen estimator, with a 95% confidence interval as shown in Fig. 9. Significant changes are observed in the most frequent systems in both regions, comprising both negative trends (~ − 2%/year) and positive trends up to 3%/year. Systems lasting up to 15 h with a maximum extent of up to 20,000 km^2 are becoming more frequent. These systems are generally less circular, warmer, and slower than the rest of the DCS population. The negative trends are observed in the upper part of the distribution and concerns systems with extents up to 200,000 km^2 and durations up to 15 h. However, this negative trend area is not significant over the Atlantic Ocean.

The largest and longest-lived systems (D > 15 h; S_{max} > 50000 km^2) show no significant trends over both Africa and the Atlantic Ocean. As shown in Supplementary Note 1, this is also the case for the Sahel region during the boreal summer. This finding contrasts with previous claims of significant changes in this region during 1983–2016 (Taylor et al. 2017). The divergence arises from differences in the datasets used, the definitions of Mesoscale

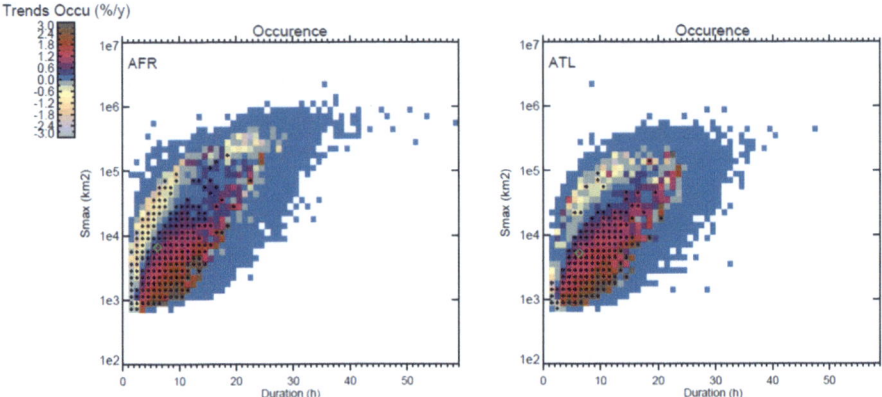

Fig. 9 Linear trend in the occurrence over the 1991–2023 period in %/year. Black dots mark the significant trends. The green diamond corresponds to the average of the 2D distribution of Fig. 2. For Africa (left) and the Atlantic region (right)

Convective Systems (MCS) employed, and the periods selected for regression analysis. In particular, the originally reported trends are still present when using the present METEO-SAT FCDR but with a muted amplitude by a factor of approximately 2. These differences and their implications are discussed in more detail in Supplementary Note 1.

4 Discussion

These changes in convective organization could either result from artifacts in the observational record or be associated with large-scale climate evolution. Both possibilities are discussed next.

4.1 Possible Artifacts in the Record in the Trend's Estimation

The analysis of a well curated DCS database over two contrasted climatological regions, built on an advanced METEOSAT record carefully homogenized over the last four decades reveals a change of *convective organization*. Over the last 32 years, the deep cloudiness over both regions shows no trends despite pronounced interannual variability in each region. The analysis of the changes of the DCS morphology indicates a shift with less frequent short lived and large systems and more frequent mid lived and small systems. *At a constant cloud fraction*, this change of occurrences in the phase space translates in a change in arrangement of the systems in complement to a change in the most frequent morphology. Changes concern both the tropical Africa and the tropical Atlantic regions with stronger significance over land.

Cold cloudiness time series do not show obvious breaks, especially when the record transitions from the MFG to the MSG era, ruling out this problem as a possible artifact source. Both the Atlantic and the African region exhibit distinct interannual variability further indicative of the absence of a strong bias in the record but leaving room for an observing system-scale remaining trend residual that could corrupt our interpretation of the system morphology changes. Formulating a new null hypothesis, by which there is no change in the system's morphology but that the record contains an artificial residual trend (RT in K/y or K/decade), we can quantitively assess the needed magnitude of such an artifact to be commensurate with the reported change in morphology. Following Fiolleau and Roca (2024), a 4-month period over tropical Africa and Atlantic Ocean is used. The TOOCAN algorithm is run on the data that are perturbed by a given amount to simulate the impact of a systematic bias on the statistics of the DCS occurrence. The sensitivity study indicates that for a + 3 K systematic bias, the population of systems lasting 10 h would decrease for all the S_{max} and the smallest system would be impacted more. A − 3 K bias would yield to similar but opposite in sign, changes in the DCS population (Fig. 10). Similarly, for the systems reaching a size of 10^4 km^2, the sign of the bias would correspond to a systematic under (or over) estimation across the various duration population. This is not consistent with our results that show positive and negative trends for short lived and large systems and mid lived and small systems, respectively. Figure 10 further indicates that over the 1991–2023 period, the magnitude of the trend is very much larger than that of a 3 K systematic bias. In other words, for the magnitude of the reported trends in the occurrence of DCS to be due to a residual trend in the METEOSAT FCDR, it would require such residual to be much larger than 3 K, or 1 K/decade. That is much larger than an order of magnitude larger than the estimated residual of 0.5 K/decade as discussed in the previous section and by consequence we rule out such a possible erroneous attribution of our reported trends.

The currently enforced quality control (QC) of the data is very conservative and could, in principle, introduce a possible source of inhomogeneity in the record. To assess the potential impact of such QC on our trend estimates, the computations are replicated for the more recent 1997–2023 period, during which data availability is consistently high. The results, shown in Figure S2, are very similar to the 1991–2023 analysis, further supporting the robustness of the observed changes in convective organization.

Additionally, these trend estimates and their significance are not sensitive to the statistical method used, whether it be ordinary least squares, nonparametric methods, bootstrapping, or the Mann–Kendall test (not shown). Ruling out artifacts in our analysis prompts a discussion of possible reasons for the observed changes in convective organization over the last few decades.

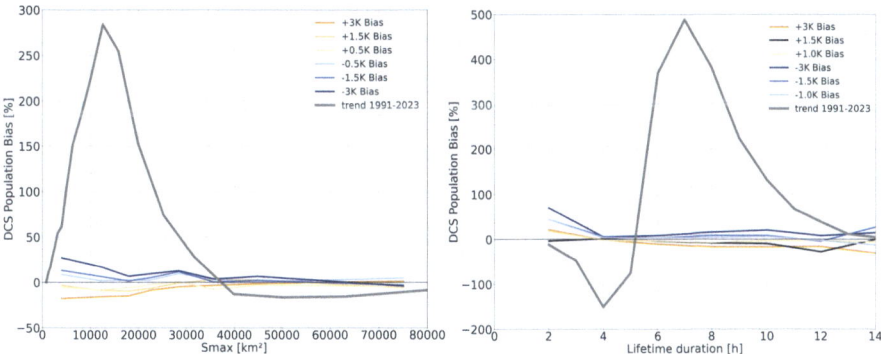

Fig. 10 Sensitivity of the DCS population occurrence to a bias in the IR brightness temperature over tropical Africa and Atlantic Ocean. (left) For the systems lasting 10 h as a function of Smax. (right) For the systems with Smax = 10^4 km^2 as a function of the duration. The thick gray lines are the 1991–2023 trends for Africa from Fig. 9

4.2 Possible Climate-based Interpretation of the Trends in Morphology

The results have been aggregated for the tropical African continent to contrast with the Atlantic Ocean. But it should be kept in mind that these bulk results for Africa are mainly influenced by changes over the Congo region given its overwhelming role in the climatology (Fig. 1). This is also, albeit less significantly, the case for the Sahel region (See Supplementary Note 1). Thermodynamic and dynamical conditions are often invoked as controlling factors or drivers of the properties of the tropical systems (Yang et al. 2017; Schiro et al. 2020; Palmer et al. 2023). Trends in meridional gradient of surface temperature have been linked to regional changes in atmospheric circulation and in the environmental conditions and further connected with convective systems changes (Taylor et al. 2017, 2018; Cook et al. 2020; Klein et al. 2021; Baidu et al. 2022). Over tropical oceans, similar drivers are noted but with also stronger emphasis on water vapor and the link with the wind shear is less conclusive (Hagos et al. 2013; Duncan et al. 2014; Galarneau et al. 2023a; Chen et al. 2023; Berthet et al. 2017). There, instability is not limited as for the land-based conditions and CAPE is always available. Deep convection can be sustained more easily thanks to the moist maritime boundary layer compared to land (Houze 2004). Despite these contrasted environmental conditions between land and ocean, our study shows similar responses over both continental Africa and the tropical Atlantic regions suggest that the change in *convective organization* is taking place at large scale. The major large scale decadal change over this region concerns the interhemispheric surface temperature gradient (Hart et al. 2019). This large-scale meridional gradient encapsulates previous regional studies showing atmospheric circulations intensification and surface temperature gradient changes. Over the Atlantic sector, it is also associated with a warming of SST of ~ 0.5K[1] over the last 30 years and a concomitant water vapor increase (Schröder et al. 2016). This relatively modest moisture increase is consistent with small changes in maximum extent and would also be compatible with extended system duration reported here.

Our analysis reveals a space/time scale dependence in the trends of DCS occurrence. The MCS subset of the DCS appears less altered than the core of the DCS distribution, indicating a need for new investigations into the drivers of these less explored systems. The extent to which the squall line (or MCS) conceptual model applies to the entire spectrum of DCS remains unclear. We can only speculate that the more frequent, not very deep, relatively small but long-lasting systems may not follow simple environmental control by dynamics. The increase in mean duration and decrease in extent suggest an enhanced atmospheric instability that sustains longer-lived systems without a significant increase in water vapor availability, which would favor larger systems. The maturity of the water and energy cycle depiction by the observing system, discussed at length in this Special Issue, along with concurrent advances in atmospheric reanalysis, now offers the opportunity to explore these speculative statements in greater detail. This will be the focus of future work.

5 Conclusions and Perspectives

A long-term analysis of convective organization was performed using a carefully homogenized four-decade-long METEOSAT archive. The analysis indicates that cold cloudiness has not exhibited any significant trends over the last 30 years. However, an investigation of the trends in DCS morphology reveals a significant shift in the occurrence of deep convective systems, characterized by a decrease in large, short-lived systems and an increase in smaller, longer-lived ones.

The possibility that these results could be artifacts of the observational record has been ruled out through various tests. Trends in the large-scale interhemispheric surface temperature gradient may be related to changes in atmospheric circulation and environmental conditions, which in turn affect the morphology of DCS and lead to changes in convective organization.

Energetics considerations suggest that the tropical wide anvil cloud cover could be reduced under strong global warming through the "stability iris" effect (Hartmann and Larson 2002; Bony et al. 2016; Ito and Masunaga 2022). During that period, global warming of 0.6 K is observed[1] that might not have triggered the proposed mechanisms. The restricted regional dimension of our analysis might prevent such an energy budget-based signal to emerge locally. Future work could consider the entire tropics for which DCS dataset 20+ years are emerging (Feng et al. 2021) or under construction (Heidinger et al. 2024). The IR-only anvil diagnostic might also suffer from limitation. The long observational record from the TRMM and GPM precipitation radar (Wang and Tang 2020) could be used in conjunction with the present METEOSAT data to better characterize the anvil cloudiness of the DCS (e.g., Elsaesser et al. 2022). Similarly, the characterization of deep convection might benefit from a multi-satellite perspective as well (Derras-Chouk and Luo 2024; Pilewskie et al. 2024; Li et al. 2024). Exploring the stability mechanism using these combined long satellite records is a venue for future work. Finally, our results highlight changes in the arrangement of the systems over the whole DCS spectrum and suggest that convective organization theory should also address this spectral dependency.

These findings have direct implications for climate modeling. Regional convection-permitting climate models, used to predict precipitation, rely on companion global parameterized models and are not directly impacted by the simulated DCS cloud shield properties. However, accurate representation of the cloud shield is crucial for kilometer-scale global climate models (Satoh et al. 2019) that requires a proper cloud-radiation feedback. Recent global simulations exhibit a significant spread in DCS morphology when compared to satellite observations (Feng et al. 2023). Our statistics and trend analysis provide an observed constraint on the evolution of DCS distribution over the last four decades. This data should be used to validate and improve the emerging multidecadal simulations using this new generation of kilometer-scale global models.

Finally, it is important to note that this active research topic will benefit in the near future from an enhanced observational basis. Field campaigns in the Atlantic ITCZ focusing on deep convection and surface interactions are planned for summer 2024: MAESTRO (S. Bony, personal communication, 2023; MAESTRO) and ORCESTRA (C. Hohenegger, personal communication, 2023). These campaigns will help understand the drivers behind the changes in DCS morphology reported here. Additionally, the existing satellite record will soon benefit from the third generation of METEOSAT (MTG) observations (Holmlund et al. 2021). The MTG's enhanced capabilities to monitor deep convection will be instrumental in better understanding the reasons for the evolution of DCS over this region.

Supplementary Information The online version contains supplementary material available at https://doi.org/10.1007/s10712-024-09862-8.

---

[1] See NOAA National Centers for Environmental information, Climate at a Glance: Global Time Series, from https://www.ncdc.noaa.gov/cag/

Acknowledgements We thank S. Cloché for her assistance. This study benefited from the IPSL mesocenter ESPRI facility, which is supported by CNRS, UPMC, Labex L-IPSL, CNES, and Ecole Polytechnique. This study was also supported by CNES and CNRS under the Megha-Tropiques program. We extend our gratitude to Louis Netz for his valuable discussions on the organization metrics and to Graeme Stephens for stimulating discussions at the workshop. Data processing was performed on the European Weather Cloud infrastructure.

Funding Open access funding provided by Université Toulouse III - Paul Sabatier.

Data Availability The TOOCAN data are available at with https://doi.org/10.15770/EUM_SEC_CLM_1005. The CACATOES data are available at with https://doi.org/10.15770/EUM_SEC_CLM_1006.

Declarations

Conflict of interest The authors do not have financial nor non-financial conflict of interest to declare.

Open Access This article is licensed under a Creative Commons Attribution 4.0 International License, which permits use, sharing, adaptation, distribution and reproduction in any medium or format, as long as you give appropriate credit to the original author(s) and the source, provide a link to the Creative Commons licence, and indicate if changes were made. The images or other third party material in this article are included in the article's Creative Commons licence, unless indicated otherwise in a credit line to the material. If material is not included in the article's Creative Commons licence and your intended use is not permitted by statutory regulation or exceeds the permitted use, you will need to obtain permission directly from the copyright holder. To view a copy of this licence, visit http://creativecommons.org/licenses/by/4.0/.

References

Alber K, Zhou L, Raghavendra A (2021) A shift in the diurnal timing and intensity of deep convection over the Congo Basin during the past 40 years. Atmos Res 264:105869. https://doi.org/10.1016/j.atmosres.2021.105869

Baidu M, Schwendike J, Marsham JH, Bain C (2022) Effects of vertical wind shear on intensities of mesoscale convective systems over West and Central Africa. Atmos Sci Lett 23:1–9. https://doi.org/10.1002/asl.1094

Becker T, Wing AA (2020) Understanding the extreme spread in climate sensitivity within the radiative-convective equilibrium model intercomparison project. J Adv Model Earth Syst 12(10):e2020MS002165. https://doi.org/10.1029/2020MS002165

Bennartz R, Schroeder M (2012) Convective activity over Africa and the tropical Atlantic inferred from 20 Years of geostationary Meteosat infrared observations. J Clim 25:156–169. https://doi.org/10.1175/2011JCLI3984.1

Berthet S, Roca R, Duvel JP, Fiolleau T (2017) Subseasonal variability of mesoscale convective systems over the tropical northeastern Pacific. Q J Royal Meteorol Soc 143(703):1086–1094. https://doi.org/10.1002/qj.2992

Biasutti M (2019) Rainfall trends in the African Sahel: characteristics, processes, and causes. Wiley Interdiscip Rev: Climate Change 10(4):e591. https://doi.org/10.1002/wcc.591

Bony S, Stevens B, Coppin D, Becker T, Reed KA, Voigt A, Medeiros B (2016) Thermodynamic control of anvil cloud amount. Proc Natl Acad Sci U S A 113:8927–8932. https://doi.org/10.1073/pnas.1601472113

Bony S, Semie A, Kramer RJ, Soden B, Tompkins AM, Emanuel KA (2020) Observed modulation of the tropical radiation budget by deep convective organization and lower-tropospheric stability. AGU Adv 1:1–15. https://doi.org/10.1029/2019av000155

Bouniol D, Roca R, Fiolleau T, Poan DE (2016) Macrophysical, microphysical, and radiative properties of tropical mesoscale convective systems over their life cycle. J Clim 29:3353–3371. https://doi.org/10.1175/JCLI-D-15-0551.1

Chen T, Zhou S, Liang C, Hagan DF, Zeng N, Wang J, Shi T, Chen X, Dolman AJ (2020) The greening and wetting of the sahel have leveled off since about 1999 in relation to SST. Remote Sens 12(17):2723. https://doi.org/10.3390/RS12172723

Chen X, Leung LR, Feng Z, Yang Q (2023) Environmental controls on MCS lifetime rainfall over tropical oceans. Geophys Res Lett 50:1–11. https://doi.org/10.1029/2023GL103267

Cheng WY, Kim D, Rowe A (2018) objective quantification of convective clustering observed during the AMIE/DYNAMO two-day rain episodes. J Geophys Res Atmos 123:10361–10378. https://doi.org/10.1029/2018JD028497

Cook KH, Liu Y, Vizy EK (2020) Congo Basin drying associated with poleward shifts of the African thermal lows. Climate Dyn 54:863–883. https://doi.org/10.1007/s00382-019-05033-3

Derras-Chouk A, Luo ZJ (2024) Revisiting Riehl and Malkus (1958) and Riehl and Simpson (1979): Charactering Tropical Hot Towers and Estimating Convective Mass Fluxes Using Geostationary Satellite Data. Surv Geophys, in press

Duncan DI, Kummerow CD, Elsaesser GS (2014) A lagrangian analysis of deep convective systems and their local environmental effects. J Clim 27:2072–2086. https://doi.org/10.1175/JCLI-D-13-00285.1

Duvel JP (1989) Convection over tropical Africa and the Atlantic Ocean during northern summer. Part I: interannual and diurnal variations. Mon Weather Rev 117:2782–2799. https://doi.org/10.1175/1520-0493(1989)117%3c2782:COTAAT%3e2.0.CO;2

Elsaesser GS, Roca R, Fiolleau T, Del Genio AD, Wu J (2022) A simple model for tropical convective cloud shield area growth and decay rates informed by geostationary IR, GPM, and Aqua/AIRS satellite data. J Geophys Res Atmos 127:1–21. https://doi.org/10.1029/2021JD035599

Feng Z, Dong X, Xi B, McFarlane SA, Kennedy A, Lin B, Minnis P (2012) Life cycle of midlatitude deep convective systems in a Lagrangian framework. J Geophys Res Atmos 117:1–14. https://doi.org/10.1029/2012JD018362

Feng Z, Leung LR, Hagos S, Houze RA, Burleyson CD, Balaguru K (2016) More frequent intense and long-lived storms dominate the springtime trend in central US rainfall. Nat Commun 7:1–8. https://doi.org/10.1038/ncomms13429

Feng Z, Leung LR, Liu N, Wang J, Houze RA Jr, Li J, Hardin JC, Chen D, Guo J (2021) A Global high-resolution mesoscale convective system database using satellite-derived cloud tops, surface precipitation, and tracking. J Geophys Res: Atmos 126(8):e2020JD034202. https://doi.org/10.1029/2020JD034202

Feng Z, Leung LR, Hardin J, Terai CR, Song F, Caldwell P (2023) Mesoscale convective systems in DYAMOND global convection-permitting simulations. Geophys Res Lett 50(4):e2022GL102603. https://doi.org/10.1029/2022GL102603

Fiolleau T, Roca R (2013) An algorithm for the detection and tracking of tropical mesoscale convective systems using infrared images from geostationary satellite. IEEE Trans Geosci Remote Sens 51:4302–4315. https://doi.org/10.1109/TGRS.2012.2227762

Fiolleau T, Roca R (2024) A database of deep convective systems derived from the intercalibrated meteorological geostationary satellite fleet and the TOOCAN algorithm (2012–2020). Earth Syst Sci Data 16(9):4021–4050. https://doi.org/10.5194/essd-16-4021-2024

Fiolleau T, Roca R, Cloché S, Bouniol D, Raberanto P (2020) Homogenization of Geostationary Infrared Imager Channels for Cold Cloud Studies Using Megha-Tropiques/ScaRaB. IEEE Trans Geosci Remote Sens 58(9):6609–6622. https://doi.org/10.1109/tgrs.2020.2978171

Futyan JM, Del Genio AD (2007) Deep convective system evolution over Africa and the Tropical Atlantic. J Clim 20:5041–5060. https://doi.org/10.1175/JCLI4297.1

Galarneau TJ Jr, Zeng X, Dixon RD, Ouyed A, Su H, Cui W (2023a) Tropical mesoscale convective system formation environments. Atmos Sci Lett 24(5):e1152. https://doi.org/10.1002/asl.1152

Guo Y, Fu Q, Leung LR, Na Y, Lu R (2023) Trends in warm season mesoscale convective systems over Asia in 2001–2020. J Geophys Res Atmos 128:1–15. https://doi.org/10.1029/2023JD038969

Hagos S, Feng Z, McFarlane S, Leung LR (2013) Environment and the lifetime of tropical deep convection in a cloud-permitting regional model simulation. J Atmos Sci 70:2409–2425. https://doi.org/10.1175/JAS-D-12-0260.1

Hart NCG, Washington R, Maidment RI (2019) Deep convection over Africa: annual cycle, ENSO, and trends in the hotspots. J Clim 32:8791–8811. https://doi.org/10.1175/JCLI-D-19-0274.1

Hartmann DL, Larson K (2002) An important constraint on tropical cloud - Climate feedback. Geophys Res Lett 29:10–13. https://doi.org/10.1029/2002GL015835

Heidinger A, Stephens G, Schulz J, John V,Meirink JF, Stengel M, Philips C, Knapp K, Roca R, Fiolleau T, Luffarelli M, L'Ecuyer T, Eliasson S (2024) The era of the GEO-RING: The Next Generation of the International Satellite Cloud Climatology Project (ISCCP-NG), submitted to BAMS

Helsel DR, Hirsch RM, Ryberg KR, Archfield SA, Gilroy EJ (2020) Statistical methods in water resources techniques and methods 4 – A3. USGS Tech Method 458

Holmlund K, Grandell J, Schmetz J, Stuhlmann R, Bojkov B, Munro R, Lekouara M, Coppens D, Viticchie B, August T, Theodore B (2021) Meteosat third generation (MTG): continuation and innovation of observations from geostationary orbit. Bull Am Meteorol Soc 102(5):E990–E1015. https://doi.org/10.1175/BAMS-D-19-0304.1

Houze RA (1993) Cloud dynamics. Edited by Academic Press. Cloud Dynamics. San Diego, USA. https://doi.org/10.1016/0377-0265(87)90017-0

Houze RA (2004) Mesoscale convective systems. Rev Geophys 42:1–43. https://doi.org/10.1029/2004RG000150

Houze RA, Betts AK (1981) Convection in GATE. Rev Geophys 19:541–576. https://doi.org/10.1029/RG019i004p00541

Hu H, Leung LR, Feng Z (2020) Observed warm-season characteristics of MCS and Non-MCS rainfall and their recent changes in the central United States. Geophys Res Lett 47:12–15. https://doi.org/10.1029/2019GL086783

Ito M, Masunaga H (2022) Process-Level assessment of the iris effect over tropical oceans. Geophys Res Lett 49:1–9. https://doi.org/10.1029/2022gl097997

Jackson B, Nicholson SE, Klotter D (2009) Mesoscale convective systems over Western Equatorial Africa and their relationship to large-scale circulation. Mon Weather Rev 137:1272–1294. https://doi.org/10.1175/2008mwr2525.1

Jirak IL, Cotton WR, McAnelly RL (2003) Satellite and radar survey of mesoscale convective system development. Mon Weather Rev 131:2428–2449. https://doi.org/10.1175/1520-0493(2003)131%3c2428:SARSOM%3e2.0.CO;2

John VO, Tabata T, Rüthrich F, Roebeling R, Hewison T, Stöckli R, Schulz J (2019) On the methods for recalibrating geostationary longwave channels using polar orbiting infrared sounders. Remote Sens 11(10):1171

Klein C, Nkrumah F, Taylor CM, Adefisan EA (2021) Seasonality and trends of drivers of mesoscale convective systems in southern West Africa. J Climate 34(1):71–87. https://doi.org/10.1175/JCLI-D-20-0194.1

Knapp KR, Ansari S, Bain CL, Bourassa MA, Dickinson MJ, Funk C, Helms CN, Hennon CC, Holmes CD, Huffman GJ, Kossin JP (2011) Globally gridded satellite observations for climate studies. Bull Am Meteorol Soc 92(7):893–907. https://doi.org/10.1175/2011BAMS3039.1

Lafore JP, Chapelon N, Diop M, Gueye B, Largeron Y, Lepape S, Ndiaye O, Parker DJ, Poan E, Roca R, Roehrig R (2017) Deep convection. Meteorol Trop West Africa: Forecast Handb. https://doi.org/10.1002/9781118391297.ch3

Li YX, Masunaga H, Takahashi H, Yu JY (2024) When, where and to what extent do temperature perturbations near tropical deep convection follow convective quasi equilibrium? Geophys Res Lett 51(11):e2024GL108233. https://doi.org/10.1029/2024GL108233

Liefhebber F, Lammens S, Brussee PW, Bos A, John VO, Rüthrich F, Onderwaater J, Grant MG, Schulz J (2020) Automatic quality control of the Meteosat First Generation measurements. Atmos Measure Tech 13(3):1167–1179. https://doi.org/10.5194/amt-13-1167-2020

Lindzen RS, Choi YS (2021) The iris effect: a review. Asia–pacific J Atmos Sci. https://doi.org/10.1007/s13143-021-00238-1

Liu C, Zipser EJ, Nesbitt SW (2007) Global distribution of tropical deep convection: Different perspectives from TRMM infrared and radar data. J Clim 20:489–503. https://doi.org/10.1175/JCLI4023.1

Mapes B, Tulich S, Lin J, Zuidema P (2006) The mesoscale convection life cycle: building block or prototype for large-scale tropical waves? Dyn Atmos Oceans 42(1–4):3–29. https://doi.org/10.1016/j.dynatmoce.2006.03.003

Mauritsen T, Stevens B (2015) Missing iris effect as a possible cause of muted hydrological change and high climate sensitivity in models. Nat Geosci 8:346–351. https://doi.org/10.1038/ngeo2414

Maybee B, Marsham J, Klein C, Parker DJ, Barton EJ, Taylor CM, Lewis H, Sanchez C, Jones RW, Warner JL (2024) Wind shear effects on entrainment in convection–permitting models influence convective storm rainfall and forcing of tropical circulation. ESS Open Archive. https://doi.org/10.22541/essoar.171536360.00696271/v1

Mohr KI, Zipser EJ (1996) Mesoscale convective systems defined by their 85-GHz Ice scattering signature: size and intensity comparison over tropical oceans and continents. Mon Weather Rev 124:2417–2437. https://doi.org/10.1175/1520-0493(1996)124%3c2417:mcsdbt%3e2.0.co;2

Morake DM, Blamey RC, Reason CJC (2021) Long-lived mesoscale convective systems over Eastern South Africa. J Clim 34:6421–6439. https://doi.org/10.1175/JCLI-D-20-0851.1

Muller C, Yang D, Craig G, Cronin T, Fildier B, Haerter JO, Hohenegger C, Mapes B, Randall D, Shamekh S, Sherwood SC (2022) Spontaneous aggregation of convective storms. Annu Rev Fluid Mech 54(1):133–157. https://doi.org/10.1146/annurev-fluid-022421-011319

Nguyen H, Duvel JP (2008) Synoptic wave perturbations and convective systems over equatorial Africa. J Clim 21:6372–6388. https://doi.org/10.1175/2008JCLI2409.1

Palmer PI, Wainwright CM, Dong B, Maidment RI, Wheeler KG, Gedney N, Hickman JE, Madani N, Folwell SS, Abdo G, Allan RP (2023) Drivers and impacts of Eastern African rainfall variability. Nature Rev Earth Environ 4(4):254–270. https://doi.org/10.1038/s43017-023-00397-x

Panthou G, Vischel T, Lebel T (2014) Recent trends in the regime of extreme rainfall in the Central Sahel. Int J Climatol 34:3998–4006. https://doi.org/10.1002/joc.3984

Pendergrass AG (2020) Changing degree of convective organization as a mechanism for dynamic changes in extreme precipitation. Curr Clim Chang Rep 6:47–54. https://doi.org/10.1007/s40641-020-00157-9

Peters O, Neelin JD (2006) Critical phenomena in atmospheric precipitation. Nat Phys 2:393–396. https://doi.org/10.1038/nphys314

Pilewskie J, Stephens GL, L'Ecuyer TS, Takahashi H (2024) A multi-satellite perspective on "hot tower" characteristics in the equatorial trough zone. Surv Geophys, submitted.

Prein AF, Liu C, Ikeda K, Trier SB, Rasmussen RM, Holland GJ, Clark MP (2017) Increased rainfall volume from future convective storms in the US. Nat Clim Chang 7:880–884. https://doi.org/10.1038/s41558-017-0007-7

Prein AF, Ge M, Valle AR, Wang D, Giangrande SE (2022) Towards a unified setup to simulate mid-latitude and tropical mesoscale convective systems at kilometer-scales. Earth Sp Sci 9:1–20. https://doi.org/10.1029/2022EA002295

Raghavendra A, Zhou L, Jiang Y, Hua W (2018) Increasing extent and intensity of thunderstorms observed over the Congo basin from 1982 to 2016. Atmos Res 213:17–26. https://doi.org/10.1016/j.atmosres.2018.05.028

Retsch MH, Jakob C, Singh MS (2020) Assessing convective organization in tropical radar observations. J Geophys Res: Atmos 125(7):e2019JD031801

Roca R, Fiolleau T (2020) Extreme precipitation in the tropics is closely associated with long-lived convective systems. Commun Earth Environ 1:1–6. https://doi.org/10.1038/s43247-020-00015-4

Roca R, Ramanathan V (2000) Scale dependence of monsoonal convective systems over the Indian Ocean. J Clim 13:1286–1298. https://doi.org/10.1175/1520-0442(2000)013%3c1286:SDOMCS%3e2.0.CO;2

Roca R, Aublanc J, Chambon P, Fiolleau T, Viltard N (2014) Robust observational quantification of the contribution of mesoscale convective systems to rainfall in the tropics. J Climate 27(13):4952–4958. https://doi.org/10.1175/JCLI-D-13-00628.1

Roca R, Fiolleau T, Bouniol D (2017) A simple model of the life cycle of mesoscale convective systems cloud shield in the tropics. J Clim 30:4283–4298. https://doi.org/10.1175/JCLI-D-16-0556.1

Satoh M, Stevens B, Judt F, Khairoutdinov M, Lin SJ, Putman WM, Düben P (2019) Global cloud-resolving models. Curr Climate Change Rep 5:172–184

Schiro KA, Sullivan SC, Kuo YH, Su H, Gentine P, Elsaesser GS, Jiang JH, Neelin JD (2020) Environmental controls on tropical mesoscale convective system precipitation intensity. J Atmos Sci 77:4233–4249. https://doi.org/10.1175/JAS-D-20-0111.1

Schmetz J, Pili P, Tjemkes S, Just D, Kerkmann J, Rota S, Ratier A (2002) An introduction to Meteosat second generation (MSG). Bull Am Meteorol Soc 83:977–992. https://doi.org/10.1175/1520-0477(2002)083%3c0977:AITMSG%3e2.3.CO;2

Schröder M, Lockhoff M, Forsythe JM, Cronk HQ, Haar THV, Bennartz R (2016) The GEWEX water vapor assessment: Results from intercomparison, trend, and homogeneity analysis of total column water vapor. J Appl Meteorol Climatol 55:1633–1649. https://doi.org/10.1175/jamc-d-15-0304.1

Schumacher C, Houze RA Jr (2006) Stratiform precipitation production over sub-Saharan Africa and the tropical East Atlantic as observed by TRMM. Q J Royal Meteorol Soc: J Atmos Sci, Appl Meteorol Phys Oceanogr 132(620):2235–2255. https://doi.org/10.1256/qj.05.121

Stephens G, Polcher J, Zeng X, Van Oevelen P, Poveda G, Bosilovich M, Ahn MH, Balsamo G, Duan Q, Hegerl G, Jakob C (2023) The first 30 years of GEWEX. Bull Am Meteorol Soc 104(1):E126–E157

Stephens GL, Shiro KA, Hakuba MZ, Takahashi H, Pilewskie JA, Andrews T, Stubenrauch CJ, Wu L (2024) Tropical deep convection, cloud feedbacks and climate sensitivity. Surveys in Geophysics. 31:1–29

Tan J, Jakob C, Rossow WB, Tselioudis G (2015) Increases in tropical rainfall driven by changes in frequency of organized deep convection. Nature 519:451–454. https://doi.org/10.1038/nature14339

Taylor CM, Belušić D, Guichard F, Parker DJ, Vischel T, Bock O, Harris PP, Janicot S, Klein C, Panthou G (2017) Frequency of extreme Sahelian storms tripled since 1982 in satellite observations. Nature 544(7651):475–478. https://doi.org/10.1038/nature22069

Taylor CM, Fink AH, Klein C, Parker DJ, Guichard F, Harris PP, Knapp KR (2018) Earlier seasonal onset of intense mesoscale convective systems in the Congo basin since 1999. Geophys Res Lett 45:13458–13467. https://doi.org/10.1029/2018GL080516

Tobin I, Bony S, Roca R (2012) Observational evidence for relationships between the degree of aggregation of deep convection, water vapor, surface fluxes, and radiation. J Clim 25:6885–6904. https://doi.org/10.1175/JCLI-D-11-00258.1

Tompkins AM, Semie AG (2017) Organization of tropical convection in low vertical wind shears: role of updraft entrainment. J Adv Model Earth Syst 9:1046–1068. https://doi.org/10.1002/2016MS000802

Vondou DA (2012) Spatio-temporal variability of western central african convection from infrared observations. Atmosphere (Basel) 3:377–399. https://doi.org/10.3390/atmos3030377

Wang T, Tang G (2020) Spatial variability and linkage between extreme convections and extreme precipitation revealed by 22-year space-borne precipitation radar data. Geophys Res Lett 47:1–10. https://doi.org/10.1029/2020GL088437

Weger RC, Lee J, Zhu T, Welch RM (1992) Clustering, randomness and regularity in cloud fields: 1 theoretical considerations. J Geophys Res Atmos 97:20519–20536. https://doi.org/10.1029/92JD02038

White BA, Buchanan AM, Birch CE, Stier P, Pearson KJ (2018) Quantifying the effects of horizontal grid length and parameterized convection on the degree of convective organization using a metric of the potential for convective interaction. J Atmos Sci 75:425–450. https://doi.org/10.1175/JAS-D-16-0307.1

Wing AA, Stauffer CL, Becker T, Reed KA, Ahn MS, Arnold NP, Bony S, Branson M, Bryan GH, Chaboureau JP, De Roode SR (2020) Clouds and convective self-aggregation in a multimodel ensemble of radiative-convective equilibrium simulations. J Adv Model Earth Syst. 12(9):e2020MS002138. https://doi.org/10.1029/2020MS002138

Yang Q, Houze RA Jr, Leung LR, Feng Z (2017) Environments of Long-Lived Mesoscale Convective Systems Over the Central United States in Convection Permitting Climate Simulations. J Geophys Res: Atmos 122(24):13–288. https://doi.org/10.1002/2017JD027033

Zhao S, Cook KH, Vizy EK (2024) Greenhouse gas-induced modification of intense storms over the west African sahel through thermodynamic and dynamic processes. Clim Dyn. https://doi.org/10.1007/s00382-024-07193-3

Publisher's Note Springer Nature remains neutral with regard to jurisdictional claims in published maps and institutional affiliations.

Surveys in Geophysics (2024) 45:1999–2048
https://doi.org/10.1007/s10712-024-09824-0

Lessons Learned from the Updated GEWEX Cloud Assessment Database

Claudia J. Stubenrauch[1] · Stefan Kinne[2] · Giulio Mandorli[1] · William B. Rossow[3] · David M. Winker[4] · Steven A. Ackerman[5] · Helene Chepfer[1] · Larry Di Girolamo[6] · Anne Garnier[4,7] · Andrew Heidinger[8] · Karl-Göran Karlsson[9] · Kerry Meyer[10] · Patrick Minnis[4,7] · Steven Platnick[10] · Martin Stengel[11] · Szedung Sun-Mack[4,7] · Paolo Veglio[5] · Andi Walther[5] · Xia Cai[7] · Alisa H. Young[12] · Guangyu Zhao[6]

Received: 25 September 2023 / Accepted: 8 January 2024 / Published online: 29 February 2024
© The Author(s) 2024

Abstract

Since the first Global Energy and Water Exchanges cloud assessment a decade ago, existing cloud property retrievals have been revised and new retrievals have been developed. The new global long-term cloud datasets show, in general, similar results to those of the previous assessment. A notable exception is the reduced cloud amount provided by the Cloud-Aerosol Lidar and Infrared Pathfinder Satellite Observation (CALIPSO) Science Team, resulting from an improved aerosol–cloud distinction. Height, opacity and thermodynamic phase determine the radiative effect of clouds. Their distributions as well as relative occurrences of cloud types distinguished by height and optical depth are discussed. The similar results of the two assessments indicate that further improvement, in particular on vertical cloud layering, can only be achieved by combining complementary information. We suggest such combination methods to estimate the amount of all clouds within the atmospheric column, including those hidden by clouds aloft. The results compare well with those from CloudSat-CALIPSO radar–lidar geometrical profiles as well as with results from the International Satellite Cloud Climatology Project (ISCCP) corrected by the cloud vertical layer model, which is used for the computation of the ISCCP-derived radiative fluxes. Furthermore, we highlight studies on cloud monitoring using the information from the histograms of the database and give guidelines for: (1) the use of satellite-retrieved cloud properties in climate studies and climate model evaluation and (2) improved retrieval strategies.

Keywords Satellite remote sensing · Atmosphere · Cloud cover · Cloud height · Cloud radiative properties · Microphysical properties · Cloud types

Extended author information available on the last page of the article

Article Highlights

- Since the first Global Energy and Water Exchanges (GEWEX) cloud assessment a decade ago, existing cloud property retrievals have been revised and new retrievals have been developed. Eleven climate data records of cloud properties are investigated, and a common database is made available
- While the spread in the absolute values between the datasets is still large, their latitudinal and seasonal variations agree well, except at the polar latitudes
- Though it is very challenging to detect small inter-annual global and zonal changes, similar geographical change patterns between the datasets demonstrate that the data can be used to study climatological change patterns, in particular those for specific cloud types
- The similar results of the two assessments indicate that further improvement, in particular on vertical cloud layering, needs combining complementary information

1 Introduction

Clouds are important regulators of the energy and water cycle. While surface observations provide the morphological structure of the cloud types, satellite observations make it possible to get an overview of the synoptic cloud fields (Fig. 1). Furthermore, the satellite observation era, beginning in the 1970s and spanning more than 40 years, allows us to monitor physical cloud properties over the entire globe. However, it is challenging to evaluate cloud properties on a global scale, as no unique dataset offers the ultimate truth, due to associated uncertainties.

The Global Energy and Water Exchanges (GEWEX) Cloud Assessment (Stubenrauch et al. 2013) provided the first coordinated inter-comparison of publicly available, global cloud products (gridded, monthly statistics). In addition to self-assessments (Annex I of Stubenrauch et al. 2012), indicating the maturity of the datasets, this inter-comparison has

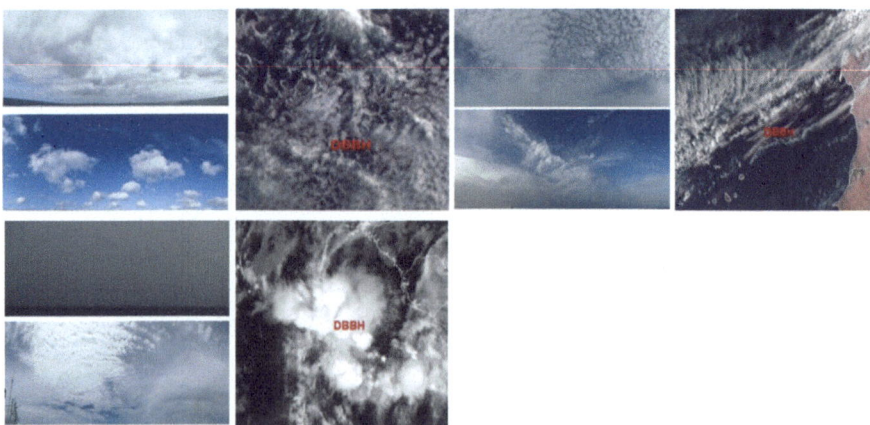

Fig. 1 Cloud types from surface observation (morning and afternoon) and synoptic cloud fields from Meteosat satellite images, for specific days in May 2023 during the science cruise M189/2 EARS. Examples show: stratocumulus fields off the coast of Namibia (top left), Cirrus and low-level clouds off the Sahara (top right) and convective cells and anvils in the ITCZ over the Atlantic (bottom). Position of the German research vessel Meteor, at 7:15 UTC:DBBH

shown how cloud property averages and distributions are affected by instrument characteristics and retrieval methods. The database associated with the GEWEX Cloud Assessment has revealed its usefulness for the assessment of new datasets. Since then, several teams have revised their retrievals, and new global long-term datasets have emerged. The updated cloud database contains data until 2020, in the same format as the original GEWEX Cloud Assessment database, with monthly statistics (averages, variability and histograms) per cloud property, at a spatial resolution of 1° latitude × 1° longitude. Since cloud property statistics are stored in files per year, we foresee a yearly extension of this database, if the cloud teams are able to provide the files.

These satellite data are very valuable for climate studies and for the evaluation of climate models, *if one is aware of the limits specific to the applied dataset*. In the following, we provide insight on how and why the retrieved cloud properties of these datasets differ. In this overview, we concentrate on cloud amount, cloud height as well as cloud radiative and microphysical properties. For the inter-comparison of averages and spatial distributions, we used the common 8-year period of 2008–2015.

After a short description of the different sensors, retrieval methods and the database in Sect. 2, we compare cloud amounts and cloud properties of the individual datasets in Sect. 3. This section also presents statistics of specific cloud types from two-dimensional histograms in the space of near-cloud-top pressure and visible cloud optical depth. As passive remote sensing only provides information on the uppermost cloud layers, Sect. 4 illustrates how complementary information can be used to estimate the full amount of all cloud layers, including the amount hidden under clouds aloft. Section 5 discusses the monitoring of cloud type occurrences with respect to time and to global surface temperature. A summary and guidelines are given in Sect. 6.

2 Sensors, Retrieval Methods and Database

The updated GEWEX cloud assessment database includes monthly cloud property statistics of eleven cloud climatologies, based on measurements from five different sensor types:

- *multi-spectral imagers*, with a maximum of 5 channels between visible (VIS) and infrared (IR), covering 0.5 µm to 12 µm, and a footprint size of about 4 km (Sect. 2.1),
- *advanced multi-spectral imagers*, like the Moderate Resolution Imaging Spectroradiometer (MODIS) with 36 channels between VIS (0.4 µm) and IR (14.5 µm) and a nadir footprint size from 250 m to 1 km (Sect. 2.2),
- *IR sounders*, with channels along the 15 µm CO_2 absorption band and a footprint size of about 13 km (Sect. 2.3)
- *multi-angle solar-spectral imager*, with 4 solar-spectral channels and a footprint size of about 1 km (Sect. 2.4)
- *active lidar*, along nadir tracks with a footprint size of about 90 m (Sect. 2.5)

The longest time series are available since the 1980s, from instruments aboard meteorological polar orbiting and geostationary satellites. Instruments operating since the 2000s have increased retrieval capabilities, in particular the spaceborne lidar and radar missions CALIPSO and CloudSat launched in 2006 (Stephens et al. 2018a, b).

All datasets of this assessment database, except ISCCP, are built from measurements aboard polar orbiting satellites. Table 1 recapitulates the datasets of this assessment

Table 1 Participating datasets, type of sensors, nominal local observation times and time period in the updated GEWEX Cloud Assessment Database

Dataset	Version	Sensor type	Observation time	Time period
ISCCP Young et al. (2018) and Rossow et al. (2022)	Version H	Multi-spectral imagers	1:30 AM/PM 7:30 AM/PM	1984–2015 1984–2015
AVHRR PATMOS-x Heidinger et al. (2012), Walther and Heidinger (2012)	Version 5	Multi-spectral imagers	1:30AM/PM 7:30 AM/PM	1983–2020 1992–2020
AVHRR-CLARA Karlsson et al. (2017)	Version 2	Multi-spectral imagers	1:30 AM/PM 7:30 AM/PM	1982–2015 1991–2015
AVHRR-Cloud_cci Stengel et al. 2020	Version 3	Multi-spectral imagers	1:30 AM/PM 7:30 AM/PM	1982–2016 1992–2016
MODIS CERES Science Team Trepte et al. (2019), Minnis et al. (2021)	Edition 4	Advanced multi-spectral imagers	1:30 AM/PM 10:30 AM	2000–2020 2002–2020
MODIS CLDPROP Frey et al. (2020); Platnick et al. (2021)	Continuity version	Advanced multi-spectral imagers	1:30 AM/PM	2002–2020
AIRS-CIRS, IASI-CIRS Stubenrauch et al. (2017) and Guignard et al. 2012	Version 2	Advanced IR sounders	1:30 AM/PM 9:30 AM/PM	2003–2018 2008–2018
MISR Di Girolamo et al. (2010)	Version 2	Multi-angle SW imager	10:30 AM/PM	2000–2017
CALIPSO-GOCCP Chepfer et al. (2010)	Version 2.9	Lidar	1:30 AM/PM	2007–2017
CALIPSO Science Team Liu et al. (2019)	Version 4.2	Lidar	1:30 AM/PM	2007–2016
IIR CALIPSO Science Team Ganier et al. (2021)	Version 4.2	Lidar/IR imager	1:30 AM/PM	2007–2016

database, with their nominal local observation times and temporal periods. The original ISCCP data, which provide cloud properties at eight specific universal times, have been processed for this database in such a way that the observation times are close to 1:30 and 7:30 local time (LT), AM and PM. Thus, with maximal four measurements per day, the analyses in this article are only able to partially resolve the diurnal cycle of the cloud properties. When presenting the optical and microphysical cloud properties as well as specific cloud types and their monitoring in Sects. 3.4 to 5, we only concentrate on analyses at one specific time of the day: 1:30 PM LT, because VIS optical depth is not available at night. This means that our analysis in the latter sections provides a snapshot and does not consider systematic diurnal variations of the specific cloud types (cf. Rossow and Schiffer 1991, 1999; Eastman et al. 2011; Eastman and Warren 2014; Rossow et al. 2022).

Furthermore, the National Oceanic and Atmospheric Administration (NOAA) satellites drifted by several hours during their life time (e.g. Foster et al. 2023) and the morning orbit time changed with advent of NOAA-17 in 2002 from 7:30 LT to 10:00 LT, which perturbs the values of the cloud properties retrieved from Advanced Very High Resolution Radiometer (AVHRR) measurements according to their diurnal variation and therefore may introduce biases in the time series (see Sect. 5).

Sections 2.1–2.7 shortly describe the different datasets. Table 2 summarizes their retrieval characteristics. Section 2.8 describes the database and gives some insight on the interpretation of the satellite-retrieved cloud properties. Not all datasets provide the complete set of variables shown in Table 3. An appendix of all acronyms is given in the supplement.

2.1 Long-Term Cloud Data from Multi-spectral Imagers

The *International Satellite Cloud Climatology Project* (***ISCCP***) uses a combination of polar orbiting and geostationary satellites in order to get a three-hourly diurnal sampling. ISCCP has recently released a new dataset (*Version H*, Rossow 2017, Young et al. 2018, Rossow et al. 2022), with a better spatial sampling (10 km instead of 30 km) and more up-to-date ancillary products (topography, land, water, snow and ice mask, atmospheric temperature-humidity profiles, ozone and aerosols). Night-time cloud properties are adjusted, based on the daytime differences between VIS/IR and IR-only results and the time interpolation of the VIS retrievals over the night-time. Changes in the polar regions are due to the removal of near-IR radiances in the retrieval over ice.

The version of ISCCP-H provided for this assessment is not the standard version as the monthly averages at four specific local times have been built from the ISCCP-HGG data, at 1° spatial resolution, given at eight specific universal times per day in such a way that the results of the two closest times were averaged. In the polar regions during periods without sunlight, only the IR-based results are available in the full ISCCP products, but in the GEWEX database, the height-stratified cloud amounts from VIS/IR analysis are not reported. Furthermore, only the 2-dimensional COD-CP and CEM-CP histograms are constructed from the initial 10 km pixels, while the 1-dimensional histograms are constructed from the 1° averages.

Measurements by the Advanced Very High Resolution Radiometers (AVHRR) aboard the NOAA and EUMETSAT polar orbiting satellites, integrated by ISCCP, have also been used on their own to produce cloud products by three different retrieval methods. The latter, listed below, are taking full advantage of all five channels and are based on very similar calibrated radiances of all available AVHRR instruments, on both

Table 2 Summary of Cloud Property Retrieval Characteristics of the Datasets Participating in the updated GEWEX Cloud Assessment Database. In bold are the cloud height parameters directly retrieved. The 3.7 micron channel has been replaced by another channel during the daylight morning period (see Sect. 2.1)

Dataset	Spatial resolution	Cloud detection	Variables	Retrieval method	Ancillary input
ISCCP-H	5 km,10 km (sampled) 50–75 km clear sky estimation	1 VIS 1 IR window time–space variances	COD, **CT**→CP, CZ	TB(11 µm)→CT, VIS→COD, CT correction for COD<15	NNHIRS retrieved *T/RH* profiles, ozone, aerosols, rad transfer + cloud layer/ particle model, surface properties (including snow/ice, topography)
			CEM	(1-exp(-COD/const(W/I)))	
			Phase (W/I)	ice: CT < 253K	
			CWP	fct(COD,phase,fixed CRE)	
PATMOS-x	1 km×4 km Sampling distance: 3 km along, 5 km across track	6 Bayesian classifiers derived from CALIPSO	CEM, **CT**→CP	Optimal estimation (11,12 µm)	NCEP reanalysis profiles (V1) *MODIS snow mask, rad. transfer + particle model (mixed habits for ice)*
			Phase (W/I)	Spectral differences	
			COD, CRE	LUT approach (0.6, 3.7 µm)	
			CWP	CLWP = fct(CODW, CREW); CWP = fct(CODI)	
AVHRR-CLARA	1 km×4 km Sampling same as PATMOS-x	Multi-spectral thresholding VIS/NIR/IR Threshold tuning against CALIPSO	**CT**→CZ,CP COD, CRE Phase (W/I)	IR matching of RTM-simulations + IR split-window histogram with ERA-interim T/WV-profiles as reference LUT approach (0.6, 3.7, 11 µm) spectral differences	ERA-interim *reanalysis profiles, rad. transfer + particle model*
			CWP	fct(COD, CRE, phase)	
AVHRR-Cloud_cci	1 km×4 km. sampling same as PATMOS-x	VIS/NIR/IR neural network	COD, **CP**, CRE, Phase→CT	Optimal estimation on VIS/NIR/IR (0.7, 0.9, 3.7, 11, 12 µm)	ERA-interim *reanalysis profiles, rad. transfer + particle model*
			CWP	fct(COD), CRE, phase)	

Table 2 (continued)

Dataset	Spatial resolution	Cloud detection	Variables	Retrieval method	Ancillary input
MODIS-CE	1 km, 4 km (sampled) 32 km clear sky estimation	multi-spectral IR/NIR/VIS (5 channels similar to VIRS)	CEM, **CT** → CZ,CP	IR split-window; lapse rate (7.1 K/km) + T profile	GMAO GEOS G5.4.1-CERES reanalysis profiles, *rad. transfer + particle model (mixed habits for ice)*
			Phase, COD, CRE	CT + LUT approach (0.6, 2.1 µm), (3.8 µm)	
			CWP	CWP = fct(COD, CRE, phase)	
MODIS-CLDPROP	1 km	multi-spectral IR/NIR/VIS (16 channels) + time-space variances	**CP**, CEM → CT	CO_2 slicing for CP < 650 hPa, TB(11 µm)	NCEP GDAS reanalysis *profiles, 16 day spectral surf. albedo climatology, rad. transfer + particle model (mixed particle habits for ice)*
			Phase (W/I)	VIS/NIR/IR spectral differences	
			COD, CRE	LUT approach (0.7, 0.9, 1.2, 1 6 µm), (2 1 µm)	
			CWP	fct(COD, phsse, CRE)	
AIRS-CIRSv2 IASI-CIRSv2	13.5 km 12 km	a posteriori: coherence of 6 spectral cloud emissivities (9–12 µm) &CEM > 0.05/0.1 (H/ML)	CEM & **CP**, CP → CT,CZ, CEM → COD	weighted χ^2 method on 8 CO_2; absorption channels, T/virt. T profile for CT/CZ; COD = −c × ln (1-CEM)	ERA-interim reanalysis *profiles, spectral surf. emissivities: IASI climatology; 4A/OP + DISORT rad. transfer + ice crystal SSPs of hex. columns and aggregates*
			Phase (W/I)	ice: CT < 250 K liquid: CT > 260 K	
			CREIH, CIWPH	LUT approach on 6 spectral emissivities (9–12 µm)	
MISR	1 km	multi-specral + angle VIS/NIR	**CZ**	stereoscopic cloud top height	
CALIPSO-GOCCP	0.09 km, sampling: 0.34 km	Lidar VIS backscatter vertical averaging	**CZ** → CT	cloud mean altitude, uppermost cloud layer (for GEWEX)	GMAO MERRA2 reanalysis profiles

Table 2 (continued)

Dataset	Spatial resolution	Cloud detection	Variables	Retrieval method	Ancillary input
CALIPSO-ST	0.09 km, sampling: 0.34 km	Lidar VIS backscatter horizontal averaging	$CZ \rightarrow CT$	cloud top, uppermost cloud layer (for GEWEX)	GMAO MERRA-2 reanalysis profiles
			Phase (W/I)	ice: 532 nm depolarization	

Table 3 Variable names of the cloud properties with statistics also distinguished by altitude (H: CP < 440 hPa, M: 440 hPa < CP < 680 hPa, L: CP > 680 hPa) and by thermodynamic phase (W: water clouds, I: ice clouds, IH: ice clouds with CP < 440 hPa) available in the updated GEWEX Cloud Assessment database

Variable	Total	H	M	L	W	I	IH
Cloud amount	CA	CAH	CAM	CAL	CAW	CAI	CAIH
Effective cloud amount	CAE	CAEH	CAEM	CAEL	CAEW	CAEI	CAEIH
Relative to total CA		CAHR	CAMR	CALR	CAWR	CAIR	CAIHR
Cloud temperature	CT	CTH	CTM	CTL	CTW	CTI	CTIH
Cloud pressure	CP						
Cloud height	CZ						
Cloud emissivity	CEM	CEMH	CEMM	CEML	CEMW	CEMI	CEMIH
Cloud optical depth	COD	CODH	CODM	CODL	CODW	CODI	CODIH
Cloud water path					CLWP	CIWP	CIWPH
Cloud effective radius					CREW	CREI	

morning (nominal equator crossing at 7h30 AM and PM LT) and afternoon (nominal equator crossing 1h30 AM and PM LT) satellites. The observation times slowly drift in time, in particular the ones of the afternoon satellite orbits (cf. Foster et al. 2023). Furthermore, the orbit time of the morning satellites changed from about 7:30 LT to about 10:00 LT with the advent of NOAA-17 in 2002, and during daytime the 3.7-µm channel was replaced by a 1.6-µm channel.

The *Pathfinder Atmospheres Extended* (**PATMOS-x**, *Version 5*) participated with the same retrieval version in the original GEWEX cloud assessment. Cloud detection is based on Bayesian classifiers derived from CALIPSO (Heidinger et al. 2012), and the retrieval is based on an optimal estimation approach (Heidinger and Pavolonis 2009). First cloud pressure (CP) and cloud emissivity (CEM) are retrieved using two IR channels at all times of day. Then, cloud optical depth (COD) and cloud particle effective radius (CRE) are obtained from solar channels during daytime so that finally cloud water path (CWP) can be derived from COD and CRE (Walther and Heidinger 2012).

The *Climate Monitoring Satellite Application Facility* (CM SAF) ***Cloud, Albedo and surface Radiation dataset*** from AVHRR data (***AVHRR-CLARA***, *Version 2*, Karlsson et al. 2017) detects clouds based on multi-spectral thresholding, which has been tuned through comparisons with cloud observations from the CALIPSO lidar. Cloud top heights are retrieved using radiance matching with radiances simulated by a radiative transfer model for opaque clouds and a split-window histogram approach for semi-transparent clouds. Cloud phase is interpreted from further spectral analysis, and a look-up table (LUT)-based retrieval of COD and CRE, following the classical Nakajima–King approach, is used. Finally, the cloud water path (CWP) is derived as a function of COD, CRE and cloud phase.

The ***ESA Cloud Climate Change Initiative*** has generated two cloud data records from AVHRR (AVHRR-AM and AVHRR-PM, *Version 3*, Stengel et al. 2020) which are combined in this study (***AVHRR-Cloud_cci*** hereafter). The cloud detection and cloud thermodynamic phase determination are based on artificial neural networks (ANNs), which were trained using data from CALIPSO. For the cloud property retrieval, the optimal-estimation-based Community Cloud retrieval for CLimate (CC4CL) was employed (Sus et al. 2018; McGarragh et al. 2018). AVHRR-Cloud_cci covers the time period 1982–2016. An extension until 2020 exists, which is of lower quality and not provided for this database.

2.2 Cloud Properties from Advanced Multi-spectral Imagers

The ***MODIS CERES Science Team*** (***MODIS-CE***) retrieval was recently upgraded (*Edition 4*, Trepte et al. 2019; Minnis et al. 2021), using new calibrations for solar channels, including the 1.24-μm channel for COD retrieval over snow and a new ice crystal reflectance model. The use of CO_2 absorbing channels and a regional lapse rate technique are applied to improve the height retrieval of high- and low-level clouds, respectively. CA, CEM, CT, COD and CRE are directly retrieved, and all other variables are deduced from these using ancillary data and parameterizations. MODIS data have been processed from instruments aboard Terra (equator crossing 10h30 AM and PM LT) and Aqua (equator crossing 1h30 AM and PM LT).

The new cloud property continuity product from the National Aeronautics and Space Administration (NASA) ***MODIS Science Team*** (***MODIS-CLDPROP***) is designed to provide continuity between MODIS and the Visible Infrared Imaging Radiometer Suite (VIIRS) with an algorithm that uses only a subset of similar spectral channels available on both imagers. Note that VIIRS is missing key atmospheric absorbing spectral channels used in the MODIS Standard Product (MOD/MYD06) cloud retrievals, in particular the 13–14 μm CO_2 slicing channels. CLDPROP (Meyer et al. 2020; Platnick et al. 2021), and the companion common algorithm continuity CLDMSK cloud mask product (Frey et al. 2020), are based on heritage algorithms used in the MODIS Collection 6.1 cloud product suite (Platnick et al. 2017) with the exception of the cloud top property datasets that are based on an optimal estimation algorithm (Heidinger and Pavolonis 2009; Heidinger et al. 2019). CLDPROP is in production for both MODIS Aqua and, to date, VIIRS on SNPP and NOAA-20. For convenience, CLDPROP in this article will also refer to the CLDMSK product and only the MODIS CLDPROP production stream is discussed. COD, CRE and CWP are retrieved using a LUT approach based on solar reflectance and midwave IR channels (Platnick et al. 2021). Only microphysical retrievals that use the MODIS 2.1-μm channel are currently reported as part of the GEWEX dataset (the CLDPROP production version includes retrievals based on other SWIR and MWIR channels). Note that the gridded sampling of COD, CRE and CWP may be a subsample of the CP, CEM, CT pixel population because retrievals of optical and microphysical properties are only attempted for masked pixels that meet stricter requirements than that of the cloud top properties algorithm. Further the optical/microphysical pixels population is partitioned into two separate datasets, a dataset for pixels that appear to be overcast at the spatial resolution of MODIS (1 km at nadir along with 250 m observations over the ocean) and a dataset for pixels that are likely to be partly cloudy (Platnick et al. 2017). As such, files named ***MODIS-CLDPROP*** provide optical/microphysical cloud properties only for the so-called overcast pixels having a successful retrieval (considered highest quality), while ***MODIS-CLDPROP_ALL*** files, also part of the updated cloud assessment database, *provide statistics that include the successful optical/microphysical partly cloudy pixel retrieval population as well.*

2.3 Cloud Properties from IR Sounders

The original GEWEX cloud assessment database includes three different IR sounder cloud climatologies: HIRS-NOAA (Wylie et al. 2005), TOVS Path-B (Stubenrauch et al. 2006) and AIRS-LMD (Stubenrauch et al. 2010).

The measurements of the High Resolution Infrared Radiation Sounder (HIRS) instruments, flown on, aboard 16 satellites from 1980 through 2015, have been recently reprocessed by Menzel et al. (2016), after having been carefully intercalibrated. The cloud retrieval method is similar to the one of HIRS-NOAA, but is now applied to HIRS measurements that have been identified as cloudy by the PATMOS-x cloud mask (Heidinger et al. 2012). This dataset is not part of the updated GEWEX cloud assessment database.

The ***CIRS (Clouds from IR Sounders)*** retrieval is an updated version of the AIRS-LMD retrieval. It has been developed for the application on any IR sounder data and has been applied so far to observations from the Atmospheric Infrared Sounder (AIRS), aboard the NASA platform Aqua, and the Infrared Atmospheric Sounding Interferometers (IASI) aboard the EUMETSAT platforms Metop-A and Metop-B (equator crossing at 9h30 AM and PM). A weighted χ^2 method (Stubenrauch et al. 1999) using eight channels along the CO_2 absorption band simultaneously provides CP and CEM. CT and CZ are derived from CP, using ancillary atmospheric T and water vapour profiles. T inversions stronger than 2 K are accounted for, setting the near cloud-top to the level of the inversion and scaling the cloud emissivity. An 'a posteriori' cloud detection relies on the spectral coherence of the retrieved cloud emissivity. The original CIRS version only kept clouds with CEM > 0.1 (Stubenrauch et al. 2017). Here we present results of CIRS *version 2*, with CEM > 0.05 for high-level clouds (as in AIRS-LMD), and CEM > 0.1 for all lower clouds (CP > 440 hPa). The synergy of AIRS-CIRS and IASI-CIRS (Feofilov et al. 2019) was achieved by using the same ancillary data, in particular atmospheric profiles of the meteorological reanalyses ERA-Interim (Dee et al. 2011). Since the production of ERA-Interim ceased in August 2019, the CIRS retrieval has been recently adapted to ancillary data of ERA5 (Hersbach et al. 2020), and the whole data record until present is being produced for AIRS and for IASI at the French data centre AERIS. This dataset will be part of this updated GEWEX cloud assessment database in the near future.

2.4 Cloud Properties from a Multi-angle Solar-Spectral Imager

The Multi-angle Imaging SpectroRadiometer (***MISR***, *version 2*) provides a Cloud Fraction by Altitude product derived from four measurements in the spectral domain between VIS and shortwave-infrared (SWIR) that are collected by nine separate cameras ranging in viewing zenith angles from 0° to 70.5° in the along-track direction (Di Girolamo et al. 2010). Cloud detection is achieved using spectral, spatial, angular signature, and stereoscopic tests, while cloud top height (CZ) is achieved using stereoscopy. This cloud height retrieval does not rely on ancillary products and is independent of radiometric calibration. The solar wavelengths are especially useful for retrieving CZ of low-level clouds, also in the presence of thin higher clouds above. A comparison with Cloud-Aerosol Transport System (CATS) lidar data (Mitra et al. 2021) has shown that MISR detects the lower cloud in a two-layered system, provided top-layer optical depth < ~ 0.3. MISR participated in the original GEWEX cloud assessment, and since then the stereoscopic technique has been improved (Mueller et al. 2013). Furthermore, the additional dataset ***MISR_RC*** introduces a *resolution-corrected cloud amount*. MISR_RC uses a machine learning technique that has been trained using collocated MISR and ASTER (15 m resolution) data to correct for the resolution bias in cloud amount caused by coarse (~ 1 km) resolution measurements. This effectively retrieves cloud amount as would be measured at 15 m resolution. Cloud amount is estimated by only counting fully cloudy footprints of a size of 15 m (Jones et al. 2012; Dutta et al. 2020) and therefore provides a lower limit on cloud amount.

2.5 Cloud Data from Active Lidar

Two products, using different approaches, are based on lidar data of the Cloud-Aerosol Lidar and Infrared Pathfinder Satellite Observation (CALIPSO) mission, and both approaches participated in the original GEWEX cloud assessment. Differences between these products in the retrieved cloud properties are related to differences in the detection and cloud–aerosol classification algorithms used. In order to be more compatible with the results from passive remote sensing, both datasets provide a version which only keeps the information of the uppermost cloud layer in the case of multiple cloud layers.

The ***GCM-Oriented CALIPSO Cloud Products*** (***CALIPSO-GOCCP***, *Version 2.9*) are derived from the CALIPSO L1 Version 4 data, after vertically averaging single-shot profile data (0.33 km along track × 0.090 km cross-track) to 0.48 km. The method used is the same as described in (Chepfer et al. 2010), using a solar noise correction for daytime data described in the first GEWEX Cloud Assessment report (Stubenrauch et al. 2012). Cloud amounts are computed as the number of cloudy profiles within a grid cell divided by the total number of profiles.

The ***CALIPSO Science Team*** (***CALIPSO-ST***, *Version 4.2*) uses horizontal averaging (5, 20 or 80 km) to detect weakly scattering, thin cirrus, while only single-shot data are used for water clouds and dense ice clouds. The 90 m receiver footprints of the CALIPSO lidar are separated by 333 m along-track (Winker et al. 2010). Cloud amounts are computed as the number of cloudy profiles within a grid cell divided by the total number of profiles. The revised dataset uses CALIPSO L2 Version 4.2 data products, which feature an improved aerosol-cloud discrimination algorithm (Liu et al. 2019), an improved ice-water classification (Avery et al. 2020) relative to the original version of CALIPSO-ST, and an improved selection of cirrus lidar ratio constrained by infrared observations (Garnier et al. 2015).

For the updated cloud assessment database, in addition to the cloud amounts of only the uppermost cloud layer (here reported as CALIPSO-ST_top), CALIPSO-ST reports the cloud amounts of all detected cloud layers within the atmospheric column (CALIPSO-ST_column), one based on only the highest cloud layer in each column having COD > 0.3 (CALIPSO-ST_passive) and one based only on cloud layers which attenuate the lidar surface return signal to the point where it can no longer be detected (CALIPSO-ST_opaque), approximately at COD > 3 (Garnier et al. 2021a). *CALIPSO-ST_passive is meant to give an indication of the effect on cloud amount due to sensitivity limits of a typical passive imager, recognizing that this COD threshold detection limit depends on passive instrument and technique, underlying surface characteristics, cloud type, aerosol loading, sun-view geometry, and thermal structure of atmosphere and surface. However, CALIPSO-ST_passive reports the actual lidar-detected cloud top altitudes (as do all CALIPSO-ST products) and therefore does not mimic the cloud heights retrieved by passive sensors* (Sect. 2.8).

2.6 Cloud Data from Combined Active Lidar and IR Imager

Since the IR spectrum is sensitive to the size of ice crystals (up to a radius of about 50 µm), an infrared imager radiometer (IIR) was developed to accompany the lidar of the CALIPSO mission. It has three channels between 8 and 12 µm and a footprint size of 1 km. The IIR_CALIPSO-ST products, new to this assessment, include IIR ice and water microphysical retrievals for single-layer scenes identified by the CALIPSO lidar. Cloud radiative temperatures are estimated from the CALIPSO lidar profiles (Garnier et al. 2021b). IIR_CALIPSO-ST provides CT and CEM, further converted to COD, but only where microphysical properties are successfully retrieved in single-layer ice or water clouds.

2.7 Cloud Data from Combined Active Radar and Lidar

While the lidar cannot probe the atmosphere below optically dense clouds, the radar cannot detect thinner layers at any level and has a reduced sensitivity near the ground (1 km), because of surface clutter. Therefore, the most complete picture of the cloud vertical structure is obtained by combining the lidar with radar measurements, both part of the A-Train mission since 2006 (Stephens et al. 2018a). Compared to passive remote sensing, which does not provide such a complete picture of vertical structure, lidar and radar sampling provides a more sophisticated vertical view but only along narrow nadir tracks. These measurements are very sparse in comparison with observations of cloud properties retrieved from passive remote sensing. This very sparse sampling leads to uncertainties in regional mean cloud properties (e.g. Astin et al. 2001; Kotarba 2022). In particular, individual radar–lidar transects need to be averaged at least over an area $10° \times 10°$ in longitude and latitude in order to achieve an accuracy of 1% (for CA) or 150 m (for CTH). The annual mean CA estimate is very sensitive to infrequent sampling, resulting in 14% or 7% average uncertainty over 1° or 2.5°, respectively.

For the comparisons in Sect. 4, we analysed CloudSat-CALIPSO GEOPROF data (*version R04*, Mace et al. 2009) of a dataset provided by J. Mace for this data assessment. It is available at https://macegroup.chpc.utah.edu/qzhang/occ/occ/occ.jsp and includes monthly statistics on maximal two vertical cloud layers, at 1° latitude × 1° longitude. Cloud top statistics has been stratified into eight categories with values of 17, 12, 8, 4.5, 2.5, 1.5, 0.75 and 0.25 km. Seven different cloud layer thicknesses are assigned with each cloud top, given by values of 0.125, 0.375, 1.00, 2.25, 4.50, 8 or 12 km. Since only maximal two cloud layers are given in this dataset, we extended the cloud layer thickness towards the surface for those cloud layers with a vertical extent of at least 8 km. For a comparison of cloud amounts stratified into high-, mid- and low- level clouds according to pressure (with boundaries at 440 and 680 hPa), polar boundaries at 3 and 6 km were increased towards 3.6 and 7.2 km at lower latitudes (35 S–35 N). Compared to version R04, the new version R05 (Mace and Zhang 2014) shows reduced false detections of the CloudSat cloud mask but at a cost of a significant loss in the true weak signals. By comparing with the CALIPSO lidar vertical feature mask, Hu et al. (2020) have improved this scheme for a future CloudSat data processing which will lead to a better detection of thin clouds.

2.8 Database

The database covers the same variables as the original GEWEX cloud assessment database (Stubenrauch et al. 2012). These variables are listed in Table 3.

In addition to averages over grid boxes of 1° latitude × 1° longitude, the database also contains monthly histograms of cloud properties given at their retrieval spatial resolution. Exceptions are the MODIS-CE and ISCCP histograms, containing values which have been first averaged over 0.2° (CERES sensor footprint) and 1° (ISCCP map grid resolution), respectively. Only the 2-dimensional COD-CP and CEM-CP histograms of ISCCP were constructed from the initial 10 km pixels.

Cloud amount (CA) is difficult to quantify at high accuracy, as it *depends on satellite sensor ability to detect optically thin clouds and partial cloud cover over the instantaneous field of view (IFOV)*. In general, for the retrieval of cloud properties it is assumed that the IFOV is completely cloud-covered. Overall, the detection thresholds should be chosen to balance between detecting and overestimating thin cirrus or broken low-level clouds (e.g. Wielicki and Parker 1992). However, there is no guarantee that this balance can be achieved to reach unbiased CA at regional scales (e.g. Zhao and Di Girolamo 2006).

Fortunately, those clouds which are the most difficult to detect are also those with the smallest radiative effect (e. g. Zhang et al. 2004), but they may still play a role in other applications (e.g. Koren et al. 2008; Pincus et al. 2012).

Height-stratified cloud amounts (CAH, CAM, CAL) give separate amounts of high-, mid- and low-level clouds. Satellite sensors observe the clouds from above, and hence, in the case of multiple layers, passive remote sensing only yields the properties of the uppermost cloud layer or of the whole cloud column. This means that the amounts of cloud top layers sum up to total cloud amount: CAH + CAM + CAL = CA, while the real amount of the underlying cloud layers may be larger due to parts hidden by the layers above. The thresholds of 440 hPa (> 6 km) and 680 hPa (< 3 km), initially defined for ISCCP (Rossow and Schiffer 1991), were motivated by cloud type classification of surface observers, adjusted to approximate the difference between cloud base and cloud top locations. These designations have been consistently followed for both versions of the GEWEX Cloud Assessment database.

The most *accurate geometrical cloud-top height* can be determined by spaceborne lidar (CALIPSO) and by a stereoscopic retrieval from multiple viewing angles (MISR). In the case of multi-layer clouds, the latter usually reports the height of the low-level clouds. Both approaches, however, come at the cost of low spatiotemporal sampling, particularly CALIPSO. Passive IR sensors are usually better in this respect, though the remote sensing techniques applied to their measurements provide a '*radiative cloud height*'. For water clouds, this is less of a problem as those clouds have in general well-defined cloud-tops. For ice clouds that have a 'fuzzy' cloud-top however, the retrieved cloud-top height may lie a few kilometres below the actual cloud-top height, depending on their extinction profile, e.g. in tropical high-level clouds (e.g. Liao et al. 1995a, b; Wylie and Wang 1997; Minnis et al. 2008; Hamann et al. 2014; Stubenrauch et al. 2010, 2017; Mitra et al. 2021). For opaque, tropical deep convective clouds, it has been demonstrated that the mean effective emission level of the MODIS 11-μm band corresponds to a level at which the cloud reaches an IR optical depth of about 0.7 ± 0.3 (Wang et al. 2014). The CIRS cloud height corresponds to a level within the cloud where the optical depth from CALIPSO reaches about 0.5 or to a level near mid-cloud for optically thin clouds (Stubenrauch et al. 2017). The difference between cloud-top height and radiative cloud height affects CAH differences between CALIPSO and the other sensors only at higher latitudes, because the fixed threshold of 440 hPa is in the tropics far below the tropopause, while at higher latitudes (in particular in winter) 440 hPa is close to the tropopause and therefore the radiative cloud height may be in the mid-level cloud category while the cloud-top is in the high-level cloud category (Fig. 6).

Other difficulties in the interpretation of retrieved cloud height arise from *differences in footprint size* (1–15 km), with larger footprints leading to a larger probability of partial overlapping multiple cloud layers and from the retrieval of different variables which represent height (cloud temperature or cloud pressure), which are then transformed to height by using *different ancillary data* (Sect. 3.2).

To diagnose cloud radiative effects, one needs in addition to cloud amount at least cloud temperature, VIS optical depth and IR emissivity. The height-stratified *effective cloud amount* (CAEH, CAEM, CAEL), defined as the cloud amount weighted by the cloud IR emissivity for the specific cloud height category, should reconcile differences in cloud amount, in particular CAH, because a smaller cloud amount due to lower sensitivity is compensated by larger cloud IR emissivity.

Cloud water path (CWP) and *effective particle radius* (CRE, averaged over a size distribution within the cloud) are the variables predicted or parameterized in climate models. These variables are then used in radiative transfer computations to obtain the *optical depth* COD. In general, at a constant CWP, the solar albedo increases with decreasing CRE. The latter may be retrieved from multi-spectral differences in the solar or thermal domain, using pre-computed look-up tables. The radiative transfer computations *need certain assumptions*, such as optical properties and crystal habit in the case of ice clouds and shape of the particle size distribution. These retrievals provide CRE at a certain depth within the cloud, which depends on the wavelengths used and on the opaqueness of the clouds. While for low-level liquid clouds in general the effective cloud droplet size decreases from top to base, the effective ice crystal size increases from top to base when sedimenting (e.g. Jensen et al. 2018).

3 Assessment of the Individual Datasets

3.1 Total and Height-Stratified Cloud Amounts

3.1.1 Averages

Figure 2 presents global averages of total and height-stratified cloud amounts of the eleven datasets participating in the updated GEWEX cloud assessment database. These averages are area-weighted, day-night averages over 1:30 AM and PM LT, except for IASI (9:30 AM and PM LT) and MISR (10:30 AM LT).

CA: Results are in general similar to those of the original GEWEX Cloud Assessment. The ***global cloud amount, CA, 0.66 ± 0.04, is slightly lower than before*** (0.68 ± 0.03), due to a much lower value from the new CALIPSO-ST version V4 (0.66 compared to 0.73 based on version V2) and due to slightly lower CA from the newly developed datasets AVHRR-CLARA and AVHRR-Cloud_cci. The decrease in CA of CALIPSO-ST is mostly due to an improved aerosol–cloud discrimination (Liu et al. 2019), which corrected for high altitude dust layers classified as cloud in earlier data versions. Keep in mind that these averages come mostly from observations at 1:30 AM and PM. Therefore, they include a

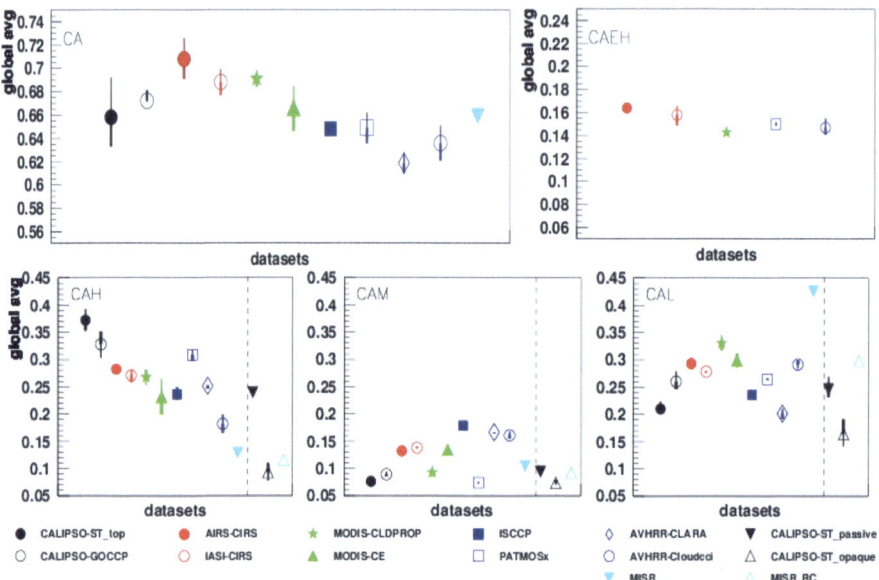

Fig. 2 Upper left panel: Global averages of total cloud amount (CA) of the eleven datasets participating in the updated GEWEX cloud assessment database. Upper right panel: Global averages of CAEH of the five datasets providing this variable. Lower panel: Global averages of height-stratified cloud amount (CAH, CAM and CAL) of the eleven datasets. CAH and CAM of clouds with COD > 0.3 and of clouds with COD > 3 from CALIPSO-ST as well as results from MISR_RC are also shown for comparison. Statistics averaged over day and night (1:30AMPM LT, except IASI (9:30AMPM) and MISR (10:30 AM)) over the common period of 2008—2015. *The colours correspond to specific instrument types (black: active lidar, red: IR sounders, green: advanced multi-spectral imagers, blue: multi-spectral imagers, cyan: solar-spectral imager). The vertical bars indicate day (thick) and night (thin) differences with respect to the daily mean*

small uncertainty due to the sparse time sampling. The CA averages at all available specific observation times available in this database are shown in Table S1 in the supplement.

In Fig. 2, we also show day–night differences, with the end of the thicker (thinner) bar indicating the daytime (night-time) value. A study with CATS lidar data (50 N–50 S, Noël et al. 2018) found that over ocean 1:30 PM and 1:30 AM correspond to the daily minimum and maximum cloud amount, respectively, while the diurnal cycle over land is slightly underestimated when using only these observation times. *Most datasets show indeed a slightly smaller global CA during day than during night.* The larger day-night spread for CALIPSO-ST is partly due to daytime noise from solar scattering, for which the CALIPSO-GOCCP data were corrected.

CAH We observe a large spread in CAH in the lower left panel of Fig. 2, where the datasets are grouped with decreasing detector sensitivity to thin high-level clouds. This spread in CAH is mainly explained by differences in instrument and retrieval method sensitivity to thin high-level clouds, including scenes with lower clouds underneath, and, in comparison with CALIPSO, by height-stratification differences according to cloud-top and radiative cloud height (see Fig. 6).

Active lidar has the highest sensitivity to thin cirrus, followed by IR sounders (with a high spectral IR resolution and no contributing VIS reflectance from underlying clouds). The IR-VIS imagers show slightly lower values of CAH (Figs. 2 and 3). While thin

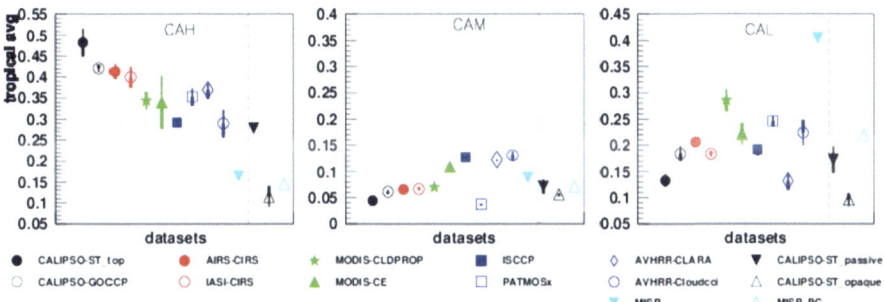

Fig. 3 Tropical (15N–15S) averages of height-stratified cloud amount (CAH, CAM and CAL) of the eleven datasets. CAH and CAM of clouds with COD > 0.3 and of clouds with COD > 3 from CALIPSO-ST as well as results from MISR_RC are also shown for comparison. Statistics averaged over day and night (1:30AMPM LT, except IASI (9:30AMPM) and MISR (10:30 AM)) over the common period of 2008—2015. *The colours correspond to specific instrument types (black: active lidar, red: IR sounders, green: advanced multi-spectral imagers, blue: multi-spectral imagers, cyan: solar-spectral imager). The vertical bars indicate day (thick) and night (thin) differences with respect to the daily mean*

cirrus can often be identified, e. g. by spectral IR differences, their height assignment is difficult as mentioned in Sect. 2.8.

In order to quantify the two categories, 1) missed thin cirrus in a single layer and 2) mis-identified thin cirrus overlying clouds at lower levels, we use information from the different CALIPSO-ST datasets and our CloudSat-CALIPSO GEOPROF analysis, summarized in supplemental Table S2. We estimate the *amount of all thin cirrus (COD < 0.3)* as the CAH difference between CALIPSO-ST_top (0.37) and CALIPSO-ST_passive (0.24). The global average is 0.13, with largest amounts in the tropics and over Antarctica, according to Figure S1a in the supplement. Indeed, the CAH of MODIS-CE, ISCCP and AVHRR-CLARA are close to the CAH of CALIPSO-ST_passive. Yet CAH of CALIPSO-ST_passive provides only a rough estimate of what a passive imager would estimate, because the accumulative CAH changes by about 0.1 for high-level clouds with COD between 0.2 and 0.3 (Balmes et al. 2019).

We further deduce from the CloudSat-CALIPSO data that 35% of all high-level clouds are single-layer clouds (Table 4). By assuming that the overlap between cirrus and lower clouds is independent of the COD of the high-level cloud, we derive

Table 4 Top panel: Global averages of height-stratified cloud amount, as seen from above (top-layer); bottom panel: column view with an illustration of 7 simplified cloud vertical structure classes, with columns including high-, mid- and low-level clouds (multiple cloud layers and contiguous cloud layers together). Amounts are reported as average from CloudSat-CALIPSO GEOPROF and CLDCLASS data (Sect. 2.7 and Oreopoulos et al. 2017). Statistics covers 82N-82S over 2007–2009 and 2007–2010, respectively. The ranges correspond to the difference between the average and the single analysis values and an additional 10% uncertainty due to the use of an earlier CALIPSO version (V3)

L	M	H	top-layer
0.24 ± 0.05	0.08 ± 0.02	0.40 ± 0.07	CloudSat-CALIPSO

1L	1M	M L	1H	H L	H M	H M L	column
0.24 ± 0.05	0.03 ± 0.01	0.06 ± 0.02	0.14 ± 0.03	0.08 ± 0.02	0.07 ± 0.03	0.12 ± 0.03	CloudSat-CALIPSO

the *amount of single-layered thin cirrus (COD < 0.3)* as 0.35 x (CAH(CALIPSO-ST_top)—CAH(CALIPSO-ST_passive)) = 0.05. The remaining amount of 0.08 (about 60%) corresponds to *thin cirrus overlying clouds at lower levels*. The latter should then correspond to the difference between the CAM derived by imager and CAM of CALIPSO-ST_passive. This seems indeed to be the case for ISCCP, AVHRR-CLARA, AVHRR-Cloud_cci and MODIS-CE.

Furthermore, we observe *large differences due to retrieval methodology*: The CALIPSO-ST horizontal averaging scheme and detection thresholding are more sensitive to subvisible cirrus than the CALIPSO-GOCCP vertical averaging, with CAH of 0.37 and 0.33, respectively. CAH deduced from AVHRR measurements even ranges from 0.18 (AVHRR-Cloud_cci) to 0.31 (PATMOS-x). The latter value is due to an overestimation at higher latitudes, in particular in the SH (see Fig. 5).

The ISCCP results do not show any day-night difference because night-time results were adjusted by adding the daytime difference of VIS/IR and IR results interpolated over the night-time. Since the combination of VIS-only information together with stereoscopy favours the height retrieval of lower clouds, CAH from MISR (0.12) is the lowest amongst the datasets and only slightly larger than the amount of opaque high-level clouds (0.09) estimated from CALIPSO-ST_opaque.

In order to mitigate the effect of the difference between cloud-top and radiative height on CAH, we investigate the tropical averages (15N-15S) of the height-stratified cloud amounts in Fig. 3. We see indeed a smaller spread in CAH than for the global averages. The relative difference in CAH between CALIPSO-ST and CIRS decreases from 24% in global average to 15% in the tropics, and CAH of CIRS and CALIPSO-GOCCP is very close. The spread within the AVHRR datasets has also decreased. CAH of ISCCP and now also AVHRR-Cloud_cci are close to CAH of CALIPSO-ST_passive. The retrieval methods of PATMOS-x and AVHRR-CLARA seem to have a higher sensitivity towards high-level clouds, close to those from MODIS.

Similar results in CAH between CIRS and MODIS-CE during night (0.29 and 0.27 global, 0.42 and 0.41 tropical, respectively) show that the sensitivity of radiometers can be enhanced by using several IR spectral channels. However, when the retrieval method includes VIS information during the day, without considering the possibility of underlying clouds, thin cirrus may be mis-identified as a more opaque lower cloud because of a large VIS reflectance from the cloud underneath.

Therefore, we recommend to develop retrieval methods (during daytime) which are able to distinguish single- and multi-layered cloud scenes. This may be achieved by first using the IR radiances and their spectral differences alone and then comparing the retrieved cloud properties to those obtained when including VIS reflectances.

ISCCP is using a two-step method (first IR and then VIS), but because of building a coherent climatology based on geostationary data back to the 1980s, it is not possible to include spectral IR differences. The MODIS-CLDPROP retrieval treats IR and VIS separately, leading indeed to CEM and COD from different methods, but the sampling of cloudy scenes is different: all cloudy scenes (0.69 over 60 N–60 S) for CEM and only 'overcast' cloudy scenes (0.47 over 60 N–60 S) for COD. Therefore, their statistics are difficult to combine, but in this database products with the same IR and VIS sampling are also provided, which will be compared in Sect. 3.4.

CAEH The effective amount of high-level clouds, CAEH, of the datasets agrees much better (Fig. 2, upper right panel) relative to CAH, with a global average of about 0.15 ± 0.01 from the five datasets, which directly retrieve CEMH. Global averages of CAE and CEM, as well as their height-stratified averages, are shown separately in supplemental Figures S2 and S3.

CAM The global amount of mid-level clouds with no higher clouds above is quite small, with smallest values from CALIPSO, PATMOS-x and MODIS-CLDPROP (0.08 ± 0.01). The CAM averages of ISCCP, AVHRR-CLARA and AVHRR-Cloud_cci are largest (0.18 ± 0.01), because they likely include thin cirrus clouds for which the height assignment is biased low. Considering the tropics (Fig. 3), also CAM of CIRS is small and very close to the values of MODIS-CLDPROP and CALIPSO, while the global average is slightly larger. This may be explained by the fact that the cloud height retrieved by CIRS has a broader distribution than MODIS-CLDPROP (see Fig. 6), most probably due to the worse spatial resolution which increases the probability of partly overlapping multiple layer clouds in the field of view, and at higher latitudes the threshold of 440 hPa puts some of these clouds into a lower cloud category.

CAL The global amount of single-layer low-level clouds is about 0.26 ± 0.05, with CALIPSO-ST_top and AVHRR-CLARA at the lower end (0.21) and MODIS, CIRS and AVHRR-Cloud_cci at the higher end (0.30 ± 0.02). Again, the global mean of CAL provided by ISCCP is close to the one of CALIPSO-ST_passive. The low CAL of AVHRR-CLARA can be explained by the misidentification between low- and mid-level clouds, in particular over land and in the stratocumulus regions (see appendix Fig. 15). The larger values may be overestimated due to partly cloudy footprints, which have been counted as overcast. CAL of MISR (0.42) is very similar to CAL_{col} (corrected for clouds underneath opaque high-level clouds, see Sect. 4). However, as stated in Sect. 2.4, MISR detects low-level clouds underneath clouds with COD < 0.3. Hence, one expects CAL of MISR to be similar to CAL of CALIPSO-ST_passive (0.25). Therefore, the combination of VIS-only reflectances together with stereoscopy seems to largely overestimate single-layer CAL. Compared to MISR, MISR_RC shows much smaller values for CAL (Figs. 2, 3). Partly cloudy fields can be identified by the ratio of CAL of MISR_RC over CAL of MISR that is smaller than 1. When considering the geographical map of this ratio in Figure S1b in the supplement, the largest changes are in the trade wind cumulus regions and over land in the tropics and subtropics.

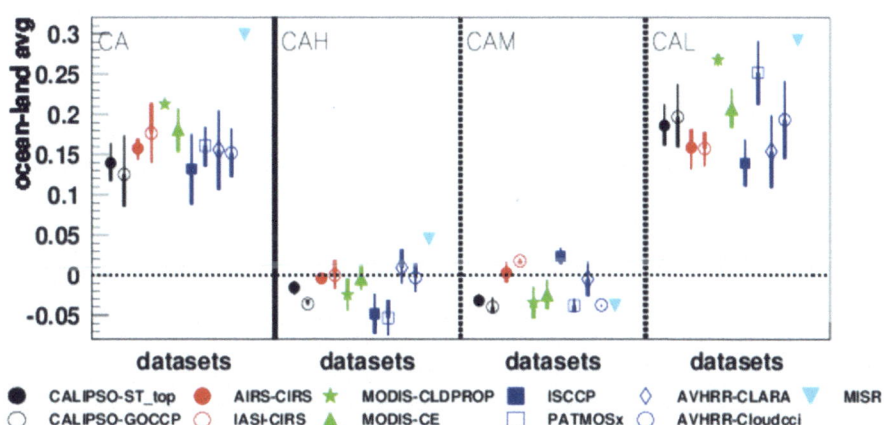

Fig. 4 Ocean minus land differences in total cloud amount (CA) and in height-stratified cloud amount of the eleven datasets participating in the updated GEWEX cloud assessment database. Statistics averaged over day and night (1:30AMPM LT, except IASI (9:30AMPM) and MISR (10:30AM)) over the common period of 2008–2015. The same land–ocean mask was used in the analysis. *The colours correspond to specific instrument types (black: active lidar, red: IR sounders, green: advanced multi-spectral imagers, blue: multi-spectral imagers, cyan: solar-spectral imager). The vertical bars indicate day (thick) and night (thin) differences with respect to the daily mean*

From Fig. 4, we deduce that the *cloud amount over ocean is larger than over land, by about 0.15 ± 0.05, which is almost entirely due to more low-level clouds over ocean*. The spread in the CAL difference is larger than 0.10. The MISR results differ from the others by a twice as large ocean-land difference in CA and a positive ocean-land difference in CAH. The geographical maps in appendix Figure 15 indicate that MISR identifies low-level clouds over ocean where other datasets identify high-level clouds, as expected since stereoscopy favours the height retrieval of lower clouds when cirrus are aloft, while PAT-MOS-x provides less low-level clouds over land at higher latitudes, because of identifying more high-level clouds there (due to a training with CALIPSO-ST which includes polar stratospheric clouds).

Instrument retrieval capabilities aside the observed spread in annual global ocean-land cloud amount differences do have other contributing factors tied to sampling differences between the instruments. This includes the time of day in which samples are collected (thus impacted by the diurnal cycle of clouds), the instrument swath (largely impacting the amount of coverage near the poles), and the reliance on sunlight of the retrieval algorithms (largely impacting large areas at mid to high latitudes of the winter hemisphere).

In order to estimate the uncertainty due to limited diurnal sampling, we also investigated the ocean-land differences given at 7:30 and 10:30 AM and PM in supplemental Figure S4. *The values averaged over AM and PM are very similar between 1:30 and 7:30–10:30 LT.* However, the day–night spread of the ocean-land difference in CA between 1:30 AM and 1:30 PM LT reaches 0.1 and originates mostly from low-level clouds. This is most probably due to their nearly opposite diurnal cycle over ocean and over land (cf. Eastman and Warren 2014; Rossow et al. 2022). The only dataset with a negligible day-night spread is MODIS-CLDPROP. Missing observation times, in particular early morning and late afternoon, lead to further uncertainty in these comparisons of about 0.02–0.05 (Rossow et al. 2022).

The spread of 0.06 in the ocean-land difference of CAH (excluding MISR) vanishes when considering CAEH (not shown). This indicates that the slightly larger CAH values over land are due to more semi-transparent cirrus.

3.1.2 Latitudinal and Seasonal Variation

Figure 5 presents the latitudinal variation of CA and CAH as well as their differences between boreal summer and boreal winter. *In general, the latitudinal variation agrees well between all datasets, except at the polar latitudes.* We concentrate our discussion in this section on CAH, because the variation of CAM and CAL is influenced by CAH and by the sensor or retrieval sensitivity of thin high-level clouds and therefore difficult to interpret. The latitudinal variation of the column amount of mid- and low-level clouds, including the amount hidden by cloud layers aloft, is discussed in Sect. 4.

The largest CA is found in the SH mid-latitudes, mostly from single-layer low-level and mid-level clouds in summer (supplement Figure S5) and from high-level (storm) clouds during winter. The seasonal oscillation of SH mid-latitude cloud tops was noted in Rossow and Schiffer (1999, Fig. 8), where it was shown that in the NH CP does not vary while CT does with air temperature but in the SH CP varies seasonally. The local peak around 5 N of the annual mean of CA is due to the high-level clouds in the Inter-Tropical Convergence Zone (ITCZ).

The peak in CAH is seen by all datasets, but with a large range from 0.35 (AVHRR-Cloud_cci) to 0.55 (CALIPSO-ST_top), and MISR with 0.2, as discussed in Sect. 3.1.1.

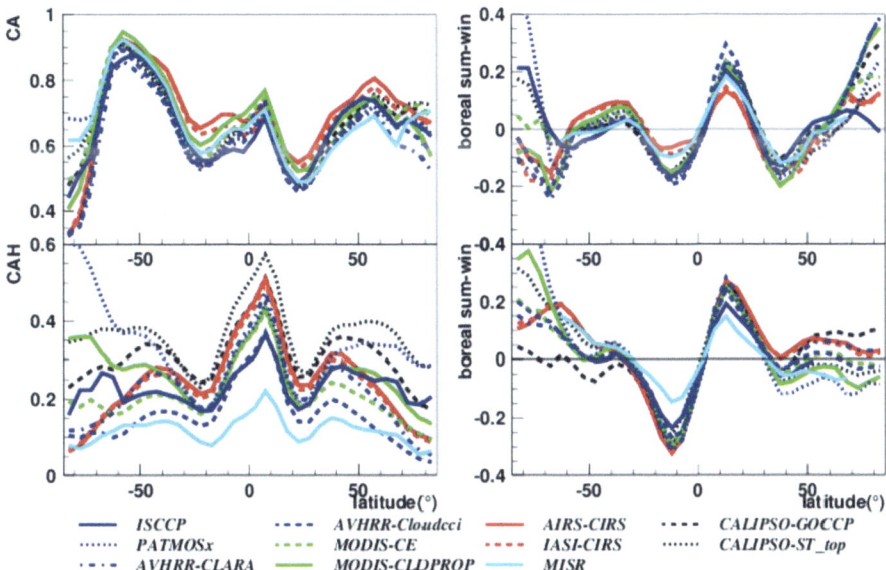

Fig. 5 Left panel: Zonal averages of CA (top) and CAH (bottom) of the eleven datasets participating in the updated GEWEX cloud assessment database. Right panel: Differences between boreal summer (June, July, August) and boreal winter (December, January, February) of CA (top) and CAH (bottom). Statistics averaged over day and night (1:30AMPM LT, except IASI (9:30AMPM) and MISR (10:30AM)) over the common period of 2008—2015. The colours correspond to specific instrument types (black: active lidar, red: IR sounders, green: advanced multi-spectral imagers, blue: multi-spectral imagers, cyan: solar-spectral imager)

The displacement of the ITCZ by about 20° between boreal summer and boreal winter is well described by all datasets. The CAH latitudinal variation of MISR is close to the one of CALIPSO-ST_opaque (not shown). Considering CAEH (supplement Figure S6), the local minima and maxima in the zonal means are at similar latitudes but they are less pronounced, and the tropical peak value lies between 0.2 and 0.35, indicating more thin cirrus in the tropics than at the other latitudes. PATMOS-x, the passive remote sensing dataset with the largest global mean CAH, agrees in the tropics with the other datasets, but has much larger values at higher latitudes. This may be related to the fact that in the training with CALIPSO polar stratospheric clouds were included. In particular, the extremely large CAH southward of 50S is identified as very thin high-level cloud (in comparison with CAEH, supplemental Figure S6).

The *spread of CAH in the polar latitudes is large, because* on one hand *passive remote sensing of clouds is less reliable over ice and snow,* and on the other hand *these regions are often covered by multi-layered clouds with thin cirrus (polar stratospheric) aloft low-level clouds*, deduced from cloud amounts of CALIPSO-ST_column (see Sect. 4).

The subtropics in both hemispheres have a local minimum in CA and in CAH. The NH mid-latitudes have a smaller CA in summer than in winter. At higher latitudes, we observe a large CAH difference between CALIPSO and all passive remote sensing datasets, while their CAM exceeds the one of CALIPSO (Figure S5). This is due to the retrieval of cloud-top height and radiative cloud height, respectively. At these latitudes, the latter may fall into a lower height category than the cloud-top height (see Sect. 3.2).

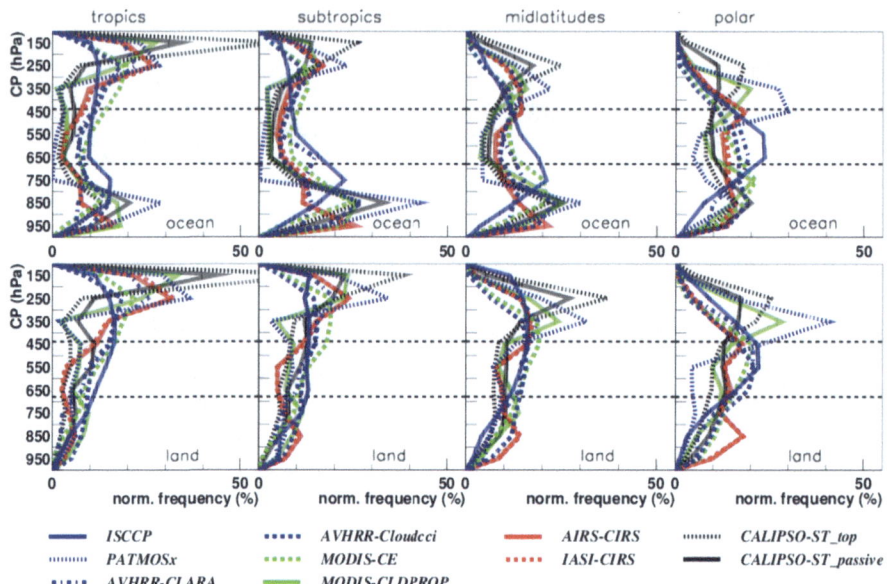

Fig. 6 Normalized frequency distributions of near-cloud-top pressure, CP, in four latitude bands (tropics: 10° N–10° S; subtropics: 10°–30°, mid-latitudes: 30°–60°; and polar: 60°–90°), separately over ocean (top) and over land (bottom). The frequency is given per 100 hPa. Statistics averaged over 1:30AM and PM LT (9:30AM and PM IASI). These normalized distributions only reflect a view from above, not considering hidden cloud layers. The colours correspond to specific instrument types (black: active lidar, red: IR sounders, green: advanced multi-spectral imagers, blue: multi-spectral imagers)

3.2 Height

Figure 6 presents the normalized frequency distributions of CP for four latitude bands, separately over ocean and over land. For the normalization, the histograms given in the database are divided by the total number of cloudy cases (sum over all events). Since the interval width of the histograms is 100 hPa, these frequencies are given per 100 hPa. These normalized distributions only reflect a view from above, not considering hidden cloud layers. We have also included the distributions of CALIPSO-ST_passive in order to illustrate the effect of sensitivity. Similar distributions of CZ are given in the supplement Figure S7, which includes MISR and CALIPSO-GOCCP.

In general, the CP distributions over ocean are bimodal, with relatively large populations of cloud-tops in the upper and in the lower troposphere. Over land, the contribution of single-layer low-level clouds is smaller, as expected from Fig. 4. *The normalized frequency of low-level clouds is highest over subtropical ocean in all datasets except for AVHRR-CLARA*.

A weaker mid-level mode appears at 5 km in the tropics in CZ for MISR and MODIS-CLDPROP and for CALIPSO-ST over land (supplemental Figure S7). A tri-modal distribution of convection with peaks at 16 km, 6 km and 2 km, corresponding to the cloud top height of Cumulonimbus, Congestus and shallow cumulus, was found by Johnson et al. (1999) with data from the Tropical Ocean Global Atmosphere Coupled Ocean–Atmosphere Response (TOGA-COARE) Experiment. This congestus peak appears at about 7 km in MODIS-CE, while the CIRS and AVHRR datasets (CLARA and Cloud_cci) show a continuous distribution, probably linked to a coarser spatial resolution. This indicates that the tropics have a wide range of cloud heights in the convective regions, with narrow convective towers and widespread anvils and thin cirrus in the upper troposphere (UT) and congestus clouds in the middle troposphere as well as shallow cumulus and stratus in the subsidence regions (see also Sect. 4).

Compared to the original GEWEX cloud assessment (Stubenrauch et al. 2013), there are *still large differences in these profiles between the different datasets, but these can be understood as follows:*

1) retrieved cloud-top (CALIPSO, MISR) versus radiative height within the cloud:

In the tropical UT, the CALIPSO-ST cloud-top height peaks are narrower and at higher altitude (by about 50 hPa or 2—2.5 km) than radiative cloud height peaks obtained from passive remote sensing, which are indeed, as expected, up to a few km below the tops. An exception is MODIS-CLDPROP in the tropics, with peak values at the same height as CALIPSO-ST, but nevertheless a broader distribution. The vertical averaging of CALIPSO-GOCCP leads to a broader CZ peak than CALIPSO-ST (supplemental Figure S7).

From the tropics towards the poles, the peaks in the UT decrease in height (increase in CP) and get broader. *While the cloud-tops determined by CALIPSO stay in the UT, the radiative cloud height starts to decrease towards the middle troposphere, which partly explains an increasing height-stratified cloud amount difference between CALIPSO and passive remote sensing towards higher latitudes.*

2) different sensitivity to thin cirrus, in particular when overlying lower clouds:

From the comparison between CALIPSO-ST_top and CALIPSO-ST_passive, we deduce a large contribution of thin cirrus in the tropics close to the tropopause, but also the polar regions have a nonnegligible contribution. The polar thin cirrus clouds often appear in combination with lower clouds (estimated by the difference of CAL between corrected CALIPSO-ST_column and CALIPSO-ST_top, see Sect. 4). Part of these polar clouds of CALIPSO are stratospheric clouds, which are most likely missed by passive remote sensing. While most datasets show a bimodal distribution also over the polar latitudes, ISCCP, AVHRR-CLARA and AVHRR-Cloud_cci show profiles with one peak in the middle troposphere.

In the tropical UT, the CP peak positions and amplitudes of CIRS, PATMOS-x and AVHRR-CLARA are quite close to each other, whereas the CP peaks decrease in amplitude, by getting broader, towards MODIS-CE (in particular during daytime, not shown), AVHRR-Cloud_cci and ISCCP. Here we have to keep in mind that ISCCP in the GEWEX database provides histograms of averages over 1°, and therefore these results are worse than using the original dataset (Rossow et al. 2022).

The CZ distribution of MISR (supplemental Figure S7, zoomed version) shows a much larger low-cloud mode than the other datasets because MISR favours the accurate height retrieval of low clouds. While the peak in the low- and mid-level cloud mode between CALIPSO-ST_passive and MISR are within their reported uncertainty, the peaks in their high-level cloud mode are ~3 km apart in the tropics (at 15.5 km and 12.5 km, respectively). This is more than 2 km larger than the uncertainty in retrieved heights of thin cirrus for CALIPSO and MISR. This, together with the very small peak amplitude of MISR, suggests that, in the tropics, there is a much higher relative frequency of occurrence of optically thin cirrus with tops at ~15.5 km compared to optically thin cirrus with tops ~12.5 km, diurnal sampling differences notwithstanding.

3) assuming fully cloudy footprints (this effect should increase with footprint size):

Over ocean, the peak positions corresponding to the single-layer low-level clouds lie around 850 hPa (or 1.5 km) and agree quite well between CALIPSO, MODIS-CE, AVHRR-Cloud_cci and PATMOS-x (and MISR in CZ, supplement Figure S7). Over tropical ocean, MODIS-CLDPROP finds low-level clouds equally frequent between 950 and 850 hPa, while *CIRS peaks at 950 hPa*, most probably due to *partly covered larger footprints, which include surface emission*. This effect is added to by the common presence of a temperature inversion over low oceanic clouds that affects the ancillary temperature profiles used to convert cloud top temperature to cloud top pressure (see point 4).

4) differences in the transformation between CP, CT and CZ:

Another reason for discrepancies in CP and CZ is linked to the use of different ancillary data, in particular atmospheric profiles, which are taken from HIRS observations by ISCCP and from different meteorological reanalyses (NCEP V1, NCEP GDAS, ERA-Interim, GMAO GEOS-CERES, GMAO MERRA-2, see Table 3) by the other datasets. CALIPSO and MISR do not use ancillary products in retrieving CZ.

However, it is interesting to note that both ISCCP and AVHRR-CLARA, which first retrieve CT and then derive CP, using different atmospheric profiles (HIRS and ERA, respectively), place low-level clouds too high in height, with broader peaks. This is particularly noticeable for AVHRR-CLARA over subtropical ocean, where a major part of the low-level clouds is assigned to mid-level clouds. This bias in low-level cloud heights may be due to confusion in how to use temperature to assign height in the case of a strong inversion at the top of the marine boundary layer (e.g. Holz et al. 2008).

3.3 Thermodynamic Phase and Radiative Properties

While cloud droplets in liquid clouds are spherical, ice crystals exist in a diversity of shapes. The refractive index of liquid and ice particles is also different. Therefore, the thermodynamic phase is important for the determination of the radiative effect of clouds as well as for the retrieval of cloud optical depth. Among other applications, cloud phase is also important for studying cloud glaciation and precipitation.

Liquid and ice clouds can be distinguished by polarization (CALIPSO), by near cloud-top temperature (ISCCP: $CTI < 253$ K and $CTW > 253$ K, CIRS: $CTI < 250$ K, $CTW > 260$ K and mixed phase clouds assumed to exist between 250 and 260 K but not considered in the I-W separation) or by use of multi-spectral information (PATMOS-x, AVHRR-CLARA, AVHRR-Cloud_cci and MODIS-CE).

Relative water (CAWR) and ice (CAIR) cloud amounts differ between the datasets, as observed in Fig. 7, but their latitudinal behaviour is similar. Comparing CAIR and CAIHR shows that while in the tropics the vast majority of the ice clouds are high-level clouds, the relative contribution of lower level ice clouds increases towards higher latitudes (in winter, supplement Figure S8).

Normalized frequency distributions of CT are compared separately for water clouds and high-level ice clouds in Fig. 8 (left). Again, the normalization is obtained by dividing the histograms by the sum of the cloudy cases building the histograms, and since the interval widths vary, each frequency has been divided so that it corresponds to a frequency per K. This is a finer binning than in CP. The distributions of IIR_CALIPSO-ST are subsampled to single-layer clouds with a successful microphysical property retrieval. For AIRS-CIRS, we show high-level ice cloud distributions for all and only for those clouds subsampled for the retrieval of bulk microphysical properties ($CTIH < 235$ K and $0.85 < CEMIH < 0.2$).

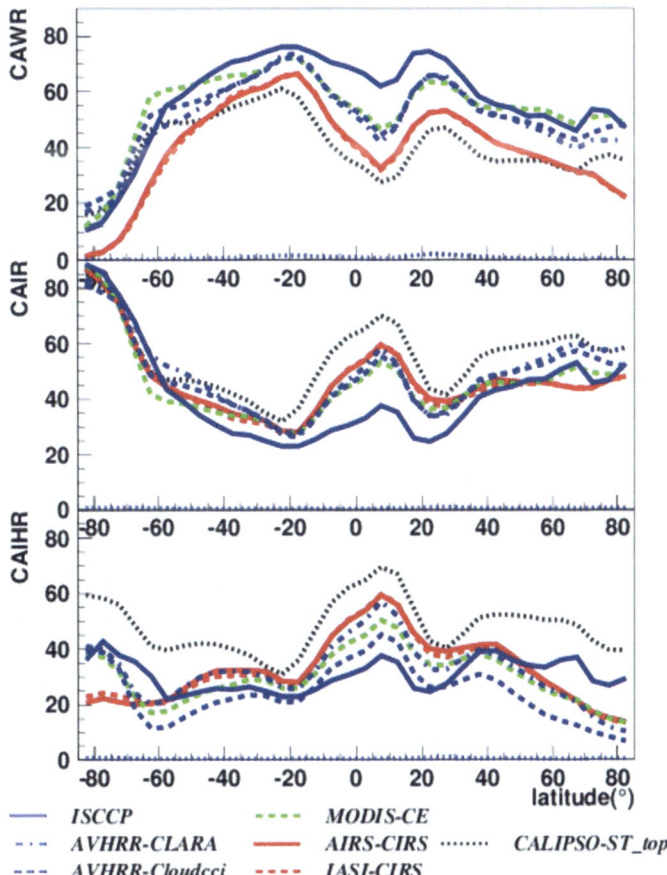

Fig. 7 Zonal variation of CAWR, CAIR and CAIHR (in %) of seven datasets participating in the updated GEWEX cloud assessment database. Note that CAWR + CAIR ≠ 100% indicates that a fraction of the clouds have an unassigned phase (which may be interpreted as mixed phase). Statistics averaged over day and night (1:30AMPM LT, except IASI (9:30AMPM)) over the common period of 2008–2015. The colours correspond to specific instrument types (black: active lidar, red: IR sounders, green: advanced multi-spectral imagers, blue: multi-spectral imagers)

CTW distributions are similar, with a peak around 285 K, slightly colder for AVHRR-CLARA and ISCCP. The CTW distribution of IIR-CALIPSO-ST shows much colder temperatures, with a peak value at 265 K, because IR microphysical retrievals tend to fail in dense and warm clouds near the surface. The CTIH distributions peak around 230 K (including IIR_CALIPSO-ST which reports radiative cloud temperatures), except for CALIPSO cloud-top temperature at 210 K and AVHRR-Cloud_cci at 240 K. Distributions of cloud-top temperatures for all high-level ice clouds (CALIPSO-ST_top) and for those with IIR_CALIPSO-ST microphysical retrievals are similar (not shown). The second peak around 195 K for ISCCP corresponds to very thin cirrus clouds, which ISCCP sets to the tropopause height. This peak is seen by CALIPSO-ST_top and not by CALIPSO-ST_passive, as expected.

CEM is retrieved at thermal wavelengths, while COD is retrieved at solar wavelengths and is therefore only available during daytime. MODIS-CLDPROP, AVHRR-Cloud_cci,

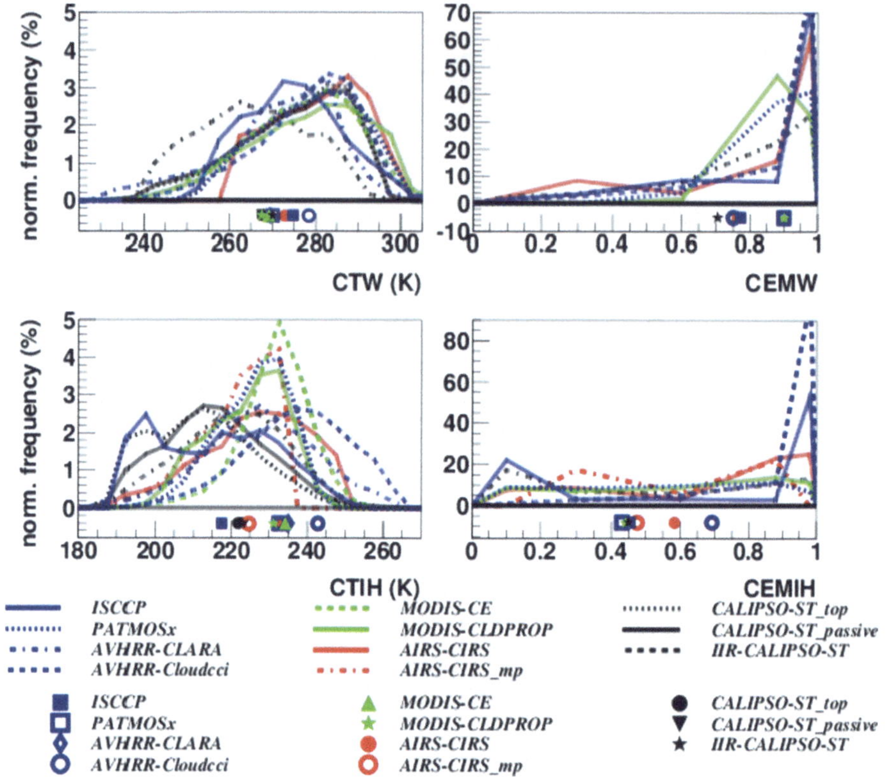

Fig. 8 Normalized frequency distributions of radiative properties (*T* and CEM) as well as their averages (below), for liquid clouds (top) and for high-level ice clouds (bottom). The frequencies of T and CEM correspond to those per 1 K and per 0.1, respectively. Statistics is averaged over 60 N–60 S, at 1:30 PM LT. The colours correspond to specific instrument types (black: active lidar, red: IR sounders, green: advanced multi-spectral imagers, blue: multi-spectral imagers)

and PATMOS-x retrieve both variables independently. ISCCP, CIRS and IIR_CALIPSO-ST only retrieve VIS COD or IR CEM, respectively, and use a conversion formula to obtain the other. AVHRR-CLARA only provides COD, and the other datasets (CALIPSO-ST, CALIPSO-GOCCP and MISR) do not provide any information on these variables.

Figure 8 (right) presents the normalized frequency distributions of CEM, separately for water clouds and high-level ice clouds. The frequencies are given per 0.1. All the CEMW distributions exhibit a peak towards large emissivity, but with a larger contribution of emissivities between 0.8 and 0.95 for MODIS-CLDPROP, PATMOS-x and IIR_CALIPSO-ST. AIRS-CIRS shows also a small contribution of emissivities smaller than 0.4. These smaller emissivities can be due to slightly overestimated cloud amount (see Sect. 3.1), and in the case of IIR_CALIPSO-ST due to the failure of the microphysical retrieval in dense clouds. The shape of the CEMIH distributions varies more. In general, there are many more semi-transparent high-level ice clouds than water clouds, as the peaks near 1 are much smaller. Only AVHRR-Cloud_cci shows a similar frequency near 1 as for CEMW and nearly no contribution of CEMIH < 0.5. Combined with the information that CTIH is also warmer than for the other datasets, the AVHRR-Cloud_cci retrieval method, based on optimal estimation with an assumption of single-layer clouds, overestimates both cloud emissivity and temperature of high-level ice clouds. The small peak of CEMIH near 0.1 of ISCCP corresponds to the very thin cirrus near the tropopause, with CTIH near 195 K, while for IIR_CALIPSO-ST these thin cirrus clouds seem to have a more variable radiative CTIH (no peak near 195 K).

Distributions of COD and bulk microphysical properties are presented and shortly discussed in the supplement (supplement Figures S9 to S12). COD distributions are difficult to present, as they are strongly skewed with some datasets showing very long tails at large COD. Though all monthly mean values are radiatively averaged, these tails yield large mean values, as for PATMOS-x, with a mean of 6.5 compared to values between 4 and 5.2 of the other datasets (supplement Figure S9). In general, the conversion of IR CEM to VIS COD underestimates the average COD, with a mean value of 2, because a CEM reaching 1 leads to a maximum COD value of about 10.

Since the distributions of all these variables are quite broad and not Gaussian, their mean values appear to be not very informative. Instead, using histograms allows one to monitor the relative changes in the distributions. This is further investigated in the next sections.

3.4 Relative Occurrence of Specific Cloud Types

The database also includes two-dimensional (2D) histograms in the CP-COD and CP-CEM space, respectively, which were designed for a more detailed analysis of various cloud types (i.e. Lau and Crane 1995, 1997; Rossow et al. 2005a; Tselioudis et al. 2013, 2021). Figure 9 presents the normalized relative occurrence frequency of cloud types distinguished by CP and COD in seven latitude bands, using the initial 7×7 intervals. Figures S13 and S14 present the CP-COD and CP-CEM distributions, respectively, regrouped into 3×4 intervals. Those distributions are more similar than the ones with the finer, original intervals.

As already seen in Fig. 6, the decrease of the tropopause height with latitude is visible in all datasets by a decrease in cloud height. An exception is the large proportion of very thin and high cirrus identified by ISCCP. These clouds are detected marginally in the IR and usually not in the VIS.

The ratio of optically thicker to optically thinner clouds increases towards higher latitudes, except for AVHRR-Cloud_cci, which has a large contribution in the middle COD range in the polar regions.

All datasets also show a minimum of single-layer mid-level clouds in the tropics and subtropics. Yet, by comparing datasets based on the same observations, we observe large differences in the distributions related to the retrieval method. They mostly come from multi-layered cloud fields: While the COD derived from VIS reflectances corresponds to the column optical depth, leading to correct TOA fluxes (Zhang and Rossow 2023), methods based on spectral IR alone identify the uppermost cloud and in this case the COD derived from the IR radiances corresponds to the COD of the uppermost detected cloud. This shows that these methods are complementary and their synergy should be used to describe the whole atmospheric cloud column (Sect. 4). The agreement between MODIS-CLDPROP and AIRS-CIRS is quite good, even if the footprint sizes of the two instruments are very different. *This indicates that the IR spectral information is more important than the spatial resolution to identify cirrus.*

Maps of the normalized occurrence of nine cloud types distinguished by COD and CP in Fig. 10 reveal in all datasets specific geographical patterns, such as the Stratocumulus

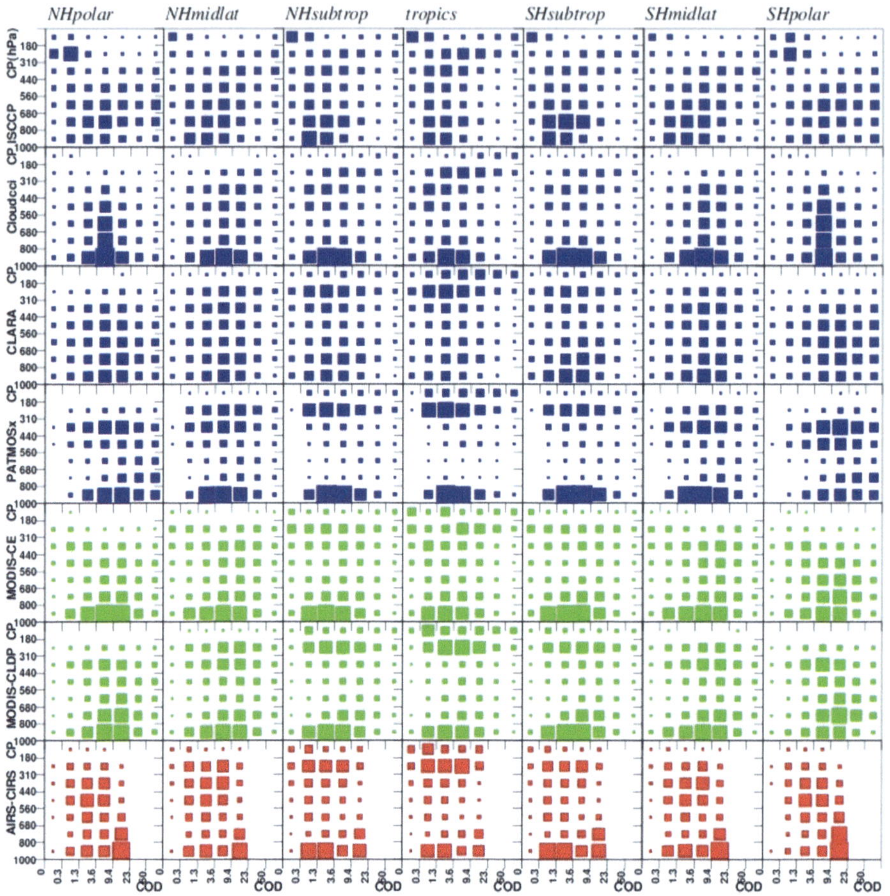

Fig. 9 Normalized occurrence frequency of cloud types distinguished by CP and COD, for 7 latitude bands. The original binning of 7 intervals in COD and in CP is presented. Statistics averaged over 1:30PM LT, from 2008 to 2015, except for MODIS-CE (2016–2020). The size of the boxes indicates the relative frequency. Long-term datasets in blue, advanced imagers in green and IR sounder in red

regions off the Western coasts in the (COD,CP) interval (2,3), tropical cirrus anvils in the (COD,CP) interval (2,1), tropical thin cirrus in the (COD,CP) interval (1,1) or midlatitude storm tracks in the (COD,CP) interval (2,2). Yet, some of these cloud fields appear in neighbouring COD-CP intervals as well. In particular, tropical thin cirrus and cirrus anvils also appear as mid-level clouds for ISCCP, AVHRR-Cloud_cci and AVHRR-CLARA. When considering a finer binning (not shown), the stratocumulus regions appear slightly higher and more opaque for CIRS and AVHRR-CLARA. For ISCCP they are shifted towards slightly smaller COD, while for PATMOS-x, AVHRR-Cloud_cci and MODIS-CLDPROP they are placed at larger CPs. These differences highlight the CP and COD uncertainties due to retrieval method and ancillary data. Thus, by regrouping the intervals, we were able to show a better agreement between the datasets.

The intervals with CP < 310 hPa are mainly populated by tropical and subtropical clouds. Many tropical thin cirrus are situated over the Warm Pool region. ISCCP also often

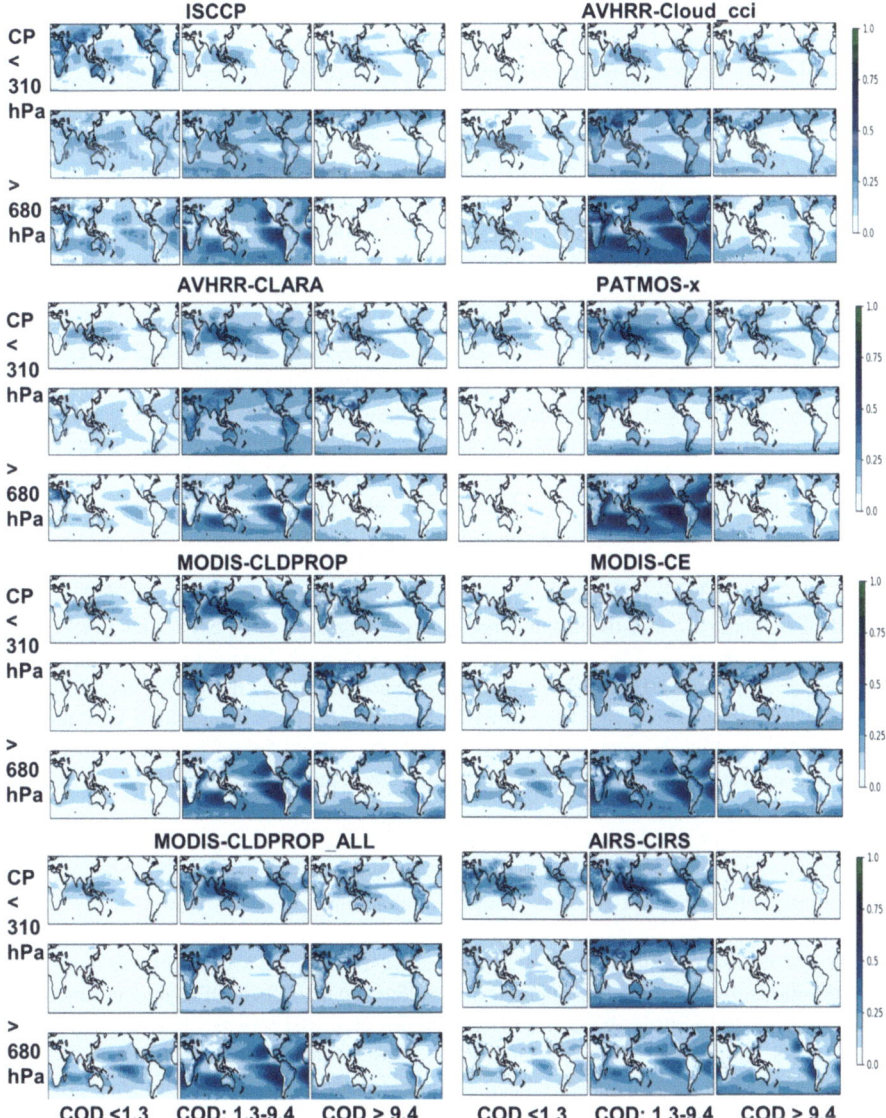

Fig. 10 Geographical distributions (60°N-60°S) of normalized occurrence frequency of nine cloud types distinguished by COD (horizontal axis) and CP (vertical axis) from ISCCP, AVHRR-Cloud_cci, AVHRR-CLARA, PATMOS-x, MODIS-CLDPROP, MODIS-CE, MODIS-CLDPROP_ALL and AIRS-CIRS. Statistics averaged over 1:30PM LT from 2008 to 2015, except for MODIS-CE (2016–2020)

identifies this type of cloud over all continents, while the Cloud_cci retrieval method misidentifies thin cirrus as thin mid-level clouds. Both datasets seem to attribute at least a part of these tropical thin cirrus to a lower altitude ((COD,CP) interval (1,2)), whereas MODIS-CLDPROP identifies few thin clouds in the mid- to low-level range. This is explained by the fact that the MODIS-CLDPROP retrieval determines optical depth only for well-retrieved clouds, which are about 69% of all retrieved clouds. The inclusion of the other clouds (MODIS-CLDPROP_ALL) shows a relative increase in the category of thin low-level clouds, as expected. Due to the large footprint size, the CIRS cloud type occurrences are slightly shifted towards lower COD intervals, compared to MODIS-CLDPROP.

4 Cloud Column Structure from Complementary Information

A complete cloud column structure is desirable for evaluating climate models (e.g. Wang and Rossow 1998). First associations between the appearance of specific cloud type combinations and specific meteorological situations were identified by surface observers (e. g. Warren et al. 1985; Hahn et al. 2001). In general, random cloud overlap is a good first approximation (e. g. Chen et al. 2000). Several studies (e.g. Mace et al. 2009; Li et al. 2015; Jing et al. 2016) found slightly larger overlap between high- and mid-level clouds, especially over land, and less overlap between high- and low-level clouds, in particular over ocean at higher latitudes. Furthermore, the vertical structure depends on cloud types linked to dynamics (Naud et al. 2008) and weather states (Tselioudis et al. 2013, 2021).

Though the combination of spaceborne radar and lidar measurements provides the most complete column view, the resulting cloud statistics still depend on methods in cloud detection and cloud-type distinction. We estimate an uncertainty due to methodology by comparing results from our analyses using the CloudSat-CALIPSO GEOPROF dataset described in Sect. 2.7 and those published using CloudSat-CALIPSO CLDCLASS data (Oreopoulos et al. 2017). The latter data are based on a feature-based cloud-type classification (Sassen and Wang 2012), starting from CloudSat-CALIPSO GEOPROF data. Besides, because CloudSat and CALIPSO are flown in a single afternoon orbit with a very narrow swath width, the biases introduced by the limited diurnal sampling produce some small systematic uncertainty in the results discussed below (see also Sect. 3.1 for ocean–land contrasts).

The scheme in Table 4, introduced by Rossow et al. (2005b), illustrates cloud vertical layer structure by distinguishing seven simplified classes, based on vertical stratification in three atmospheric layers (H, M, L as described before), and associates these classes to top and column view.

The global averages summarized in Table 4 correspond to the average of two independent estimations, from our analysis using the CloudSat-CALIPSO GEOPROF data described in Sect. 2.7 and those given in Table 2 of Oreopoulos et al. 2017 using CloudSat-CALIPSO CLDCLASS data. These CloudSat-CALIPSO datasets did not use the latest version of CALIPSO data (V4.2) with improved aerosol–cloud discrimination and thus show a larger total cloud amount (0.72) than the one from CALIPSO-ST_top (0.66, Sect. 3.1). Therefore, the uncertainty coming from the retrieval method, and given in Table 4, has been estimated as the difference between the average and the single analysis values to which a 10% uncertainty has been added according to the difference between the CALIPSO versions.

The *radar–lidar data include the cloud layers hidden by those aloft*, and we can compute the amount of all associated mid-level and low-level clouds within the atmospheric column by combining the following cloud vertical structure classes to:

$$M_{col} = 1M + ML + HM + HML \text{ and } L_{col} = 1L + ML + HL + HML,$$

with global average $CAM_{col} = 0.27 \pm 0.06$ and $CAL_{col} = 0.50 \pm 0.10$ (CloudSat-CALIPSO).

Since the *lidar cannot probe the atmosphere below optically dense clouds*, the CALIPSO-ST_column data underestimate CAM_{col} and CAL_{col}, with global averages of

0.18 and 0.30, respectively. We have therefore corrected these values for the amounts hidden by opaque cloud layers aloft, in guidance with the CloudSat-CALIPSO GEOPROF analysis, by using complementary information from CALIPSO-ST_opaque as well as overlap assumptions and formulas given in Appendix section "Combining CALIPSO-ST_column and CALIPSO-ST_opaque". These corrections lead to global average $CAM_{col} = 0.24 \pm 0.04$ and $CAL_{col} = 0.39 \pm 0.06$ (CALIPSO-ST_overlap), in closer agreement with the results from CloudSat-CALIPSO above than the initial CALIPSO-ST_column values.

As the instantaneous coverage of CALIPSO is very sparse, we also have used complementary information from passive remote sensing to estimate a complete cloud column structure. Therefore, we have combined CIRS data which identify cirrus also in the case of underlying clouds, once with data from MODIS-CE and once with data from AVHRR-Cloud_cci, both having a better spatial resolution and using additional VIS information during day which is helpful for the detection of low-level clouds.

For each 1° grid box, we choose the maximum CAH and the minimum CAL between the two considered datasets (CIRS and MODIS-CE or CIRS and AVHRR-Cloud_cci), and then, by applying cloud overlap assumptions and formulas given in Appendix section "Complementary passive remote sensing", we deduce global averages $CAM_{col} = 0.25 \pm 0.03$ and $CAL_{col} = 0.48 \pm 0.05$ (complementary passive), in good agreement with CloudSat-CALIPSO.

A complete 3D description of clouds is also necessary in order to derive radiative heating / cooling profiles, essential for a better understanding of cloud-radiative feedbacks. While the terrestrial (LW) fluxes act primarily on the atmosphere, the solar (SW) fluxes primarily impact the surface energy budget. ISCCP provides radiative flux products (Rossow and Zhang 1995; Zhang and Rossow 2023). For their computation, a cloud vertical layer (CVL) model accounts for cloud layer overlap by assigning specific layer structures to each of the ISCCP cloud types. These types are defined by three intervals of CP and three intervals of COD (similar to the 2D histograms in Sect. 3.4), separately for each thermodynamic phase. The original model, based on the ISCCP-D version and a cloud layer climatology derived from radiosonde humidity profiles (Rossow et al. 2005b), was recently refined by combining the ISCCP-H version and CloudSat-CALIPSO GEOPROF data (Zhang and Rossow 2023). Since this model also corrects the ISCCP cloud type biases linked to multiple cloud layers, it can only be applied to ISCCP or a dataset with similar biases.

We have slightly modified the model developed for ISCCP-D and evaluated with CloudSat-CALIPSO GEOPROF data by Rossow and Zhang (2010), so that we can use the statistics within the intervals of the 2D CP-COD histograms provided in our database: We moved the lower COD threshold from 2.4 to 1.3 for ML as well as the upper COD thresholds from 2.4 to 3.6 for 1H and HL and from 9.4 to 23 for 1 M. Furthermore, we changed the thresholds to define HM from (COD 6.3–23) to (COD 3.6–9.4), and we do

Table 5 Modified Cloud vertical layer model for ISCCP cloud types as function of CP and COD. The original model (Rossow et al. 2005b) assumes ISCCP mid-level clouds (440–680 hPa) as 1 M, HL, HL, ML, instead of HL, ML, ML, ML / 1 M in the four COD intervals, respectively

< 440 hPa	1H	1H	HM	HML
440–680 hPa	HL	ML	ML	ML 1 M
> 680 hPa	1L	1L	1L	1L
CP / COD range	0–1.3	1.3–3.6	3.6–9.4	> 9.4 > 23

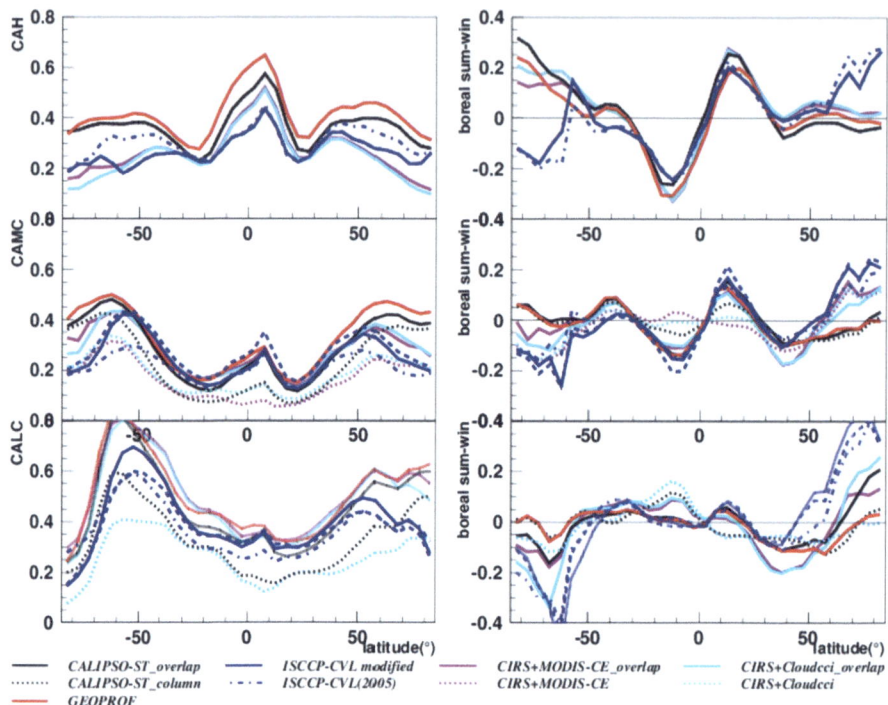

Fig. 11 Left panel: Zonal mean values of CAH, CAM_{col} and CAL_{col} of GEOPROF, CALIPSO-ST_overlap, complementary passive (CIRS & MODIS-CE_overlap and CIRS & AVHRR-Cloud_cci_overlap) and of ISCCP corrected with different CVL models. CAM and CAL of CALIPSO-ST_column and complementary passive are also shown. These are uncorrected, while the overlap values include all clouds hidden by clouds aloft. Right panel: Differences of CAH, CAM_{col} and CAL_{col} between boreal summer (June, July, August) and boreal winter (December, January, February). Statistics averaged 1:30AMPM LT (except ISCCP, 1:30PM LT, during daytime) over the common period of 2008–2015 (except GEOPROF 2007–2009)

not distinguish between multiple layers of HML (COD 2.4–6.3) and one contiguous layer HxMxL (COD > 23), and simply define HML/HxMxL by COD > 9.4. This modified CVL model is summarized in Table 5.

We have applied this model on the statistics of the CP-COD histograms to deduce global averages CAH = 0.29, CAM_{col} = 0.23 ± 0.03 and CAL_{col} = 0.41 ± 0.05 (ISCCP-CVL).

The latter values agree well with the others noted above.

Figure 11 compares the latitudinal variation of CAH, CAM_{col} and CAL_{col} from Cloud-Sat-CALIPSO GEOPROF, CALIPSO-ST_column, CALIPSO-ST_overlap (derived from column and opaque), complementary passive (CIRS-MODIS-CE and CIRS-AVHRR-Cloud_cci) and ISCCP-CVL. Zonal means of their differences between boreal summer and boreal winter are also reported.

Since CAH of the complementary datasets is the maximum of both, which is most often CIRS, it is relatively close to CALIPSO-ST_top, with less agreement towards the poles due to different height categories according to retrieval of radiative versus cloud-top height (Sect. 3.2). The bias-corrected CAH of ISCCP-CVL also agrees well with the other datasets, but the peak in the tropics is still slightly lower. CAH of GEOPROF is slightly larger than of CALIPSO-ST_top, because the latter has a reinforced aerosol detection.

The amounts of all mid- and low-level clouds within the atmospheric column, CAM_{col} and CAL_{col}, agree very well among the datasets. Only ISCCP indicates a slightly smaller CAL_{col} at higher latitudes, particularly between 50 and 65°S. In this region the ISCCP CAL_{col} also differs more with respect to the assumptions in the two CVL models (new one and the one of Rossow and Zhang 2005b). The comparison of CALIPSO-ST_overlap (solid) with the original CAM and CAL of CALIPSO-ST_column (dotted) demonstrates that the correction of CALIPSO-ST_column due to partial obscuration by opaque clouds aloft is not negligible, particularly for low-level clouds. By comparing CAL (dotted) and CAL_{col} (solid) of the complementary datasets, we deduce that *about half of the low-level clouds are obscured by clouds aloft.*

The seasonal differences are now easier to interpret: CAM_{col} and even CAL_{col} follow the displacement of the ITCZ, because the high-level convective clouds in the ITCZ fill the whole atmospheric column and the cloud fields around convection are complex with several layers of clouds. At the high latitudes, we observe more low-level clouds in summer than in winter. This effect is much smaller when considering only single-layer clouds (CAL, Figure S5).

5 Monitoring of Cloud Amount and of Relative Occurrence of Specific Cloud Types

Some of these global cloud climate data records now cover up to 40 years. Time series of the AVHRR-based datasets have already been compared with ISCCP by Karlsson et al. (2018). They found a weak negative trend in global cloud amount from all datasets over the period 1984–2009, but also noticed clear signs of various discontinuities related to satellite shifts and orbital drift effects. They state that only after the introduction of the AVHRR/3 sensors from 2001 onwards, all AVHRR Climate Data Records show more consistent results among them, for the afternoon orbits. In the supplemental Figure S15, we present the time series of the global mean CA, separately for all four observation times given in this database. It is interesting to note that the satellite drifts of the afternoon satellites are the most strongly pronounced at 1h30 PM LT, in particular for PATMOS-x.

Marvel et al. (2015) developed a multi-variate fingerprint that captures coherent, externally forced cloud changes in order to reduce the time to detect these changes. By using ISCCP and PATMOS-x data, they could still demonstrate the poleward migration in the major latitudinal features of total cloud amount. Bender et al. (2012) and Norris et al. (2016) removed artefacts of ISCCP and PATMOS-x by filtering problematic regions, spurious viewing geometry effects and specific cloud types, or by using empirical corrections, respectively. Their analyses reveal large-scale patterns of cloud change between the 1980s and the 2000s consistent with poleward retreat of mid-latitude storm tracks, expansion of subtropical dry zones, and increasing height of the highest cloud tops at all latitudes.

So far, the original ISCCP provides the best temporal sampling. Since our database does not completely resolve the diurnal cycle with four measurements per day, we concentrate in the following only on variabilities at 1h30 PM LT, as for the analysis of the specific cloud types in Sect. 3.4. Recently, Liu et al. (2023) found weak opposing trends of CA over land and over ocean under global warming, but only by using an EOF analysis which was able to separate the large signals from El-Niño-Southern-Oscillation-associated (ENSO-associated) modes in the 42 years of meteorological reanalysis data.

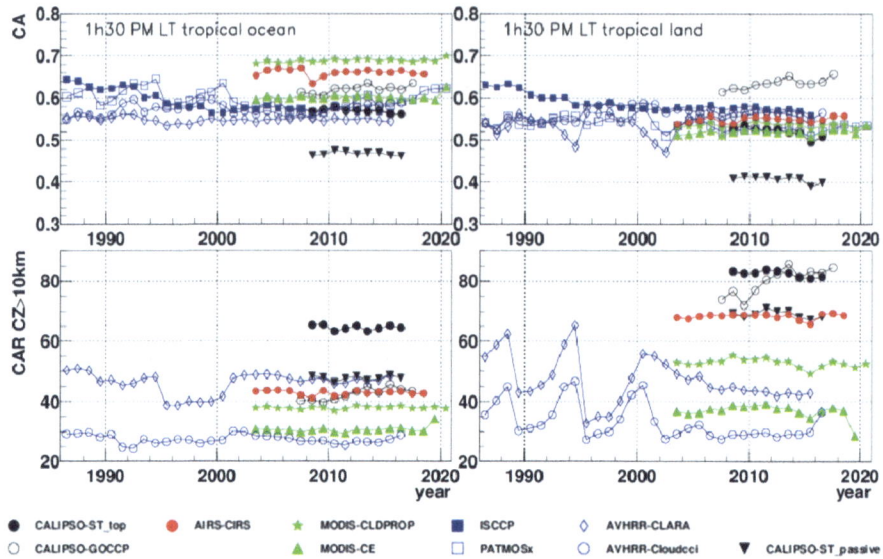

Fig. 12 Time series of cloud amount (top) and relative proportion of clouds with CZ > 10 km (bottom), over tropical ocean (left) and tropical land (right). Statistics at 1:30PM LT over 25N-25S. The colours correspond to specific instrument types (black: active lidar, red: IR sounders, green: advanced multi-spectral imagers, blue: multi-spectral imagers)

5.1 Interannual Variability

The upper panel of Fig. 12 shows annual values of cloud amount for all datasets from 1986 to 2020, averaged over the latitudinal band 25 N–25 S, separately over ocean and over land. Before 2003, the AVHRR-based time series show inter-annual variations which do not coincide, without any clear trends. The three AVHRR cloud datasets use the same radiance calibration, but at this particular time of day the cloud detection of PATMOS-x (ocean) and CLARA (land) seems to be more sensitive to changes in observation time due to orbital drifts than the one by Cloud_cci. ISCCP, mostly based on geostationary imager data, shows a nearly linear decrease in cloud amount from 1986 to 2000, both over ocean and over land. The negative trend has already been thoroughly investigated for the earlier ISCCP version (Appendix 2 by W. B. Rossow in Stubenrauch et al. 2012) and also for the current ISCCP version (Rossow et al. 2022): this systematic slow decrease of CA may be slightly overestimated but it cannot be explained alone by spurious effects linked to calibration, viewing geometry changes, changes in land–ocean and day-night sampling. This conclusion is also supported by a quantitative agreement between the TOA radiative fluxes, calculated from ISCCP cloud properties and those determined from ERBS observations (Zhang et al. 2004; Norris 2005). The global cloud amount (Figure S15) has a very similar behaviour. Concerning the newer instruments, it is interesting to note that both CALIPSO datasets show a slightly different behaviour, in particular over land. The time series show in general a flat behaviour, which shows that one needs a good filtering in order to observe any changes due to specific climate modes.

In order to illustrate another application of the data, we derive the relative proportion of clouds above 10 km from the CZ histograms. The histograms allow us to define

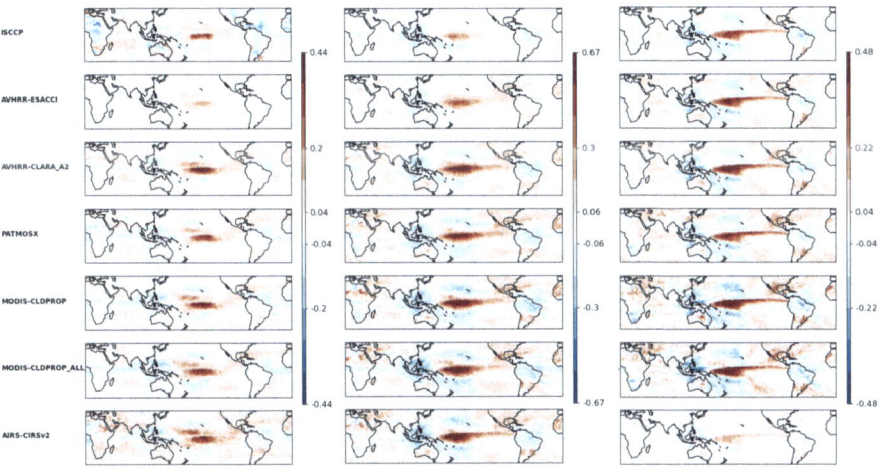

Fig. 13 Geographical maps of linear regression slopes between monthly mean anomalies in amount of thin cirrus (left), cirrus (middle) and high opaque clouds (right), *relative to all clouds*, and global mean surface temperature anomalies. Units are in fractional change per °C warming. Cloud data are from ISCCP, AVHRR-Cloud_cci, AVHRR-CLARA, PATMOS-x, MODIS-CLDPROP_ALL, MODIS-CLDPROP and AIRS-CIRS (from top to bottom), and surface temperature from the GISS Surface Temperature Analysis, during the period 2003–2015, 40N-40S

cloud types by different thresholds, depending on the subject of study. The lower panel of Fig. 12 shows again that the data are less variable after 2003. Both CLARA and Cloud_cci have large variations over land, most probably due to the NOAA afternoon satellite drifting. In addition, CLARA also displays an abrupt decrease over ocean in 1995, due to an increase in semi-transparent clouds (see next section and supplement Figure S17). Again, the behaviour of the two CALIPSO datasets is different over land, with CALIPSO-ST exhibiting less variability.

IASI-CIRS and MISR results are presented together with the other morning measurements in Figure S16 in the supplement. Except PATMOS-x, which shows a drop in CA after 2010, the data from the morning satellites are again quite stable after 2002.

5.2 Anomaly Patterns with Respect to Global Surface Temperature Changes

Since the AVHRR data records are less consistent before 2000 and since the observational period of the new satellite era is still too short to directly study long-term cloud variability related to climate warming, an alternative approach is to analyse cloud type variability with respect to inter-annual global mean surface temperature anomalies. For this analysis, we use deseasonalized global monthly mean surface temperature anomalies from the GISS Surface Temperature Analysis (GISTEMP Team 2023; Lenssen et al. 2019).

We use the 2D histograms in the CP-COD space to select upper tropospheric cloud types with CP < 310 hPa and distinguish these further into thin cirrus, cirrus and high opaque clouds, using the same three COD ranges as in Fig. 10. Then, we determine geographical change patterns in amount of these cloud types, relative to all clouds, with respect to global surface temperature: $d(CA_{type}/CA)/dT_s$. These derivatives are computed for each grid box of 1° latitude × 1° longitude, by linear regression between their anomalies

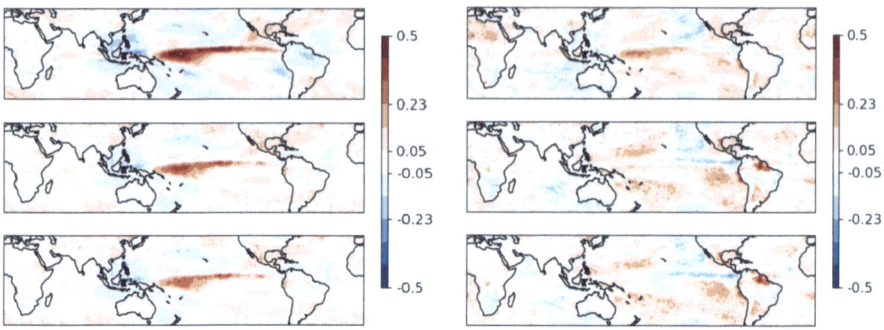

Fig. 14 Geographical maps of linear regression slopes between monthly mean anomalies in CAM_{col} (left) and CAL_{col} (right) and global mean surface temperature anomalies. Units are in fractional change per °C warming. The column amounts have been obtained from complementary information, as discussed in Sect. 4, using the CVL model for ISCCP (top), and using AIRS-CIRS combined with AVHRR-Cloud_cci (middle) or MODIS-CE (bottom). Surface temperature from the GISS Surface Temperature Analysis, during the period 2003–2015

as in (Stubenrauch et al. 2017). Results are presented in Fig. 13 for the seven datasets which provide these 2D histograms over the common time period 2003–2015. All datasets show very similar geographical change patterns, with an increase per °C of warming at the equator in the tropical Pacific and a decrease over Indonesia, a pattern similar to ENSO (e.g. Stephens et al. 2018b). Because of the large footprint (15 km), the change in amount of high opaque clouds identified by AIRS-CIRS corresponds to the one of large convective areas. Their amount is much smaller than the one of convective areas identified by a smaller spatial resolution (Fig. 10). Smaller changes in cirrus by AVHRR-Cloud_cci and ISCCP can be explained by misplacing their heights. We also observe an increase in cirrus and thin cirrus in the subtropics. This may be in relation to a hypothesized widening of the Hadley cell with global warming (e.g. Staten et al. 2018). The relative amount of thin cirrus also increases around the equator, close to the increase of high opaque clouds. This can be reconciled with the findings that tropical mesoscale convective systems get higher (e. g. Stubenrauch et al. 2021) and the area of surrounding thin cirrus gets larger with their height (Protopapadaki et al. 2017). This signal decreases with decreasing sensitivity to thin cirrus, the lowest signal being given by AVHRR-Cloud_cci.

The different geographical change patterns of high opaque clouds and thin cirrus lead to variations in atmospheric heating and cooling gradients which then influence the large-scale circulation, as has already been investigated by Slingo and Slingo (1991).

It is interesting to note that the change patterns are similar when exploring a longer time period (2003–2018 for MODIS-CLDPROP and AIRS-CIRS, and 1995–2015 for the AVHRR datasets and ISCCP), as shown in supplemental Figure S17, except for AVHRR-CLARA exhibiting artificial increases of cirrus and thin cirrus from 1995 to 2001.

Changes in column cloud amount of mid- and low-level clouds with respect to global surface temperature are displayed in Fig. 14. We have used the complementary information and computations discussed in Sect. 4. The change patterns, both in CAM_{col} and in CAL_{col}, are very similar among the different methods. The increase in column mid-level clouds in the Pacific along the equator is coherent with the change pattern of high opaque cloud amount. This indicates increasing tropical deep convective cloud amount in the tropical Pacific close to the equator, with warming. All data agree that the column low-level cloud amount decreases in the stratocumulus regions off the Californian coast and off the west coast of Australia, while it increases off the west coasts of South America and Africa as well as north and south of the increasing deep convective band in the Western Pacific.

To summarize: It is very challenging to detect small inter-annual global and zonal changes with these data, in particular with long-term datasets, which have undergone orbit drifts, instrument changes and orbit changes. However, similar geographical change patterns between the different datasets demonstrate that the data can be used to study climatological change patterns, in particular, those for specific cloud types (considering the sensitivity of a specific dataset).

6 Summary and Outlook

Since the original GEWEX cloud assessment a decade ago, existing cloud property retrievals have been revised and/or temporally extended, and additional retrievals from new sensor data have emerged.

In general, results have remained similar to those of the previous assessment. A notable exception is a reduced cloud amount derived by CALIPSO-ST, which is associated with an improved distinction between aerosol and clouds. Therefore, the global mean cloud amount 0.66 ± 0.04, estimated from the eleven datasets of the updated GEWEX cloud assessment database, is slightly lower than before (0.68 ± 0.03). Both these estimations have additional uncertainties due to their limited diurnal sampling.

The absolute values of high-level cloud amount depend on instrument and retrieval sensitivity to thin cirrus, varying between 0.12 (MISR, visible information only) and 0.37 (CALIPSO-ST, active lidar). By exploiting the different CALIPSO data in combination with CloudSat-CALIPSO, we estimated the amount of thin cirrus with COD < 0.3 roughly to 0.13, of which about 60% occur with lower clouds underneath, when assuming that multiple cloud layers are COD-independent of the higher cloud. This larger contribution compared to a much earlier result of about 30% obtained for cirrus with COD < 1, using HIRS observations (Jin and Rossow 1997), can be partly explained by the different spatial resolutions (1° compared to 17 km) and by the assumption of COD-independent overlap made in our estimation. When the retrieval method includes IR and VIS information during day, without considering the possibility of underlying clouds, thin cirrus may be mis-identified as a more opaque lower cloud because of the large VIS reflectance from the cloud underneath.

While the absolute values of the retrieved variables show systematic differences due to retrieval method, instrument sensitivity, spectral and ancillary data and footprint size, their geographic, synoptic and seasonal variations agree well. Only the polar regions remain a challenge for passive remote sensing. Another issue is the difference between the radiative cloud height, retrieved by passive remote sensing (except stereoscopic height retrieval by MISR) and the cloud top height from active remote sensing. At higher latitudes, the radiative height starts to decrease towards the middle troposphere, while the cloud top height still stays in the category of the upper troposphere.

The three datasets based on AVHRR observations differ strongly. Two of these datasets have been recently reprocessed: PATMOS-x (Foster et al. 2023) and AVHRR-CLARA (Karlsson et al. 2023). Both new versions show a global CA increased by 0.03, with PATMOS-x changing from 0.65 to 0.68 and AVHRR-CLARA from 0.62 to 0.65. The AVHRR-Cloud_cci dataset is not continued.

For the determination of cloud impacts on the radiative heating profile not only altitude-stratified cloud amount but also associated values of attenuation (optical depth, emissivity) and composition (particles size, phase) matter. The presented diversity of these cloud properties is also associated with the strengths and limitations of the retrieval methods and assumptions, so that a combination of sensor capabilities is recommended.

In this article, we have highlighted how to use this updated GEWEX cloud assessment database, which includes many variables and monthly statistics in the form of averages and histograms. In particular, the two-dimensional histograms in the CP-COD or CP-CEM space allow a more detailed cloud type specification. Differences in the shape of the normalized distributions indicate the CP and COD uncertainties due to retrieval method and ancillary data. By regrouping to coarser intervals, we were able to show a better global agreement between the datasets, which indicates that the uncertainties are larger than the original intervals given in the CP-COD and CP-CEM histograms.

In spite of large uncertainties, these data are very valuable for cloud climate and process studies via analyses which cluster statistics in the CP-COD space, leading to the definition of Weather States or Cloud Regimes (e.g. Rossow et al. 2005a; Tselioudis et al. 2013; Gryspeerdt and Stier 2012; Stachnik et al. 2013; Luo et al. 2017; Tan and Oreopoulos 2019) or which construct cloud systems from adjacent measurements (e.g. Protopapadaki et al. 2017; Stubenrauch et al. 2021, 2023). Furthermore, by including radiative fluxes, as for example directly available for ISCCP (Zhang and Rossow 2023) and AVHRR-Cloud_cci (Stengel et al. 2020), cloud radiative effects may be quantified. The latter data have been used by Philipp et al. (2020) to analyse Arctic feedback mechanisms between sea ice and low-level clouds.

While passive remote sensing can address radiative (climate) impacts at the top of the atmosphere (Zhang et al. 2004), those in the atmosphere and at the surface require information on cloud-base altitude and cloud amounts hidden by higher clouds aloft, which only space-borne active radar–lidar remote sensing can offer. Still, the uncertainty in the retrieval method applied to these active data is about 10 to 20%. We have shown that insights into these properties are possible by combining complementary passive sensor data (like IR sounder and IR-VIS imager data) and an empirical overlap model or by applying a bias correction and vertical layer model specifically developed for ISCCP (Zhang and Rossow 2023) to the 2D histograms of ISCCP. Furthermore, these cloud datasets can be combined with other measurements for diagnostic purposes.

For IR-VIS radiometers, we recommend developing retrieval methods (during daytime) which are able to distinguish single- and multi-layered cloud scenes. This may be achieved by first using the IR radiances and their spectral differences alone and then comparing the retrieved cloud properties to those obtained when including VIS reflectances (e.g. Chang and Li 2005). Machine-learning methods that use VIS-IR radiances, single-layer retrievals, and ancillary data as input are another effective approach to identify overlapped clouds (e.g. Sun-Mack et al. 2023) and retrieve some of their properties (e.g. Minnis et al. 2019). It is also beneficial to combine radiances of different instruments on the same platform for a more sophisticated retrieval. Still, combining different sensors might come at a cost, e.g. in terms of spatial coverage (swath width) or temporal consistency (instrument issues). First steps in this direction have recently emerged, including the new version of the PATMOS-x cloud retrieval (Foster et al. 2023), which uses combined AVHRR and HIRS measurements, and a new cloud retrieval technique for two-layered cloud systems using the fusion of MISR VIS-stereoscopy and MODIS IR radiances (Mitra et al. 2023).

Reliable Climate Data Records require calibration between instruments and homogeneous ancillary data. EUMETSAT and NOAA/NESDIS are collaborating on generating a new GEO-RING set of data based on the recent improvements in the geostationary imager capabilities (from 2018 onwards). A next generation of the International Satellite Cloud Climatology Project (ISCCP-NG) is being developed as an application of this new GEO-RING data. The AVHRR retrieval teams in the USA and Europe are already generating prototype cloud products from ISCCP-NG. This data will have spatial resolution of $0.05°$ and temporal resolution of 30 min. The final resolutions are still to be determined. The goal is to have ISCCP-NG generating routine data once the Third Generation of Meteosat (MTG) has finished its first year of operations.

Furthermore, we strongly recommend the use of complementary information as much as possible. ISCCP employs a cloud vertical layer model based on complementary information from radar–lidar (CloudSat-CALIPSO) to compute the cloud-induced radiative fluxes, and we have proposed a scheme which combines IR sounder and VIS-IR imager cloud amounts and an empirical cloud overlap scheme to get a full cloud amount within the atmospheric column. This method assumes increasing overlap with increasing cloud emissivity of the clouds aloft and decreasing vertical distance. Both approaches should be applied on instantaneous data, but we saw also an improvement in comparisons with the CloudSat-CALIPSO data when applied to the monthly data.

As published before, global and zonal long-term trends are only coherent among the AVHRR datasets after 2002, when all cloud amounts become more stable. The decrease in total cloud amount from ISCCP from 1986 to 2018 can be explained by changes in particular cloud types (Rossow et al. 2022; Zhang and Rossow 2023). The more capable satellite sensors available since 2000 do not show any statistically significant trends in global and zonal means. Yet, regional changes in the relative occurrence of specific cloud types as well as in column low-level cloud amount can be well identified with respect to global surface temperature anomalies.

Appendix 1: Geographical Patterns of the Individual Datasets

The geographical distributions of annual averages of CAH, CAM, CAL and clear sky amount (1-CA) are displayed in Fig. 15 for each of the twelve individual datasets.

While the absolute values differ, the geographical patterns obtained from these data are very similar. All data sources show the Inter-Tropical Convergence Zone (ITCZ) as the most prominent feature in CAH. Other important features are large CAL in the stratocumulus regions off the western coasts in the subtropics and over ocean in the mid-latitudes, especially in the SH. The clear sky amount (1-CA) distributions also agree well and show prominent features over the deserts and the subtropical ocean, in particular where one expects open cell cumulus clouds. Another common feature comprises the storm tracks in the zonal westerlies, which are evinced by the relatively large CAH and CAM means. In general, we observe more CAM at high than at low latitudes. This effect is stronger for the passive remote sensing, because when the tropopause is lower (down to 300 hPa) the radiative cloud height is closer to the constant threshold of 440 hPa and therefore some of the high-level clouds fall into the category of mid-level clouds (see Sect. 3.2).

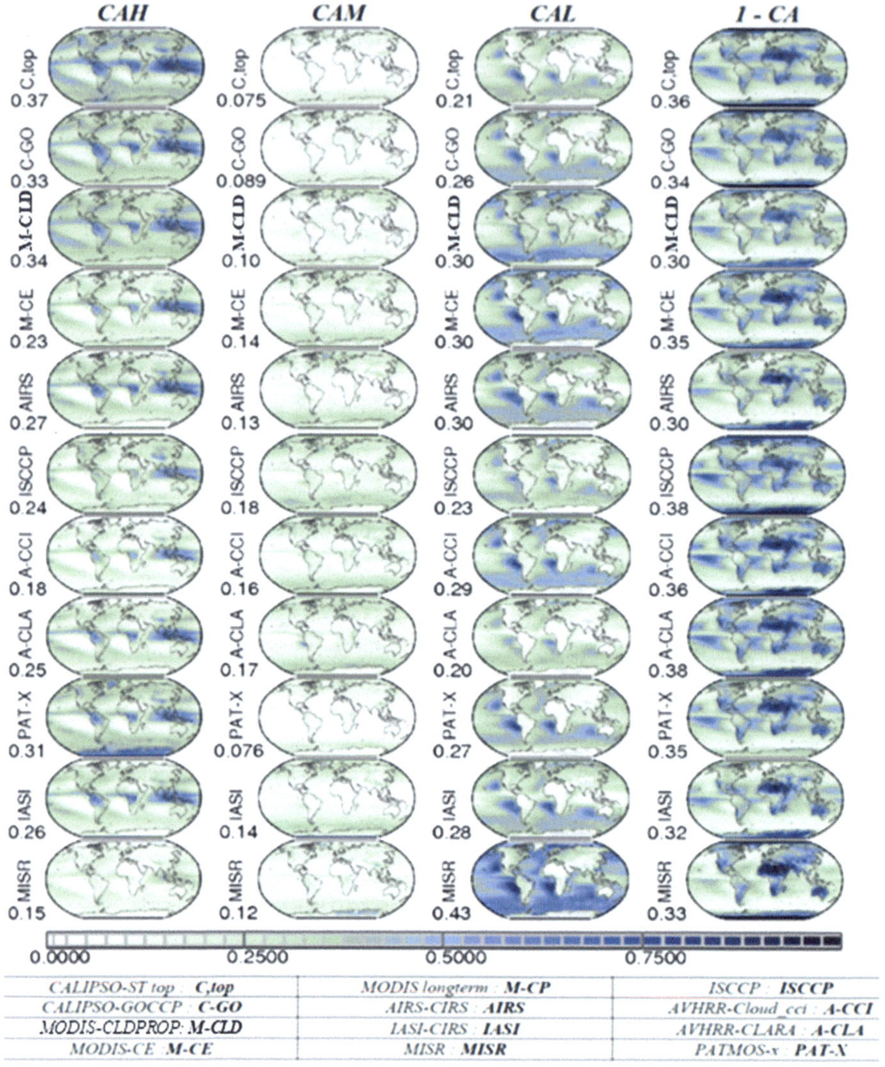

Fig. 15 Geographical maps of annual averages of CAH (column 1), CAM (column 2), CAL (column 3) and (1-CA) (column 4) from eleven datasets of the updated GEWEX cloud assessment database. Statistics 2008–2015. The values to the lower left indicate global (area-weighted) averages

Appendix 2: Estimation of Cloud Column Structure from Complementary Information

Combining CALIPSO-ST_column and CALIPSO-ST_opaque

Since the lidar cannot probe the atmosphere below optically dense clouds, the CALIPSO-ST_column dataset underestimates CAM_{col} and CAL_{col} by about 25–30%, in comparison to the results from combined CloudSat–CALIPSO data in Table 6. The

Table 6 Global averages of CAM and CAL, including the parts hidden by clouds aloft, separately over ocean and over land. Compared are results from CloudSat-CALIPSO (CC) (analysis described in Sect. 2.6 and by Oreopoulos et al. 2017), from CALIPSO-ST_column and CALIPSO-ST_overlap, and from complementary passive remote sensing (CP)

	ocean			Land		
	CC	C	CP	CC	C	CP
CAM(column)		0.16			0.23	
CAM_{col}	0.26 ± 0.06	0.19/**0.23**/0.25	0.25/0.25	0.31 ± 0.04	0.26/**0.29**/0.31	0.26/0.26
CAL(column)		0.37			0.15	
CAL_{col}	0.58 ± 0.08	0.44/**0.50**/0.53	0.55/0.54	0.33 ± 0.06	0.19/**0.29**/0.32	0.32/0.30

CC: averaged results from CloudSat GEOPROF-lidar and CloudSat CLDCLASS-lidar analyses

C: CALIPSO-column CAL, CAM

C: CAL_{col}, CAM_{col} corrected for clouds hidden by opaque ones aloft under three different overlap assumptions (random, **random to maximum** and maximum overlap)

CP: complementary passive remote sensing (CIRS-MODIS-CE), under two different overlap assumptions (random and **increasing overlap with CEMH**)

Fig. 16 Scheme of cloud overlap, with parts hidden from CALIOP by opaque cloud layers aloft

latter are given as averages from the results of 1) Oreopoulos et al. (2017), using cloud top and base heights from CloudSat-CALIPSO CLDCLASS data (Sassen and Wang 2012) and transforming these to pressure coordinates via ECMWF auxiliary data, and of 2) our analysis of CloudSat-CALIPSO GEOPROF data described in Sect. 2.6. The 10–15% differences in Table 6 may be due to the use of CP and CZ for the height-stratification and due to the different definition of cloud types in the original data.

We estimate CAM_{col} and CAL_{col} by correcting the CALIPSO-ST_column amounts (CAM(column) and CAL(column)) for those hidden by opaque clouds aloft (Fig. 16). CAH(opaque) (0.09) contains the vertical structures HM and HML, and CAM(opaque) (0.07) corresponds to ML (Table 4). We use formulas similar to those of Ham et al. (2021)

$$CAM_{col} = a\,CAH(opaque) + CAM(column)/(1 - (1-a)\,CAH(opaque)) \text{ and}$$
$$CAL_{col} = a\,(CAH(opaque) + CAM(opaque))$$
$$+ CAL(column)/(1 - (1-a)(CAH(opaque) + CAM(opaque)))$$

under three different overlap assumptions: (1) random ($a = 0$), (2) *between random and maximum* ($a = 0.75$), and (3) maximum ($a = 1$). The computation of CAL_{col} is similar, but it considers high- and mid-level opaque clouds aloft with the assumption of minimum

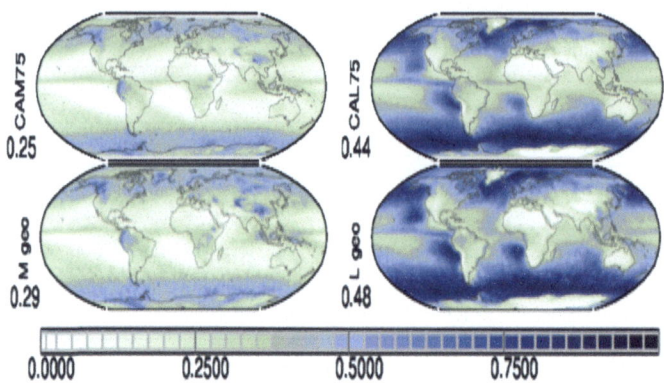

Fig. 17 Geographical patterns of annual averages of CAM_{col} (left) and CAL_{col} (right), deduced from the above formulas applied on CALIPSO-ST (top) and from CloudSat-CALIPSO GEOPROF (bottom). Statistics 2007–2009. The values to the lower left indicate global (area-weighted) averages

overlap between these two. The uncertainties in CAM_{col} and CAL_{col} linked to overlap assumptions are about 10 to 15%, according to Table 6.

The maps in Fig. 17 reveal that the CALIPSO-ST CAM_{col} and CAL_{col} averages are very similar to their CloudSat-CALIPSO GEOPROF counterparts. The larger global averages can be explained by an earlier CALIPSO version used in these GEOPROF data. Strictly speaking, this method should be applied to the instantaneous gridded data. Nevertheless, it also works well on the monthly data, as demonstrated in Fig. 17.

Complementary Passive Remote Sensing

To estimate CAM_{col} and CAL_{col} from passive remote sensing, we need a consistent definition for cloud layer height regimes (CAH, CAM and CAL) and assumptions about how they overlap. To minimize mis-location of thin clouds into lower height regimes, a 'best estimate' is suggested by *combining the advantages of IR sounders* (with best IR spectral resolution and hence a good sensitivity to thin cirrus: e. g. AIRS, IASI), *and the advantages of multi-spectral sensors* (with a better spatial resolution for the retrieval of smaller scale low-level cloud properties: e. g. MODIS-CE, AVHRR_Cloud_cci).

Among the two complementary datasets we choose CAH, CAM and CAL for each 1° grid cell at a specific observation time as follows:

$$CAH = \max(CAH1, CAH2) \quad CAL = \min(CAL1, CAL2)$$

$$CAM = \max(CAH1 + CAM1, CAH2 + CAM2) - CAH$$

For estimates of the complete (including hidden) cloud amount at mid- (CAM_{col}) and low- (CAL_{col}) levels, we have developed an empirical formulation that considers the cloud emissivity of the high- (CEMH) and mid- (CEMM) level clouds aloft, by assuming that the probability of cloud overlap increases with increasing cloud emissivity and it decreases with increasing vertical distance:

$$CAM_{col} = aCAH + CAM/(1-(1-a)CAH) \text{ with } a = CEMH**2$$
$$CAL_{col} = bCAH + cCAM + CAL/(1-(1-b)CAH - (1-c)CAM).$$
$$\text{with } c = CEMM**2, b = 0.75 \, CEMH**2$$

These formulas allow again random ($a, b, c = 0$) to maximum cloud layer overlap ($a, b, c = 1$). By comparing geographical distributions with CloudSat-CALIPSO GEOPROF results we developed an empirical formula, assuming *increasing overlap with increasing emissivity of the higher clouds aloft* ($a = CEMH^2$) *and decreasing overlap with increasing vertical distance between the cloud layers* (factor 0.75). These formulas lead to the global averages $CAM_{col} = 0.26$ and $CAL_{col} = 0.48$.

This 'best estimate' is not perfect, but by considering the complementary advantages of the two instruments, it gives an idea what to expect at the best from passive remote sensing.

In order to evaluate the stability of the cloud column structure from complementary passive remote sensing, we have also combined AIRS-CIRS with AVHRR-Cloud_cci. CAH of AVHRR-Cloud_cci is underestimated (Fig. 1), but the combination of both, built in the same manner as above, leads to results very similar to the ones from the CIRS–MODIS-CE combination, with global averages $CAM_{all} = 0.25$ and $CAL_{all} = 0.47$.

Strictly speaking, this method should be applied to the instantaneous gridded data, but again the results show that this approximation on monthly data compares well to the column cloud amounts from the active sensors (Fig. 11).

Supplementary Information The online version contains supplementary material available at https://doi.org/10.1007/s10712-024-09824-0.

Acknowledgements This article was proposed to be part of the theme "Challenges in Understanding the Global Water Energy Cycle and its Changes in Response to Greenhouse Gas Emissions" within the framework of the workshop held at the International Space Science Institute (ISSI) in Bern, Switzerland (26–30 September 2022). The authors want to thank Jay Mace for the help with the specific CloudSat-CALIPSO GEOPROF dataset used for this study, as well as Genevieve Sèze for discussions on the article. Furthermore, we want to thank three anonymous reviewers for their very valuable comments which makes this manuscript easier to follow.

Author's Contribution All authors read and approved the final manuscript.

Funding Open Access funding enabled and organized by Projekt DEAL. This study was partly supported by the Centre National de la Recherche Scientifique (CNRS), the Centre National d'Etudes Spatiales (CNES).

Data availability This updated GEWEX cloud assessment database is distributed by the French data centre AERIS at https://gewexca.aeris-data.fr/.

Declarations

Conflict of interest The authors declare that they have no competing interests.

Open Access This article is licensed under a Creative Commons Attribution 4.0 International License, which permits use, sharing, adaptation, distribution and reproduction in any medium or format, as long as you give appropriate credit to the original author(s) and the source, provide a link to the Creative Commons licence, and indicate if changes were made. The images or other third party material in this article are included in the article's Creative Commons licence, unless indicated otherwise in a credit line to the material. If material is not included in the article's Creative Commons licence and your intended use is not permitted by statutory regulation or exceeds the permitted use, you will need to obtain permission directly from the copyright holder. To view a copy of this licence, visit http://creativecommons.org/licenses/by/4.0/.

References

Astin I, Di Girolamo L, Van de Poll HM (2001) Baysian confidence intervals for true fractional coverage from finite transect measurements: implications for cloud studies from space. J Geophys Res 106:17303–17310. https://doi.org/10.1029/2001JD900168

Avery MA, Ryan RA, Getzewich BJ, Vaughan MA, Winker DM, Hu Y, Garnier A, Pelon J, Verhappen CA (2020) CALIOP V4 cloud thermodynamic phase assignment and the impact of near-nadir viewing angles. Atmos Meas Tech 13:4539–4563. https://doi.org/10.5194/amt-13-4539-2020

Balmes KA, Fu Q, Thorsen TJ (2019) Differences in ice cloud opticaldepth from CALIPSO andground-based Raman lidar at the ARMSGP and TWP sites. J Geophys Res Atmos 124:1755–1778. https://doi.org/10.1029/2018JD028321

Bender FA-M, Ramanathan V, Tselioudis G (2012) Changes in extratropical storm track cloudiness 1983–2008: observational support for a poleward shift. Clim Dyn 38:2037–2053. https://doi.org/10.1007/s00382-011-1065-6

Chang F-L, Li Z (2005) A new method for detection of cirrus overlapping water clouds and determination of their optical properties. J Atmos Sci 62:3993–4009. https://doi.org/10.1175/JAS3578.1

Chen T, Zhang Y, Rossow WB (2000) Sensitivity of atmospheric radiative hetaing rate profiles to variations of cloud layer overlap. J Clim 13:2941–2959. https://doi.org/10.1175/1520-0442(2000)013%3c2941:SOARHR%3e2.0.CO;2

Chepfer H, Bony S, Winker D, Cesana G, Dufresne JL, Minnis P, Stubenrauch CJ, Zeng S (2010) The GCM oriented CALIPSO cloud product (CALIPSO-GOCCP). J Geophys Res 115:D00H16. https://doi.org/10.1029/2009JD012251

Dee DP, Uppala SM, Simmons AJ, Berrisford P, Poli P, Kobayashi S, Andrae U, Balmaseda MA, Balsamo G, Bauer P, Bechtold P, Beljaars AC, van de Bergn M, Bidlot L, Bormann J, Delsol NC, Dragani R, Fuentes M, Geer AJ, Haimberger L, Healy SB, Hersbach H, Holm EV, Isaksen L, Kallberg P, Kohler M, Matricardi M, McNally AP, Monge-Sanz BM, Morcrette J-J, Park B-K, Peubey C, de Rosnay P, Tavolato C, Thepaut J-N, Vitart F (2011) The ERA-Interim reanalysis: configuration and performance of the data assimilation system. Q J R Meteorol Soc 137:553–597. https://doi.org/10.1002/qj.828

Dutta S, Di Girolamo L, Dey S, Zhan Y, Moroney CM, Zhao G (2020) The reduction in near-global cloud cover after correcting for biases caused by finite resolution measurement. Geophys Res Lett 47:e2020GL090313. https://doi.org/10.1029/2020GL090313

Eastman R, Warren SG (2014) Diurnal cycles of cumulus, cumulonimbus, stratus, stratocumulus, and fog from surface observations over land and ocean. J Clim 27:2386–2404. https://doi.org/10.1175/JCLI-D-13-00352.1

Eastman R, Warren SG, Hahn CJ (2011) Variations in cloud cover and cloud types over the ocean from surface observations, 1954–2008. J Clim 24:5914–5934. https://doi.org/10.1175/2011JCLI3972.1

Feofilov AG, Stubenrauch CJ (2019) Diurnal variation of high-level clouds from the synergy of AIRS and IASI space-borne infrared sounders. Atmos Chem Phys 19:13957–13972. https://doi.org/10.5194/acp-19-13957-2019

Foster MJ, Phillips C, Heidinger AK, Borbas EE, Li Y, Menzel WP, Walther A, Weisz E (2023) PATMOS-x version 6.0: 40 years of merged AVHRR and HIRS global cloud data. J Clim 36:1143–1160. https://doi.org/10.1175/JCLI-D-22-0147.1

Frey RA, Ackerman SA, Holz RE, Dutcher S, Griffith Z (2020) The continuity MODIS-VIIRS cloud mask. Remote Sens 12:3334

Garnier A, Pelon J, Vaughan MA, Winker DM, Trepte CR, Dubuisson P (2015) Lidar multiple scattering factors inferred from CALIPSO lidar and IIR retrievals of semi-transparent cirrus cloud optical depths over oceans. Atmos Meas Tech 8:2759–2774. https://doi.org/10.5194/amt-8-2759-2015

Garnier A, Pelon J, Pascal N, Vaughan MA, Dubuisson P, Yang P, Mitchell DL (2021a) Version 4 CALIPSO imaging infrared radiometer ice and liquid water cloud microphysical properties: part II: results over oceans. Atmos Meas Tech 14:3277–3299. https://doi.org/10.5194/amt-14-3277-2021

Garnier A, Pelon J, Pascal N, Vaughan MA, Dubuisson P, Yang P, Mitchell DL (2021b) Version 4 CALIPSO imaging infrared radiometer ice and liquid water cloud microphysical properties: part I: the retrieval algorithms. Atmos Meas Tech 14:3253–3276. https://doi.org/10.5194/amt-14-3253-2021

Di Girolamo L, Menzies A, Zhao G, Mueller K, Moroney C, Diner DJ (2010) MISR Level 3 cloud fraction by altitude theoretical basis, JPL D-62358, Jet Propulsion Laboratory, Pasadena, CA, p 24. Available at https://eospso.gsfc.nasa.gov/sites/default/files/atbd/MISR_CFBA_ATBD.pdf

GISTEMP Team (2023) GISS surface temperature analysis (GISTEMP), version 4. NASA Goddard Institute for Space Studies. Dataset accessed 2023-12-21 at data.giss.nasa.gov/gistemp/

Gryspeerdt E, Stier P (2012) Regime-based analysis of aerosol-cloud interactions. Geophys Res Lett 39:21802. https://doi.org/10.1029/2012GL053221

Guignard A, Stubenrauch CJ, Baran AJ, Armante R (2012) Bulk microphysical properties of semi-transparent cirrus from AIRS: a six year global climatology and statistical analysis in synergy with geometrical profiling data from CloudSat-CALIPSO. Atmos Chem Phys 12:503–525. https://doi.org/10.5194/acp-12-503-2012,2012

Hahn CJ, Rossow WB, Warren SG (2001) ISCCP cloud properties associated with standard cloud types identified in individual surface observations. J Clim 14:11–28. https://doi.org/10.1175/1520-0442(2001)014%3c0011:ICPAWS%3e2.0.CO;2

Ham S-H, Kato S, Rose FG, Loeb NG, Xu K-M, Thorsen T, Basilovich MG, Sun-Mack S, Chen Y, Miller WF (2021) Examining cloud macrophysical changes over the Pacific for 2007–17 using CALIPSO, CloudSat, and MODIS observations. J Appl Meteor Climatol 60:1105–1126. https://doi.org/10.1175/JAMC-D-20-0226.1

Hamann U, Walther A, Baum B, Bennartz R, Bugliaro L, Derrien M, Francis PN, Heidinger A, Joro S, Kniffka A, Le Gléau H, Lockhoff M, Lutz H-J, Meirink JF, Minnis P, Palikonda R, Roebeling R, Thoss A, Platnick S, Watts P, Wind G (2014) Remote sensing of cloud top pressure/height from SEVIRI: analysis of ten current retrieval algorithms. Atmos Meas Tech 7:2839–2867. https://doi.org/10.5194/amt-7-2839-2014

Heidinger AK, Pavolonis MJ (2009) Gazing at cirrus clouds for 25 years through a split window. Part I: methodology. J Appl Meteor Climatol 48:1100–1116. https://doi.org/10.1175/2008JAMC1882.1

Heidinger AK, Evan AT, Foster MJ, Walther A (2012) A naive Bayesian cloud detection scheme derived from CALIPSO and applied within PATMOS-x. J Appl Meteorol Climatol 51:1129–1144. https://doi.org/10.1175/JAMC-D-11-02.1

Heidinger AK, Bearson N, Foster MJ, Li Y, Wanzong S, Ackerman S, Holz RE, Platnick S, Meyer K (2019) Using sounder data to improve cirrus cloud height estimation from satellite imagers. J Atmos Ocean Technol 36:1331–1342. https://doi.org/10.1175/JTECH-D-18-0079.1

Hersbach H, Bell B, Berrisford P et al (2020) The ERA5 global reanalysis. Q J R Meteorol Soc 146:1999–2049. https://doi.org/10.1002/qj.3803

Holz RE, Ackerman SA, Nagle FW, Frey R, Dutcher S, Kuehn RE, Vaughan MA, Baum B (2008) Global MODIS cloud detection and height evaluation using CALIOP. J Geophys Res 113:D00A19. https://doi.org/10.1029/2008JD009837

Hu X, Ge J, Li Y, Marchand R, Huang J, Fu Q (2020) Improved hydrometeor detection method: an application to CloudSat. Earth Space Sci 7:e2019EA000900. https://doi.org/10.1029/2019EA000900

Jensen EJ, van den Heever SC, Grant LD (2018) The life cycles of ice crystals detrained from the tops of deep convection. J Geophys Res Atmos 123:9624–9634. https://doi.org/10.1029/2018JD028832

Jin Y, Rossow WB (1997) Detection of cirrus overlapping low-level clouds. J Geophys Res 102:1727–1737. https://doi.org/10.1029/96jd02996

Jing X, Zhang H, Peng J, Li J, Barker H (2016) Cloud overlapping parameter obtained from CloudSat/CALIPSO dataset and its application in AGCM with McICA scheme. Atmos Res 170:52–65. https://doi.org/10.1016/j.atmosres.2015.11.007

Johnson RH, Rickenbach TM, Rutledge SA, Ciesielski PE, Schubert WH (1999) Trimodal characteristics of tropical convection. J Clim 12:2397–2418. https://doi.org/10.1175/1520-0442(1999)012%3c2397:TCOTC%3e2.0.CO;2

Jones AL, Di Girolamo L, Zhao G (2012) Reducing the resolution bias in cloud fraction from satellite derived clear-conservative cloud masks. J Geophys Res 117:D12201. https://doi.org/10.1029/2011JD017195

Karlsson K-G, Devasthale A (2018) Inter-comparison and evaluation of the four longest satellite-derived cloud climate data records: CLARA-A2, ESA Cloud CCI V3, ISCCP-HGM, and PATMOS-x. Remote Sens 10:1567. https://doi.org/10.3390/rs10101567

Karlsson K-G, Anttila K, Trentmann J, Stengel M, Meirink J-F, Devasthale A, Hanschmann T, Kothe S, Jääskeläinen E, Sedlar J, Benas N, van Zadelhoff G-J, Schlundt C, Stein D, Finkensieper S, Håkansson N, Hollmann R, Fuchs P, Werscheck M (2017) CLARA-A2: CM SAF cLoud, Albedo and surface RAdiation dataset from AVHRR data-edition 2, satellite application facility on climate monitoring. https://doi.org/10.5676/EUM_SAF_CM/CLARA_AVHRR/V002

Karlsson K-G, Stengel M, Meirink JF, Riihelä A, Trentmann J, Akkermans T, Stein D, Devasthale A, Eliasson S, Johansson E, Håkansson N, Solodovnik I, Benas, N, Clerbaux N, Selbach N, Schröder M, Hollmann R (2023) CLARA-A3: the third edition of the AVHRR-based CM SAF climate data record on clouds, radiation and surface albedo covering the period 1979 to 2023. Earth Syst Sci Data Discuss [preprint]. https://doi.org/10.5194/essd-2023-133 (in review)

Koren I, Oreopoulos L, Feingold G, Remer LA, Altaratz O (2008) How small is a small cloud? Atmos Chem Phys 8:3855–3864. https://doi.org/10.5194/acp-8-3855-2008

Kotarba AZ (2022) Errors in global cloud climatology due to transect sampling with the CALIPSO satellite lidar mission. Atmos Res 279:106379. https://doi.org/10.1016/j.atmosres.2022.106379

Lau N-C, Crane MW (1995) A satellite view of the synoptic-scale organization of cloud properties in mid-latitude and tropical circulation systems. Mon Weather Rev 123:1984–2006. https://doi.org/10.1175/1520-0493(1995)123%3c1984:asvots%3e2.0.co;2

Lau N-C, Crane MW (1997) Comparing satellite and surface observations of cloud patterns in synoptic-scale circulations. Mon Weather Rev 125:3172–3189. https://doi.org/10.1175/1520-0493(1997)125%3c3172:csasoo%3e2.0.c0;2

Lenssen N, Schmidt G, Hansen J, Menne M, Persin A, Ruedy R, Zyss D (2019) Improvements in the GISTEMP uncertainty model. J Geophys Res Atmos 124:6307–6326. https://doi.org/10.1029/2018JD029522

Li J, Huang J, Stamnes K, Wang T, Lv Q, Jin H (2015) A global survey of cloud overlap based on CALIPSO and CloudSat measurements. Atmos Chem Phys 15:519–536. https://doi.org/10.5194/acp-15-519-2015

Liao X, Rossow WB, Rind D (1995a) Comparison between SAGE II and ISCCP high-level clouds, part I: global and zonal mean cloud amounts. J Geophys Res 100:1121–1135. https://doi.org/10.1029/94JD02429

Liao X, Rossow WB, Rind D (1995b) Comparison between SAGE II and ISCCP high-level clouds, part II: locating cloud tops. J Geophys Res 100:1137–1147. https://doi.org/10.1029/94JD02430

Liu Z, Kar J, Zeng S, Tackett J, Vaughan M, Avery M, Pelon J, Getzewich B, Lee K-P, Magill B, Omar A, Lucker P, Trepte C, Winker D (2019) Discriminating between clouds and aerosols in the CALIOP version 4.1 data products. Atmos Meas Tech 12:703–734. https://doi.org/10.5194/amt-12-703-2019

Liu H, Koren I, Altaratz O, Chekroun MD (2023) Opposing trends of cloud coverage over land and ocean under global warming. Atmos Chem Phys 23:6559–6569. https://doi.org/10.5194/acp-23-6559-2023

Luo Z, Anderson RC, Rossow WB, Takahashi H (2017) Tropical cloud and precipitation regimes as seen from near-simultaneous TRMM, CloudSat and CALIPSO observations and comparison with ISCCP. J Geophys Res Atmos 122:5988–6003. https://doi.org/10.1002/2017JD026569

Mace GG, Zhang Q (2014) The CloudSat radar–lidar geometrical profile product (RL-GeoProf): updates, improvements, and selected results. J Geophys Res Atmos 119:9441–9462. https://doi.org/10.1002/2013JD021374

Mace GG, Zhang Q, Vaughan M, Marchand R, Stephens G, Trepte C, Winker D (2009) A description of hydrometeor layer occurrence statistics derived from the first year of merged Cloudsat and CALIPSO data. J Geophys Res 114:D00A26. https://doi.org/10.1029/2007JD009755

Marvel K, Zelinka M, Klein SA, Bonfils C, Caldwell P, Doutriaux C, Santer BD, Taylor KE (2015) External influences on modeled and observed cloud trends. J Clim 28:4820–4840. https://doi.org/10.1175/JCLI-D-14-00734.1

McGarragh GR, Poulsen CA, Thomas GE, Povey AC, Sus O, Stapelberg S, Schlundt C, Proud S, Christensen MW, Stengel M, Hollmann R, Grainger RG (2018) The Community Cloud retrieval for CLimate (CC4CL): part 2: the optimal estimation approach. Atmos Meas Tech 11:3397–3431. https://doi.org/10.5194/amt-11-3397-2018

Menzel WP, Frey RA, Borbas EE, Baum BA, Cureton G, Bearson N (2016) Reprocessing of HIRS satellite measurements from 1980 to 2015: development toward a consistent decadal cloud record. J Appl Meteor Climatol 55:2397–2410. https://doi.org/10.1175/JAMC-D-16-0129.1

Meyer K, Platnick S, Holz R, Dutcher S, Quinn G, Nagle F (2020) Derivation of shortwave radiometric adjustments for SNPP and NOAA-20 VIIRS for the NASA MODIS-VIIRS continuity cloud products. Remote Sens 12:4096. https://doi.org/10.3390/rs12244096

Minnis P, Yost CR, Sun-Mack S, Chen Y (2008) Estimating the physical top altitude of optically thick ice clouds from thermal infrared satellite observations using CALIPSO data. Geophys Res Lett 35:L12801. https://doi.org/10.1029/2008GL033947

Minnis P, Sun-Mack S, Yost CR, Chen Y, Smith WL Jr, Chang F-L, Heck PW, Arduini RF, Trepte QZ, Ayers K, Bedka K, Bedka S, Brown RR, Heckert E, Hong G, Jin Z, Palikonda R, Smith R, Scarino B, Spangenberg DA, Yang P, Xie Y, Yi Y (2021) CERES MODIS cloud product retrievals for edition 4, part I: algorithm changes to CERES MODIS. IEEE Trans Geosci Remote Sens 59:2744–2780

Minnis P, Sun-Mack S, Smith WL Jr, Hong G, Chen Y (2019) Advances in neural network detection and retrieval of multilayer clouds for CERES using multispectral satellite data. In: Proceedings of the SPIE conference remote sensing clouds and the atmosphere. XXIV, Strasbourg, France, Sept 9–12, 11152, p 12. https://doi.org/10.1117/12.2532931

Mitra A, Di Girolamo L, Hong Y, Zhan Y, Mueller KJ (2021) Assessment and error analysis of Terra-MODIS and MISR cloud-top heights through comparison with ISS-CATS lidar. J Geophys Res Atmos 126:e2020JD034281. https://doi.org/10.1029/2020JD034281

Mitra A, Loveridge JR, Di Girolamo L (2023) Fusion of MISR stereo cloud heights and terra-MODIS thermal infrared radiances to estimate two-layered cloud properties. J Geophys Res Atmos 128:e2022JD038135. https://doi.org/10.1029/2022JD038135

Mueller K, Maroney C, Jovanovic V, Garay MJ, Muller J-P, Di Girolamo L, Davies R (2013) MISR level 2 cloud product algorithm theoretical basis, JPL D-73327, Pasadena, CA, p 61. Available at https://eospso.gsfc.nasa.gov/sites/default/files/atbd/MISR_L2_CLOUD_ATBD-1.pdf

Naud CM, Del Genio A, Mace GG, Benson S, Clothiaux EE, Kollias P (2008) Impact of dynamics and atmospheric state on cloud vertical overlap. J Clim 21:1758–1770. https://doi.org/10.1175/2007JCLI1828.1

Noël V, Chepfer H, Chiriaco M, Yorks J (2018) The diurnal cycle of cloud profiles over land and ocean between 51° S and 51° N, seen by the CATS spaceborne lidar from the International Space Station. Atmos Chem Phys 18:9457–9473. https://doi.org/10.5194/acp-18-9457-2018

Norris JR (2005) Multidecadal changes in near-global cloud cover and estimated cloud cover radiative forcing. J Geophys Res 110:D08206. https://doi.org/10.1029/2004JD005600

Norris JR, Allen RJ, Evan AT, Zelinka MD, O'Dell CW, Klein SA (2016) Evidence for climate change in the satellite cloud record. Nature 536:72–75. https://doi.org/10.1038/nature18273

Oreopoulos L, Cho N, Lee D (2017) New insights about cloud vertical structure from CloudSat and CALIPSO observations. J Geophys Res Atmos 12:9280–9300. https://doi.org/10.1002/2017JD026629

Philipp D, Stengel M, Ahrens B (2020) Analyzing the arctic feedback mechanism between sea ice and low-level clouds using 34 years of satellite observations. J Clim 33:7479–7501. https://doi.org/10.1175/JCLI-D-19-0895.1

Pincus R, Platnick S, Ackerman SA, Hemler RS, Hofmann RJP (2012) Reconciling simulated and observed views of clouds: MODIS, ISCCP, and the limits of instrument simulators. J Clim 25:4699–4720. https://doi.org/10.1175/JCLI-D-11-00267.1

Platnick S, Meyer KG, Yang P, Ridgway WL, Riedi JC, King MD, Wind G, Amarasinghe N, Marchant B, Arnold GT et al (2017) The MODIS cloud optical and microphysical products: collection 6 updates and examples from Terra and Aqua. IEEE Trans Geosci Remote Sens 55:502–525. https://doi.org/10.1109/TGRS.2016.2610522

Platnick S, Meyer K, Wind G, Holz RE, Amarasinghe N, Hubanks PA, Marchant B, Dutcher S, Veglio P (2021) The NASA MODIS-VIIRS continuity cloud optical properties products. Remote Sens 13:2. https://doi.org/10.3390/rs13010002

Protopapadaki ES, Stubenrauch CJ, Feofilov AG (2017) Upper tropospheric cloud systems derived from IR sounders: properties of Cirrus Anvils in the tropics. Atmosph Chem Phys 17:3845–3859. https://doi.org/10.5194/acp-17-3845-2017

Rossow WB (1989) Measuring cloud properties from space: a review. J Clim 2:201–213. https://doi.org/10.1175/1520-0442(1989)002%3c0201:mcpfsa%3e2.0.co;2

Rossow WB, Schiffer RA (1991) ISCCP cloud data products. Bull Am Meteorl Soc 72:2–20. https://doi.org/10.1175/1520-0477(1991)072%3c0002:icdp%3e2.0.co;2

Rossow WB, Schiffer RA (1999) Advances in understanding clouds from ISCCP. Bull Am Meteorl Soc 80:2261–2287. https://doi.org/10.1175/1520-0477(1999)080%3c2261:aiucfi%3e2.0.co;2

Rossow WB, Zhang Y-C (1995) Calculation of surface and top-of-atmosphere radiative fluxes from physical quantities based on ISCCP datasets: 2. Validation and first results. J Geophys Res 100:1167–1197. https://doi.org/10.1029/94JD02746

Rossow WB, Zhang Y-C (2010) Evaluation of a statistical model of cloud vertical structure using combined CloudSat and CALIPSO cloud layer profiles. J Clim 23:6641–6653. https://doi.org/10.1175/2010JCLI3734.1

Rossow WB, Tselioudis G, Polak A, Jakob C (2005a) Tropical climate described as a distribution of weather states indicated by distinct mesoscale cloud property mixtures. Geophys Res Lett 32:L21812. https://doi.org/10.1029/2005GL024584

Rossow WB, Zhang Y-C, Wang J (2005b) A statistical model of cloud vertical structure based on reconciling cloud layer amounts inferred from satellites and radiosonde humidity profiles. J Clim 18:3587–3605. https://doi.org/10.1175/JCLI3479.1

Rossow WB, Knapp KR, Young AH (2022) International satellite cloud climatology project: extending the record. J Climate 35:141–158. https://doi.org/10.1175/jcli-d-21-0157.1

Rossow WB (2017) Climate data record (CDR) Program climate algorithm theoretical basis document of international satellite cloud climatology project (ISCCP) cloud properties. CDRP-ATBD-0872.

Available at https://www.ncei.noaa.gov/pub/data/sds/cdr/CDRs/Cloud_Properties-ISCCP/AlgorithmDescription_01B-29.pdf

Sassen K, Wang Z (2012) The clouds of the middle troposphere: composition, radiative impact, and global distribution. Surv Geophys 33:677–691. https://doi.org/10.1007/s10712-011-9163-x

Slingo JM, Slingo A (1991) The response of a general circulation model to cloud longwave radiative forcing. II: further studies. Q J R Meteorol Soc 117:333–364

Stachnik JP, Schumacher C, Ciesielski PE (2013) Total heating characteristics of the ISCCP tropical and subtropical cloud regimes. J Clim 26:7097–7116. https://doi.org/10.1175/JCLI-D-12-00673.1

Staten PW, Lu J, Grise KM, Davis SM, Birner T (2018) Re-examining tropical expansion. Nat Clim Change 8:768–775. https://doi.org/10.1038/s41558-018-0246-2

Stengel M, Stapelberg S, Sus O, Finkensieper S, Würzler B, Philipp D, Hollmann R, Poulsen C, Christensen M, McGarragh G (2020) Cloud_cci advanced very high resolution radiometer post meridiem (AVHRR-PM) dataset version 3: 35-year climatology of global cloud and radiation properties. Earth Syst Sci Data 12:41–60. https://doi.org/10.5194/essd-12-41-2020

Stephens GL, Winker D, Pelon J, Trepte C, Vane D, Yuhas C, L'Ecuyer T, Lebsock M (2018a) CloudSat and CALIPSO within the a-train: ten years of actively observing the earth system. Bull Am Meteorol Soc 99:569–581. https://doi.org/10.1175/BAMS-D-16-0324.1

Stephens GL, Hakuba MZ, Webb M, Lebsock M, Yue Q, Kahn BH, Hristova-Veleva S, Rapp A, Stubenrauch C, Elsaesser GS, Slingo J (2018b) Regional intensification of the tropical hydrological cycle during ENSO. Geophys Res Lett 45:4361–4370. https://doi.org/10.1029/2018GL077598

Stubenrauch CJ, Chédin A, Armante R, Scott NA (1999) Clouds as seen by infrared sounders (3I) and imagers (ISCCP): part II a new approach for cloud parameter determination in the 3I algorithms. J Clim 12:2214–2223. https://doi.org/10.1175/1520-0442(1999)012%3c2214:CASBSS%3e2.0.CO;2

Stubenrauch CJ, Chédin A, Rädel G, Scott NA, Serrar S (2006) Cloud properties and their seasonal and diurnal variability from TOVS Path-B. J Clim 19:5531–5553. https://doi.org/10.1175/JCLI3929.1

Stubenrauch CJ, Cros S, Guignard A, Lamquin N (2010) A six-year global cloud climatology from the atmospheric InfraRed sounder aboard the AQUA satellite: statistical analysis in synergy with CALIPSO and CloudSat. Atmos Chem Phys 10:7197–7214. https://doi.org/10.5194/acp-10-7197-2010

Stubenrauch CJ, Rossow WB, Kinne S, Ackerman S, Cesana G, Chepfer H, Di Girolamo L, Getzewich B, Guignard A, Heidinger A, Maddux BC (2013) Assessment of global cloud datasets from satellites: project and database initiated by the GEWEX radiation panel. B Am Meteorol Soc 94:1031–1049. https://doi.org/10.1175/BAMS-D-12-00117.1

Stubenrauch CJ, Feofilov AG, Protopapadaki SE, Armante R (2017) Cloud climatologies from the infrared sounders AIRS and IASI: strengths and applications. Atmos Chem Phys 17:13625–13644. https://doi.org/10.5194/acp-17-13625-2017

Stubenrauch CJ, Caria G, Protopapadaki SE, Hemmer F (2021) The effect of tropical upper tropospheric cloud systems on radiative heating rate fields derived from synergistic A-train satellite observations. Atmos Chem Phys 21:1015–1034. https://doi.org/10.5194/acp-21-1015-2021

Stubenrauch CJ, Mandorli G, Lemaitre E (2023) Convective organization and 3D structure of tropical cloud systems deduced from synergistic A-Train observations and machine learning. Atmos Chem Phys 23:5867–5884. https://doi.org/10.5194/acp-23-5867-2023

Stubenrauch CJ, Rossow WB, Kinne S, GEWEX Cloud Assessment Team (2012) Assessment of Global Cloud Datasets from Satellites, A Project of the World Climate Research Programme Global Energy and Water Cycle Experiment (GEWEX) Radiation Panel, WCRP report, p 180. Available at chrome-extension://efaidnbmnnnibpcajpcglclefindmkaj/https://www.wcrp-climate.org/documents/GEWEX_Cloud_Assessment_2012.pdf and at http://climserv.ipsl.polytechnique.fr/gewexca/

Sun-Mack S, Minnis P, Chen Y, Hong G, Smith WL Jr (2023) Identification of ice-over-water multilayer clouds using an artificial neural network with multispectral satellite data. Atmos Meas Tech Disc. https://doi.org/10.5194/egusphere-2023-2804

Sus O, Stengel M, Stapelberg S, McGarragh G, Poulsen C, Povey AC, Schlundt C, Thomas G, Christensen M, Proud S, Jerg M, Grainger R, Hollmann R (2018) The Community Cloud retrieval for CLimate (CC4CL): part 1: a framework applied to multiple satellite imaging sensors. Atmos Meas Tech 11:3373–3396. https://doi.org/10.5194/amt-11-3373-2018

Tan J, Oreopoulos L (2019) Subgrid precipitation properties of mesoscale atmospheric systems represented by MODIS cloud regimes. J Clim 32:1797–1812. https://doi.org/10.1175/JCLI-D-18-0570.1

Trepte QZ, Minnis P, Sun-Mack S, Yost CR, Chen Y, Jin Z, Chang F-L, Smith WL Jr, Bedka KM, Chee TL (2019) Global cloud detection for CERES 4 using Terra and Aqua MODIS data. IEEE Trans Geosci Remote Sens 57:9410–9449. https://doi.org/10.1109/TGRS.2019.2926620

Tselioudis G, Rossow WB, Zhang Y, Konsta D (2013) Global weather states and their properties from passive and active satellite cloud retrievals. J Clim 26:7734–7746. https://doi.org/10.1175/JCLI-D-13-00024.1

Tselioudis G, Rossow WB, Jakob C, Remillard J, Tropf D, Zhang Y-C (2021) Evaluation of clouds, radiation, and precipitation in CMIP6 models using global weather states derived from ISCCP-H cloud property data. J Clim 34:7311–7324. https://doi.org/10.1175/jcli-d-21-0076.1

Walther A, Heidinger AK (2012) Implementation of the daytime cloud optical and microphysical properties algorithm (DCOMP) in PATMOS-x. J Appl Meteorol Climatol 51:1371–1390. https://doi.org/10.1175/JAMC-D-11-0108.1

Wang J, Rossow WB (1998) Effects of cloud vertical structure on atmospheric circulation in the GISS GCM. J Clim 11:3010–3029. https://doi.org/10.1175/1520-0442(1998)011%3c3010:eocvso%3e2.0.co;2

Wang C, Luo ZJ, Chen X, Zeng X, Tao W-K, Huang X (2014) A physically based algorithm for non-blackbody correction of cloud top temperature and application to convection study. J Appl Meteorol Climatol 53:1844–1857. https://doi.org/10.1175/JAMC-D-13-0331.1

Warren SG, Hahn CJ, London J (1985) Simultaneous occurrence of different cloud types. J Clim Appl Meteorol 24:658–667. https://doi.org/10.1175/1520-0450(1985)024%3c0658:SOODCT%3e2.0.CO;2

Wielicki BA, Parker L (1992) On the determination of cloud cover from satellite sensors: the effect of sensor spatial resolution. J Geophys Res 97:12799–12823. https://doi.org/10.1029/92jd01061

Winker DM, Pelon J, Coakley JA Jr et al (2010) The CALIPSO mission: a global 3D view of aerosols and clouds. Bull Am Meteorol Soc 91:1211–1229. https://doi.org/10.1175/2010BAMS3009.1

Wylie DP, Wang P-H (1997) Comparison of cloud frequency data from the high-resolution infrared radiometer sounder and the stratospheric aerosol and gas experiment II. J Geophys Res 102:29893–29900. https://doi.org/10.1029/97JD02360

Wylie DP, Jackson DL, Menzel WP, Bates JJ (2005) Trends in global cloud cover in two decades of HIRS observations. J Clim 18:3021–3031. https://doi.org/10.1175/JCLI3461.1

Young AH, Knapp RK, Inamdar A, Hankins W, Rossow WB (2018) The international cloud climatology project H-series climate data record product. Earth Syst Sci Data 10:583–593. https://doi.org/10.5194/essd-10-583-2018

Zhang Y, Rossow WB (2023) Global radiative flux profile data set: revised and extended. J Geophys Res Atmos 128:e2022JD037340. https://doi.org/10.1029/2022JD037340

Zhang Y, Rossow WB, Lacis AA, Oinas V, Mishchenko MI (2004) Calculation of radiative fluxes from the surface to top of atmosphere based on ISCCP and other global data sets: refinements of the radiative transfer model and the input data. J Geophys Res 109:D19105. https://doi.org/10.1029/2003JD004457

Zhao G, Di Girolamo L (2006) Cloud fraction errors for trade wind cumuli from EOS-Terra instruments. Geophys Res Lett 33:L20802. https://doi.org/10.1029/2006GL027088

Publisher's Note Springer Nature remains neutral with regard to jurisdictional claims in published maps and institutional affiliations.

Authors and Affiliations

Claudia J. Stubenrauch[1] · Stefan Kinne[2] · Giulio Mandorli[1] · William B. Rossow[3] · David M. Winker[4] · Steven A. Ackerman[5] · Helene Chepfer[1] · Larry Di Girolamo[6] · Anne Garnier[4,7] · Andrew Heidinger[8] · Karl-Göran Karlsson[9] · Kerry Meyer[10] · Patrick Minnis[4,7] · Steven Platnick[10] · Martin Stengel[11] · Szedung Sun-Mack[4,7] · Paolo Veglio[5] · Andi Walther[5] · Xia Cai[7] · Alisa H. Young[12] · Guangyu Zhao[6]

✉ Stefan Kinne
Stefan.Kinne@mpimet.mpg.de

[1] Laboratoire de Météorologie Dynamique / Institut Pierre-Simon Laplace, (LMD/IPSL), Sorbonne Université, Ecole Polytechnique, CNRS, 4 Place Jussieu, 75252 Paris Cedex 05, France

2 Max Planck Institute for Meteorology, Bundesstr 53, 20146 Hamburg, Germany

3 Franklin, USA

4 NASA Langley Research Center, Hampton, VA 23681, USA

5 Space Science and Engineering Center, University of Wisconsin—Madison, 1225W Dayton, Madison, WI 53706, USA

6 Department of Atmospheric Sciences, University of Illinois at Urbana–Champaign, 1301 W. Green Str., Urbana, IL 61820, USA

7 Analytical Mechanics Associates, 21 Enterprise Parkway Suite 300, Hampton, VA 23666, USA

8 NOAA/NESDIS/STAR, 1225W Dayton, Madison, WI 53706, USA

9 Swedish Meteorological and Hydrological Institute, Folksborgsvaegen 17, 60176 Norrköping, Sweden

10 Earth Sciences Division, NASA Goddard Space Flight Center, Bldg 33, Greenbelt, MD 20771, USA

11 Deutscher Wetterdienst, Frankfurter Str 135, 63067 Offenbach am Main, Germany

12 NOAA Great Lakes Environmental Research Lab, 4840S State Rd, Ann Arbor, MI 48108, USA

The manufacturer's authorised representative in the EU is Springer Nature Customer Service Centre GmbH, Europaplatz 3, 69115 Heidelberg, Germany. If you have any concerns regarding our products, please contact ProductSafety@springernature.com

Printed and bound by CPI Group (UK) Ltd, Croydon, CR0 4YY

26/03/2026

02078939-0006